Vertebrate Conservation and Biodiversity

TOPICS IN BIODIVERSITY AND CONSERVATION

Volume 5

The titles published in this series are listed at the end of this volume.

Vertebrate Conservation and Biodiversity

Edited by

David L. Hawksworth

and

Alan T. Bull

Reprinted from *Biodiversity and Conservation*, volume 16:4 (2007)

Springer

A C.I.P. Catalogue record for this book is available from the library of Congress.

ISBN 978-1-4020-6319-0 (HB)
ISBN 978-1-4020-6320-6 (e-book)

Published by Springer,
P.O. Box 17, 3300 AA Dordrecht, The Netherlands.

www.springer.com

Printed on acid-free paper

Contents

Vertebrate Conservation and Biodiversity

Introduction

Vertebrate Conservation and Biodiversity

This book brings together a selection of original studies submitted to *Biodiversity and Conservation* addressing aspects of the conservation of vertebrates and biodiversity, especially those in terrestrial habitats[1]. Vertebrates are, along with plants, the best-known and most intensively studied components of biological diversity on Earth. As such, studies on vertebrates may be expected to provide models for other groups, but at the same time they pose particular problems because of their relative mobility, which can be transcontinental in migratory birds. In addition, many mammals and fish are also subject to extensive human exploitation for food or sport.

Two contributions consider the issue of habitat modification in relation to birds, mammals and also reptiles, one carried out in East African coastal forests, and the other a planted tree crop in Nicaragua; in both cases primary and relict forests were critical to the survival of the most specialized species.

The articles concerned with mammals cover the conflict caused by large carnivores attacking livestock in Africa, mammals in ranches and other habitats in the Brazilian Cerrado, mammals in the Caxiuanã National Forest in Brazil (monitored using a camera phototrapping programme), koalas and a wombat in Australia, the maned wolf in the Brazilian Cerrado and Pantanal, bats in Madagascar, marmosets in Brazil, and endemic primates in Kenya. Problems of surveying carnivores at different scales are revealed by the comparison of different methods in the Mediterranean.

The contributions on birds included here fall into two main categories: the relationships of the avifauna to habitat types and changes, and studies on particular species. In the former category are studies comparing abundance and diversity in grassland and non-grassland habitats in Hungary, birds in wetland fragments in the neotropics, responses to prescribed burning in prairies in the USA, relationships to human activities in the Córdoba Mountains of Argentina, and ones in montane forests of different degrees of modification in Ecuador. The studies on particular species presented encompass ones on the Lower Grey Shrike in France and Spain, the migratory Lesser Grey Shrike in Europe, Great Spotted Woodpeckers in Switzerland, the Sooty Shearwater on the South American coast, the Sierra Madre Sparrow in México, and Blue-winged Macaws in Brazil. Also included is a critical study of bird endemism and biodiversity in the different subregions of China which may help direct conservation efforts in the country.

---

[1] A series of contributions including ones on fish was included in a previous volume of collected papers in this series, *Marine, Freshwater, and Wetlands Biodiversity Conservation* [Topics in Biodiversity and Conservation Vol. 4], edited by D. L. Hawksworth & A. T. Bull (Springer, Dordrecht, 2006; 399 pp., ISBN 10: 1 4020 5733 4, 13: 978 1 4020 5633).

Reptiles and amphibians include many of the most endangered vertebrates, with anuran decline being an issue of ongoing major world concern. The situation with anurans is exemplified here by a study of 131 species in the Brazilian Cerrado leading to the design of a network of conservation areas. Other investigations reported on these groups here include ones on amphibian diversity hot-spots in the Tanzanian mountains, characterization of factors favouring these organisms in a National Park in Portugal, and the decline of Loggerhead Turtles on a beach in Turkey. The issue of exploitation is also represented by reports on the reptile skin trade in México, and the sale of snakes in markets and stores in Brazil as pets, sources of traditional medicines, or uses in religious rites. Another aspect is the effect of introduced species on others, exemplified here by the threat posed to native species by the Boa constrictor on a Mexican island.

This series of studies is presented together here to provide an indication of current research activities, that will be of value to students undertaking courses in aspects of biodiversity and conservation. They can be viewed as a series of case studies that will expose students to primary research being conducted now. As such they will complement the necessarily less-detailed specific information in textbooks and secondary review articles.

DAVID L. HAWKSWORTH
Editor-in-Chief Biodiversity and Conservation
Universidad Complutense de Madrid
22 March 2007

Biodivers Conserv (2007) 16:827–839
DOI 10.1007/s10531-006-9107-9

ORIGINAL PAPER

Responses of plant and bird communities to prescribed burning in tallgrass prairies

Fred Van Dyke · Jamie D. Schmeling · Shawn Starkenburg · Sung Heun Yoo · Peter W. Stewart

Received: 29 December 2005 / Accepted: 7 August 2006 / Published online: 21 November 2006

Abstract Historic losses and fragmentation of tallgrass prairie habitat to agriculture and urban development have led to declines in diversity and abundance of plants and birds associated with such habitat. Prescribed burning is a management strategy that has potential for restoring and rejuvenating prairies in fragmented landscapes, and through such restoration, might create habitat for birds dependent upon prairies. To provide improved data for management decision-making regarding the use of prescribed fire in tallgrass prairies, we compared responses of plant and bird communities on five burned and five unburned tallgrass prairie fragments at the DeSoto National Wildlife Refuge, Iowa, USA, from 1995 to 1997. Overall species richness and diversity were unaffected by burning, but individual species of plants and birds were affected by year-treatment interactions, including northern bobwhite (*Colinus virginianus*) and ring-necked pheasant (*Phasianus colchicus*), which showed time-delayed increases in density on burned sites. Analyses of species/area relationships indicated that, collectively, many small sites did make significant contributions to plant biodiversity at landscape levels, supporting the overall conservation value of prairie fragments. In contrast, most birds species were present on larger sites. Thus, higher biodiversity in bird communities which contain area-sensitive species might require larger sites able to support larger, more stable populations, greater habitat heterogeneity, and greater opportunity for niche separation.

Keywords DeSoto National Wildlife Refuge · Grassland birds · Grassland plants · Prairie restoration · Prescribed burning · Tallgrass prairie

F. Van Dyke (✉) · J. D. Schmeling · S. H. Yoo · P. W. Stewart
Department of Biology, Wheaton College, 501 College Avenue, Wheaton, IL 60187, USA
e-mail: fred.g.vandyke@wheaton.edu

S. Starkenburg
Department of Botany and Plant Pathology, Oregon State University, Corvallis, OR 97331, USA

Introduction

The structure and function of grasslands worldwide have been disrupted by habitat fragmentation, the removal of native grazers and altered fire frequency (Samson and Knopf 1994). Such changes threaten the integrity of native grassland communities of plants and animals in many ways and at many levels (Janzen 1983; Collins 2000; Cully and Michaels 2000). Habitat fragmentation alters size, spacing and context of habitat patches, which can result in an increase in the local rate of extinction of plant and animal species by reducing fecundity, population size and colonization of species from similar habitats (Noss and Csuti 1992). Compounding the effects of fragmentation, extirpation of fire and loss of populations of native grazers can lead to encroachment of woody vegetation and loss of grassland (Leach and Givnish 1996; Collins et al. 1998).

In addition to effects on plant communities, the use of prescribed fire to enhance grasslands is of increasing relevance as regional losses, isolation and fragmentation of grasslands have led to corresponding declines in local population densities of many species of grassland-dependent birds (Askins 1993; Vickery et al. 1994). Area-sensitive species in grasslands are especially vulnerable to these effects (Herkert 1994a, b) which can create specific physical, environmental and ecological changes (Yahner 1988) that lower habitat quality for area-sensitive species that have been historical specialists of grassland habitats. Prescribed burning cannot change the size, shape, or area of a site, but it can alter vegetation density, vegetation structure and habitat heterogeneity. Thus, prescribed burning can be a management tool for providing appropriate habitat to vegetation-sensitive species.

With these concerns in mind, we examined the effects of prescribed burning on plant and bird communities in tallgrass prairie fragments in order to determine (1) changes in the structure of vegetation and abundance of plant species following burning in tallgrass prairie habitats; (2) the response of individual species of resident grassland-dependent breeding birds; and (3) the effects of prescribed burning on overall plant and avian community species richness and diversity. The questions we wished to answer were (1) would differences in management on fragmented prairie habitat lead to different communities of plants and birds associated with such habitat and (2) would differences in management treatments be effective in moving communities of low species diversity toward higher levels of biodiversity. Our null hypothesis was that there would be no differences in characteristics of these communities on burned versus unburned sites.

Study area

The DeSoto National Wildlife Refuge (DNWR) was established along the Missouri River in Iowa and Nebraska (USA) in 1959 to conserve wetlands used as resting and staging areas for migrating waterfowl. In addition to wetlands (approximately 1000 ha), DNWR's most abundant habitats are forests (1350 ha) and native tallgrass prairie (665 ha). As a result of settlement and conversion of prairie communities in this region to agriculture and pasture lands beginning in the second half of the 19th century, no large blocks of native tallgrass prairie vegetation remained by the time DNWR was created. Given the rarity of tallgrass prairie in the upper Midwest, prairies at DNWR was re-established intentionally on marginal croplands within refuge borders to conserve this historic native habitat and its biological diversity. Tallgrass prairie exists today at DNWR in fragments of 3–10 ha that are actively managed by prescribed burning every third year.

Initial conversions of croplands into prairie were made on smaller sites, sites with configurations that made cultivation difficult, sites with historically poor crop yields, sites with potentially high wildlife value, or sites with a combination of these traits. Restoration efforts began with a single site in 1965 that was planted with one species, big bluestem (*Andropogon gerardii*), a C_4 grass often considered an indicator of tallgrass prairies. Six additional sites were converted in 1972–1974 with the planting of 3–5 species of warm season C_4 grasses but no forb species. In 1980–1983, five more sites were converted with plantings of six species of grasses and up to five species of forbs. Thus, all early restoration efforts were characterized by low species diversity (1–11 species). Management following plantings was site-specific according to site characteristics and management objectives, but typically involved burning and mowing at 3–4 year intervals.

By the 1990s, vegetation in these tallgrass prairie fragments at DNWR was dominated by big bluestem, Indiangrass (*Sorghastrum nutans*), switchgrass (*Panicum virgatum*) and various species of goldenrod (mainly *Solidago canadensis*). The most abundant bird species were common yellowthroats (*Geothlypis trichas*) and field sparrows (*Spizella pusilla*). The general status of sites had been monitored since restoration, but quantitative measurements of plant and animal community composition were limited. Our study was the first comprehensive quantitative examination of such restored sites at DNWR and the first direct comparison of plant and bird communities on burned and unburned sites there.

Materials and methods

Sampling of plant and bird communities

In January 1995, 10 independent noncontiguous sites of native tallgrass prairie at DNWR were selected for evaluation as statistical sampling units. All sites possessed similar physical characteristics and site histories, and all were within a relatively narrow size range (3.0–9.3 ha) typical of prairie fragments on the refuge. Five sites were designated randomly for burning while the other five were left untreated. Burning was completed on all sites between 22 April and 11 May in 1995. Fires were ignited at multiple points along a prescribed line using kerosene drip torches and completed in 1–2 h. Each site retained its treatment designation for sampling through 1997, but burned sites were burned only in 1995 and were not reburned in subsequent years, an approach that permitted us to track changes on these sites subsequent to burning.

Regeneration on burned sites was rapid and vegetation was re- established by the time sampling began in June. Vegetation in each prairie fragment was examined using five 50 × 50 m macroplots sufficient to cover all parts of the site. On all sites, macroplots were nonoverlapping and covered the entire prairie fragment so that all parts of the site were sampled and total diversity of the site's plant community could be assessed. Six 50-m belt transects were established perpendicular to the baseline of each macroplot at random intervals. Plant species' composition and percent ground cover were determined in five 25.4 × 50.8 cm microplots placed randomly along each belt (30 microplots/macroplot). Estimates of cover were used as the primary index of plant abundance and were made using a visually calibrated frame. Presence of individual species was used to determine species richness. Species diversity was determined by using the Shannon Index (H') (Shannon and Weaver 1949),

$$H' = -\sum_i (p_i \ln p_i),$$

where p_i is the proportion of individuals of the ith species. Species dominance on each site was determined using the Simpson index (C) (Pielou 1975), which measures dominance as the probability that two randomly selected individuals from a community will belong to the same species and is calculated as

$$C = \sum_i p_i^2.$$

From 20 May through 10 July 1995–1997, the density and diversity of resident breeding bird species was determined by a census of singing males by using the spot mapping method (International Bird Census Committee 1970). Each site was visited 10–12 times each year. Birds were designated as grassland residents if they established home ranges that were confined to the site or, if extending beyond the site, remained within adjacent grassland habitat. We also included two nonterritorial species, brown-headed cowbird (*Molothrus ater*) and mourning dove (*Zenaida macroura*), as members of the resident grassland community, as well as American goldfinch (*Carduelis tristis*), a species that typically established territories subsequent to our sampling period. Because these species invested most of their time in grassland fragments, we included them in the resident bird community. In these species, we censused males within the grassland whether vocalizing or not. Composite results of all visits were used to determine the abundance of individual species, species richness and diversity.

Numerical and statistical analyses

We compared community species richness, diversity, dominance and species cover (plants) and density (resident birds) through a repeated measures analysis of variance, with years as repeated measures, to determine whether these variables were independent of treatment. Plant cover estimates were arcsine-transformed before analysis. $P \leq 0.05$ was considered significant. For species that had significant treatment–year interactions, we conducted unpaired t-tests to determine in which years treatment affected species abundance. Bonferroni corrections were applied to account for the multiple years being compared. To determine the cumulative effect of all sites on the species richness of both plant and bird communities, we performed a rarefaction analysis (Krebs 1989) that provided an estimate of the expected number of species present per sampled area.

We measured similarity of plant communities between plots of different treatments and years using the Jaccard index, a measure of association (similarity) between communities based on differences in species presence and absence. The Jaccard Index provides an easily interpreted measure of association between communities but should not be analyzed using standard analysis of variance

(Dyer 1978). To extend analysis of the Jaccard Index across multiple factors, we analyzed species composition of plant communities between treatments and among years using the linear model proposed by Dyer (1978). The baseline similarity estimated by the model is a measure of the average Jaccard similarity between sites in the same treatment and year. The coefficient estimated for the effect of treatment provides a measure of change in similarity between treatments in the same year. The coefficient estimated for the effect of year provides a measure of change in similarity on the same site in different years. Based on a permutation test applied separately to treatment and year factors (Edgington 1995), we evaluated whether communities from different treatments or years were dissimilar to one another. The impact of each species on the Jaccard computations also was estimated.

Results

Species richness, diversity and dominance of tallgrass plant and bird communities were independent of treatment (Tables 1, 2). Overall, species richness and diversity in plant and bird communities were low on all sites. Year effects were significant influences on species richness in both plant and bird communities (Table 1) and affected species diversity and dominance in bird communities (Table 2). Resident bird communities were lower in richness and diversity and higher in dominance in 1995 compared to subsequent years. Proportional treatment differences in species richness were greatest immediately after burning (1995) in both plants and birds, but diminished in subsequent years.

Plant communities at DNWR contained few species on individual sites (range 8–24). Forty-six percent of ground cover was contributed by one species of grass, *A. gerardii*, and one species of forb, *S. canadensis*. Rarefaction analysis revealed that the entire array of sites collectively contributed many more species than any individual site, with the upper asymptote of expected number of species in any given year (20–30 species) not being approached until sampling area exceeded 30 ha (Fig. 1), more than three times the area of the largest individual site.

One plant species (*Cassia fasciculata*) showed sensitivity to treatment effects, five (*A. gerardii, C. fasciculata, P. virgatum, Schizachyrium scoparium* and *S. nutans*) to year effects (Table 3) and three (*C. fasciculata, Polygonum pensylvanicum* and *S. scoparium*) to treatment–year interactions ($P = < 0.01, 0.03, 0.02$, respectively).

Table 1 Effect of management treatment and year on species richness in communities of plants and resident birds in tallgrass prairie fragments at DNWR, Iowa, USA, 1995–1997

Community	1995			1996			1997			*P* (Treatment)	*P* (Year)
	Mean	SE	Range	Mean	SE	Range	Mean	SE	Range		
Plants	14.4	1.6	8–22	21.2	1.2	14–26	19.6	1.5	13–25	0.63	< 0.01
Burned	16.2	2.2	11–22	20.6	2.0	14–24	20.2	2.3	13–25	–	–
Untreated	12.6	2.2	8–19	21.8	1.5	18–26	19.0	2.1	14–25	–	–
Resident Birds	7.8	1.0	3–14	10.2	0.4	8–12	10.4	0.7	7–14	0.09	< 0.01
Burned	9.0	1.6	4–14	10.6	0.5	9–12	11.8	0.8	10–14	–	–
Untreated	6.6	1.0	3–8	9.8	0.5	8–11	9.0	0.7	7–11	–	–

Means of species richness compared through repeated measures (year) analysis of variance

Table 2 Effect of management treatment and year on species diversity (Shannon index) and dominance (Simpson index) of plant and resident grassland bird communities in burned and untreated tallgrass prairies at DNWR, Iowa, USA, 1995–1997

Community	1995				1996				1997				*P* (Treatment)	*P* (Year)
	Burned		Untreated		Burned		Untreated		Burned		Untreated			
	Mean	SE	Mean	SE	Mean	SE	Mean	SE	Mean	SE	Mean	SE		
Plants														
Diversity	1.87	0.13	1.82	0.12	1.95	0.09	1.92	0.07	1.85	0.12	1.80	0.13	0.75	0.24
Dominance	0.22	0.03	0.21	0.03	0.21	0.02	0.23	0.02	0.24	0.03	0.26	0.05	0.79	0.19
Resident birds														
Diversity	1.57	0.13	1.27	0.12	1.74	0.07	1.72	0.17	1.91	0.06	1.57	0.09	0.11	< 0.01
Dominance	0.27	0.02	0.35	0.03	0.26	0.03	0.26	0.05	0.22	0.03	0.28	0.02	0.23	0.04

Means of species diversity and dominance compared through repeated measures (year) analysis of variance

C. fasciculata, a N-fixing legume, grew only on burned sites and only 1–2 years after burning occurred. *A. gerardii* declined in cover on burned and unburned sites with time. *S. scoparium* and *S. nutans* declined in coverage on untreated sites over time. Among species that demonstrated treatment–year interactive effects, *C. fasciculata* had greater coverage on burned sites in 1996 ($t_8 = -3.4$, $P = 0.03$, Bonferroni corrected). Other treatment differences in individual years among species were not significant ($P \geq 0.43$, *d.f.* = 8, all cases, Bonferroni corrected). In community structure, graminoid, shrub and total plant cover were affected by year but not treatment (Table 3). Graminoid and total plant cover declined, and shrub cover increased, with time on burned and unburned sites. Heights of individual species were not directly measured, but we observed visually similar plant height and stratification on burned and unburned sites, supporting the inference from coverage data that community structure was similar in both treatments.

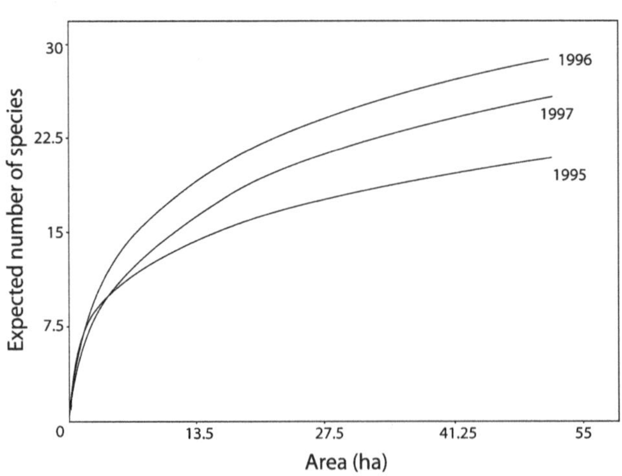

Fig. 1 Rarefaction analysis estimating the expected number of plant species per sampled area in tallgrass prairies at DNWR, Iowa, USA, 1995–1997. Different years plotted separately

Table 3 Percent ground coverage of 10 most common plants in burned and untreated tallgrass prairies at DNWR, Iowa, USA, 1995–1997

Community	1995				1996				1997				*P* (Treatment)	*P* (Year)
	Burned		Untreated		Burned		Untreated		Burned		Untreated			
	Mean	SE	Mean	SE	Mean	SE	Mean	SE	Mean	SE	Mean	SE		
Andropogon gerardii	11.0	4.5	9.9	2.0	9.8	4.3	7.7	2.6	5.1	1.1	3.8	1.1	0.75	< 0.01
Asclepias spp.	1.2	0.5	0.8	0.5	2.2	0.5	1.0	0.6	0.9	0.5	0.2	0.1	0.10	0.07
Bouteloua curtipendula	1.1	1.1	3.8	2.2	4.6	3.2	4.7	3.5	4.0	2.4	3.2	2.2	0.74	0.42
Carex spp.	0.3	0.1	0.3	0.3	0.2	0.1	0.5	0.3	0.1	< 0.1	0.9	0.7	0.48	0.75
Cassia fasciculata	0.0	0.0	0.0	0.0	3.1	1.2	0.0	0.0	0.7	0.4	0.0	0.0	0.01	< 0.01
Panicum virgatum	5.1	1.6	9.6	2.7	5.1	1.5	7.5	2.2	3.4	1.0	5.5	1.3	0.20	< 0.01
Polygonum pensylvanicum	0.9	0.7	0.7	0.5	0.8	0.7	1.7	1.3	0.4	0.3	2.2	1.6	0.56	0.70
Schizachyrium scoparium	2.9	1.9	5.1	0.9	1.3	0.8	2.3	0.9	1.8	0.9	1.1	0.8	0.43	0.01
Solidago canadensis	16.2	5.4	10.9	5.1	12.7	4.5	10.5	3.4	13.0	4.2	15.5	4.9	0.88	0.33
Sorghastrum nutans	6.7	2.2	5.3	1.6	3.6	1.0	1.6	0.4	4.5	1.9	1.5	0.6	0.28	< 0.01
Total grass	29.7	3.7	35.0	3.4	27.2	5.1	25.4	4.0	20.0	2.7	17.0	2.4	0.99	< 0.01
Total forb	25.5	7.4	15.3	4.3	22.6	6.3	16.0	4.3	17.6	5.3	20.8	5.9	0.66	0.80
Total shrub	0.0	0.0	0.0	0.0	1.5	0.9	1.0	0.3	0.7	0.4	0.6	0.4	0.77	< 0.01
Total plant cover	55.2	5.0	50.3	4.8	51.4	6.1	42.4	2.5	38.4	5.1	38.3	4.1	0.38	0.01

Means of cover of different treatments compared through repeated measures (year) analysis of variance. Percent cover arcsine-transformed for analysis

Plant communities on individual sites were dissimilar in year-to-year comparisons but not in comparisons between sites of different treatments. The average similarity of plots within the same treatment and year (baseline coefficient) was 0.43 (SE = 0.01). Burned and unburned sites contained essentially the same species with little decrease in similarity between treatments (treatment coefficient = – 0.009, SE = 0.007, $P = 0.45$), but site similarity decreased over time (year coefficient = – 0.086, SE = 0.008, $P < 0.01$). More than 70% of dissimilarity between sites was contributed by just 10 species (Table 4).

Burning had no effect on the total density of resident breeding birds or on the densities of individual species (Table 5). In contrast to plant communities, most species of birds were present on most sites. Fourteen species established territories on at least one site in every year, a total only slightly higher than the average number of species found on untreated sites (9–10) and burned sites (10–12) in 1996 and 1997. In 1996, the year of highest resident bird density, rarefaction analysis revealed an upper asymptote of 12 species at 10 ha (Fig. 2), a species–area relationship approximated on most of the larger sites. In years of lower population densities, upper asymptotes of expected numbers of species were associated with much larger areas (30–40 ha).

In population density, treatment effects were not statistically significant in any species. It is noteworthy, however, that field sparrow averaged twice the density on untreated sites as on burned sites throughout the study (Table 5). This difference was marginally nonsignificant ($F_1 = 4.6$, $P = 0.06$). However, our small sample size, combined with consistently large differences in density on different treatments in every year, makes it inappropriate for us to assert a conclusion of "no effect" of prescribed burning in this species. Rather, it raises the possibility of a biologically important effect that was not detected due to low statistical power associated with the test.

Total bird density and densities of seven species were affected by year, suggesting that annual populations of resident breeding birds on these sites varied significantly. In affected species, densities were lowest in 1995 and, in five of these, highest in 1996, as was total density. Six species showed the greatest proportional treatment differences in 1995, immediately after burning. One songbird, eastern kingbird (*Tyrannus tyrannus*) ($F_2 = 4.9$, $P = 0.02$), and two upland game birds, northern bobwhite

Table 4 Relative contribution to dissimilarity index of 10 plant species in plots sampled at DNWR, Iowa, USA, 1995–1997

Species	(%)
Graminoids	
Andropogon gerardii	4.76
Bouteloua curtipendula	4.29
Carex spp.	5.79
Panicum virgatum	10.91
Schizachyrium scoparium	8.15
Sorghastrum nutans	10.91
Forbs	
Apocynum cannabinum	3.84
Asclepias spp.	9.48
Polygonum pensylvanicum	3.41
Solidago canadensis	10.91
Total	72.45

Table 5 Densities (males/10 ha) of 12 most common resident grassland bird species in burned and untreated tallgrass prairies at DNWR, Iowa, USA, 1995–1997

Community	1995				1996				1997				*P* (Treatment)	*P* (Year)
	Burned		Untreated		Burned		Untreated		Burned		Untreated			
	Mean	SE	Mean	SE	Mean	SE	Mean	SE	Mean	SE	Mean	SE		
American Goldfinch	0.4	0.3	< 0.1	< 0.1	2.1	0.7	1.9	0.5	2.3	0.9	3.1	0.8	0.92	< 0.01
Brown-headed Cowbird	0.1	0.1	0.0	0.0	1.7	0.5	2.4	1.0	1.4	0.4	1.5	0.3	0.60	< 0.01
Common Yellowthroat	2.1	0.9	4.9	1.0	16.4	4.0	17.0	7.3	9.6	2.5	7.2	1.7	0.92	< 0.01
Dickcissel	4.3	1.3	1.9	1.2	3.2	1.1	2.6	0.8	1.9	0.3	2.0	1.2	0.44	0.27
Eastern Kingbird	0.2	0.1	0.1	0.1	0.9	0.2	2.3	0.6	0.8	0.3	0.6	0.2	0.29	< 0.01
Field Sparrow	2.5	0.8	5.8	1.3	5.0	0.9	10.1	2.6	3.8	0.6	7.0	2.2	0.06	0.03
Grasshopper Sparrow	0.2	0.2	0.2	0.2	0.2	0.1	0.0	0.0	0.3	0.2	< 0.1	< 0.1	0.24	0.66
Northern Bobwhite	0.2	0.1	0.2	0.1	0.8	0.3	0.7	0.3	1.4	0.3	0.4	0.1	0.14	0.01
Red-winged Blackbird	1.6	0.8	0.9	0.6	2.4	0.9	2.4	2.0	1.2	0.7	1.0	1.0	0.82	0.09
Ring-necked Pheasant	0.3	0.1	0.2	0.1	1.7	0.5	1.7	0.4	2.0	0.6	0.2	0.1	0.21	< 0.01
Vesper Sparrow	0.3	0.3	0.3	0.3	0.2	0.2	1.7	1.0	0.3	0.1	0.1	< 0.1	0.37	0.14
Western Meadowlark	0.1	0.1	< 0.1	< 0.1	0.5	0.5	0.1	0.1	0.2	0.2	0.0	0.0	0.41	0.29
Total Density	13.0	2.5	15.0	1.7	35.6	6.6	43.0	8.2	25.8	3.9	23.6	3.6	0.68	< 0.01

Means of species densities of different treatments compared through repeated measures (year) analysis of variance

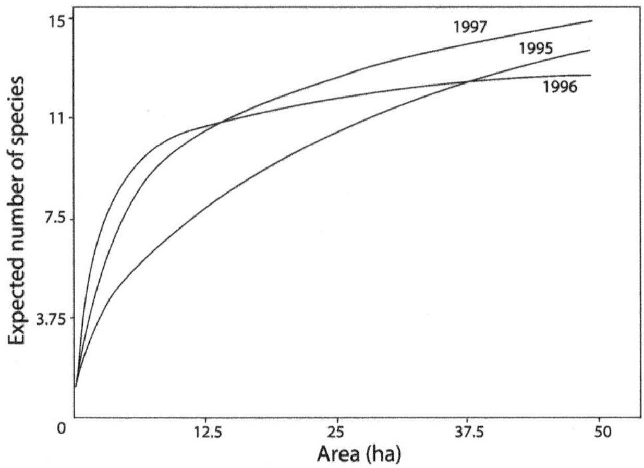

Fig. 2 Rarefaction analysis estimating the expected number of grassland resident bird species per sampled area in tallgrass prairies at DNWR, Iowa, USA, 1995–1997. Different years plotted separately

(*Colinus virginianus*) (F_2 = 3.9, P = 0.04) and ring-necked pheasant (*Phasianus colchicus*) (F_2 = 8.3, P < 0.01), showed significant treatment–year interactions. Among these species, the density of bobwhite was 3.5 times higher (t_8 = – 3.2, P = 0.04, Bonferroni corrected) and the density of pheasant 10 times higher (t_8 = – 3.0, P = 0.05, Bonferroni corrected) on burned sites two years after burning compared to untreated sites. Densities were not different between treatments in earlier years, suggesting a time lag in the effect of burning in these species.

Discussion

Factors affecting responses in plant communities

The low diversity and abundance of plant species in tallgrass prairie fragments at DNWR are, in part, reflections of site-specific management histories. Initial re-establishment of grasslands at DNWR did not use the historical array of plant species that were typical of tallgrass prairie systems before settlement and cultivation, but only a small number of native species that could be inexpensively and efficiently planted. Prairie fragments at DNWR retain the influence of their initial plantings, and their low levels of species diversity also are representative of the depauperate condition of their surrounding landscape, one that lacks larger prairie remnants and diverse seed banks.

In our study, only *Cassia fasciculata*, a N-fixing legume, was exclusive to burned sites. Burning is known to increase forb biodiversity in N-limited systems by benefiting species of N-limited legumes through increased N availability in soils (Dudley and Lajtha 1993; Towne and Knapp 1996). The fact that burning added only one such species in our study is yet further evidence of the impoverished state of local seed banks.

Factors affecting responses in bird communities

Like the simplified communities of plants they inhabited, communities of native grassland-dependent birds on burned and unburned sites also possessed low species diversity. Patterns of spatial distribution and community organization on these sites supported the long-held hypothesis that the persistence and success of species in grassland habitat are related directly to the ecological differences (i.e., niche separation) between them (Cody 1968). The small size of these fragments, coupled with their low diversity of plant species, probably reduced habitat heterogeneity and contributed to reduced opportunity for niche separation, resulting in low species diversity.

At DNWR, area sensitivity, another important constraint on bird species diversity, might be mediated by density. Rarefaction analysis suggested that, at high densities (1996), most species were present on larger sites. This pattern of occupancy is consistent with predictions of the Ideal Free Distribution theorem (Rosenzweig 1991), which asserts that strength of habitat selection declines with increasing population density. Our results also are consistent with species-specific null models which suggest that, at low densities, a species' probability of occurrence decreases with decreasing field size (Horn et al. 2000). Such considerations might explain why rarefaction analysis showed fewer species associated with areas in the size range of our sites (3–10 ha) in a year of low population densities (1995) than in years of higher population densities (1996 and 1997).

Both species of upland game birds, northern bobwhite and ring-necked pheasant, experienced increases in density on burned sites by the second year after burning, but not before. These species are ground-nesters that begin incubating in April at DNWR, before spring burning occurred. Thus, spring burning in 1995 destroyed some nests, making its immediate effects detrimental. However, increased densities on burned sites after 2 years suggested that longer-term effects of burning might be beneficial. Although we cannot identify the mechanism of such increases given the apparent similarity of plant communities on burned and unburned sites, such patterns of population change should alert managers to the fact that the effects of burning on individual species might involve time lags that would not be detected without ongoing, longer-term monitoring.

Management implications

Plant communities

Our results demonstrated that, in plant communities, many small sites do make cumulative contributions to landscape biodiversity. Thus, there is value in managing such small sites effectively and intensively, and they should not be neglected in an overall strategy of plant conservation.

At DNWR, most grass cover on burned sites was contributed by just five species of C_4 grasses (*Andropogon gerardii*, *Bouteloua curtipendula*, *Panicum virgatum*, *Schizachyrium scoparium* and *Sorghastrum nutans*). These species benefit from early spring burning on these sites, but their resulting dominance does not encourage the establishment of more diverse and historically representative plant communities. If historic plant diversity is the normative management goal, managers might achieve greater success through summer burns rather than fall or spring burns (Howe 1994a, b), and by using burns less frequently rather than more frequently.

Bird communities

Given that burning had no effect on species richness and diversity of resident grassland birds, species–area relationships might be more important determinants of these variables, especially given the demonstrated sensitivity to area in some species (Herkert 1994a, b; Vickery et al. 1994; Swengel 1996). It is likely that grassland sites need to be more than 50 ha in size, preferably at least 200 ha, if they are to support a high level of avian biodiversity (Vickery et al. 1994). Field sparrow, the only species suggesting sensitivity to treatment in our study, also was the only species known to be insensitive to area (Vickery et al. 1994). Managers might have greater success in enhancing bird species diversity, as well as increasing densities of individual species, by acquiring and restoring larger sites rather than investing in intensive management of smaller sites.

Although our results demonstrated that many small sites can contribute to increased landscape diversity of plant communities, the collective contribution of small sites was less pronounced for bird communities, particularly at higher population densities. At DNWR, it appears that adding additional small sites would provide little increase in total bird species diversity. If this is the case, managers should use small fragments to benefit selected species while simultaneously working to acquire and restore sites > 200 ha in size in order to develop larger and more diverse avian communities.

Acknowledgements Funding was provided through the USFWS Cooperative Cost Share Program, the USFWS Nongame Bird Research Program and Northwestern College (Iowa, USA). We thank the staff of DNWR, especially G. Gage, S. Van Riper, M. Buskey and M. Sheets who assisted in selection and preparation of sites and provided vehicles for travel in the field. R. M. King, USFS, provided expertise in statistical analysis and suggestions to improve preliminary versions of this manuscript. L. K. Page and J. A. Darragh, Wheaton College (Illinois, USA), offered comments on a preliminary draft of the manuscript.

References

Askins RA (1993) Population trends in grassland, shrubland, and forest birds in eastern North America. Curr Ornithol 11:1–34

Cody ML (1968) On the methods of resource division in grassland bird communities. Am Nat 102:107–147

Collins SL (2000) Disturbance frequency and community stability in native tallgrass prairie. Am Nat 155:311–325

Collins SL, Knapp AK, Briggs JM, Steinauer EM (1998) Modulation of diversity by grazing and mowing in native tallgrass prairies. Science 280:745–747

Cully JF Jr, Michaels HL (2000) Henslow's sparrow habitat associations on Kansas tallgrass prairie. Wilson Bull 112:115–123

Dudley JL, Lajtha K (1993) The effect of prescribed burning on nutrient availability and primary production in sandplain grasslands. Am Midl Nat 130:286–298

Dyer DP (1978) An analysis of species dissimilarity using multiple environmental variables. Ecology 59:117–125

Edgington ES (1995) Randomization tests, 3rd edn. Marcel Dekker, New York, USA

Herkert JR (1994a) The effects of habitat fragmentation on Midwestern grassland bird communities. Ecol Appl 4:461–471
Herkert JR (1994b) Status and habitat selection of the Henslow's sparrow in Illinois. Wilson Bull 106:35–45
Horn DJ, Fletcher RJ Jr., Koford RL (2000) Detecting area sensitivity: a comment on previous studies. Am Midl Nat 144:28–35
Howe HF (1994a) Response of early- and late-flowering plants to fire season in experimental prairies. Ecol Appl 4:121–133
Howe HF (1994b) Managing species diversity in tallgrass prairie: assumptions and implications. Conserv Biol 8:691–704
International Bird Census Committee (1970) Recommendations for an international standard for a mapping method in bird census work. Swedish National Scientific Research Council, Stockholm. Bull Ecol Res Comm 9:49–52
Janzen DH (1983) No park is an island: increase in interference from outside as park size decreases. Oikos 41:402–410
Krebs CJ (1989) Ecological methodology. Harper Collins, New York, USA
Leach MK, Givnish TJ (1996) Ecological determinants of species loss in remnant prairies. Science 273:1555–1558
Noss R, Csuti B (1992) Habitat fragmentation. In: Meffe GK, Carroll CR (eds) Principles of conservation biology. Sinauer Associates, Sunderland, Massachusetts, USA, pp 269–304
Pielou EC (1975) Ecological diversity. Wiley, New York, USA
Rosenzweig ML (1991) Habitat selection and population interactions: the search for mechanism. Am Nat 137:S5–S28
Samson F, Knopf F (1994) Prairie conservation in North America. BioScience 44:418–421
Shannon CE, Weaver W (1949) The mathematical theory of communication. University of Illinois Press, Urbana, USA
Swengel SR (1996) Management responses of three species of declining sparrows in tallgrass prairie. Bird Conserv Int 6:241–253
Towne EG, Knapp AK (1996) Biomass and density responses in tallgrass prairie legumes to annual fire and topographic position. Am J Bot 83:175–179
Vickery PD, Hunter ML, Melvin SM (1994) Effects of habitat area on the distribution of grassland birds in Maine. Conserv Biol 8:1087–1097
Yahner RH (1988) Changes in wildlife communities near edges. Conserv Biol 2:333–339

Biodivers Conserv (2007) 16:841–856
DOI 10.1007/s10531-006-9090-1

ORIGINAL PAPER

Population trends and spatial synchrony in peripheral populations of the endangered Lesser grey shrike in response to environmental change

David Giralt · Francisco Valera

Received: 9 November 2005 / Accepted: 7 July 2006 / Published online: 11 November 2006

Abstract Regional synchronization in species dynamics as well as particular ecological and demographic characteristics of peripheral populations poses special challenges for conservation purposes, particularly under the current scenario of global climate change. Here, we study the population trend and spatial synchrony of several peripheral populations of the endangered Lesser grey shrike *Lanius minor* at the western limit of its breeding range (southern France and northeast Spain). In an attempt to ascertain the effect of environmental change on the decline of the species we also look for evidence of climate changes in the breeding and wintering area of this shrike and related effects on vegetation by using the normalized difference vegetation index (NDVI). We found that the interannual fluctuations of the peripheral populations in France and Spain are strongly correlated, therefore suggesting that their decline can be under the influence of a common factor. We obtained clear evidence of climatic change (an increased thermal oscillation) in one peripheral population that could have resulted in a decrease of the NDVI index in the area. Our study finds correlational evidence that climatic variables in the breeding area may account for fluctuations in abundances of some populations and that environmental conditions experimented by some population could influence the fate of the neighboring populations. Our results indicate that the studied peripheral populations are spatially synchronized, so that conservation efforts should be applied at a large-scale encompassing all the isolated populations at the western border of the range of the species in the Mediterranean area.

D. Giralt
Centre Tecnològic Forestal de Catalunya, 25280 Solsona, Spain
e-mail: david.giralt@ctfc.es

D. Giralt
Departament de Biología Animal, Universitat de Barcelona, Facultat de Biología, 08028 Barcelona, Spain

F. Valera (✉)
Estación Experimental de Zonas Áridas (CSIC), General Segura 1, Almeria E-04001, Spain
e-mail: pvalera@eeza.csic.es

Keywords *Lanius minor* · Mediterranean region · Population trend · Range periphery · Spatial synchrony

Introduction

Understanding the spatial distribution of organisms is a crucial issue in population dynamics (Tilman and Kareiva 1997) and has important consequences for more applied sciences such as conservation biology, where strategies for long-term species conservation depend on present and future patterns of abundance (Lawton et al. 1994; Brown et al. 1995; Mehlman 1997; Williams et al. 2003). One major topic in this field is the comprehension of the abundance distribution within the range of a given species and the roles of density-dependent and density-independent processes in determining the variability of population abundances (Hengeveld and Haeck 1982; Brown 1984; Royama 1992; Mehlman 1997; Williams et al. 2003). It is well-known that peripheral populations, frequently relatively small and isolated from central populations (Lawton 1993; Lesica and Allendorf 1995), are likely to experience different regimes of natural selection than central ones, since the relative importance of abiotic and biotic factors on distribution patterns and population limitation are likely to change according to the position within the geographical range (Randall 1982; Hoffmann and Blows 1994; Brown et al. 1995; Williams et al. 2003). There is evidence supporting the fact that environmental changes and abiotic, density-independent factors, like weather, have a higher influence on demographic rates and produce greater fluctuations in peripheral populations than in central ones (Hoffmann and Blows 1994; Brown et al. 1995; Curnutt et al. 1996; Williams et al. 2003), probably because closeness to the edge of range usually indicates poorer environmental conditions for a species (Brown 1984; Brown et al. 1995). Thus, in the present scenario of climate change (Parmesan et al. 1999; Hughes 2000), we would expect peripheral populations of organisms to be under a stronger influence of environmental changes than populations closer to the core of their range.

An additional factor of uttermost importance in the conservation of peripheral populations and/or of rare or endangered species is spatial synchrony in the dynamics of local populations (Kendall et al. 2000). Spatial synchrony refers to coincident changes in the abundance or other time-varying characteristics of geographically disjunct populations (Liebhold et al. 2004). Evidence for widespread spatial synchrony in population fluctuations has been found in a variety of organisms (Paradis et al. 2000; Kendall et al. 2000), and some studies have found that weather is a likely candidate as a synchronizing factor (Paradis et al. 2000; Williams et al. 2003).

Climate per se is a major determinant of geographical distribution for many organisms and recent climate warming has been shown to affect the distribution of different species (Thomas and Lennon 1999; Parmesan et al. 1999; Hughes 2000). However, the potential impact of a change in environmental suitability on abundance within the range of a given species has received relatively little attention (Mehlman 1997; Williams et al. 2003). This is partly because there is generally little comprehensive information on the distribution of abundance within the range of a species and because long-term estimates of densities over wide geographic areas are uncommon (Williams et al. 2003).

In this paper, we study the pattern of population synchrony of three peripheral populations of the Lesser grey shrike *Lanius minor*, and the relationship between population fluctuations and weather. The Lesser grey shrike is highly endangered throughout Europe, having declined markedly in abundance and range in the last decades (Lefranc and Worfolk 1997). Relict populations of the species in the western limit of its breeding range are known to exist since long (Lefranc 1995). Such peripheral populations, now restricted to very small areas in southern France and northeast Spain, have been reported to decline in the last decades (Lefranc 1995; Giralt and Bota 2003, Giralt 2004). Overall, there is almost no information on which factors may be responsible for the general decline of the species (Lefranc and Worfolk 1997; Krištín et al. 2000). Some authors have pointed out loss of habitat (agricultural intensification) and adverse weather (Lefranc 1995, 1997; Lefranc and Worfolk 1997; Isenmann and Debout 2000) whereas others suggest that adverse circumstances on the wintering grounds or during migration may account for the decline (Herremans 1997a, 1998a). However, no specific work has investigated the causes underlying the decrease of the species. Whereas there is not much information on this shrike (but see Lefranc 1995; Herremans 1997a; Lefranc and Worfolk 1997; Isenmann et al. 2000; Krištín et al. 2000), the distribution of abundance of the species in its westernmost range and changes in the last decade are available, what provides a valuable opportunity for studying population fluctuations and general patterns of changes in abundance in response to environmental changes. Therefore we aim to: (i) evaluate the spatial scale of synchrony in fluctuations of several peripheral populations of this species, (ii) assess the role of climate on the decline of this shrike in an attempt to contribute to the conservation of this particular species. Additionally we aim to illustrate the challenges associated to the conservation of peripheral populations of endangered birds.

Methods

Study species and study area

The Lesser grey shrike is a socially monogamous long-distance migratory passerine whose breeding range is limited to warmer parts of Eurasia, spreading over 6,000 km from west to east (Cramp and Perrins 1993). The westernmost points reached by this species lie in southern France and northeast Spain (Lefranc and Worfolk 1997), at the farthest extreme of its migratory route. This shrike produces a single brood per season, although replacement clutches can be produced after nest failure. Birds arrive on the breeding grounds in Europe during May. In the study area the main egg-laying activity takes place from late May to early June and the main fledging period encompasses from late June to early July (Isenmann and Debout 2000; pers. obs.).

Two main breeding areas existed in Spain until recently (Giralt and Bota 2003). The breeding area in Girona (42° 16′ 42.84″ N 3°07′ 21.66″ E, Catalonia) lay in the Natural Park of Aiguamolls de l'Empordá (protected area since 1983) and its periphery, where the species bred for the last time in 2001 (Fig. 1). The breeding area in Lleida (Catalonia) is 220 km southwest from the first one, and consists of two nuclei 5,5 km apart from each other (Fig. 1). A third, relict nucleus remains in Aragón (35 km west of Lleida).

The breeding population of the study species in France is located in two main breeding areas, Montpellier (43°30′ 44.66″ N 3°39′ 15.26″ E, Hérault) (Isenmann et al. 2000) and Aude (43°15′ 53.48″ N 3°08′ 51.78″ E, departments of Aude and Hérault, Bara 1995), 60 km from each other and 180 and 120 km from the breeding nucleus in Girona, respectively (Fig. 1). Some isolated pairs (3 in 2002 and 1–2 in 2004) still breed in Vaunage (Gard) (Labouyrie 2003; pers. comm.).

The non-breeding range of the Lesser grey shrike is about one tenth of the size of its breeding range. During the non-breeding season, the world population concentrates in the southern African thornbelt, mainly in the Kalahari basin (Herremans 1997a, 1998a), spreading mostly over Botswana, Namibia and north of South Africa. Birds occupy their final non-breeding destination in January–March and almost all individuals have disappeared from the winter quarters by the end of April (Herremans 1997b).

Abundance data

Data on the past distribution of the species in the Iberian Peninsula have been collected from old and recent literature and personal communications. Long-term data on breeding population come from Girona where the population has been monitored during 1989–1997 (except 1992) by the staff of the Natural Park, and by D.G. from 1998 to 2004. Data on the current distribution of the species in Lleida have been collected on the basis of censuses performed by D.G. during the entire breeding season (15 May–10 July) for the period 2001–2004.

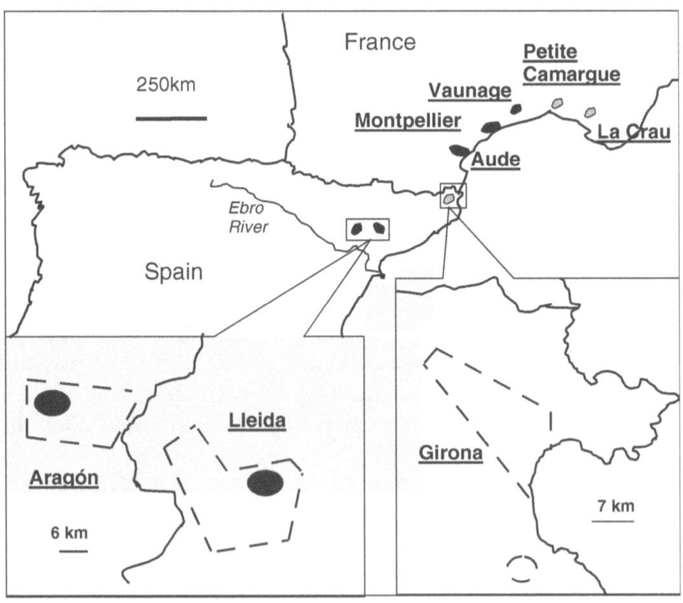

Fig. 1 Contraction of the breeding range of the French and Spanish populations of Lesser grey shrike. Current breeding areas (in black), past breeding locations (shaded areas) and contraction of the range (discontinuous lines) in the study areas are shown

Information about the two main areas in France was collected from Bara (1995), Isenmann et al. (2000), Donck and Bara (2001) and Rufray and Rousseau (2004) and kindly completed by Isenmann, Lefranc and Rufray. Censuses in the population in Aude started in 1992 (Bara 1995). The breeding population in Montpellier was discovered in 1995 (Béchet et al. 1995) and censused from that year onwards.

Abundance data from Spain and France are the result of censuses specifically undertaken to monitor the breeding population of this species. They were performed from territory establishment until fledging. Thus, such information reflects reliable actual counts.

Population trends, fluctuations and spatial synchrony

We studied the occurrence and intensity of population synchrony by correlating the fluctuations in the time series of the French and the Spanish (only Girona) populations (Paradis et al. 2000). We used the software program TRIM (Pannekoek and van Strien 2003), designed to analyze time series of counts with missing observations by using Poisson regressions, that produces estimates of yearly indices and trends.

First, we studied the trend of the French population (Aude and Montpellier) from 1993 to 2004 (two missing data—1993 and 1994—for the population in Montpellier) and of the population in Girona from 1993 to 2002 (although the species bred there for the last time in 2001 we assigned 1 pair to 2002 to allow a better comparison between both populations—see below). For these purposes we used a linear trend model with stepwise selection of change-points. Change-points are moments in time (i.e. years) where the slope parameter changes. This model provides an overall trend as well as selects specific time points (i.e. years) when the slope parameters differ significantly from the ones obtained before and after that time point. Possible violations to the assumption of Poisson distributions due to overdispersion or serial correlation were corrected with the methods implemented in TRIM.

Then, we explored whether the fluctuations of the Spanish (Girona) and French population (Aude and Montpellier) are interrelated by examining the relationship between the annual departures from the long-term trends of each population. For this we first calculated the trend for each population from 1993 to 2002 and the yearly deviations from each linear trend by using the time effects model implemented in TRIM. Since null values are not admitted when using the time effects model we restricted the study period until 2002 (breeding in Girona did not occur from 2002 onwards, see results), and considered one pair to breed in 2002 after checking that the linear trend obtained was similar to the one estimated with the linear trend model run with zero breeding pairs in 2002. In contrast to the linear trend model this one calculates separate parameters for each year and estimates yearly deviations from the linear trend. This model provides the best estimates for deviations of the general trend (van Strien, pers. comm.). Finally, we correlated the yearly deviations from the linear trends found in France and Girona.

Meteorological variables

We used meteorological data from the meteorological stations closest to each breeding population: Mauguio–Montpellier (Montpellier), Béziers (Aude) and Aiguamolls de l'Empordà (Girona), all of them less than 15 km from the respective breeding sites. Weather records consisted in daily rainfall (mm) during

January–July for the period 1989–2001 for Girona and maximum and minimum daily temperatures (°C) during May–June (encompassing most of the breeding period of the study species) for the period 1989–2002 for Aude, Montpellier and Girona. Daily thermal oscillation was calculated as the difference between the latter variables. Mean temperatures were calculated as the average of daily maximum and minimum temperatures.

Rainfall in Botswana and Namibia (October–February) was also gathered from the Tyndall Center for Climate Change Research (data set TYN CY 1.1, Mitchell et al. 2002). Since rainfall varies widely in and around the Kalahari we compared seasonal rainfall (October–February) during the years 1989–2000 with the mean rainfall for those months in the same areas for the larger period 1902–2000 (data set TYN CY 1.1). We got an average of 502.93 mm. We defined a threshold value of 150 mm below and above the long-term mean to classify years of the period 1989–2000 into 'low', 'average' and 'high' rainfall years (thresholds of 352.93 and 652.93 mm, respectively). Similar classifications have been applied in other studies (see, for instance, Wiegand et al. 1999, Tews and Jeltsch 2004).

Normalized difference vegetation index (NDVI)

The amount and vigor of vegetation at the land surface was estimated by means of the NDVI. This index, based on satellite images indicating the condition of rainfall-dependent vegetation in time, is strongly correlated with the fraction of photosynthetically active radiation absorbed by vegetation (see Asrar et al. 1984; Prince and Justine 1991; Myneni et al. 1997 for more details about the index and Sanz et al. 2003 for a similar use of the index).

NDVI data corrected by surface topography, land-cover type, presence of clouds and solar zenith angle were provided by Clark Labs in IDRISI format as world monthly images at spatial resolution of 0.1° in a 0–255 scale values. Using IDRISI32 software, we obtained mean NDVI values for the period May–June (at 1-month interval) from 1988 to 2000 for the square areas sized 0.25° occupied by breeding populations in Montpellier (E 3°30′–3°45′ N 43°25′–43°40′), Aude (E 3°–3°15′ N 43°10′–43°25′) and Girona (E 3°–3°15′ N 42°10′–42°25′) and for the period January–March (when most birds occupy their final wintering destination) from 1988 to 2000 for the wintering area (E 18°–28°, S 20°–27°). The selected wintering area matches the Kalahari basin defined by Herremans (1997a, 1998a) as the core area for the Lesser grey shrike.

Statistical analyses

Separate stepwise multiple regressions were performed to determine the effect of climatic variables (thermal oscillation) and NDVI indexes on the population size of each of the three peripheral areas. Thermal oscillations in the three study areas during May–June were the independent variables for the first set of analyses. For the second set we used the mean NDVI index during May–June of each study area as well as the mean NDVI index for January-March in the wintering grounds in Africa. Parametric tests were used where the assumptions for normality were met. In some cases transformations were used to meet the requirements for normality. Otherwise non-parametric tests were used. Statistical analyses were carried out with the

STATISTICA 6.0 package (StatSoft Inc. 2001). Unless otherwise stated means and standard errors are offered and two-tailed tests used.

Results

Contraction of the breeding range

The geographic range of the species at the southwestern limit of its distribution has contracted dramatically. The breeding nucleus in Girona, that held in 1989 around 50% of the estimated breeding population in Spain, contracted progressively until it became extinct in 2002 (Fig. 1). Although the breeding population in Lleida seems relatively stable during the last years (1.4, 2.4, 1.4 and 2.3 breeding pairs/km^2 for 2001–2004) the breeding range has contracted ca. 50% in last decade due to the progressive disappearance of breeding sites to the south and west of the current breeding area (Fig. 1). As a whole, the contraction of the range in Catalonia (Lleida and Girona) has been estimated at –68% between 1984 and 2002 (Giralt 2004).

The breeding area in Aragón, with ca. 2–7 pairs in the last 3 years, has also contracted during the last decade after disappearance of the easternmost breeding sites (Giralt and Bota 2003; Albero and Rivas, pers. comm.) (Fig. 1).

The French population has also declined in range due to the loss of breeding localities during the 90s in Gard (Petite Camargue) and Bouches du Rhône (La Crau), a stronghold of the species in the 70s (Lefranc 1999; Donck and Bara 2001; Labouyrie 2003) (Fig. 1). More recently, a 50% reduction of the number of pairs in the main nuclei (Aude and Montpellier) since 2002 (Rufray and Rousseau 2004) has contributed to the contraction of the range (Fig. 1).

Population trend and spatial synchrony in peripheral populations

Breeding populations of the Lesser grey shrike have been decimated along the southwestern range of its distribution (Fig. 2).

The decline of the species in France (Montpellier and Aude) fits a linear model (Likelihood ratio = 1.83, df = 10, $P = 0.99$) with a significant decreasing slope of 8.25% per year for the period 1993–2004 (Overall Multiplicative Slope imputed with intercept = 0.917, SE = 0.0069, $P < 0.05$). Particularly marked decreases in the trend occur between 1997–1998, 1999–2000, and 2001–2002 (Tables 1 and 3). During 2002–2004 there was a constant yearly decrease of 75% that has resulted in the lowest ever numbers of the French population (Fig. 2).

The decline of the Spanish population in Girona during the period 1993–2002 also fits a linear model (Likelihood ratio = 0.03, df = 1, $P = 0.85$) with a significant decreasing slope of around 14% per year for the period 1993–2002 (Overall Multiplicative Slope imputed with intercept = 0.856, SE = 0.013, $P < 0.05$) (Fig. 2). Similar to the French population, this tendency is not constant and several significant change-points can be distinguished (Table 2). Sharp decreases occur between 1993–1994, 1995–1996, and 1997–1998 whereas increases occurred between 1994–1995, 1996–1997 and 1998–1999 (Tables 2 and 3). During 1999–2001 there was a constant yearly decrease of 76% that, in fact, led to the extinction of the species in Girona from 2002 onwards (Table 3, Fig. 2).

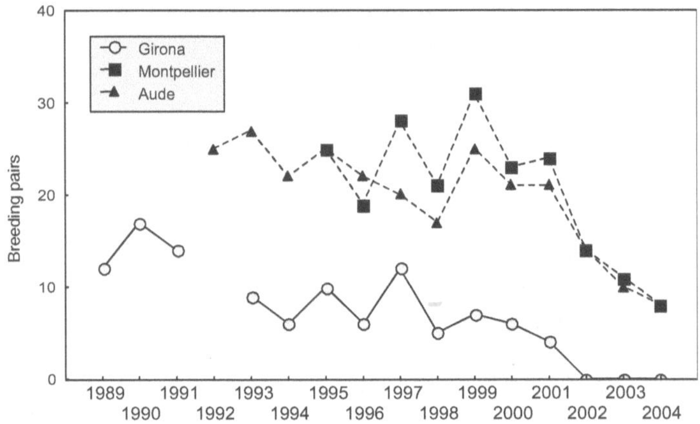

Fig. 2 Population trend of the Lesser grey shrike in Girona (Spain) (open circles) and two French populations (filled symbols) during the period 1989–2004

Table 1 Wald-tests for the significance between time-points of changes in the slope of the population trend of the Lesser grey shrike in France (Montpellier and Aude) during the period 1993–2004

Change-point	Wald-test	df	*P*
1993	2.80	1	0.09
1994	1.75	1	0.18
1995	2.64	1	0.10
1996	5.05	1	0.02
1997	6.29	1	0.01
1998	14.77	1	0.0001
1999	18.09	1	0.0000
2000	2.92	1	0.08
2001	9.59	1	0.002
2002	1.81	1	0.17

Annual deviations of the Spanish (Girona) and French population (Aude and Montpellier) from their respective long-term trends during 1993–2002 are strongly correlated (Pearson correlation, $r = 0.70$, $P = 0.024$, $n = 10$) (Fig. 3).

NDVI index and climate change in breeding and wintering grounds

The average of the mean temperature for May–June has not changed significantly in the period 1989–2002 in any of the studied locations (Girona, Aude, and Montpellier) (Pearson correlations, $P > 0.10$ and $n = 14$ for all cases). However, the average

Table 2 Wald-tests for the significance between time-points of changes in the slope of the population trend of the Lesser grey shrike in Girona (Spain) during the period 1993–2002

Change-point	Wald-test	df	*P*
1993	9.19	1	0.002
1994	14.59	1	0.0001
1995	21.38	1	0.0000
1996	26.11	1	0.0000
1997	52.98	1	0.0000
1998	24.20	1	0.0000
1999	11.64	1	0.0006
2001	14.91	1	0.0001

Table 3 Parameter estimates of the trend for each time interval (defined after significant change-points) referred to the French and Spanish population size in the previous interval

French population			Spanish population		
From–up to	Multiplicative slope	Standard error	From–up to	Multiplicative slope	Standard error
1993–1994	0.81	0.099	1993–1994	0.67	0.089
1994–1995	1.07	0.11	1994–1995	1.67	0.21
1995–1996	0.82	0.07	1995–1996	0.60	0.07
1996–1997	1.17	0.10	1996–1997	2.00	0.25
1997–1998	0.79	0.07	1997–1998	0.41	0.05
1998–1999	1.47	0.13	1998–1999	1.45	0.20
1999–2000	0.78	0.06	1999–2001	0.76	0.05
2000–2001	1.02	0.09			
2001–2002	0.62	0.06	2001–2002	0.23	0.06
2002–2004	0.75	0.05	–	–	–

thermal oscillation for the same season (May–June) and period has increased dramatically in Girona ($r = 0.90$, $P < 0.0001$) whereas there is no significant change either in Montpellier ($r = -0.39$, $P = 0.17$) or Aude ($r = -0.16$, $P = 0.58$). Precipitation (total amount during January–July) in Girona has not changed during 1989–2001 ($r = -0.30$, $P = 0.32$).

Climatic changes seem to have influenced the vegetation in Girona, where the mean NDVI index for May–June has decreased significantly during 1989–2000 ($r = -0.62$, $P = 0.033$, $n = 12$). We found non-significant negative correlations in the mean NDVI index for the same season (May–June) and period in the other study areas (Montpellier, $r = -0.16$; Aude, $r = -0.07$, $P > 0.50$ and $n = 12$ for all cases).

Overall, the NDVI index recorded during 1989–2000 for each area correlates with the one in the other areas (Pearson correlations, $P < 0.01$ in all cases).

An analysis of rainfall in the wintering grounds (October–February) during the period 1989–2000 suggests that it has not changed significantly ($r = 0.10$, $P = 0.76$, $n = 12$). A comparison of precipitations during October–February for each of the years in that period with the average precipitation (502.93 mm) and the threshold values (352.93 and 652.93 mm) (see Methods) for the longer series 1902–2000 suggests that only two seasons in the period 1989–2000 (1991/1992—273.1 mm—and 1994/1995—311.4 mm) can be classified as dry whereas the remaining seasons have average precipitations. Accordingly, the mean NDVI index for January-March does not show any trend along 1988–2000 ($r = 0.11$, $P = 0.72$, $n = 13$).

Population declines and climatic variables

Variation in the number of breeding pairs in Girona during the period 1989–2002 (except 1992) can be explained by climatic variables (i.e. thermal oscillation during May–June) in the local area. A multiple regression analysis provided a significant model (Table 4) where only thermal oscillation in Girona proved significant ($\beta = -0.92$, $P = 0.0007$), suggesting that the larger the thermal oscillation, the less birds in Girona (Table 4). Repeating the analysis with the NDVI values for the breeding locations and the wintering grounds (period 1989–2000) as independent variables yields a significant model (Table 4) where only the NDVI in Aude is excluded. The strongest effect was found for the NDVI in Girona

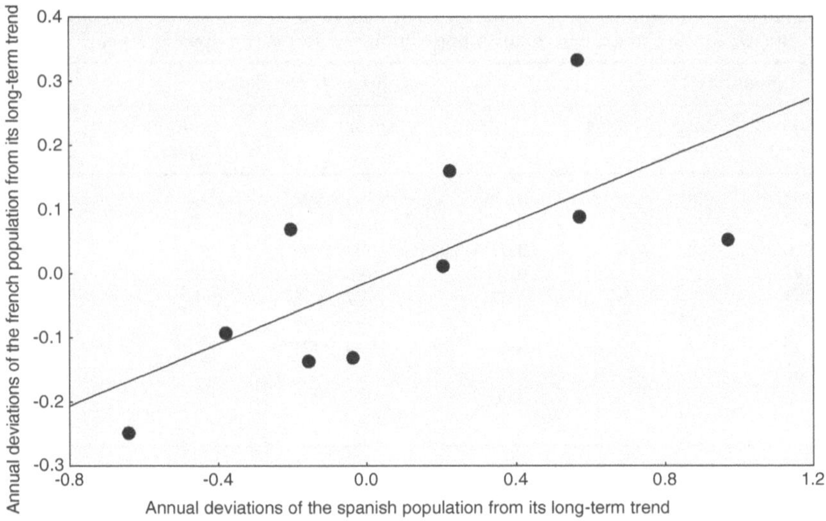

Fig. 3 Correlation between the annual deviations from the long-term trend calculated separately for the breeding populations of Lesser grey shrike in Girona (Spain) and France (Montpellier and Aude) from 1993 to 2002

(β = 1.44, P = 0.006), the other significant variables being the NDVI in Montpellier (β = –1.22, P = 0.009) and in the wintering grounds (β = –0.78, P = 0.01) (Table 4).

The fluctuations in the number of breeding pairs in Aude during the period 1992–2002 can be explained by a marginally significant model (P = 0.056, Table 4) where

Table 4 Results of stepwise multiple regression analyses with number of breeding Lesser grey shrike pairs in three peripheral populations (Girona, Aude and Montpellier) as dependent variables and (i) climatic variables (mean thermal oscillation during May–June), (ii) vegetation variables (mean NDVI index during May–June for the breeding locations and mean NDVI index during January–March for the wintering area) in each of these locations as independent ones. β coefficients, P values and statistics of each model are shown

Location	Thermal oscillation			Vegetation variables (NDVI)			
	β coeff.	P	Model		β coeff.	P	Model
Girona							
Girona	–0.92	<0.001	$F_{3,9}$ = 8.5,	Girona	1.44	<0.01	$F_{3,7}$ = 6.3,
Montpellier	–0.38	0.17	P < 0.01,	Montpellier	–1.22	<0.01	P = 0.02,
Aude	0.25	0.33	R^2 = 0.74	Wintering area	–0.78	0.01	R^2 = 0.73
Aude							
Girona	–0.71	<0.05	$F_{2,8}$ = 4.2, P = 0.056,	Wintering area	–0.42	0.25	$F_{1,7}$ = 1.5, P = 0.25,
Montpellier	0.29	0.28	R^2 = 0.51				R^2 = 0.17
Montpellier							
Montpellier	–0.39	0.33	$F_{1,6}$ = 1.1, P = 0.33, R^2 = 0.15	No variable entered			

only thermal oscillation in Girona had a significant effect (β = −0.71, P = 0.021). Remarkably, thermal fluctuations in Aude did not enter the model. Variations in the breeding population in Aude during 1992–2000 could not be explained by the NDVI indexes in the breeding and wintering areas (non-significant model, Table 4).

Finally, we did not find any significant model explaining variation in the breeding population in Montpellier either with thermal oscillation (period 1995–2002, Table 4) or with NDVI indexes (period 1995–2000, no variable entered the model).

Discussion

Both the breeding range and the abundance of the Lesser grey shrike at the southwestern limit of its distribution have drastically decreased in the last decades. The French population has been decimated in a few years with a yearly decline of 8.25% during 1993–2004. A stronger yearly decline of 14% during 1993–2002 in the population in Girona resulted in the extirpation of the species from the northeast portion of its range in Spain. The extinction of the species in this protected area in 2002 seems durable and this may very well be the fate for the French population in the near future (Rufray and Rousseau 2004).

Several reasons have been argued to explain the decline of the Lesser grey shrike, most of them pointing out at large-scale processes (like agricultural intensification or climatic changes) acting on the breeding areas (Lefranc and Worfolk 1997; Isenmann and Debout 2000) or on the winter quarters (Herremans 1998a). However, to our knowledge, no specific study has been done in this regard. Our results show correlated fluctuations of the French and the Spanish population in the last decade (suggesting that their decline can be under the influence of a common factor) together with concomitant climate changes, what suggests that climate may play a role in the decline of the species (see Schaub et al. 2005). We found clear evidence of climatic change in our study area in Girona (a drastic increase in thermal oscillation during May–June) in the last years, what agrees with other studies performed in the area (Piñol et al. 1998; Peñuelas et al. 2002; Gordo et al. 2005). Specifically, Piñol et al. (1998) described an increase in temperature in northeastern Spain resulting from increased maximum but not minimum temperatures (1910–1994 period) or from the increase of both maximum and minimum temperatures (1968–1994 period). Such changes may have probably resulted in the observed decrease of the NDVI index in the area, which suggests less vigorous and abundant vegetation. For the two other study areas in France we found no significant trends either in climatic variables or vegetation index during the studied period (May–June). However, there is strong evidence supporting climate change in those areas too. Spring temperature (March–May) rose significantly at Montpellier during the period 1984–1992 (Chuine et al. 1998) and in this same area the flowering date occurred at progressively earlier dates during this period (Osborne et al. 2000).

Climatic conditions can directly influence survival and reproduction in birds (Ringsby et al. 2002 and references therein). Moreover, the peripheral location of these breeding populations could enhance the negative effect of climate change since density-independent factors, like weather, have a high influence on demographic rates in border areas (Hoffmann and Blows 1994; Mehlman 1997; Williams et al. 2003). Thus, it could be that shrikes at the border of the range are more sensitive to these changes than populations at the core of the area. In line with this, the only comparative data available come from a Slovak population, connected to the main breeding area, which keeps a stable and dense population since years (Krištín 1995; Krištín et al. 2000).

A non-mutually exclusive explanation is that climate changes may have altered the availability/quality of basic resources (like habitat quality or food abundance) that are known to largely influence the size of peripheral populations (McArdle et al. 1990; Williams et al. 2003). In fact, we found a positive relation between the breeding population in Girona and the NDVI index that, in turn, decreases with time. Alterations in northeast Spain in the last decades (like increased aridity—Piñol et al. 1998—or altered life cycles of plants and insects—Peñuelas et al. 2002) due to climatic changes may have resulted in a decrease of food availability and/or in a decoupling between bird physiology and their food supply (Crick et al. 1997; Visser et al. 1998; Sanz et al. 2003) that could influence the reproduction of this predator specialized on large phytophagous insects (Krištín 1995).

Alternatively, human-related activities could account for the decline of the species via habitat modification. Widespread agricultural intensification has been occurring in western Europe since the sixties. However, the protected area in Girona and its periphery has not experienced appreciable changes in the last decade (Romero and Martí, pers. comm.) and the breeding areas in France have not seemingly suffered substantial alteration so as to produce the observed drastic decline of the breeding population (Rufray and Rousseau 2004; Lefranc, pers. comm.).

Factors like drought or habitat changes on the wintering grounds could also account for the decline of the species (Herremans 1997a, 1998a). We did not find any significant trend in precipitation in the main wintering area, what agrees with previous findings (see Herremans 1998b and references therein). Whereas we did not detect differences in NDVI index either, qualitative changes in vegetation cover types in the Kalahari basin due to human and natural causes do have occurred (Ringrose et al. 2002), including an increasing prevalence of thorn trees and widespread bush encroachment. The Lesser grey shrike prefers in the wintering grounds open habitat with limited presence of trees, scattered low bushes and low grass cover that favors prey visibility (Herremans 1997a, 1998a). We found that increases in NDVI in Africa influenced negatively the breeding population in Girona. Thus, it is likely that human factors and/or a series of years with average precipitations could deteriorate the structural qualities of the habitat for the Lesser grey shrike (e.g. bush encroachment caused by overgrazing, increase of grass cover) and thus reduce food availability either directly or indirectly by reducing prey visibility (Herremans 1998a, b).

Our results have two remarkable aspects. One is the spatial synchrony of the studied populations, which can be caused by similar extrinsic effects operating on different populations (Kendall et al. 2000; Williams et al. 2003; Liebhold et al. 2004). Weather has been frequently reported as a likely synchronizing factor (Paradis et al. 2000; Williams et al. 2003). For the population in Girona we detected a strong relationship between climatic variables and fluctuations in population abundances. The lack of correlation for other populations can be due to inaccuracies of our climatic data (e.g. the period when temperature or any other environmental variable influences habitat quality could be different to the one here considered), to the interaction between climate and local factors (Ringsby et al. 2002) or to the limited sample size. However, both our results and evidence from other studies (Chuine et al. 1998, Osborne et al. 2000) suggest that the studied populations are under similar changing environmental conditions and, therefore, common climatic conditions could be one factor accounting for the observed population declines.

A second relevant aspect arising from our study is that the circumstances experimented by one population could influence the fate of neighboring populations. Variations in the population size of Aude are correlated with environmental variables in Girona. Similarly, the NDVI in Montpellier correlated negatively with the population size in Girona. The migration pattern of the species implies that birds have to pass through the breeding areas in France to reach the ones in Spain. It is therefore likely that these populations are closely interconnected, so that individuals produced in a population recruit into neighboring populations in the next breeding season. It could well be that some birds born in Spain could decide to breed in France the next year (philopatry in juveniles of this species is known to be low, Krištín et al. in press). On the other side, favorable conditions in, let's say, Montpellier, could influence birds' decision to stay there rather than continue on migration to Girona. Curnutt et al. (1996) showed that source-sink dynamics can explain variability at the periphery of a species range where population size is governed largely by migration rather than by reproduction and survival (Curnutt et al. 1996). The role of each of the studied populations in this dynamic remains to be investigated.

Conservation implications

Species with synchronous populations are thought to confront greater risk of extinction because density crashes can occur simultaneously in all populations (Heino et al. 1997; Palmqvist and Lundberg 1998). Faced to the widespread extinction of this species in the southwestern border of its range, it is crucial to identify the causes of the decline and the ecological mechanisms involved and to establish conservation measurements. Whereas the occurrence of a major factor accounting for the decline of this shrike would make its conservation easier, this is unlikely given its migratory nature and the variety of environmental conditions they have to face at different locations throughout the year (Schaub et al. 2005). Our study adds to the preservation of this species by identifying some issues of practical importance to conservation biology. The parallel fluctuations of the French and Spanish (Girona) populations suggest common causes for the decline of the species in this area. Our results suggest that climatic change, acting directly or indirectly on the different populations and probably moderated by local conditions (Osborne et al. 2000; Ringsby et al. 2002) can be one of such causes. However, the correlational nature of this study limits our ability to determine causal factors. More effort is needed to identify other possible factors and the exact mechanisms by which climate may influence population dynamics of this species so that mitigation activities can be implemented.

Although climate change can be observed globally, the magnitude of changes and their effects on organisms is likely to vary considerably within a given region (Osborne et al. 2000). It is therefore crucial to preserve those populations/areas where the study species seems to be stable (e.g. Lleida). However, conservation actions could prove useless if applied separately for some population given that the fates of these populations seem to be interconnected. In line with this more information is needed about the degree of interconnection among peripheral populations and between these and more central ones, and whether the spatial synchrony here reported occurs at a larger spatial scale.

Acknowledgements We are indebted to J.J. Sanz for helping us with NDVI data and for providing useful advises and improving earlier versions of this work. We also thank P. Isenmann, X. Rufray, N. Lefranc, F. Labouyrie, J.C. Albero, J.L. Rivas, J. Estrada and Natural Park of Aiguamolls staff (J. Martí, E. Streich and S. Romero) for kindly providing us unpublished information and to Arco van Strien for his very kind assistance concerning the use of TRIM. Anders P. Møller, T. Szep, M. Herremans, L. Brotons and S. Mañosa provided valuable suggestions and unpublished information. We thank J. Castelló, A. Bonan and L. Juliá for their assistance in the field and G. Bota for his collaboration at the beginning of the project. Two anonymous referees contributed to improve the manuscript. We are grateful to the owners of the properties for all the facilities given and to the people who offered information about the past and current distribution of the species. This research has been funded by the Departament de Medi Ambient i Habitatge (Generalitat de Catalunya) (2001–2002), REGSEGA (2003–2004) and the Fundació Territori i Paisatge (1999). FVH was supported by a I3P contract funded by the European Union and by the Programa de Ayudas para el Retorno de Investigadores de la Junta de Andalucía.

References

Asrar G, Fuchs M, Kanemasu ET, Hatfield JL (1984) Estimating absorbed photosynthetic radiation and leaf area index from spectral reflectance in wheat. Agron J 76:300–306

Bara T (1995) La population de Pies-Grièches à Poitrine Rose *Lanius minor* dans la Basse Plaine de l'Aude. Alauda 63:191–198

Béchet A, Isenmann P, Mauffrey JF (1995) Un deuxième site de nidification de la Pie-grièche à poitrine rose *Lanius minor* en Languedoc. Alauda 63:243–244

Brown JH (1984) On the relationship between abundance and distribution of species. Am Nat 124:255–279

Brown JH, Mehlman DW, Stevens GC (1995) Spatial variation in abundance. Ecology 76:2028–2043

Chuine I, Cour P, Rousseau DD (1998) Fitting models predicting dates of flowering of temperate-zone trees using simulated annealing. Plant Cell and Environ 21:455–466

Cramp S, Perrins CM (eds) (1993) The birds of the western palearctic, vol VII. Oxford University Press, Oxford and New York

Crick HQP, Dudley C, Glue DE, Thomson DL (1997) UK birds are laying eggs earlier. Nature 388:526

Curnutt JL, Pimm SL, Maurer BA (1996) Population variability of sparrows in space and time. Oikos 76:131–144

Donck G, Bara T (2001) Oiseaux Nicheurs rares en France en 1999: Pie Grièche à Poitrine Rose *Lanius minor*. Ornithos 8–4:131

Giralt D, Bota G (2003) Alcaudón chico, *Lanius minor*. In: Martí R, del Moral JC (eds), Atlas de las Aves Reproductoras de España. Dirección General de Conservación de la Naturaleza-Sociedad Española de Ornitología, Madrid, pp 532–533

Giralt D (2004) Trenca *Lanius minor*. In: Estada J, Pedrocchi V, Brotons L, Herrando S (eds), Atles dels ocells nidificants de Catalunya 1999–2002. Institut Català D'Ornitologia (ICO)/ Lynx edicions, Barcelona, pp 478–479

Gordo O, Brotons L, Ferrer X, Comás P (2005) Do changes in climate patterns in wintering areas affect the timing of the spring arrival of trans-Saharan migrant birds?. Glob Change Biol 11:12–21

Heino M, Kaitala V, Ranta E, Lindström J (1997) Synchronous dynamics and rates of extinction in spatially structured populations. Proc R Soc Lond Ser B 264:481–486

Hengeveld R, Haeck J (1982) The distribution of abundance. I. Measurements. J Biogeogr 9:303–316

Herremans M (1997a) Habitat segregation of male and female Red-backed Shrikes *Lanius collurio* and Lesser Grey Shrikes *Lanius minor* in the Kalahari basin, Botswana. J Avian Biol 28:240–248

Herremans M (1997b) Lesser Grey Shrike *Lanius minor*. In: Harrison JA, Allan DG, Underhill LG, Herremans M, Tree AJ, Parker V, Brown CJ (eds), The atlas of southern African birds, vol II. Birdlife South Africa Johannesburg, Johannesburg pp 406–407

Herremans M (1998a) Monitoring the world population of the Lesser Grey Shrike (*Lanius minor*) on the non-breeding grounds in southern Africa. J Ornithol 139:485–493
Herremans M (1998b) Conservation status of birds in Botswana in relation to land use. Biol Conserv 86:139–160
Hoffmann AA, Blows MW (1994) Species borders: ecological and evolutionary perspectives. Trends Ecol Evol 9:223–227
Hughes L (2000) Biological consequences of global warming: is the signal already apparent?. Trends Ecol Evol 15:56–61
Isenmann P, Debout G (2000) Vineyards harbour a relict population of Lesser Grey Shrike (*Lanius minor*) in Mediterranean France. J Ornithol 141:435–440
Isenmann P, Debout G, Lepley M (2000) La Pie-Grièche à Poitrine Rose *Lanius minor* nicheuse à Montpellier. Alauda 68:123–131
Kendall BE, Bjornstad ON, Bascompte J, Keitt TH, Fagan WF (2000) Dispersal, environmental correlation and spatial synchrony. Am Nat 155:628–636
Krištín A (1995) Why the Lesser Grey Shrike (*Lanius minor*) survives in Slovakia: food and habitat preferences and breeding biology. Folia Zool 44:325–334
Krištín A, Hoi H, Valera F, Hoi C (2000) Breeding biology and breeding success of the Lesser Grey Shrike (*Lanius minor*) in a stable and dense population. Ibis 142:305–311
Krištín A, Hoi H, Valera F, Hoi C Philopatry, Dispersal Patterns and Nest-site Reuse in Lesser Grey Shrike (*Lanius minor*). Biodivers Conserv DOI 10.1007/s10531-006-9019-8
Labouyrie F (2003) Status de la Pie-Grièche à poitrine rose *Lanius minor* en Vaunage, Gard. Meridionalis 5:54–60
Lawton JH (1993) Range, population abundance and conservation. Trends Ecol Evol 8:409–413
Lawton JH, Nee S, Letcher AJ, Harvey PH (1994) Animal distributions: patterns and processes. In: Edwards PJ, May RM, Webb NR (eds), Large-scale ecology and conservation biology. Blackwell Scientific, Oxford England, pp 41–58
Lefranc N (1995) Decline and current status of the Lesser Grey Shrike (*Lanius minor*) in Western Europe. In: Reuven R, Lohrer FE (eds), Shrikes (*Lanius*) of the World: Biology and Conservation. Proceedings of the Western Foundation for Vertebrate Zoology 6:93–97
Lefranc N (1997) Shrikes and the farmed landscape in France. In: Pain DJ, Pienkowski MW (eds), Farming and birds in Europe, Academic Press, London, pp 236–268
Lefranc N (1999) Les Pies-grièches *Lanius sp.* en France: rèpartition et statut actuels, histoire récente, habitats. Ornithos 6:58–82
Lefranc N, Worfolk T (1997) Shrikes, A Guide to the Shrikes of the World. Sussex
Lesica P, Allendorf FW (1995) When are peripheral populations valuable for conservation? Conserv Biol 9:753–760
Liebhold A, Koenig WD, Bjornstad ON (2004) Spatial synchrony in population dynamics. Annu Rev Ecol Evol Syst 35:467–490
McArdle BH, Gaston KJ, Lawton JH (1990) Variation in the size of animal populations: pattern, problems, and artifacts. J Anim Ecol 59:439–454
Mehlman DW (1997) Change in avian abundance across the geographic range in response to environmental change. Ecol Appl 7:614–624
Mitchell TD, Hulme M, New M (2002) Climate data for political areas. Area 34:109–112
Myneni RB, Keeling CD, Tucker CJ, Asrar G, Nemani RR (1997) Increased plant growth in the northern high latitudes from 1981 to 1991. Nature 386:698–702
Osborne CP, Chuine I, Viner D, Woodward FI (2000) Olive phenology as a sensitive indicator of future climatic warming in the Mediterranean. Plant Cell Environ 23:701–710
Palmqvist E, Lundberg P (1998) Population extinctions in correlated environments. Oikos 83:359–367
Pannekoek J, van Strien AJ (2003) TRIM 3 Manual. TRends and Indices for Monitoring Data. Research paper n° 0102 Voorburg, The Netherlands: Statistics Netherlands. (Available freely at http://zeus.nyf.hu/~szept/trim.htm)
Paradis E, Baillie SR, Sutherland WJ, Gregory RD (2000) Spatial synchrony in populations of birds: effects of habitat, population trend, and spatial scale. Ecology 81:2112–2125
Parmesan C, Ryrholm N, Stefanescus C, Hill JK, Thomas CD, Descimon H, Huntley B, Kaila L, Killberg J, Tammaru T, Tennent WJ, Thomas JA, Warren M (1999) Poleward shifts in geographical ranges of butterfly species associated with regional warming. Nature 399:579–583
Peñuelas J, Filella I, Comás P (2002) Changed plant and animal life cycles from 1952 to 2000 in the Mediterranean region. Glob Change Biol 8:531–544

Piñol J, Terradas J, Lloret F (1998) Climate warming, wildfire hazard, and wildfire occurrence in coastal eastern Spain. Clim Change 38:345–357
Prince SD, Justine CO (1991) Coarse resolution remote sensing of the Sahelian environment. Int J Remote Sens 12:1133–1421
Randall MGM (1982) The dynamics of an insect population throughout its altitudinal distribution: *Coleophora alticolella* (Lepidoptera) in northern England. J Anim Ecol 51:993–1016
Ringrose S, Chipanshi AC, Matheson W, Chanda R, Motoma L, Magole I, Jellema A (2002) Climate and human-induced woody vegetation changes in Botswana and their implications for human adaptation. Environ Manage 30:98–109
Ringsby TH, Saether BE, Tufto J, Jensen H, Solberg EJ (2002) Asynchronous spatiotemporal demography of a house sparrow metapopulation in a correlated environment. Ecology 83:561–569
Royama T (1992) Analytical population dynamics. Chapman and Hall, London, UK
Rufray X, Rousseau E (2004) La Pie-Grièche à poitrine rose (*Lanius minor*): une fin annnoncée. Ornithos 11:36–38
Sanz JJ, Potti J, Moreno J, Merino S (2003) Climate change and fitness components of a migratory bird breeding in the Mediterranean region. Glob Change Biol 9:461–472
Schaub M, Kania W, Koppen U (2005) Variation of primary production during winter induces synchrony in survival rates in migratory white storks *Ciconia ciconia*. J Anim Ecol 74:656–666
StatSoft Inc (2001) STATISTICA for Windows. StatSoft, Inc
Tews J, Jeltsch F (2004) Modelling the impact of climate change on woody plant population dynamics in South African savanna. BMC Ecology 4:17. doi: 10.1186/1472-6785-4-17
Thomas CD, Lennon JJ (1999) Birds extend their ranges northwards. Nature 399:213
Tilman D, Kareiva P (eds), (1997) Spatial ecology: the role of space in population dynamics and interspecific interactions. Princeton University Press, Princeton NJ
Visser ME, van Noordwijk AJ, Tinbergen JM, Lessells CM (1998) Warmer springs lead to mistimed reproduction in great Tits (*Parus major*). Proc R Soc Lond B 265:1867–1870
Wiegand K, Jeltsch F, Ward D (1999) Analysis of the population dynamics of desert-dwelling *Acacia* trees with a spatially-explicit computer simulation model. Ecol Modell 117:203–224
Williams CK, Ives AR, Applegate RD (2003) Population dynamics across geographical ranges: time-series analyses of three small game species. Ecology 84:2654–2667

Biodivers Conserv (2007) 16:857–870
DOI 10.1007/s10531-006-9094-x

ORIGINAL PAPER

Monitoring mammals in the Caxiuanã National Forest, Brazil – First results from the Tropical Ecology, Assessment and Monitoring (TEAM) program

Simone de Souza Martins · James G. Sanderson · José de Sousa e Silva-Júnior

Received: 14 July 2005 / Accepted: 7 July 2006 / Published online: 11 November 2006

Abstract The need for long-term biodiversity monitoring using standardized protocols led to the creation of the Tropical Ecology, Assessment and Monitoring (TEAM) initiative. At some 50 field stations in tropical forests around the world, TEAM will monitor various taxa such as ants, birds, butterflies, medium and large terrestrial mammals, primates, litter fall, and trees, as well as landscape change in nine tropical biodiversity hotspots and three tropical wilderness areas. The TEAM terrestrial mammal program calls for using a grid of camera phototraps to monitor long-term trends in densities and occupancy rates of species that can or cannot be uniquely identified, respectively. We describe the TEAM camera phototrapping program and provide results for the first TEAM site–Caxiuanã National Forest in northern Brazil. An intensive one year camera trapping effort was carried out to determine which months were most suitable for long-term monitoring. Fifteen species of medium and large terrestrial mammals and two large birds were recorded, including three xenarthrans, five carnivores, one perissodactyle, three artiodactyles, two rodents, and one marsupial. The medium and large terrestrial mammal diversity was well represented during two consecutive wet and dry months, respectively. We also recorded activity patterns for all species photographed by our camera traps more than 10 times.

Keywords Brazil · Camera trapping · Monitoring · TEAM program · Tropical rainforest

---

S. S. Martins · J. S. Silva-Júnior
Zoology Department, Museu Paraense Emilio Goeldi, Campus de Pesquisa, Av Perimetral 1901, Terra Firma, Belém, Pará Cep: 66.077-530, Brasil

J. G. Sanderson (✉)
Tropical Ecology, Assesement and Monitoring, Center for Applied Biodiversity Science, Conservation International, 1919 M Street NW, Washington, DC 20036, USA
e-mail: j.sanderson@conservation.org

Introduction

At the Rio Earth Summit in 1992 over 150 heads of state from around the world signed the Convention on Biological Diversity (CBD). The CBD was the world's first comprehensive international agreement committing governments to sustainable use and conservation of biological resources. More than 175 governments have now ratified the CBD that obligates the Parties to slow biodiversity loss by (1) establishing a system of protected areas giving special consideration to threatened species and ecosystems, (2) creating economic incentives for sustainable use of biodiversity, (3) evaluating development projects to understand their impacts on biodiversity, and (4) protecting indigenous people's rights to traditionally utilize and manage their natural resources.

To effectively evaluate progress in satisfying CBD goals monitoring biodiversity is critical. In 1996 CBD Parties were encouraged to set measurable targets to achieve conservation and sustainable use objectives. Formal agreement was reached to establish a core set of biodiversity indicators that could be monitored. To track biodiversity trends and threats that may be national or global in scope requires a long-term biodiversity monitoring program that has at its heart a standardized monitoring procedure for collecting quantitative data.

Monitoring natural populations, habitats, and threats is vital to assessing impacts on natural landscapes (Balmford et al. 2003; Kremen et al. 1994; ter Keurs and Meelis 1986). Monitoring is the repeated collection of data in the same area for the same time period to analyze changes in species' populations (Comiskey et al. 2001; Thompson et al. 1998). Monitoring can serve as an early warning system to alert managers that changes in biodiversity may require changes in management schemes to promote the long-term maintenance of biodiversity (Balmford et al. 2003; Yoccoz et al. 2001; Hellawell 1991). To be effective, field data must be collected in a rigorous, consistent manner according to standard, accepted protocols (Comiskey et al. 2001; Debinski and Humphrey 1997). Recently, increased emphasis has been placed on the standardization of data collection methodologies to enable comparisons between different projects (Henschel and Ray 2003). The need for a network of long-term biodiversity monitoring sites using standardized protocols led to the creation of the Tropical Ecology, Assessment and Monitoring, or TEAM, program.

TEAM was established in 2002 with a grant from the Gordon and Betty Moore Foundation to Conservation International. TEAM's mission is to monitor long-term trends in biodiversity through a network of 50 or more tropical forest field stations located in nine tropical forested hotspots (Atlantic Forest, East and West Africa, Indonesia, Madagascar, Philippines, MesoAmerica and Caribbean, Southeast Asia, and Tropical Andes, and three wilderness areas (Amazônia, Central Africa, and Papua New Guinea) (Myers et al. 2000), TEAM will provide an early warning system on the status of biodiversity that can effectively guide conservation actions. The primary goal of monitoring a broad range of biodiversity indicators is to detect changes and trends that differ significantly from normal and natural fluctuation.

In addition to the list of standard variables to be monitored across all sites, each TEAM field station is also be expected to monitor site-specific variables that are of local or regional importance. These variables include, for example, threatened or endangered species, economically important species, and rare and unique habitat types. Currently, TEAM monitoring protocols are being implemented at 3 sites in the Amazon wilderness of Brazil, and one site each in the Atlantic Forest and MesoAmerica hotspots.

TEAM is a ten-year program that will hopefully function in perpetuity once results show the value of long-term globally collected, standardized data analyzed by local professional biologists and other scientists. TEAM has chosen to work in already established research stations in tropical forests managed primarily for the conservation of biodiversity. However, TEAM might also serve as a model monitoring program for other tropical CBD Parties as one step towards satisfying the 1996 agreements. The goal of implementing standardized protocols for monitoring biodiversity globally could thus be realized.

Monitoring medium and large terrestrial mammals

Monitoring is the application of a systematic data collection program repeatedly executed in the same place during a specific time. A statistically robust methodology that gives repeatable results under the same conditions is essential. The number of individuals of all species using a certain area within each habitat type is a powerful measure of biodiversity but is generally and practically unattainable. However, when the number of individuals of certain focal species within a specific area can be statistically estimated year after year population trends above background noise can be recognized and documented (Karanth and Nichols 1998). For other species occupancy rates can be computed (MacKenzie et al. 2006).

Because every component of biodiversity cannot be monitored, TEAM, with help from international scientists, chose to monitor several core components of biodiversity in tropical evergreen forests. Often socio-economic factors lead to unsustainable pressure on certain terrestrial medium and large mammals. Thus, TEAM's monitoring program does not involve all mammal fauna but instead targets medium and large mammals most likely to respond to specific threats such as landscape scale changes, direct and indirect hunting pressure, or changes in ecological processes such as fire and climate change (Lambeck 1997). Some species lend themselves to individual identification and also serve as focal species for monitoring programs (Lambeck 1997). Spotted cats of various sizes, and large herbivores such as tapirs and elephants serve as focal species since they are sensitive to a variety of threats. Ideal for monitoring are threatened mammal species, particularly those on the IUCN Red List (Lawler et al. 2003).

No method can provide a measure of density, i.e., the number of individuals per unit area, unless (1) individuals can be identified, and (2) the size of the area being sampled can be determined. Determining the size of the area sampled depends upon knowledge of the home range of the species under study, and this requires validation by fieldwork (Wemmer et al. 1996). Because individuals of the focal species chosen by TEAM can be uniquely identified their population densities can be monitored through time (Karanth and Nichols 1998; Trolle and Kéry 2003). Data on non-focal species can be used to estimate siteoccupancy rates (MacKenzie et al. 2006).

For large areas with focal species the entire area generally cannot be monitored simultaneously in its entirety. In addition, individuals have different probabilities of being detected within the sampled area during a specific time interval and not all individuals are detected during each time interval. Thus, a subset of the entire area must be monitored and inferences about densities and abundances of the entire area can be inferred using appropriate, well documented, and proven mathematical software (Karanth and Nichols 2002; MacKenzie et al. 2006). Moreover, monitoring continuously in time is unnecessary. If detection rates are low then species' relative abundances can be estimated.

Major challenges must be overcome to monitor carnivores and other shy species. Carnivores, particularly those in tropical forests, are usually elusive and not easily observed by humans. Some are nocturnal or move about the landscape using dense cover. Typically, carnivores range widely and occur infrequently over large parts of their home range (Sunquist and Sunquist 2001). Their population densities are usually low making direct observation methodologies unreliable. The basic ecology of carnivores makes their populations inherently difficult to monitor. However, despite these challenges, population and habitat information on carnivores provides robust information on biodiversity (Balmford et al. 2003).

Camera phototrapping

Automatic cameras have been used to capture photographs of wildlife for at least 100 years (Chapman 1927; Nesbit 1926; Shiras 1906). Remote surveying and monitoring of terrestrial and some arboreal mammals has been enabled by a new generation of camera phototraps and well developed capture-recapture models (Trolle 2003; Karanth and Nichols 2002; Jones and Raphael 1993; Mace et al. 1990; Joslin 1977). Population estimates can now be made for individually identifiable species by photographing both sides of individuals simultaneously and occupancy rates can be calculated for other species (MacKenzie et al. 2006). For instance, Karanth and Nichols (2002) estimated tiger densities in four national parks in India, and Trolle and Kéry (2003) estimated ocelot densities in an area of the Pantanal, Brazil. Carbone et al. (2001) suggested that camera phototraps and home range information could be used to estimate densities of animals that cannot be individually identified but this remains contraversial. Concurrently, prey populations can also be monitored with camera phototraps.

The first TEAM station was established in the Caxiuanã National Forest, an important Conservation Unit in the Xingu-Tocantins basin, in December, 2002. Caxiuanã National Forest is also a RAINFOR site where important ecological and biogeochemical consequences of environmental change are being monitored to understand their effects on forest biomass and dynamics (Malhi et al. 2002). The results obtained by RAINFOR and TEAM are complementary and comprehensive, potentially allowing environmental impacts to be understood more thoroughly. Here we report on the first year of camera trapping results obtained at Caxiuanã National Forest.

Methods

Study area

Located in Melgaço, Pará, Brazil, Caxiuanã National Forest is 330,000 ha, the largest protected area between the Tocantins and Xingu rivers. The landscape is primary rainforest with 85% terra firme, 12% flooded igapó and várzea, and 3% secondary forest (Lisboa et al. 1997). The forest is dominated by Sapotaceae (72 spp.), Chrysobalanaceae (59 spp.), and Lauraceae (46 spp.) (Silva and Almeida, personal communication). The Ferreira Penna Scientific Station, located at 1°42′30″ S, 51°31′45″ W, has an area of 33,000 ha operated by IBAMA (Instituto Brasileiro do Meio Ambiente e dos Recursos Naturais Renováveis) and Museu Paraense Emílio Goeldi, Belém (Lisboa 1997).

The climate is humid tropical with two seasons: dry (June–December) and wet (January–May). The medium annual temperature is 26°C (Oliveira et al. 2002). Rainfall peaks in March with 379.8 mm and declines to a minimum of 50.7 mm October. Total yearly rainfall is approximately 1400 mm (Costa and Moraes 2002).

Serveral groups of vertebrates including reptiles (Ávila-Pires and Hoogmoed 1997; Bernardi et al. 2002; Estupiñán-T et al. 2002), birds (Silva et al. 2001; Silva and Pimentel Neto 1997), bats (Marques-Aguiar and Aguiar 2002), and primates (Jardim 1997; Veracini 2002, 1997; Bobadilla 1998; Pina 1999; Souza 1999; Tavares 1999) have been the subject of previous studies.

The local population of about 283 people (2003) is distributed in three villages within the reserve (Silveira et al. 2002). People live simply subsisting on small-scale farming, forest extraction, fishing, and hunting (Ferraz et al. 2002). The impact on the natural environment has not been investigated.

Time and equipment

From December 2002–November 2003 we placed a pair of camera traps at 12 sites 3–4 km apart to cover approximately 70 km^2. Sites chosen were active animal paths and locations with a significant number of animal tracks. Each of 24 camera traps was run continuously, day and night, for 12 mon or until it failed. After taking a photograph, each camera trap was forced to wait 20 s (the so-called *latency time*) before another photograph could be taken.

We used Cam Trakker^{TM} passive camera traps (CamTrak South, Inc., Watkinsville, GA 30667, USA, http://www.camtrakker.com). The passive system is activated by heat-in-motion within an opportunity cone that begins at the camera trap in a 4 cm circle that widens to a 2 m circle 8 m from the camera trap. The 35 mm camera with automatic focus within the camera trap photographed a larger frame that was encompassed by the opportunity cone of the sensor. When the heat-in-motion censor was trigger, a 35 mm camera took a photograph 0.6 s later so that most subjects were properly centered. Day and time were recorded on each photograph.

Data analysis

For those species whose individuals can be uniquely identify, density estimates are possible. For others site occupancy rates can be estimated. If detection probabilities are low then estimating relative abundances must be considered. To estimate relative abundance of medium and large terrestrial mammals and terrestrial birds, we analyzed the camera trapping data in three different ways. First, we assumed that only one individual male or female per species (or group in the case of peccarys, coatimundis and birds) could be photographed at each camera location during a 24 h period beginning at midnight (unless we could determine without doubt otherwise). That is, if a deer of unknown sex was photographed at a single site 3 times during a

24 h period beginning at midnight then this was counted as a single photograph of the same deer.

We identified all felids uniquely from their spot patterns or unique body characters (Karanth and Nichols 1998; Trolle and Kéry 2003). We used this information to re-estimate relative abundance.

The most conservative estimate of relative abundance is to count the photograph of a single species at each camera trap site once and only once for the length of the study. This redefines a camera trapping occasion to be the length of the study. That is, if paca were photographed at each of 8 sites multiple times during the study then each of 8 sites was recorded to have photographed a single paca during the length of the study. Thus, no camera trap could record more than one occurrence of a species unless individuals could be determined.

For each species we recorded the total number of pictures, the number of pictures per day, and the relative percent number of pictures of each species as surrogate for the percent relative abundance of each species. We also recorded activity patterns for those species photographed at least 10 times (Fig. 1).

Results of continuous monitoring were used to determine which months were most representative of the mammalian fauna so that monitoring could continue during just those months, thus reducing the cost of the monitoring program.

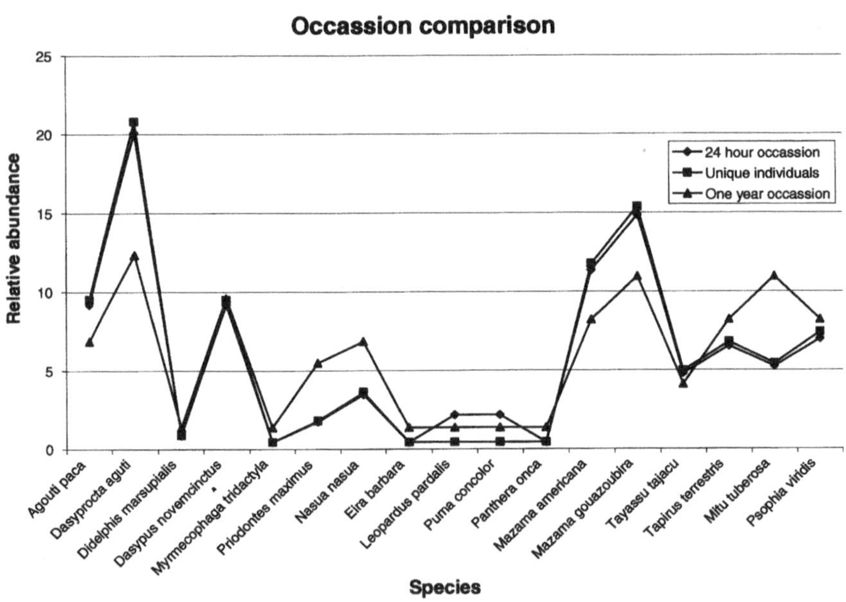

Fig. 1 The number of days in a camera trapping occasion (the number of days used to record detection or non-detection) did not effect the relative abundance of species. With several species were we able to identify individuals and thus avoid repeatedly counting them but this also had little effect on relative abundance. Shown is a comparison of the relative abundance of each species using a 24 h camera trapping occasion, non-repeated counting of unique individuals, and a one year camera trapping occasion

Results

During one year of camera trapping at 12 sites, 348 photographs were recorded. Of these photographs 12 days or times could not be deciphered leaving 336 useful photographs. Approximately 1/3 of the photographs taken at each site during a 24 h period beginning at midnight (a camera trapping *occasion* of 24 h) were of the same species. Assuming that repeated photographs of the same species during an occasion were the same individual (unless we could determine otherwise) 236 photographs of 15 medium and large mammals in six orders and 2 terrestrial birds were recorded during 2838.35 camera trap days (Table 1). Detection rates at our camera traps was less than 2/3 animal per day for all 12 camera trapping sites thus precluding the estimation of reasonably low variance occupancy rates.

Using each focal species unique marks and body characteristics we recorded one photograph of a male jaguar, three photographs of the same male puma, and five photographs of the same male ocelot.

Relative abundances did not change significantly when an occasion was defined as a 24 h period beginning at midnight (Table 1), when individuals could be determined (Table 2), or most conservatively when an occasion was defined as the entire study period (Table 3–4).

Activity patterns for species recorded at least 10 times revealed that *Agouti paca* and *Dasypus novemcinctus* were nocturnal, *Dasyprocta leporina, Tayassu tajacu* and

Table 1 During one year of camera trapping at 12 sites, 336 useful photographs were recorded

Species	Common name	Total	24 h	Pics/day	Pct. rel.
Agouti paca	paca	23	21	0.0074	9.17
Dasyprocta aguti	red-romped aguti	65	46	0.0162	20.09
Didelphis marsupialis	common opossum	2	2	0.0007	0.87
Dasypus novemcinctus	nine-banded armadillo	24	21	0.0074	9.17
Dasypus sp.	unknown armadillo	4	3	–	–
Myrmecophaga tridactyla	giant anteater	1	1	0.0004	0.44
Priodontes maximus	giant armadillo	4	4	0.0014	1.75
Nasua nasua	South American coati	22	8	0.0028	3.49
Eira barbara	tayra	1	1	0.0004	0.44
Leopardus pardalis	ocelot	6	5	0.0018	2.18
Puma concolor	puma	5	5	0.0018	2.18
Panthera onca	jaguar	1	1	0.0004	0.44
Mazama americana	red brocket deer	35	26	0.0092	11.35
Mazama gouazoubira	gray brocket deer	42	34	0.0120	14.85
Tayassu tajacu	collared peccary	24	11	0.0039	4.80
Tapirus terrestris	brazilian tapir	25	15	0.0053	6.55
Mitu tuberose	razor-billed curassow	29	12	0.0042	5.24
Psophia viridis	grey-winged trumpeter	23	16	0.0056	6.99
		336	232	0.0807	100

Approximately 1/3 of the photographs taken at a single site during a 24 h period beginning at midnight (a camera trapping *occasion*) were of the same species. Assuming that repeated photographs of the same species during an occasion were the same individual (unless we could determine otherwise) 232 photographs of 16 medium and large mammals and 2 terrestrial birds were recorded. In four photographs we were unable to identify which species of armadillo was present. The relative abundance of each species was determined by dividing the total number of each species photographed by the total number of all species photographed during a 24-h occasion. Total pictures per day (Pics/day), and percent relative abundance (Pct. rel.) for each species are given. All 12 camera trap sites recorded approximately one photograph each day.

Table 2 For each of three species of felids we were able to individually identify each individual

Species	Total pictures	Pics/day	Pct. rel. abd.
Agouti paca	21	0.0074	9.50
Dasyprocta aguti	46	0.0162	20.81
Didelphis marsupialis	2	0.0007	0.90
Dasypus novemcinctus	21	0.0085	9.50
Myrmecophaga tridactyla	1	0.0004	0.45
Priodontes maximus	4	0.0014	1.81
Nasua nasua	8	0.0028	3.62
Eira barbara	1	0.0004	0.45
Leopardus pardalis	1	0.0004	0.45
Puma concolor	1	0.0004	0.45
Panthera onca	1	0.0004	0.45
Mazama Americana	26	0.0092	11.76
Mazama gouazoubira	34	0.0120	15.38
Tayassu tajacu	11	0.0039	4.98
Tapirus terrestris	15	0.0053	6.79
Mitu tuberosa	12	0.0042	5.43
Psophia viridis	16	0.0056	7.42
	221	0.0779	100

This reduced the relative abundance of each felid species and increased the relative abundance of all other species. Total pictures, pictures per day (Pics/day), and percent relative abundance (Pct. rel. abd.) for each species are given

Table 3 The most conservative estimate of relative abundance is to assume that if at each site a species was photographed then the same individual of that species was photographed repeatedly (unless we could determine otherwise) during the entire camera trapping period

Species	Total	Pics/day	Pct. rel. abd.
Agouti paca	5	0.0018	6.85
Dasyprocta aguti	9	0.0032	12.33
Didelphis marsupialis	1	0.0004	1.37
Dasypus novemcinctus	7	0.0025	9.59
Myrmecophaga tridactyla	1	0.0004	1.37
Priodontes maximus	4	0.0014	5.48
Nasua nasua	5	0.0018	6.85
Eira barbara	1	0.0004	1.37
Leopardus pardalis	1	0.0004	1.37
Puma concolor	1	0.0004	1.37
Panthera onca	1	0.0004	1.37
Mazama americana	6	0.0021	8.22
Mazama gouazoubira	8	0.0028	10.96
Tayassu tajacu	3	0.0011	4.11
Tapirus terrestris	6	0.0021	8.22
Mitu tuberosa	8	0.0028	10.96
Psophia viridis	6	0.0021	8.22
	73	0.0257	100

For instance, this assumption says that no more than 12 A. *paca* (one per camera trap site) could be photographed during the entire period. This assumption significantly reduces the number of unique photographs but does not alter significantly percent relative abundance. Total pictures, pictures per day (Pics/day), and percent relative abundance (Pct. rel. abd.) for each species are given

the two bird species *Mitu tuberosa*, and *Psophia viridis* were diurnal, and *Mazama americana* and *M. gouazoubira* were continuously active. *Tapirus terrestris* were active nearly continuously but no photographs were obtained between 8:30 and noon

Table 4 Using a 24 h camera trapping occasion, activity patterns for species photographed at least 10 times are given as the number of photographs in each two hour period beginning at midnight

Time	*A. paca*	*D. aguti*	*D. nov.*	*M. amer.*	*M. gou.*	*T. terrestris*	*T. tajacu*	Birds
0–2	1		4	2	1	2		
2–4	6		5	3	3			
4–6	2		3	4		2		
6–8		12		3	11		2	10
8–10		9	1	1	4	1	2	3
10–12		2			5			3
12–14		4		2	4	1	2	4
14–16		5		1	3	1	2	3
16–18		7	1		2		1	5
18–20	6	7	2	4	1	2	2	
20–22	4		4	5		5		
22–24	2		1	1		1		

D. nov. ≡ Dasypus novemcinctus, *M. amer.* ≡ Mazama americana, *M. gou.* ≡ *M. gouzaoubira. Mitu tuberosa* and *Psophia viridis* showed similar activity patterns and so are combined into Birds. Note that *A. paca* began their nocturnal foraging at 19:16 and *D. aguti* were active until 18:49 so that these species showed no activity overlap. With some exceptions, activity patterns are generally clear

Table 5 For each month (J ≡ January, etc.) and for each species the total number of photographs using a 24 h camera trapping occasion is given

Species	N	D	J	F	M	A	M	J	J	A	S	O	N	D	Sum
Agouti paca		1	3		3		6	5	2	1					21
Dasyprocta agouti		4	5	1	12	6		7	4	1	1	3	2		46
Dasypus novemcinctus	2				8	4				1			3	3	21
Didelphis marsupialis								2							2
Eira Barbara													1		1
Leopardus pardalis			1	1			2	1							5
Mazama americana			5	1	4	1	3	2	4	1	3	1		1	26
Mazama gouazoubira	1	2	10	1	2	5	1			3		5	3	1	34
Myrmecophaga tridactyla					1										1
Nasua nasua	1	1	2	2			1							1	8
Panthera onca											1				1
Priodontes maximus		1	1	1								1			4
Puma concolor					2	2	1								5
Tapirus terrestris			2	1	2		2	1	1	4			1	1	15
Tayassu tajacu				1	7	1						1		1	11
Mitu tuberosa	1	2	3	2	1		1							2	12
Psophia vividis	1	4		1	5	1	1	1	1	1					16
Total records	6	15	32	12	47	20	18	19	12	12	5	11	10	10	229
Richness	5	7	9	10	11	7	9	7	5	7	3	5	5	7	17

Camera traps recorded the most photographs and highest species richness in March during the wet season and the lowest number in September during the dry season. Note that two terrestrial bird species, *Mitu tuberosa* and *Psophia vividis* were not detected in the dry season. The wet months of mid March through mid May, and the dry months of mid September to mid November were selected for long-term monitoring. Only *Didelphis marsupialis* was not recorded during either of these seasons suggesting that a resting period, perhaps at a wallow, might occur during this time (Table 4).

Based on the number of photographs and the diversity of species, two wet months and two dry months were selected for future long-term monitoring (Table 5).

Discussion

Despite being one of the largest protected areas in oriental Amazônia, much information regarding Caxiuanã National Forest remains unknown. Of the mammalian fauna, order Chiroptera was sampled reasonably well by Marques-Aguiar and Aguiar (2002), order primates has received greater attention (Bobadilla 1998; Bobadilla and Ferrari 2000; Jardim 1997; Jardim and Oliveira 1997, 2002; Moegenburg and Jardim 2002; Pina 1999; Pina et al. 2002; Souza 1999; Souza et al. 2002; Tavares 1999; Tavares and Ferrari 2002), and other mammals have been inferred to be present based on interviews and observations of hunters (Lisboa et al. 2002; Santana and Lisboa 2002). As a result of the present study the presence of 11 non-volant mammal species has been confirmed. Moreover, four additional non-volant mammals have been added to the list of mammals: ocelot *(Leopardus pardalis)* and giant armadillo *(Priodontes maximus)* were recorded for the first time, and two deer species *(Mazama americana* and *M. gouazoubira)* were positively identified. According to IBAMA (2003), five of the photographed species are vulnerable: *Myrmecophaga tridactyla, P. maximus, L. pardalis, Puma concolor* and *Panthera onca.*

Our results did not offer any surprises in medium and large mammalian species composition. Voss and Emmons (1996) suggested that xenarthran, carnivore, and ungulate faunas are remarkably uniform throughout greater Amazônia. For these taxa, the species present in our area were a subset of those from 10 other sites in Amazônia. What is surprising is the relatively low number of photographs and the relatively low *apparent* densities we recorded. For instance, we expected to record species such as deer, pacas, and agutis at each camera trap site. In fact, deer, pacas, and agutis were recorded at 5, 7, and 9 sites, respectively, out of 12 possible sites. Moreover, at sites that successfully recorded these species, the number of records was relatively low suggesting these species were foraging widely. Supporting evidence of this hypothesis comes from recording only 3 individual cats, one each of 3 species, during the entire year in our study area of 70 km^2. In contrast, in a three month study in Brazil's Pantanal Trolle and Kéry (2003) obtained 54 pictures of approximately 11 ocelots in 10 km^2.

Species' relative abundance did not vary greatly when the number of days in a camera trapping occasion was changed from a single 24-h period to one year (Tables 2–5). We do not recommend defining a camera trapping occasion as one year because this would preclude density studies. Our point is that relative abundances are fairly robust to changes in the number of days used to define a camera trapping occasion.

Because we only recorded one individual each of three focal species where we could identify individuals we were not able to estimate densities for these species. Indeed, carnivores and many other species appeared to be relatively sparse at Caxiuanã National Forest compared to other habitats e.g. Pantanal (Trolle and Kéry 2003); and Espírito Santo (Srbek-Araujo and Chiarello 2005) in Brazil. More data on rainfall, seasonal flooding, tree growth rates, hunting pressure, and continued monitoring of the medium and large mammals will help us determine why the mammalian fauna is apparently sparse.

As expected the two terrestrial birds, *Mitu tuberosa* and *Psophia viridis*, as well as peccary and aguti were exclusively diurnal. The paca was crepuscular. Moreover, paca and aguti were photographed at the same camera trap sites but during periods of the day. Closer analysis of the 2 h period from 1800 to 2000 showed that no agutis were active after 18:49 while no pacas were active before 19:16. This evidence suggests these guild members were avoiding competing for resources.

The TEAM camera trapping monitoring effort at Caxiuanã National Forest is a continuing effort that will enable year-to-year comparisons of richness, relative abundance, activity patterns, and rates of change in site occupancy (MacKenzie et al. 2003, 2006). Individual carni vores that were identified were too few in number to permit density estimates (Karanth and Nichols 1998). That such low numbers of carnivores might persist remains to be discovered.

Acknowledgements Support for the TEAM program was provided with a grant from the Gordon and Betty Moore Foundation to Conservation International. We thank Francisco Braga (Calafate), Gustavo Fonseca, Magalli Henriques, Thomas Lacher, Juarez Pezutti, and Ima Vieira. Maurício Barbanti Duarte identified the *Mazama* species. We also thank two anonymous reviewers for their valuable comments and suggestions.

References

Ávila-Pires TCS, Hoogmoed MS (1997) The Herpetofauna. In: Lisboa PLB (ed) Caxiuanã. Museu Paraense Emílio Goeldi, Belém, Pará, Brasil, pp 389–401

Balmford A, Green RE, Jenkins M (2003) Measuring the changing state of nature. Trends Ecol Evol 18:326–336

Bernardi JAR, Rufino N, Costa GNR, Rocha RA-T (2002) Répteis. In: Lisboa PLB (ed) Caxiuanã: Populações Tradicionais, Meio Físico e Diversidade Biológica. Museu Paraense Emílio Goeldi, Belém, Pará, Brasil, pp 533–540

Bobadilla UL (1998) Abundância, tamanho de agrupamento e uso do hábitat por Cuxiús de Uta Hick *Chiropotes satanas utahicki* Hershkovits, 1985 em dois sítios na Amazônia Oriental: Implicações para a conservação. Masters Dissertation, Universidade Federal do Pará, Belém, Brasil. 62 p

Bobadilla UL, Ferrari SF (2000) Habitat use by *Chiropotes satanas utahicki* and syntopic platyrrhines in eastern Amazonia. Am J Primatol 50:215–224

Carbone C, Christie S, Conforti K, Coulson T, Franklin N, Ginsberg JR, Griffiths M, Holden J, Kawanishi K, Kinnaird M, Laidlaw R, Lynam A, Macdonald DW, Martyr D, McDougal C, Nath L, O'Brien T, Seidensticker J, Smith DJL, Sunquist M, Tilson R, Wan Shahrudden WN (2001) The use of photographic rates to estimate densities of tigers and other cryptic mammals. Anim Conserv 4:75–79

Chapman FM (1927) Who treads our trails? Natl Geogr Mag 52:331–345

Comiskey JA, Dallmeier F, Alonso A (2001) Framework for assessment and monitoring of biodiversity. In: Levin SA (ed) Encyclopedia of Biodiversity, Vol 3. Academic Press, San Diego, California, pp 63–73

Costa JPR, de Moraes JC (2002) Médias mensais de variáveis meteorológicas. In: Lisboa PLB (ed) Caxiuanã: Populações Tradicionais, Meio Físico e Diversidade Biológica. Museu Paraense Emílio Goeldi, Belém, Pará, Brasil, pp 225–232

Debinski DM, Humphrey PS (1997) An integrated approach to biological diversity assessment. Nat Areas J 17:355–365

Estupiñán-T RA, Bernardi JAR, Galatti U (2002) Fauna Anura. In: Lisboa PLB (ed) Caxiuanã: Populações Tradicionais, Meio Físico e Diversidade Biológica. Museu Paraense Emílio Goeldi, Belém, Pará, Brasil, pp 541–553

Ferraz MG, Lisboa PLB, Ninni K, Santana MRA, Andrade MS (2002) Programa Floresta-Modelo de Caxiuanã: desenvolvimento sustentável de comunidades ribeirinhas. In: Lisboa PLB (ed) Caxiuanã: Populações Tradicionais, Meio Físico e Diversidade Biológica. Museu Paraense Emílio Goeldi, Belém, Pará, Brasil, pp 127–163

IBAMA, (2003). http://www.ibama.gov.br/fauna/extincao

Jardim MMA (1997) Estratégias de forrageamento e uso de espaço por *Alouatta belzebul* (Primates, Cebidae) na Estação Científica Ferreira Penna, Melgaço, Pará, Masters Dissertation, Universidade Federal do Rio Grande do Sul, Porto Alegre, Brasil. p 120.

Jardim MMA, Oliveira LF (2002) Vocalização de um grupo de guaribas *(Alouatta belzebul)*. In: Lisboa PLB (ed) Caxiuanã: Populações Tradicionais, Meio Físico e Diversidade Biológica. Museu Paraense Emílio Goeldi, Belém, Pará, Brasil, pp 697–703

Jardim MMA, Oliveira LF (1997) Uso do espaço de *Alouatta belzebul* (Primates, Cebidae) em função da temporalidade de recursos. In: Lisboa PLB (ed) Caxiuanã. Museu Paraense Emílio Goeldi, Belém, Pará, Brasil, pp 417–436

Jones LLC, Raphael MG (1993) Inexpensive camera systems for detecting martens, fishers, and other animals: guidelines for use and standardization. General Technical Report PNW-GTR-306. U.S. Department of Agriculture, Forest Service, Pacific Northwest Research Station. Portland, Oregon

Joslin P (1977) Night stalking: setting a camera 'trapline' for nocturnal carnivores. Photo Life 7:34–35

Hellawell JA (1991) Development of a rational for monitoring. In: Goldsmith B (ed) Monitoring for conservation and ecology. Chapman and Hall, London, UK, pp 12–16

Henschel P, Ray J (2003) Leopards in African rainforests: Survey and monitoring techniques. Wildlife Conservation Society, New York

Karanth KU, Nichols JD (1998) Estimation of tiger densities in India using photographic captures and recaptures. Ecology 79:2852–2862

Karanth KU, Nichols JD (eds) (2002) Monitoring tigers and their prey. Centre for Wildlife Studies, India

Kremen C, Merenlender AM, Murphy DD (1994) Ecological monitoring: A vital need for integrated conservation and development programs in the tropics. Conserv Biol 8:388–397

Lambeck RJ (1997) Focal species: a multi-species umbrella for nature conservation. Conserv Biol 11:849–856

Lawler JJ, White D, Sifneos JC, Master LL (2003) Rare species and the use of indicator groups for conservation planning. Conserv Biol 17:875–882

Lisboa PLB (1997) A Estação Científica Ferreira Penna/ECFPn. In: Lisboa PLB (ed) Caxiuanã. Museu Paraense Emílio Goeldi, Belém, Pará, Brasil, pp 23–49

Lisboa PLB, Gomes IA, Lisboa RCL, Urbinati CV (2002) Parte III–O estilo amazônico de sobreviver: manejo dos recursos naturais. In: Lisboa PLB (ed) Natureza, Homem e manejo de Recursos Naturais na Região de Caxiuanã, Melgaço. Museu Paraense Emílio Goeldi, Belém, Pará, pp 43–170

Lisboa PLB, Silva ASL, Almeida SS (1997) Florística e Estrutura dos Ambientes. In: Lisboa PLB (ed) Caxiuanã. Museu Paraense Emílio Goeldi, Belém, Pará, Brasil, pp 163–193

Mace R, Manley T, Aune K (1990) Use of systematically deployed remote cameras to monitor grizzly bears. Montana Department of Fish, Wildlife, and Parks, Kalispell, Montana, 25 pp

MacKenzie DI, Nichols JD, Royle JA, Pollock KP, Bailey LL, Hines JE (2006) Occupancy estimation and modeling. Elsivier Academic Press, NY

MacKenzie DI, Nichols JD, Hines JE, Knutson MG, Franklin AB (2003) Estimating site occupancy, colonization, and local extinction when a species is detected imperfectly. Ecology 84:2200–2207

Malhi Y, Phillips OL, Lloyd J, Baker T, Wright J, Almeida S, Arroyo L, Frederiksen T, Grace J, Higuchi N, Killeen T, Laurance WF, Leaño C, Lewis S, Meir P, Monteaguudo A, Neill D, Núñez Vargas P, Panfil SN, Patiño S, Pitman N, Quesada CA, Rudas-L1 A, Salomão R, Saleska S, Silva N, Silveira M, Sombroek WG, Valencia R, Vãsquez Martínez R, Vieira ICG, Vinceti B (2002) An international network to monitor the structure, composition and dynamics of Amazônian forests (RAINFOR). J Veg Sci 13:439–450

Marques-Aguiar SA, Aguiar GFdS (2002) Interações de quirópteros em ecossistemas tropicais: perspectiva de estudo em Caxiuanã. In: Lisboa PLB (ed) Caxiuanã: Populações Tradicionais, Meio Físico e Diversidade Biológica. Museu Paraense Emílio Goeldi, Belém, Pará, Brasil, pp 651–668

Moegenburg SM, Jardim MAG (2002) Utilization of açaí (Euterpe oleracea Mart) fruit and fruit patches by fruit-eating birds. In: Lisboa PLB (ed) Caxiuanã: Populações Tradicionais, Meio Físico e Diversidade Biológica. Museu Paraense Emílio Goeldi, Belém, Pará, Brasil, pp 641–650

Myers N, Mittermeier RA, Mittermeier CG, Fonseca GAB, Kent J (2000) Biodiversity hotspots for conservation. Nature 403:853–858

Nesbit W (1926) How to Hunt with the Camera. EP Dutton & Co., NY

Oliveira MCF, Costa ACL, Costa JPR, Palheta MCP, Pereira MGP (2002) Comportamento dos elementos meteorológicos. In: Lisboa PLB (ed) Caxiuanã: Populações Tradicionais, Meio Físico e Diversidade Biológica. Museu Paraense Emílio Goeldi, Belém, Pará, Brasil, pp 217–224

Pina ALCB (1999) Dinâmica sócio-ecológica em uma população de guaribas-das-mãos-vermelhas *(Alouatta belzebul)* na Estação Cientifica Ferreira Penna, Pará, Brazil. *Masters Dissertation*, Universidade Federal do Pará, Belém. p 89.

Pina ALCB, Souza LL, Ferrari SF (2002) Spacing patterns of *Alouatta belzebul* groups. In: Lisboa PLB (ed) Caxiuanã: Populações Tradicionais, Meio Físico e Diversidade Biológica. Museu Paraense Emílio Goeldi, Belém, Pará, Brasil, pp 685–695

Santana MRA, Lisboa PLB (2002) Estratégias de sobrevivência. In: Lisboa PLB (ed) Caxiuanã: Populações Tradicionais, Meio Físico e Diversidade Biológica. Museu Paraense Emílio Goeldi, Belém, Pará, Brasil, pp 95–125

Shiras G (1906) Photographing wild game with flashlight and camera. The National Geographic Magazine

Silva JMC, Pimentel Neto DC (1997) As Aves. In: Lisboa PLB (ed) Caxiuanã. Museu Paraense Emílio Goeldi, Belém, Pará, Brasil, pp 403–415

Silva MNF, Rylands AB, Patton JL (2001) Biogeografia e conservação da mastofauna na floresta amazônica brasileira. In: Capobianco JPR, Veríssimo A, Moreira A, Saweyr D, Santos I, Pinto LP (eds) Biodiversidade na Amazônia Brasileira: Avaliação de Ações Prioritárias para a Conservação. Uso Sustentável e Repartição de Benefícios, Belém, Pará, Brasil, pp 110–131

Silveira IM, Kern DC, Quaresma HDAB (2002) Reconstruindo uma ocupação. In: Lisboa PLB (ed) Caxiuanã: Populações Tradicionais, Meio Físico e Diversidade Biológica. Museu Paraense Emílio Goeldi, Belém, Pará, Brasil, pp 59–76

Souza LL (1999) Comportamento alimentar e dispersão de sementes por guaribas *(Alouatta belzebul)* na Estação Científica Ferreira Penna, (Caxiuanã/Melgaço/Pará). *Masters Dissertation*, Universidade Federal do Pará and Museu Paraense Emílio Goeldi, Belém, Brazil. p 168

Souza LL, Pina ALCB, Ferrari SF (2002) Diet of the red-handed howler monkey *(Alouatta belzebul)*. In: Lisboa PLB (ed) Caxiuanã: Populações Tradicionais, Meio Físico e Diversidade Biológica. Museu Paraense Emílio Goeldi, Belém, Pará, Brasil, pp 669–683

Srbek-Araujo AC, Chiarello AG (2005) Is camera-trapping an efficient method for surveying mammals in Neotropical forests? A case study in south-eastern Brazil. J Trop Ecol 21:121–125

Sunquist ME, Sunquist FC (2001) Changing landscapes: consequences for carnivores. In: Gittleman JL, Funk SM, Macdonald D, Wayne RK (eds) Carnivore Conservation. Cambridge University Press, Cambridge, pp 399–418

Tavares LI (1999) Estratégias de forrageio de um grupo silvestre de sagüi-branco (*Callithrix argentata* Linnaeus 1771) na Estação Científica Ferreira Penna–Pará. *Masters Dissertation*, Universidade Federal do Pará, Belém, Brazil. p 89.

Tavares LI, Ferrari SF (2002) Diet of the silvery marmoset *(Callithrix argentata)*. In: Lisboa PLB (ed) Caxiuanã: Populações Tradicionais, Meio Físico e Diversidade Biológica. Museu Paraense Emílio Goeldi, Belém, Pará, Brasil, pp 705–717

ter Keurs WJ, Meelis E (1986) Monitoring the biotic aspects of our environment as a policy instrument. Environ Monit Assess 7:161–168

Thompson WL, White GC, Gowan GC (1998) Monitoring Vertebrate Populations. Academic Press, San Diego

Trolle M (2003) Mammal survey in the southeastern Pantanal, Brazil. Biodivers Conserv 12:823–836

Trolle M, Kéry M (2003) Estimation of ocelot density in the Pantanal using capture-recapture analysis of camera trapping data. J Mammal 84:607–614

Veracini C (1997) O comportamento alimentar de *Callithrix argentata* (Linnaeus, 1771) (Primata, Callithrichidae). In: Lisboa PLB (ed) Caxiuanã. Museu Paraense Emílio Goeldi, Belém, Pará, Brasil, pp 437–446

Veracini C (2002) Ecologia alimentar e o uso dos habitats de *Saguinus midas niger*. In: Lisboa PLB (ed) Caxiuanã: Populações Tradicionais, Meio Físico e Diversidade Biológica. Museu Paraense Emílio Goeldi, Belém, Pará, Brasil, pp 719–734

Voss RS, Emmons LH (1996) Mammalian diversity in neotropical lowland rainforests: a preliminary assessment. Bulletin of the American Museum of Natural History. Number 230. New York

Wemmer C, Kunz TH, Lundie-Jenkins G, McShea WJ (1996) Mammalian sign. In: Wilson DE, Cole FR, Nichols JD, Rudran R, Foster MS (eds) Measuring and monitoring biological diversity. Standard methods for mammals, pp 157–176

Yoccoz NG, Nichols JD, Boulinier T (2001) Monitoring of biological diversity in space and time. Trends Ecol Evol 16:446–453

Biodivers Conserv (2007) 16:871–881
DOI 10.1007/s10531-006-9135-5

ORIGINAL PAPER

Grassland versus non-grassland bird abundance and diversity in managed grasslands: local, landscape and regional scale effects

Péter Batáry · András Báldi · Sarolta Erdős

Received: 15 February 2006 / Accepted: 2 October 2006 / Published online: 17 November 2006

Abstract Declines of West European farmland birds have been associated with intensive agricultural practices, while in Central and Eastern European countries grasslands still harbour a diverse and unique bird community. However, in these countries comparative studies on the effects of agricultural intensity on biodiversity are virtually missing. We compared bird communities of paired extensively and intensively grazed cattle pastures in three different regions of the Hungarian Great Plain. The influence of grazing intensity, landscape and regional effects were tested on the abundance and species richness of two ecological groups of bird species (grassland and non-grassland birds), as well as on the abundance of the three commonest grassland bird species (Skylark, Yellow wagtail, Corn bunting) in linear mixed models. We found significant effects of grazing intensity on the abundance of grassland birds, which were more abundant on the extensive sites, whereas no effects were found on non-grassland birds. This could be explained by a closer dependence of grassland birds on grasslands for nesting and foraging, whereas non-grassland birds only used grasslands opportunistically for foraging. Landscape effect was shown on grassland bird abundance, but not on non-grassland birds. The regions did affect only the species richness of grassland birds. At species level, the effect of management was significant for the three commonest grassland species, which were more abundant on the extensive fields in all regions. Additionally, on Skylark abundance landscape and regional effects were also shown. These findings suggest that conservation of biodiversity in agricultural systems requires the consideration of landscape perspective to apply the most adequate management.

P. Batáry (✉) · S. Erdős
Department of Zoology, Hungarian Natural History Museum, Ludovika tér 2, Budapest H-1083, Hungary
e-mail: batary@nhmus.hu

A. Báldi
Animal Ecology Research Group of the Hungarian Academy of Sciences and the Hungarian Natural History Museum, Ludovika tér 2, Budapest H-1083, Hungary

Keywords Grazing · Species richness · Skylark · Yellow wagtail · Corn bunting · Boundaries

Introduction

In central Europe, natural ecosystems are almost absent and most nature reserves require management, which is probably the most important local factor influencing their natural values (Martin and van der Straaten 2004; Tscharntke et al. 2005). These man-made reserves include large proportions of grassland, which are among the most threatened habitats by agricultural intensification (Donald et al. 2002; Tscharntke et al. 2005). Low-intensity grassland systems have developed a characteristic and diverse fauna and flora, but their substitution by high input, simplified systems has been associated with a decline in biodiversity (Stoate et al. 2001). In recent decades, major declines and range contractions of European and North American farmland birds have been reported in relation to changes in agricultural practices (e.g. Tucker and Heath 1994; Siriwardena et al. 1998; Hole et al. 2002; Murphy 2003; Burfield and van Bommel 2004; Gregory et al. 2005).

Agri-environmental schemes of the European Common Agricultural Policy are considered an essential tool for stopping the decline of biodiversity in farmland areas. However, the review of Kleijn and Sutherland (2003) gave contrasting results in the case of birds. Out of 19 studies examining the response of birds, four showed significant increases in species richness or abundance, two showed decreases and nine showed both increases and decreases as compared to controls. In the Central and Eastern European (CEE) countries, which recently joined the EU, the agri-enviromental schemes have just been introduced; nevertheless, well-designed studies now can serve to provide baseline data to assess effectiveness in the future (Kleijn and Sutherland 2003). Although the schemes perhaps could help in conservation of biodiversity of agricultural areas, but the larger part of subsidies of the Common Policy of the EU is spent on area based intensive agriculture, which probably will extend the homogenisation process of agricultural landscapes to CEE countries (Donald et al. 2002). These all call the attention of researchers on the need of comparative studies of agricultural intensification in Eastern and Central Europe as well.

Agricultural intensification has led to decline in biodiversity not only due to increased and widespread use of pesticides and machinery, but also due to the loss of non-crop habitats and habitat fragmentation. Semi-natural grasslands, which are important bird habitats especially in Hungary, can be easily transformed to arable lands (Pywell et al. 2002; Lundstrom-Gillieron and Schlaepfer 2003). Habitat fragmentation has serious effects on avian nesting success at several spatial scales, from edges to patches and landscapes (reviews in Paton 1994; Wiens 1995; Bender et al. 1998; Stephens et al. 2003; Batáry and Báldi 2004). However, there is little knowledge about biodiversity-driven agroecosystem functioning and the relative importance of local and landscape management for agricultural landscapes (Tscharntke et al. 2005). Furthermore, because of strong and widespread effects of habitat fragmentation, there is increasing need for multi-scale studies, where local and landscape scale effects are studied together. However, simultaneous studies of the effects of local-scale management and landscape structure are comparatively rare. Brotons et al. (2005) investigated the effect of adjacent agricultural habitat on the distribution of passerines in natural grasslands. Based on their results, they suggest that the management of the context in which steppe habitat is imbedded will significantly affect the distribution and abundance of this group of passerines and determine both habitat quality and conservation value at a landscape scale.

We selected extensively and intensively grazed pastures in three regions of Hungarian Great Plain (those can be categorised as solonchak alkali steppes, solonetz alkali steppes and wet meadows). In the year of the study, in 2003 were the first supports by National Agri-Environment Schemes given to farmers. None of our extensively grazed fields were supported, however all of them met the requirements of the National Agri-Environment Schemes (Ángyán et al. 1999). The three regions differ in landscape structure (from simple through intermediate to complex). Our aims were to test the influence of a local factor (grazing intensity), landscape and regional effects on two ecological groups of bird species (grassland and non-grassland birds) and on the most frequent species (Skylark *Alauda arvensis*, Yellow wagtail *Motacilla flava*, and Corn bunting *Miliaria calandra*) of the Hungarian Great Plain.

Methods

Twenty-one pairs of bird census sites were selected on grasslands with high and low grazing pressure in three distinct biogeographic regions of the Hungarian Great Plain. The three regions differ in their landscape structure, although grasslands are the most extensive land-use type (over 60%) in all regions. One of our study areas is situated in the Heves Landscape Protection Area in Eastern Hungary. This region ('Simple' region) has the most simple landscape structure with the largest, least fragmented grassland patches and is dominated by mosaic-complex of dry and wet alkali grasslands and marshes on solonetz soil. An other region is situated parallel to the river Danube, in the Kiskunság National Park (KNP), has an intermediate landscape structure, and contains secondary Pannonic alkali steppe vegetation on solonchak soils ('Intermediate' region). The third region (also in the KNP) is more heterogeneous, has several marshy patches and woodlots in the grasslands, so it has the most complex landscape structure ('Complex' region). For detailed area description see Báldi et al. (2005).

We established seven pairs of 12.5 ha plots, usually square in shape, in the extensively and intensively grazed grasslands in all regions (altogether 21 pairs of fields). The extensive field and intensive field in a pair had the same soil type and groundwater level and were situated in similarly structured landscapes, therefore the effects of confounding environmental variables were diminished. Cattle density was about 0.5 cows/ha on extensive and >1 cow/ha on intensive fields. None of the fields were fertilised, cut or re-seeded. Some of the study fields were not exactly 12.5 ha. In these cases we cut down the hang out piece on a randomly chose side of the larger field, because of the paired study design. This made possible to compare equally sized parts of a pair.

Breeding birds were surveyed using the territory mapping approach (Bibby et al. 1992). Altogether we made four censuses, two in April and two in May of 2003. Censuses were carried out under good weather conditions (no wind and rain), from sunrise to 9–10 a.m. The extensive and intensive fields of each pair were censused in the same morning by the same observer. The order in which sites were sampled was changed in the consecutive censuses. All observations registered by sight or sound were plotted on maps of the fields. Birds just flying through and not foraging in flight were excluded from the analysis. Territories were then drawn around complementary observations made during the four visits. Nest sites were allocated to the site where the observation most indicative of a territory was made (e.g. singing or displaying male, actual nests). To draw a territory in the case of breeding birds, we took two observations within about 10–20 m, with exception of late migrants (e.g. Grasshopper Warbler *Locustella naevia*, where observations occurred only in May). In the case of territories located along the borders of study plots, we counted them, if we got at least one contact within the plot. Species whose territories are larger than the target plots (e.g. great bustard *Otis tarda*, raptors) or that use the plots for feeding and clearly nested outside them (e.g. some small passerines such as tree sparrow *Passer montanus*, white stork *Ciconia ciconia* and corvids) were also mapped. Two occurrences of them meant one territory independently of the number of individuals and of their places on the map.

We divided the species for habitat specialist and generalist, because specialist species usually react in a different way to factors than generalist species (e.g. Siriwardena et al. 1998; Magura et al. 2000; Imbeau et al. 2003; Virkkala et al. 2004). We considered those species as habitat specialists (i.e. grassland birds), which breed on the ground of grasslands, whereas the remaining was considered as habitat generalist (non-grassland birds) (Table 1). Our previous analysis on the effect of grazing intensity on birds showed that species richness might increase due to increasing grazing pressure—a rather unexpected result (Báldi et al. 2005). However, this result was based solely on the local scale (grazing effect), and we explained it with the changing landscape structure of study fields. This paper in contrast to the earlier one is based on territories of grassland and non-grassland bird species and not on occurrences of all bird species, uses a pair-wise statistical design and a multi-scale approach, thus more comprehensive.

We used aerial photographs (Institute of Geodesy, Cartography and Remote Sensing; Air project 2000; 0.5 m/pixel resolution) from the study fields to digitise land-use types. First we searched the middle point of the 12.5 ha field and around this centre we used a buffer of 500 m radius. We digitised the following land-use types: (1) grasslands; (2) arable fields; (3) forests; (4) built-up areas; (5) marshes, reeds and bogs and (6) ditches, streams and lakes. Area percentages, mean areas and patch densities were measured for land-use types within the buffer. Further we also measured the total length of boundaries within buffers. All fields were situated only on grasslands and did not contain boundaries inside at all.

We analysed the local effect (extensive or intensive grazing regime), the effect of landscape and region and the interaction between management and landscape on the species number and on the abundance of grassland and non-grassland bird species in linear mixed models with the Restricted Maximum Likelihood method. To control for confounding effects of region on landscape influences (regions were selected based on their landscape structure, but differed also in soil type and vegetation structure), we built models with and without landscape factors. As landscape variable the total length of all boundaries was used, which reflects well the landscape complexity of the three regions (mean of total length of all boundaries within 500 m buffer areas in 'Simple' region: 4.74 km; in 'Intermediate' region: 6.25 km; in 'Complex' region: 8.53 km). The cause, why this variable was used is that the amount of boundaries or edges is one of the most important factors determining the nest success of ground nesting birds (Batáry and Báldi 2004). This landscape variable

Table 1 List of grassland and non-grassland species

Grassland species	Non-grassland species
Black-tailed godwit *Limosa limosa*	Barn swallow *Hirundo rustica*
Corn bunting *Miliaria calandra*	Bee-eater *Merops apiaster*
Curlew *Numenius arquata*	Common black-headed gull *Larus ridibundus*
Grasshopper warbler *Locustella naevia*	Common buzzard *Buteo buteo*
Great bustard *Otis tarda*	Cuckoo *Cuculus canorus*
Lapwing *Vanellus vanellus*	Great white egret *Casmerodius albus*
Montagu's harrier *Circus pygargus*	Greylag goose *Anser anser*
Partridge *Perdix perdix*	Hooded crow *Corvus corone cornix*
Pheasant *Phasianus colchicus*	Hoopoe *Upupa epops*
Quail *Coturnix coturnix*	Kestrel *Falco tinnunculus*
Redshank *Tringa tetanus*	Lesser grey shrike *Lanius minor*
Skylark *Alauda arvensis*	Magpie *Pica pica*
Stone curlew *Burhinus oedicnemus*	Mallard *Anas plathyrhynchos*
Stonechat *Saxicola torquata*	Marsh harrier *Circus aeruginosus*
Tawny pipit *Anthus campestris*	Nightingale *Luscinia megarhynchos*
Whinchat *Saxicola rubetra*	Red-backed shrike *Lanius collurio*
Yellow wagtail *Motacilla flava*	Red-footed falcon *Falco vespertinus*
	Roller *Coracias garrulus*
	Rook *Corvus frugilegus*
	Sand Martin *Riparia riparia*
	Savi's warbler *Locustella luscinioides*
	Sedge warbler *Acrocephalus schoenobaenus*
	Starling *Sturnus vulgaris*
	Tree sparrow *Passer montanus*
	White stork *Ciconia ciconia*
	Wood pigeon *Columba palumbus*

correlates with most other important landscape metrics—e.g. positively with total patch density ($r_{42} = 0.895$, $P = 0.001$), negatively with area and percent of grassland ($r_{42} = -0.810$, $P = 0.001$ and $r_{42} = -0.475$, $P = 0.001$) and positively with habitat diversity ($r_{42} = 0.454$, $P = 0.003$). Models contained only management, landscape, region and interaction between management and landscape as fixed factors and pair as random factor. Further we used the same models for the most abundant species, namely for Skylark, Yellow wagtail and Corn bunting as well. In the case of two latter species, data of one region was not included in the models, because both species occurred nearly only in two regions (for Yellow wagtail the 'Intermediate' region, for Corn bunting the 'Simple' region was not included into the models).

Results

Overall 748 bird territories belonging to 43 species were recorded at the 42 study sites. Intensively and extensively grazed fields had nearly the same species richness (13 grassland and 22 non-grassland species on intensively grazed grasslands, while 14 grassland and 22 non-grassland species on extensively grazed grasslands).

Investigating management (extensive versus intensive grazing) and regional effects on species richness, we showed both effects on grassland birds, but no effect on non-grassland birds (Table 2). The species richness of grassland birds was significantly higher on extensively grazed fields, and the 'Intermediate' region proved to be the most species rich. When we included landscape (total length of boundaries) in the models, no effect was found either on grassland or non-grassland birds (Table 2).

Abundance of grassland species was higher on extensive than on intensive fields (Table 2; Fig. 1), in contrast to the abundance of non-grassland species. In the case of grassland birds, management effect was shown in both models (with and without including landscape). Further, investigating the significant landscape effect, the abundance of grassland species was negatively related to total length of boundaries (Table 2; Fig. 2; intensive fields: $r_{21} = -0.328$, $P = 0.146$; extensive fields: $r_{21} = -0.462$, $P = 0.035$). No effect of region and no interaction effect were found on the abundance of grassland and non-grassland bird species.

At the species level, the effect of management was significant for the commonest species, the Skylark (in both models), which was more abundant on the extensive fields in all regions (Table 2; Fig. 3a). Additionally, Skylark abundance was also negatively related to total length of boundaries and occurred more frequently in the 'Simple' and 'Intermediate' region than in the 'Complex' region (Table 2; Figs. 3a and 4; intensive fields: $r_{21} = -0.456$, $P = 0.038$; extensive fields: $r_{21} = -0.713$, $P = 0.001$). The effect of management and landscape on Skylark abundance was also manifested by a significant interaction between management and landscape—Skylark abundance was more sensitive to boundary length on extensively grazed fields than intensive fields (Fig. 4). In the case of two other frequent species management effects were found, the abundance of these species tended to be higher on the extensive fields than on the intensive ones (Table 2; Fig. 3b, c). Further, no landscape effects were shown for the Yellow wagtail and the Corn bunting, but in the case of Corn bunting a significant management × landscape effect was found (Table 2).

Table 2 Linear mixed models on the effects of management (intensive versus extensive grazing [M]), of landscape (total length of boundaries in surrounding landscape [L]) and of region [R] on bird communities and species of Hungarian grasslands. Model 1 contains management and region, while in model 2 landscape and management × landscape interaction were included. Bold *P*-values indicate significant effects

	Model 1				Model 2							
	M		R		M		L		R		M × L	
	F	*P*	*F*	*P*	*F*	*P*	*F*	*P*	*F*	*P*	*F*	*P*
Species richness												
Grassland species	6.936	**0.016**	4.338	**0.029**	0.419	0.524	0.135	0.716	2.548	0.103	0.088	0.770
Non-grassland species	1.337	0.261	0.481	0.626	0.098	0.758	2.120	0.154	1.281	0.299	0.006	0.940
Territories												
Grassland species	40.030	**0.001**	1.492	0.251	7.613	**0.012**	4.671	**0.038**	0.492	0.619	0.215	0.647
Non-grassland species	1.333	0.262	0.119	0.888	0.001	0.984	1.427	0.240	0.557	0.581	0.189	0.667
Skylark	9.033	**0.007**	11.354	**0.001**	9.734	**0.005**	5.088	**0.030**	8.154	**0.003**	4.534	**0.045**
Yellow wagtail	11.039	**0.003**	1.616	0.510	0.913	0.350	2.402	0.132	2.306	0.157	0.346	0.562
Corn bunting	1.174	0.294	0.325	0.616	7.499	**0.014**	1.937	0.178	0.044	0.8401	0.470	**0.005**

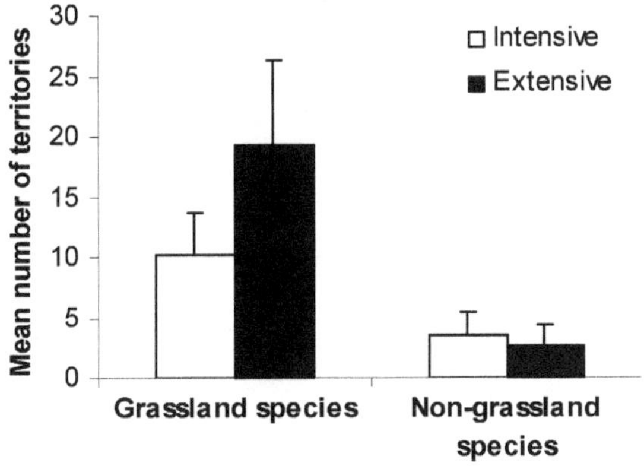

Fig. 1 Mean territory number of grassland (breeds on the ground of grasslands) and non-grassland bird species of intensively and extensively grazed grasslands of Hungarian Great Plain ($n = 42$ fields); bars show 1 SE

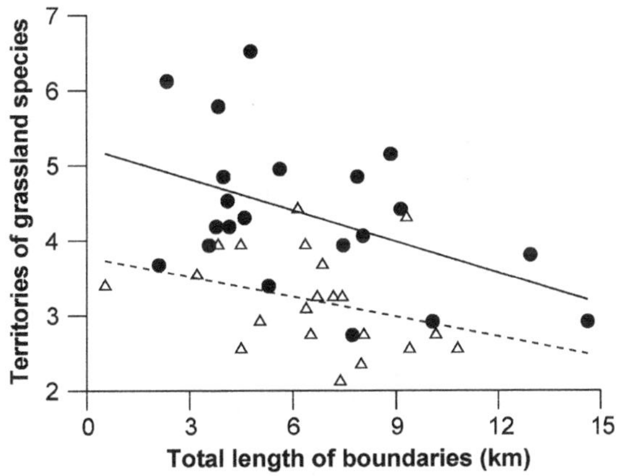

Fig. 2 Correlation between the total length of boundaries (km) and mean territories of grassland bird species on differently managed grasslands (△ intensive fields, ● extensive fields). Dashed line indicates correlation for intensive fields, while continuous line indicates correlation for extensive fields

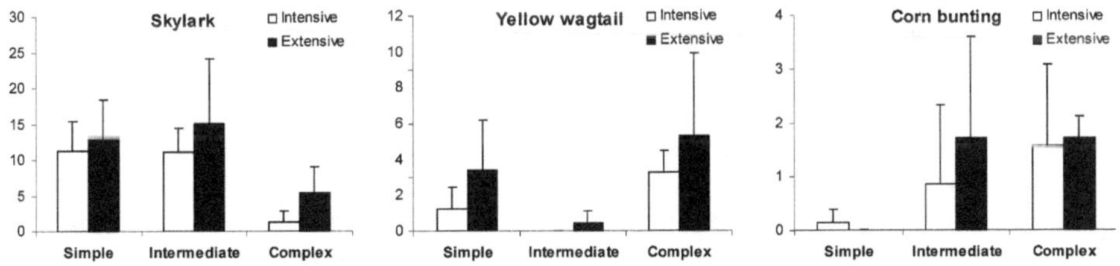

Fig. 3 Mean individual number of the most abundant grassland bird species on Hungarian grasslands. Landscape complexity increases from the 'Simple' region to 'Complex' region. Filled bars indicate extensive grazing, open bars indicate intensive grazing. (**a**) Skylark, (**b**) Yellow wagtail, (**c**) Corn bunting

Discussion

The relative importance of local farming management (intensive versus extensive grazing), landscape context and region was analysed on grassland and non-grassland bird diversity. In the case of community analysis region had no confounding effects on landscape influences. In recent decades great attention have been paid to the

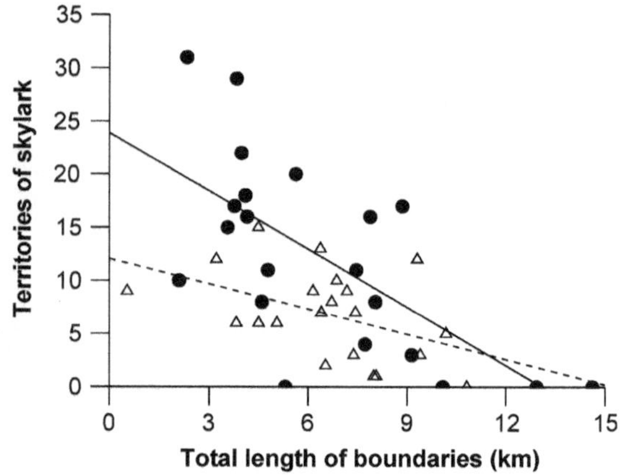

Fig. 4 Correlation between the total length of boundaries and territories of Skylark on differently managed grasslands (△ intensive fields, ● extensive fields). Dashed line indicates correlation for intensive fields, while continuous line indicates correlation for extensive fields

population declines of farmland birds attributed to intensive agricultural management (Vickery et al. 2001). Grazing generally has negative effect on bird species richness or abundance (Dobkin et al. 1998; Fuller and Gough 1999; Verhulst et al. 2004; Maron and Lill 2005). In the present study we showed effect of management on species richness and abundance of true grassland birds with higher species richness and abundance on the extensive sites, while on non-grassland species no effect was found. This could be explained probably by that true grassland birds are more specialised on grassland habitats than non-grassland birds, which only feed and not nest there. Grazing can impact on bird populations through changes in vegetation structure, food resources and predation pressure (Vickery et al. 2001). Alteration of the vegetation structure will affect the suitability of the sward for nesting and feeding (Milsom et al. 1998). Intensive grazing may increase nest losses due to predation and trampling (Ammon and Stacey 1997; Wilson et al. 1999; Pavel 2004).

Tscharntke et al. (2005) concluded in their recent review that agri-environment schemes need to broaden their perspective and to take the different responses to schemes in simple (high impact) and complex (low impact) agricultural landscapes into account. Furthermore, Benton et al. (2003) reviewed that several studies have shown heterogeneity to be associated with diversity. In the present study we found landscape effect on grassland bird abundance, however with increasing heterogeneity (increasing total length of boundaries) the abundance of grassland birds declined. The conclusion of Benton et al. (2003) is probably valid only in highly managed regions. Here we have to emphasise that our study sites contained more than 60% grasslands in all regions and generally are less intensively managed (non-fertilised and pesticide free) than in Western Europe. This, and other studies warn that the understanding of biodiversity in agricultural landscapes need a more comprehensive approach (Kleijn and Báldi 2005; Tscharntke et al. 2005). In contrast to our results, the model of Virkkala et al. (2004) explained a moderate proportion of the variation in the total density of farmland birds in the landscape. Söderström et al. (2001) emphasised the importance of landscape composition for mobile organisms such as birds and found that species richness of grassland birds decreased with increasing proportion of urban elements and arable fields in a 1000 m landscape area centred on each pasture. In a similar multi-scale study, like ours, examining the farmland management on assemblages of grassland wintering birds in Portugal,

Moreira et al. (2005) found that species richness was primarily influenced by landscape context, whereas field management mostly determined abundance. Finally, we could separate regional and landscape effects with building models with and without a landscape metric. The results showed that the regional effect on species richness was not due to landscape complexity differences between regions, but probably differences in e.g. soil type and/or vegetation structure and composition.

At species level our analyses were limited to the most abundant birds (Skylark, Yellow wagtail and Corn bunting), which are important contributors of the Hungarian grassland bird assemblages. All three species underwent smaller or larger declines during the recent decades mostly in West Europe, but the key eastern populations remained stable (Siriwardena et al. 1998; Brickle et al. 2000; Burfield and van Bommel 2004; Newton 2004; Gregory et al. 2005). In our study significant management effect was shown for all species, while landscape effect was only found in the case of Skylark. For Skylark regional effect was not separable from landscape effect, which at species level is not surprisingly, because species react individually for landscape, management, vegetation, etc. (Bradbury et al. 2004). Our results confirm that Skylarks avoid smaller fields (Donald et al. 2001; Perkins et al. 2000; Moreira et al. 2005). This is consistent with the theory that abundant generalist species should be less affected by fragmentation than (habitat) specialist species (Braschler and Baur 2005). The significant interaction between management and landscape reflect that the increased Skylark density was confined to extensively grazed fields, of which surrounding landscapes contained less boundaries.

The ongoing changes in the agriculture threaten the rich eastern European ecosystems. However, the agri-environmental schemes open up new views to protect the biodiversity there. But as other researchers, we also have to emphasise that conservation of biodiversity and ecosystem services in agricultural systems requires a landscape perspective (Bengtsson et al. 2003; Tscharntke et al. 2005). Finally, we conclude that both local management and landscape structure has significant effects on grassland bird abundance, but not on non-grassland abundance when analysed together, and that such effects depend on the ecology of each bird species.

Acknowledgements We are indebted to Prof M. Díaz for valuable comments on the manuscript. We are grateful to A. Bankovics, B. Kancsal, T. Kisbenedek and L. Molnár for help with bird counting. Staff of the Kiskunság and Bükk National Parks gave valuable help. We thank the National Parks for permissions, the Kiskunság National Park for making available the aerial photographs and landowners for allowing us to work on their fields. The study was supported by the EU-funded project 'EASY' (QLK5-CT-2002-01495) and partly by the Faunagenesis project (NKFP 3B023-04). A. B. was a Bolyai Research Fellow of the Hungarian Academy of Sciences.

References

Ammon EM, Stacey PB (1997) Avian nest success in relation to past grazing regimes in a montane riparian system. Condor 99:7–13

Ángyán J, Fésűs I, Podmaniczky L, Tar F, Madarassy A (1999) Nemzeti Agrár-környezetvédelmi Program. Ministry of Agriculture and Rural Development, Budapest

Báldi A, Batáry P, Erdős S (2005) Effects of grazing intensity on bird assemblages and populations of Hungarian grasslands. Agric Ecosyst Environ 108:251–263

Batáry P, Báldi A (2004) Evidence of an edge effect on avian nest success: a meta-analysis. Conserv Biol 18:389–400

Bender DJ, Contreras TA, Fahrig L (1998) Habitat loss and population decline: a meta-analysis of the patch size effect. Ecology 79:517–533
Bengtsson J, Angelstam P, Elqvist T, Emanuelsson U, Folke C, Ihse M, Moberg F, Nystrom M (2003) Reserves, resilience and dynamic landscapes. Ambio 32:389–396
Benton TG, Vickery JA, Wilson JD (2003) Farmland biodiversity: is habitat heterogeneity the key? Trends Ecol Evol 18:182–188
Bibby CJ, Burgess ND, Hill DA (1992) Bird Census Techniques. Academic Press, London
Bradbury RB, Browne SJ, Stevens DK, Aebischer NJ (2004) Five-year evaluation of the impact of the Arable Stewardship Pilot Scheme on birds. Ibis 146:171–180
Braschler B, Baur B (2005) Experimental small-scale grassland fragmentation alters competitive interactions among ant species. Oecologia 143:291–300
Brickle NW, Harper DGC, Aebischer NJ, Cockayne SH (2000) Effects of agricultural intensification on the breeding success of Corn buntings *Miliaria calandra*. J Appl Ecol 37:742–755
Brotons L, Wolff A, Paulus G, Martin J-L (2005) Effect of adjacent agricultural habitat on the distribution of passerines in natural grasslands. Biol Conserv 124:407–414
Burfield I, van Bommel F (2004) Birds in Europe: population estimates, trends and conservation status. BirdLife International, Cambridge
Dobkin DS, Rich AC, Pyle WH (1998) Habitat and avifaunal recovery from livestock grazing in a riparian meadow system of the Northwestern Great Basin. Conserv Biol 12:209–221
Donald PF, Green RE, Heath MF (2001) Agricultural intensification and the collapse of Europe's farmland bird populations. Proc Roy Soc Lond B 268:25–29
Donald PF, Pisano G, Rayment MD, Pain DJ (2002) The Common Agricultural Policy, EU enlargements and the conservation of Europe's farmland birds. Agric Ecosyst Environ 89:167–182
Fuller RJ, Gough SJ (1999) Changes in sheep numbers in Britain: implications for bird populations. Biol Conserv 91:73–89
Gregory RD, van Strien A, Vorisek P, Meyling AWG, Noble DG, Foppen RPB, Gibbons DW (2005) Developing indicators for European birds. Philos T Roy Soc B 360:269–288
Hole DG, Whittingham MJ, Bradbury RB, Anderson GQA, Lee PLM, Wilson JD, Krebs JR (2002) Widespread House Sparrow extinctions – Agricultural intensification is blamed for the plummeting populations of these birds. Nature 418:931–932
Imbeau L, Drapeau P, Mokkonen M (2003) Are forest birds categorised as "edge species" strictly associated with edges? Ecography 26:514–520
Kleijn D, Sutherland WJ (2003) How effective are European agri-environment schemes in conserving and promoting biodiversity? J Appl Ecol 40:947–969
Kleijn D, Báldi A (2005) Effects of set-aside land on farmland biodiversity: comments on van Buskirk and Willi. Conserv Biol 19:963–966
Lundstrom-Gillieron C, Schlaepfer R (2003) Hare abundance as an indicator for urbanisation and intensification of agriculture in Western Europe. Ecol Model 168:283–301
Magura T, Tóthmérész B, Bordán Z (2000) Effects of nature management practice on carabid assemblages, Coleoptera: Carabidae. in a non-native plantation. Biol Conserv 93:95–102
Maron M, Lill A (2005) The influence of livestock grazing and weed invasion on habitat use by birds in grassy woodland remnants. Biol Conserv 124:439–450
Martin D, van der Straaten J (2004) Cultural landscapes and land use. Kluwer Academic Publishers, Dordrecht
Milsom TP, Ennis DC, Haskell DJ, Langton SD, McKay HV (1998) Design of grassland feeding areas for waders during winter: the relative importance of sward, landscape factors and human disturbance. Biol Conserv 84:119–129
Moreira F, Beja P, Morgado R, Reino L, Gordinho L, Delgado A, Borralho R (2005) Effects of field management and landscape context on grassland wintering birds in Southern Portugal. Agric Ecosyst Environ 109:59–74
Murphy MT (2003) Avian population trends within the evolving agricultural landscape of eastern and central United States. Auk 120:20–34
Newton I (2004) The recent declines of farmland bird populations in Britain: an appraisal of causal factors and conservation actions. Ibis 146:579–600
Paton PWC (1994) The effect of edge on avian nest success: How strong is the evidence? Conserv. Biol 8:17–26
Pavel V (2004) The impact of grazing animals on nesting success of grassland passerines in farmland and natural habitats: a field experiment. Folia Zool 53:171–178

Perkins AJ, Whittingham MJ, Bradbury RB, Wilson JD, Morris AJ, Barnett PR (2000) Habitat characteristics affecting use of lowland agricultural grassland by birds in winter. Biol Conserv 95:279–294

Pywell RF, Bullock JM, Hopkins A, Walker KJ, Sparks TH, Burke MJW, Peel S (2002) Restoration of species-rich grassland on arable land: assessing the limiting processes using a multi-site experiment. J Appl Ecol 39:294–309

Siriwardena GM, Baillie SR, Buckland ST, Fewster RM, Marchant JH, Wilson JD (1998) Trends in the abundance of farmland birds: a quantitative comparison of smoothed common bird census indices. J Appl Ecol 35:24–43

Söderström B, Svensson B, Vessby K, Glimskar A (2001) Plants, insects and birds in semi-natural pastures in relation to local habitat and landscape factors. Biodiv Conserv 10:1839–1863

Stephens SE, Koons DN, Rotella JJ, Willey DW (2003) Effects of habitat fragmentation on avian nesting success: a review of the evidence at multiple spatial scales. Biol Conserv 115:101–110

Stoate C, Boatman ND, Borralho RJ, Carvalho CR, de Snoo GR, Eden P (2001) Ecological impacts of arable intensification in Europe. J Environ Manage 63:337–365

Tscharntke T, Klein AM, Kruess A, Steffan-Dewenter I, Thies C (2005) Landscape perspectives on agricultural intensification and biodiversity – ecosystem service management. Ecol Lett 8:857–874

Tucker GM, Heath MF (1994) Birds in Europe: Their Conservation Status. BirdLife International, Cambridge

Verhulst J, Báldi A, Kleijn D (2004) Relationship between land-use intensity and species richness and abundance of birds in Hungary. Agric Ecosyst Environ 104:465–473

Vickery JA, Tallowin JR, Feber RE, Asteraki EJ, Atkinson PW, Fuller RJ, Brown VK (2001) The management of lowland neutral grasslands in Britain: effects of agricultural practices on birds and their food resources. J Appl Ecol 38:647–664

Virkkala R, Luoto M, Rainio K (2004) Effects of landscape composition on farmland and red-listed birds in boreal agricultural-forest mosaics. Ecography 27:273–284

Wiens JA (1995) Habitat fragmentation – Island vs. landscape perspectives on bird conservation. Ibis 137:97–104

Wilson JD, Morris AJ, Arroyo BE, Clark SC, Bradbury RB (1999) A review of the abundance and diversity of invertebrate and plant foods of granivorous birds in northern Europe in relation to agricultural change. Agric Ecosyst Environ 75:13–30

Biodivers Conserv (2007) 16:883–912
DOI 10.1007/s10531-006-9022-0

Biogeographic patterns of the East African coastal forest vertebrate fauna

Ermias T. Azeria · Isabel Sanmartín · Stefan Ås · Allan Carlson · Neil Burgess

Received: 17 June 2005 / Accepted: 27 February 2006 / Published online: 28 March 2007

Abstract The archipelago-like coastal forest of East Africa is one of the highest priority ecosystems for biodiversity conservation worldwide. Here we investigate patterns of species richness and biogeographic distribution among birds, mammals and reptiles of these forests, using distribution data obtained from recently published reviews and information collated by the WWF Eastern Africa Coastal Forest Ecoregion Programme. Birds and mammals species were divided into forest specialists and generalists, and forest specialist reptiles into 'coastal' and 'forest' endemics. The species richness of birds and generalist mammals increased with area, and is probably a result of area-dependent extinction. Only in birds, however, species richness increased with decreasing isolation, suggesting possible isolation-dependent colonization. Forest diversity, associated to altitudinal range, is important for specialist birds and mammals, whose species richness increased with wider altitudinal range. The number of relict coastal endemic and forest endemic reptiles was higher in forests with wider altitudinal ranges and on relatively higher altitude, respectively. Such forests have probably provided a suitable (and perhaps stable) environment for these species through time, thus increasing their persistence. Parsimony

E. T. Azeria · A. Carlson
Department of Conservation Biology, Swedish University of Agricultural Sciences, 7002, SE-750 07 Uppsala, Sweden

I. Sanmartín · S. Ås
Evolutionary Biology Centre, Uppsala University, Norbyvägen. 18D, SE-752 36 Uppsala, Sweden

A. Carlson
WWF-Sweden, Ulriksdals Slott, SE-170 81 Solna, Sweden

N. Burgess
WWF-USA Conservation Science Programme, 1250 24th St NW, Washington, DC, USA

N. Burgess
Conservation Biology Group, Zoology Department, Cambridge University, CB2 3EJ Cambridge, UK

E. T. Azeria (✉)
Department of Biology, McMaster University, 1280 Main St. West, Hamilton, ON L8S 4K1, Canada
e-mail: Ermias.Azeria@nvb.slu.se

analysis of distributions (PAD) and cluster analyses showed geographical distance and general ecological similarity among forests as a determinant factor in bird distribution patterns, with compositional similarity decreasing with increasing inter-forest distance. Compositional similarity patterns of mammals among the forests did not show a strong geographical correspondence or a significant correlation with inter-forest distance, and those of reptiles were not resolved, with very low similarity levels among forest faunas. Our results suggest that the relative importance (and causal relationship) of forest attributes affecting the distribution of the East African coastal forest vertebrate fauna varies depending on life history traits such as dispersal ability and forest specialization. The groupings in PAD are partly congruent with some of the previous classifications of areas of endemism for this region, supporting the 'naturalness' of these regions.

Keywords Biogeography · Conservation · East Africa coastal forests · Fragmentation · Species–area relationship · Isolation · Altitude · Parsimony analysis of distributions · Relict fauna

Introduction

Fragmentation of once continuous habitats entails habitat loss, degradation and increased isolation, and ultimately the loss of biodiversity (Andrén 1994; Cornelius et al. 2000; Brooks et al. 2002). Ecologically important processes during fragmentation are similar to those influencing biotas of 'true' islands (MacArthur and Wilson 1967). The species richness in the remnant fragments, according to MacArthur and Wilson (1967), is primarily the result of a dynamic balance between two opposing processes, area-dependent extinction and isolation-dependent colonization. Species richness is therefore expected to increase with increasing area and/or decreasing isolation from the species source pool (MacArthur and Wilson 1967; Rosenzweig 1995; Whittaker 1998).

However, fragments may also differ in ecological factors other than area and isolation (e.g. Diamond 1981; Stuart 1981), as well as community dynamics in such fragments may depart from colonization–extinction equilibrium (Whittaker 1998; Lomolino 2000). For example, forest fragments may span smaller or larger altitudinal ranges, and forests incorporating a greater altitudinal range are expected to cover a larger span of habitats and hence harbour more species (e.g. Kingdon and Howell 1993; Ås et al. 1997). Similarly, ecological factors that determine species number are also determinants in species composition (e.g. Whittaker 1998; Lomolino 2000). For example, species composition of communities is expected to be more similar among geographically closer fragments than among more distant ones because of more habitats in common and higher rate of 'inter-fragment' dispersal (Power 1975; Nekola and White 1999; Ricklefs and Lovette 1999; Morand 2000; Azeria 2004).

However, the relative importance of these factors in determining species distribution throughout the fragmented habitat differs among taxa, depending on life history traits such as dispersal ability (birds vs. reptiles) and level of dependence on the forest resources (specialists vs. generalists) (Ricklefs and Lovette 1999; Gascon et al. 1999; Watson 2002). Moreover, historical factors (e.g. past distribution and fragmentation), origin of the biota (e.g. relict species vs. species derived from the surrounding matrix), and species geographical ranges have a profound effect on species distribution patterns (e.g. Ricklefs and Schluter 1993; Welter-Schultes and Williams 1999; Jetz and Rahbek 2002; Watson 2002; Jetz et al. 2004).

The coastal forests of Eastern Africa form an archipelago of forests extending along the coastal plain of East Africa from southern Somalia to northern Mozambique. Located within the so-called ''Swahili regional centre of endemism and Swahili-Maputaland regional transition zone'' (Clarke 2000), they form one of the major centres of endemism in Africa (Burgess and Clarke 2000; Brooks et al. 2001). The large number of endemic species, high biodiversity, and concentration of rare and threatened taxa make the coastal forests of East Africa one of the highest priority ecosystems for conservation in Africa and globally (Hawthorne 1993; Burgess and Clarke 2000; Myers et al. 2000; Brooks et al. 2001, 2002; Burgess et al., 2004). Despite their biological importance, however, the unique fauna and flora of these forests are currently threatened by human disturbance, through increasing fragmentation and forest degradation (e.g. Hawthorne 1993; Brooks et al. 2002).

The current view is that their present archipelagic distribution is the remnant of a once continuous forest that extended over tropical Africa in the Early Tertiary. This later became fragmented by the gradual desiccation of the continent that started in the Miocene and, more recently, by increasing human habitat destruction (Moreau 1933; deMonocal 1995; Clarke 2000b; Clarke and Karoma 2000 and references therein). Most coastal forest endemics have a narrow distributional range, often exhibiting single-site endemism or disjunct distributional patterns (Burgess et al. 1998). They do not seem to be part of a recent local radiation, but show closer relationships to West African lineages, or represent 'ancient lineages' that have gone extinct elsewhere (Burgess et al. 1998; Burgess 2000). Rodgers (2000) suggests that the ancient forest was probably heterogeneous, with patchy ecological conditions, even before the onset of fragmentation. Therefore, the variation in distribution patterns between forest fragments that we observe today is probably the result of both recent ecological and historical factors.

The last decade has witnessed an increase in the number of biogeographic studies on the coastal forest biota (Burgess et al. 1998; Burgess and Clarke 2000, and references therein). Most of these studies have focused on studying distributional patterns of endemic species, by simply plotting numbers of coastal forest endemics onto a map, which are then used to define 'centres of endemism' for conservation policies (Burgess et al. 1998; Burgess and Clarke 2000; but see de Klerk et al. 2002b). However, more analytical approaches are now available that can be used to visualize biotic relationships among areas. Cluster analysis based on similarity indices has been the traditional method to group areas based on their compositional similarity. An expansion of parsimony analysis of endemicity (PAE, Rosen 1988) that takes into account an ecological interpretation, parsimony analysis of distributions (PAD, Trejo-Torres and Ackerman 2001), has recently arisen as an alternative to phenetic multivariate methods in biogeographic analysis. Like cluster analysis, PAD uses presence/absence matrices to construct branching diagrams of area relationships, and can be used as a method to find hierarchical structure in biogeographical data, i.e., nested sets of area relationships (Trejo-Torres and Ackerman 2001; Cavieres et al. 2002; García-Barros et al. 2002).

In this study, we will use PAD to analyse biogeographical affinities among the coastal forests using recent data on species distributions of birds, mammals, and reptiles, and compare those groupings with current definitions of 'centres of endemism' (Burgess et al. 1998; Burgess 2000). Moreover, we will use cluster analysis to group areas based on their compositional similarity, and test whether this similarity can be explained by 'inter-forest dispersal'. Finally, we will analyse the effects of ecological attributes of the forest, such as forest size, isolation, median altitude, and elevation range, in the variation of species richness between forest fragments. Because different taxa may respond differently to ecological attributes, we adopt here a comparative approach and investigate species richness and compositional similarity patterns among forests across three vertebrate groups: birds vs. mammals vs. reptiles. We also examine differences in species richness between habitat generalists and forest specialists. This approach will give us an insight into the effect of life history characteristics on current species patterns, and help us to understand differences in forests affinities between taxonomic groups, which could then be used for conservation strategies.

Materials and methods

The study system

The East African coastal forests (Fig. 1) are a set of tiny forests found on the coastal belt of East Africa, and often imbedded within a much larger habitat mosaic of farmland, savannah-woodland, and thickets. The forests themselves cover a total area of 3,172 sq. km, consisting of over 260 forest patches varying in size and degree of isolation; most of them are less than 500 ha in size (Burgess and Clarke 2000). The forest belt extends from southern Somalia south to Mozambique (i.e. between 1° N to 25° S and 34°–41° E), distributed over six countries: Somalia, Kenya, Tanzania, Mozambique, Zimbabwe, and Malawi. The limits to the coastal forest area are set by rainfall (decreasing to the north), seasonality (increasing to the south), and by altitude, increasing to the west, where they can extend from sea-level to a maximum altitude of 1,100 m, depending on local ecological conditions. Coastal forests also may be found in areas with a low mean annual rainfall (< 470 mm), growing along river banks (Burgess et al. 1998; Clarke 2000a, b).

The word 'coastal forest' in itself is a broad term used to define a mosaic of forest types including the typical semi-evergreen and evergreen dry forest, variant type and sub-type (scrub forest, Brachystegia forest) and transitional vegetation formations (e.g. riverine, swamp, Afromontane transition forests) (for a definition, see Hawthorne 1993; Rodgers 2000; Clarke 2000a, and references therein).

Data treatment

Presence–absence matrices of species by forests were obtained from three recent reviews of the East African coastal forest fauna: birds (Mlingwa et al. 2000), mammals (Burgess et al. 2000), and reptiles (Broadley and Howell 2000). The three reviews are based on published and unpublished data collected during several research programs that have been active in the coastal forests of East Africa during the last decade (see the data sources for detailed references). These sources were updated in 2004 using information collected for the WWF Eastern Africa Coastal Forest Ecoregion Programme (WWF-EARPO 2002; Anonymous 2003; Burgess et al. 2004; WWF unpublished). Although survey effort has not been fully standardized between sites, the time spent in each forest was considerable (tens to hundreds of man days) and thus we consider that the species lists available are adequate to detect broad biogeographical groupings between forests. Data on birds and mammals are the most comprehensive, particularly for birds, which can be sampled in mist nets and

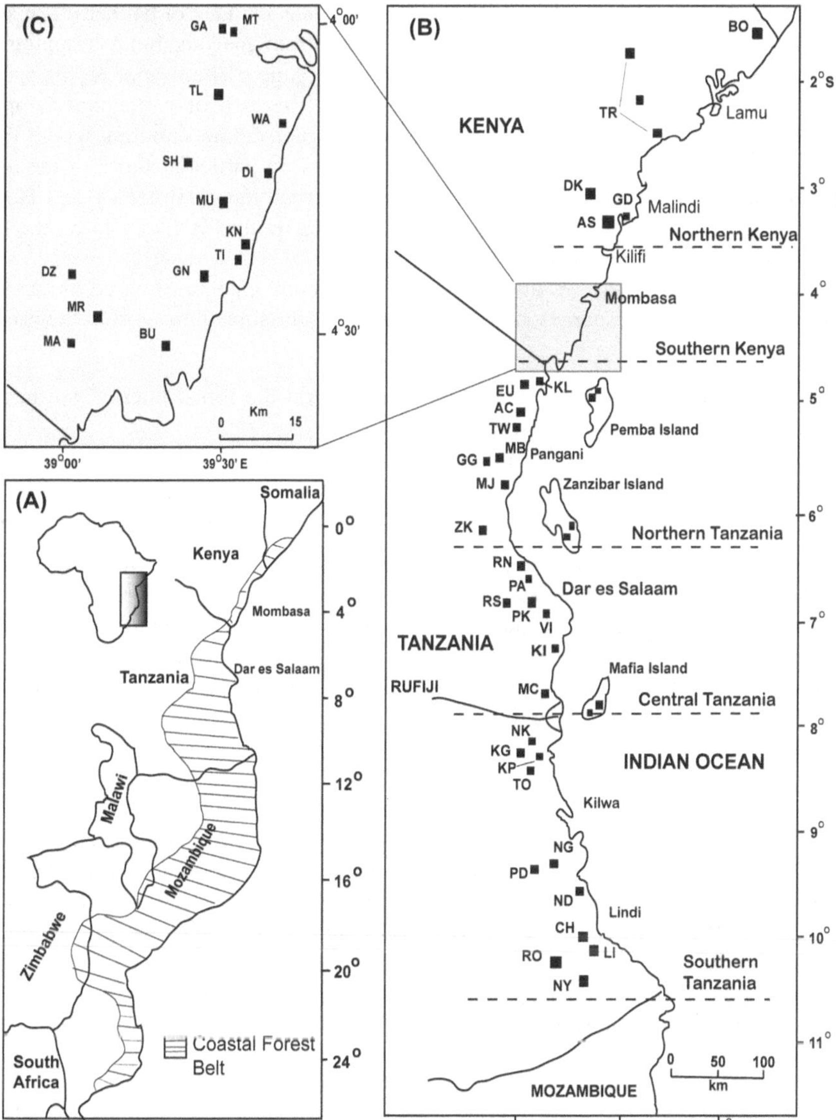

Fig. 1 Map of study area showing (**A**) the general location of the East African coastal forests; (**B**) Location of the coastal forest fragments included in this study; (**C**) Details of southern Kenya indicating the location of the coastal forest fragments studied in this region. Adapted from Burgess et al. (1998). Abbreviations (north to south): BO = Boni; TR = Tana river primate and delta; DK = Dakatcha; GD = Gede; AS = Arabuko Sokoke; GA = Gandini; MT = Mtswakara; TL = Teleza; WA = Waa; DI = Diani; SH = Shimba Hills; MU = Muhaka; KN = Kinondo, TI = Timbwa; GN = Gongoni; DZ = Dzombo; MR = Mrima; MA = Marenji; BU = Buda; EU = Eastern Usambara lowland; KL = Kilulu; AC = Amboni Caves; TW = Tongwe; MB = Msumbugwe; GG = Gendagenda; MJ = Mkwaja; ZK = Zaraninge-Kiono; RN = Ruvu North; PA = Pande; RS = Ruvu South; PK = Pugu and Kazimzumbwi; VI = Vikindu; KI = Kisiju; MC = Mchungu; NK = Namakutwa; KG = Kiwengoma; KP = Kitope; TO = Tong'omba; NG = Ngarama; PD = Pindiro; ND = Ndimba; CH = Chitoa; RO = Rondo; LI = Litipo; NY = Nyangamara. Other localities not shown in the map but included in the study are Jubba River (southern Somalia), Kimboza (Tanzania), and forests in the offshore islands of Pemba Island (PI), Zanzibar Island (ZI), and Mafia Island (MI)

observed directly and recorded through calls. For mammals the lack of bat netting at some sites means that all bat assemblage data were excluded from analysis, but as trapping was undertaken at all sites, rodents and shrews have been included. The lists of reptile species per forest are likely to be the least complete, due to the lack of a standard sampling protocol and difficulty of observation and trapping. Further reptile sampling would likely increase the number of 'coastal endemic' reptile species, in particular those of the dense forest and forest edge, many of which are fossorial herpetofauna (Broadley and Howell 2000). Hence the biogeographical zonation obtained for reptiles is likely to be the least robust to the addition to new data.

Within each faunal group, we distinguished two different categories based on the level of dependence on forest resources, and according to the classifications used in the original studies:

(a) 'Forest specialists' are species strictly dependent on the forest interior, so they are likely to disappear if the forest is modified.
(b) 'Forest generalists' are species that depend upon undisturbed forests for some of their resources, for example nesting sites in birds, but which are also able to live at the edge of the forest or in modified, managed, or secondary forests.

The original reptile study (Broadley and Howell 2000) considered two different categories: 'coastal endemics'—species which are endemic to the coastal forests, and 'forest endemics'—species that are also found in the nearby Eastern Arc Mountain forests. We have grouped these two categories under specialists in our study. No generalist species were considered in Broadley and Howell's (2000) analysis but non-forest species made up for a great part of the total reptile diversity.

Species-richness pattern

In all, a total of 163 species and 53 forests found in Somalia, Kenya and Tanzania (Table 1) were included in our study (Fig. 1). Data on area size (minimum forest area), and median and range of altitude for each forest fragment were obtained from Burgess and Clarke (2000). When a habitat is patchily distributed in the landscape, measuring the degree of isolation is not straightforward. In our analysis, we have adopted an approach commonly used in metapopulation studies. Thus, the 'isolation index' for each forest was computed after Hanski and Thomas (1994), as follows:

$$Ii = 1/\max 10\Sigma Aj \ \exp^{-\alpha Dij},$$

where Aj (in km^2) is the area of the neighbouring forest, Dij is the distance (in km) from the study or focal forest i to the neighbouring forest j (including forests not included in our study), and α (alpha) is the effect of distance on isolation. The first ten larger values of the expression $Aj \exp^{-\alpha\, Dij}$ were taken. Three different alpha values were tried (0.1, 0.2, and 0.5), arranged by decreasing weight given to long distances. Birds did show a significant correlation between species richness and isolation for all three values of alpha, but no significant correlation was found for either mammals or reptiles. An alpha value of 0.2 was chosen for birds and reptiles because it also gave the best fit for the normality assumption of residuals. For mammals, an alpha value of 0.5 showed the best fit. Inter-forest distance was estimated from the coordinates (centre) of each forest. Larger values of the isolation index "*Ii*" indicate higher isolation of the forest.

Table 1 Number of vertebrate species and number of forest fragments (with at least one confirmed species presence) of the East African coastal forest included in this study, according to taxa and specialization level

Taxonomic groups	Specialization groups	No. of forests	No. of species	Species richness (Mean ± SD)
Birds	Specialists	43	35	11.49 ± 6.28
	Generalists	43	48	23.44 ± 7.46
	Total	43	83	34.93 ± 13.11
Mammals	Specialists	30	23	4.07 ± 1.98
	Generalists	30	43	10.17 ± 6.07
	Total	30	66	14.23 ± 7.41
Reptiles	Specialists	22 (3)	37	3.52 ± 2.68
	Coastal endemic	18 (7)	23	1.88 ± 1.88
	Forest endemic	19 (6)	14	1.58 ± 1.60

Number of species does not include other species that are present in coastal forests but absent in forests included in this study (see Tables 5–7 for a complete list). Species richness (mean ± standard deviation) is shown. Numbers in brackets are forests with no recorded species

The correlation between each of the variables – area, altitude, and isolation – and the species richness of the three vertebrate group across forests was studied first by using simple correlation and regression analyses, and finally by multiple regression analysis (MINITAB v. 13.31). We initially examined three simple regression/correlation functions: untransformed (linear), semi-log and log–log. Log-transformation of area and isolation better approximated the normality distribution of values and had also a slightly better fit than the other functions. In a few cases, however, the linear model showed a better fit (see Results). Median altitude and altitudinal range were fitted to a linear function, but partial correlations were used because the two variables were significantly correlated to each other ($r = 0.73$–0.95, $P < 0.001$ for all three groups).

Biogeographical affinities

Cluster analysis

Cluster analysis is a method for grouping areas according to their overall biotic similarity (Van Tongeren 1995). Similarity distances between areas are calculated based on a wide array of similarity indices (e.g., Jaccard, Sorensen), and dendrograms are produced from the resulting similarity matrices. We use two similarity indices in our analysis. The 'Sorensen's similarity index' is calculated as $Sor=2C/(A+B)$, where C is the number of species common to both areas, and A and B are the total number of species occurring in each of the areas (Krebs 1999). This index takes the total species number of the two areas (the denominator) into account, so similarity values strongly depend on the difference in species richness between the two areas. The second index, the 'Simpson's coefficient of similarity', is not sensitive to differences in species numbers (Patterson and Brown 1991), but gives more weight to the number of shared species. Simpson's similarity index is computed as: $Simp=C/N_{small}$, where N_{small} is the number of species in the area with the lowest species number. An UPGMA dendrogram was constructed based on the Sorensen's similarity matrices. Additionally, a Mantel test of correlation between matrices was used to find whether compositional similarity (number of shared species) between each pair of forest fragments was correlated to the geographic distance between fragments. The significance of the correlation was tested against random values from 1,000 simulations, using the program PC-ORD (McCune and Mefford 1997). Inter-forest distance ranged from 1.80 km (Mtswakara to Gandini) to 1,300 km (Jubba to Rondo).

Parsimony analysis of distribution

PAD (Trejo-Torres and Ackerman 2001, also called PASA in Trejo-Torres and Ackerman 2002) is an ecological extension of the parsimony analysis of endemicity, PAE (Rosen 1988). PAE is a method to group areas based on their shared taxa according to the most parsimonious solution (criterion of simplicity or 'parsimony') (Rosen 1988; Morrone 1994). Taxa that are restricted to two or more of the studied areas are considered in PAE as analogous to shared derived characters (synapomorphies) in phylogenetic analysis, indicating a unique biotic history among the areas analysed (but see Brooks and Van Veller 2003). PAD is identical to PAE in execution but the interpretation of the results is more ecological. PAD cladograms are considered as hypotheses of biotic similarities between areas, and the terminal dichotomies in PAD are interpreted as areas that share the maximum biogeographic affinity (Trejo-Torres and Ackerman 2001; García-Barros et al. 2002).

A data matrix of forest by species was constructed for each group of taxa (see Table 1 for number of forests and species included in the analysis; Appendices 5–6 for the complete data matrices). A hypothetical, 'primitive' area coded 'all zeros' was incorporated to the matrix as an outgroup to root the trees. This ensures that the biogeographic relationships depicted in the PAD cladogram are mainly based on the appearance (gain) of new species (1) by the area, instead of on taxa losses (0). The same reasoning is used in cluster analysis to favour similarity indices that give more value to presence than absence data (Trejo-Torres and Ackerman 2002).

In PAD, single-area endemics and cosmopolitan species (species that are present in all areas) do not provide information about area relationships and are usually removed from the analysis (Trejo-Torres and Ackerman 2001; García-Barros et al. 2002); this is analogous to removing autapomorphies from a phylogenetic analysis. However, an area that does not share species with other areas but contains many single-area endemics can still be interesting from a conservation viewpoint. Therefore, we decided to include all the original species in our PAD analysis. This also makes the results from PAD comparable with those derived from the cluster analysis, which is based on complete biotas. All species usually receive the same weight in PAD, irrespective of their distributional ranges. However, species belonging to different guilds (e.g., specialists vs. generalists) should be differently treated (e.g. Trejo-Torres and Ackerman 2001). In our analysis, forest specialist species were given a weight of 2 to emphasize their dependence to the forest ecosystem, and therefore their importance as biotic markers. We compared these results with those of an unweighted analysis to analyse the effects of weighting in PAD. In all cases, the trees from the weighted analyses were longer but considerably better resolved than those from the unweighted analysis (not shown). For the cluster analysis, there was no difference between the weighted and unweighted matrices, so only the unweighted results are given.

The data matrix was analysed under the maximum parsimony criterion using the heuristic search in PAUP 4.0b10 (Swofford 2002). When more than one most parsimonious tree (MPT) was found, both strict and Adams consensus trees were computed to reflect the common groupings. The strict consensus tree reflects only those groupings that are present in all MPT trees. The Adams consensus is particularly useful in recovering common structure when incongruence is mainly caused by a few unstable taxa (Page 1998); for example forests that occupy very different positions among the trees.

For comparative purposes, each taxonomic group (birds, mammals, and reptiles) was analysed separately. However, an alternative approach was tried in which we combined all datasets in one single analysis. The objective was to examine patterns of biogeographical affinity defined by the correlation between all three groups. Because the original studies did not include the same set of areas (only 16 forests were common to the three studies), we compared only forests for which we have data for at least two of the groups studied: 29 forests in all. This means that several forests lack data for part of the taxa, but this is not necessarily an impediment for a simultaneous analysis: a direct relationship between missing data and ambiguity of results has yet to be proved (Kearney 2002). However, conclusions from this analysis should be taken with caution.

Results and discussion

Ecological determinants of species richness

Species richness of birds, both specialist and generalist, was higher in larger and less isolated forests (Table 2, see also Mlingwa et al. 2000). Moreover, the number of both specialists and generalists also increased with increasing altitudinal range (Table 2). Similarly, when forests area, isolation and altitudinal range were entered into multiple regression models, all had a significant explanatory effect for the species richness of the specialists (Multiple $r^2 = 0.66$, $F = 25.93$, d.f. = (3, 39), $P < 0.0001$), whereas altitude range was only marginally significant ($P = 0.064$) for species richness of the generalist birds (Multiple $r^2 = 0.55$, $F = 15.96$, d.f. = (3, 39), $P < 0.0001$).

Species richness of generalist mammals was positively correlated to forest area, whereas the number of specialist mammals increased with increasing altitudinal range (Table 2). Similarly, in the multiple regression analyses only area had a significant ($P < 0.01$) effect on species richness of generalist mammals (Multiple $r^2 = 0.393$, $F = 15.96$, d.f. = (4, 25), $P < 0.05$). For forest specialist mammals, only the multiple regression model with median and altitude range was marginally significant (Multiple $r^2 = 0.162$, $F = 2.6$, d.f. = (2, 27), $P < 0.10$), and only altitude range had a significant effect ($P < 0.05$) in the model.

Species richness of forest specialist reptiles was higher in forests with higher median altitude and wider altitude range if univariate correlations were considered (median: $r = 0.55$, $P < 0.01$; range: $r = 0.56$, $P < 0.01$), but no significant correlations were found after partialling out (controlling) the effect of the two altitude components (Table 2). When

Table 2 Relationship between species richness of vertebrates in the East African coastal forests in relation to area, isolation, and altitude

Taxa	Groups	Log (Area) (r^2)	Log (Isolation) (r)	Altitude	
				Range (r_p)	Median (r_p)
Birds	Specialists	**0.33*****	**−0.39****	**0.51*****	0.09
	Generalists	**0.33*****	**−0.34*****	**0.44****	−0.02
Mammals	Specialists	0.09	−0.29[u]	**0.37***	−0.25
	Generalists	**0.30****	−0.26[u]	0.23	−0.13
Reptiles	Specialists	0.08	−0.19[u]	0.214	−0.07
	Coastal endemic	0.124[a, u]	−0.15[u]	**0.52****	−0.39[a]
	Forest endemic	0.02	−0.17	−0.30	**0.43***

r = Pearson correlation coefficient, r^2 = simple linear regression coefficient. For median and range of altitude, partial correlations (r_p) were calculated because the two variables were highly correlated to each other. Significant correlations are shown in bold. 'U': linear form of the predictor was used for the analyses

***$P < 0.001$; **$P < 0.01$; *$P < 0.05$; $^{a}P < 0.10$

the two groups of specialist reptiles ('forest' and 'coastal' endemics) were considered separately, they showed significant correlations to different components of altitude. Species richness of coastal endemic reptiles tended to increase with increasing altitudinal range, whereas that of forest endemics showed a positive correlation to the median altitude (Table 2). Moreover, coastal endemic reptiles showed a marginally significant correlation to the median altitude (of negative sign) and to area size. A similar relationship was obtained in the multivariate regression analysis. Only altitude range was significantly ($P < 0.05$) correlated to species richness of coastal endemics (Multiple $r^2 = 0.433$, $F = 3.82$, d.f. = (4, 20), $P < 0.05$), whereas median altitude was the only significantly ($P < 0.05$) correlated variable to species richness of 'forest endemics' (Multiple $r^2 = 0.374$, $F = 2.98$, d.f. = (4, 20), $P < 0.05$). None of the mammal or reptile groups displayed a significant relationship to measure of isolation between forest fragments (Table 2).

The species–area relationship found here for birds and generalist mammals is consistent with the predictions of the Island Biogeography Theory (MacArthur and Wilson 1967; Rosenzweig 1995), and it is probably an effect of area-dependent extinction. It also corroborates earlier findings for various East African forest ecosystems (Diamond 1981; Stuart 1981; Newmark 2002). The general lack or the existence of a weak species–area relationship in reptiles could be due to several reasons, such as the long-term isolation and cycles of environmental fluctuation of the system, anthropogenic disturbances that could have led to a non-equilibrium condition (e.g., Whittaker 1998), or could simply being the result of a sampling artefact or undersampling of this group.

The fact that only birds showed a significant negative relationship between species richness and isolation is probably related to their higher dispersal ability. Mobile organisms like birds are expected to be able to disperse across gaps of unsuitable habitat and are therefore more likely to benefit from the 'rescue effect' (Brown and Kodric-Brown 1977), and from decolonization processes. On the other hand, isolation may not be an explanatory factor of species richness variability in groups of organisms with weak dispersal ability (Lomolino et al. 1989). In our study, this applies to forest specialist mammals and to endemic reptiles, in particular (see also Table 3), which are known for their poor ability to disperse (e.g., Foufoupolos and Ives 1999; Morand 2000). Thus, the extinction–colonization equilibrium (MacArthur and Wilson 1967) may not apply for these groups.

Altitude is an important determinant of species richness in terrestrial communities. A coastal forest spanning over a wide altitudinal range probably has a more diverse environment than one with a narrower range (Kingdon and Howell 1993; Jetz and Rahbek 2002). For example, altitudinal range was found to be one of the core predictors of species richness for range-restricted sub-Saharan birds (Jetz and Rahbek 2002). As expected in habitat specialists (e.g. Ricklefs and Lovette 1999), habitat diversity (i.e. altitude range) influenced the species richness of forest specialist birds and mammals, and also that of coastal endemic reptiles. The pattern shown by the relict coastal endemic reptiles may also reflect historical refugee opportunities, i.e. a more stable environment provided by the altitudinal range (topographic heterogeneity). Concentrations of range-restricted relict endemics on some localities may reflect low rates of extinction due to high environmental stability (e.g. Fjeldså and Lovett 1997; Fjeldså et al. 1997; de Klerk et al. 2002a). However, a correspondence between centres of endemism and altitudinal range measured at a larger scale (one degree latitude × one degree longitude) for sub-Saharan bird fauna was interpreted as topographic heterogeneity reflecting "historical opportunities for allopatric speciation" (Jetz et al. 2004). Such historical interpretation of the relationship between altitude range and number of endemics, in this case as reflecting opportunity for speciation

Table 3 Similarity in species composition between East African coastal forests in relation to geographical distance between forests

		Sorensen	Simpson
Birds	Specialists	**−0.25****	**−0.225****
	Generalists	**−0.18***	**−0.29*****
Mammals	Specialists	0.059	0.008
	Generalists	0.061	−0.015
Reptiles	Specialists	−0.044	−0.043
	Coastal endemics	−0.12	−0.14
	Forest endemics	0.006	−0.023

A Mantel test based on Sorensen's and Simpson's similarity indexes was used to test the significance of the correlation. Significant correlations are shown in bold

***$P < 0.001$; **$P < 0.01$; *$P < 0.05$

or as a refugee, may be further resolved by including information on the phylogenetic relationships among the endemic species.

Interestingly, species richness of forest endemic reptiles was higher in forests with a high median altitude. This is probably because forest endemics include mountain species that are mainly restricted to higher altitude forests (e.g. Tongwe, Tong'omba, and Rondo). Coastal forests with higher median altitude are expected to be better developed, i.e., moist forests with tall trees because they receive more rainfall (e.g. Hawthorne 1993; Lowe and Clarke 2000), and are probably less subjected to human disturbance (Clarke 2000b) than low lying forests. This may also have a historical explanation in that high altitude forests are ancient forests that may have survived during cycles of arid climate and coastal inundations, and possibly acted as refugia harbouring species that went extinct in low-lying forests during those periods (see also Burgess et al. 1998).

Biogeographical affinities among the East African coastal forest faunas

A parsimony analysis of bird distributions based on 43 coastal forest fragments resulted in two most parsimonious trees (MPTs, L = 588, CI = 0.20, RI = 0.57), whose strict consensus is shown in Fig. 2A. The Adams consensus was identical to the strict consensus. Forest characterizations given below are based on descriptions from Mlingwa et al. (2000).

The most distal clade of the cladogram (A) includes large or well-studied forests in northern Kenya (e.g., Arabuko-Sokoke). These forests are the sister-group to a second clade (B) formed by the sacred 'Kaya' forests in southern Kenya (e.g., Diani), which are protected under tribal laws. The sister-group to these Kenyan forests (A + B) is a clade (c), comprising the Pugu Hills and the East Usambara lowland forests in northern Tanzania, and its sister-clade (D), composed of forests in southern Tanzania corresponding to the 'Lindi' local centre of endemism and nearby forests (e.g., Litipo, Ngarama) (Burgess 2000), as well as four other forests in northern Tanzania (e.g., Gendagenda). The forest in the first three clades (A–C) belong to the so-called 'Usambara-Kwale' local centre of endemism and to other 'minor' centres of endemism (e.g., Arabuko-Sokoke) recognized in previous studies (Burgess et al. 1998; Burgess 2000, Appendix Table 4). Diverging basally to the whole distal clade (Tana to Msubugwe) are forests that correspond geographically to central Tanzania (e.g., Pande, Vikindu; Fig. 1).

The distal part of the cladogram explained above indicates that compositional similarity of bird assemblages between forest fragments is spatially autocorrelated, i.e. geographically close forests have similar bird faunas. The cluster tree in Fig. 2B also shows a

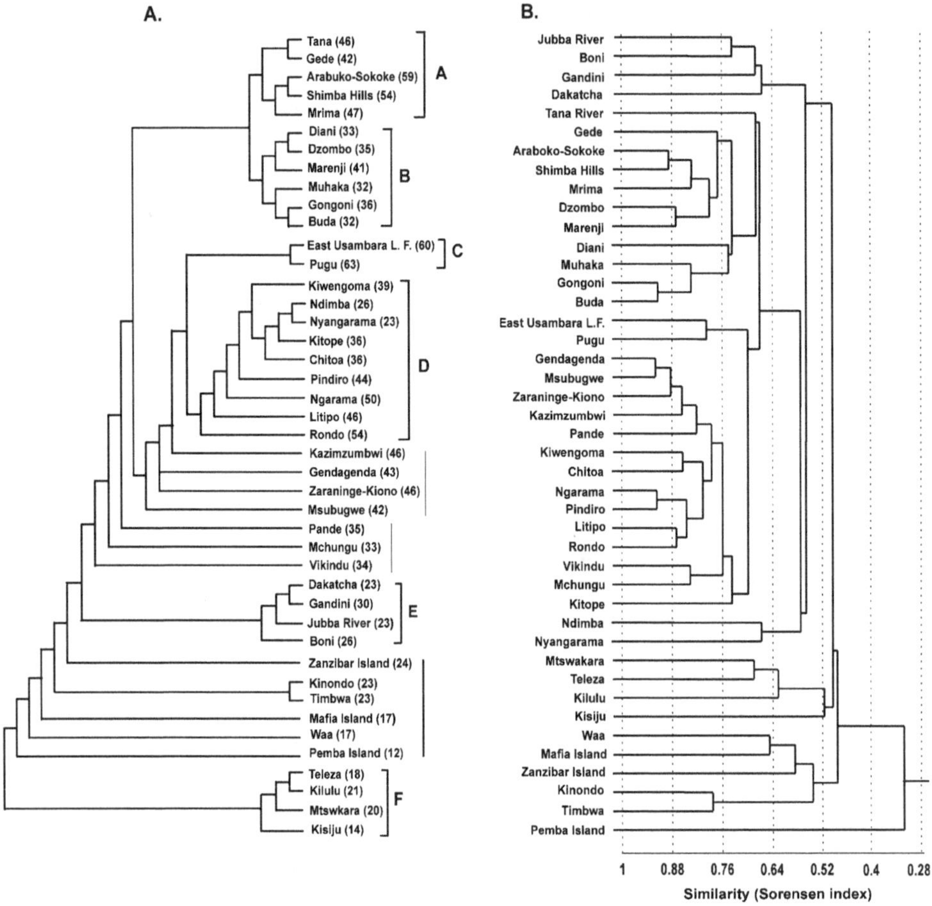

Fig. 2 Biogeographic affinities among 43 East African coastal forests based on the distributions of specialist and generalist birds. (**A**) Strict consensus of two MPTs (L = 588, CI = 0.20, RI = 0.57) obtained from a parsimony analysis of species distributions; number of species in each forest fragment is given between brackets. (**B**) UPGMA tree from a cluster analysis based on Sorensen's similarity index

similar pattern, with many of the higher-level clusters grouping forests that are geographically close. Moreover, these clusters are joined by high Sorensen's similarity values (SS = 0.7–0.8), indicating a high degree of similarity among their forest bird faunas (Fig. 2B). This is further corroborated by the Mantel test (Table 3), which shows a general decrease of compositional similarity with increasing geographic distance (see also the negative correlation between isolation and species richness in Table 2). These results suggest that geographical distance is one of the main factors determining biogeographical affinities among bird communities (see also Stuart 1981; Morand 2000; Azeria 2004).

However, this geographic pattern can also be explained by other ecological factors such as forest habitat type, since forests of similar type are usually found closer to each other. For example, bird faunas of the well developed but fairly small forests in clade B showed closer affinity to other, large forests within the main forest block (A) than to similar sized and geographically close scrubby forests (e.g., Waa, Timbwa). Similarly, other studies in coastal forests of south Kenya have found forest structure to be more important than area as determinant of the composition of understory bird fauna (Mlingwa et al. 2000). Thus, type and structure of the forests influence the species richness and composition of bird communities. However, finer details of forest structure type, such as tree height or cover of forest canopy, may not always be spatially autocorrelated, depending on local aspects such as topography and associated soil structure, or anthropogenic effects (e.g., Lowe and Clarke 2000).

The remaining groups at the base of the tree do not follow so clearly a geographical pattern but are better explained by similarities in vegetation and other ecological factors. For example, clade (E) is composed of a tight group of forests in the northern limit of the coastal forest belt, in southern Somalia and northern Kenya, consisting of relatively dry and low forests (e.g., Jubba) or small and dense woodland forests (e.g., Gandini). Diverging basal to this clade are several depauperated forests, either because of their island condition (e.g., Pemba) or because they correspond to very small, scrubby forest fragments (< 0.5 sq. km) such as Waa. The last clade (F) is formed by a group of geographically scattered but ecologically similar forests (e.g., Teleza), which are described as scrubby forests associated with low rainfall and/or poor soil conditions (Mlingwa et al. 2000). These forests have a low number of species and seem to be subordinates (in terms of species diversity) to the other Tanzanian and Kenyan forests (i.e., poor subsets of these forests). All the affinities among the depauperated bird faunas signal a possible future dire scenario if the diversity and heterogeneity of the coastal forests is lost by further fragmentation and degradation (Lowe and Clarke 2000). Thus, the bird faunas of the degraded or scrubby forests within the main forest block (e.g. Waa) resembled those of the dry forests and woodlands near the northern border of the forest chain, which are marginal in terms of biodiversity and affinity to the coastal forest.

The PAD analysis of mammals resulted in two MPTs (L = 259, CI = 0.33, RI = 0.52). The strict and Adams consensus of these trees had the same topology shown in Fig. 3A. The hierarchical grouping of forest fragments does not seem to follow a strict geographical gradient. The first clade (A) is formed by two groups of forests located at opposite ends of the coastal forest belt, i.e. forests found in the southern Somalia–northern Kenya region (Jubba to Arabuko-Sokoke) and forests found in southern Tanzania (Litipo to Chitoa). The forests in this clade belong to the 'Lindi' centre of endemism and other 'minor' centres of endemism (e.g., Tana-River). The sister-group to this clade (B) is a group of forests in northern Tanzania, which roughly corresponds to the 'Usambara-Kwale' and the 'Pugu Hills' centres of endemism, but which, as in birds, also includes other forests in-between (e.g., Zaraninge-Kiono). Therefore, the geographical pattern is opposite to that found in birds, although the composition of the individual clades is similar.

Basal to these clades, the cladogram shows several geographically dispersed forests (Tong'omba to Pemba). Most of the forests are degraded (e.g., Vikindu), and isolated by their island (Mafia, Pemba) or geographical location (e.g., Kimboza). Unlike in birds, the grouping of these forests is not only based on their poor species diversity but also on the presence of idiosyncratic species, which have restricted or disjunct distributions within the coastal forest range. For example, Kiwengoma, a relatively species-rich forest, is basal to species-poor forests, such as Kisiju (see Fig. 3A), and six of its species have idiosyncratic distribution, e.g., the forest generalist *Chlorocebus aethiops*. Like in PAD, the groupings in the cluster analysis showed also a lack of geographical correspondence (Fig. 3B). Moreover, the Mantel test (Table 3) indicated that composition similarity was not spatially autocorrelated in mammals, i.e., geographically close forests do not necessarily share similar faunas. Moreover, all high-level clusters in mammals are joined by relatively low similarity values (SS = 0.4–0.6, Fig. 3B), suggesting low similarity among their mammal

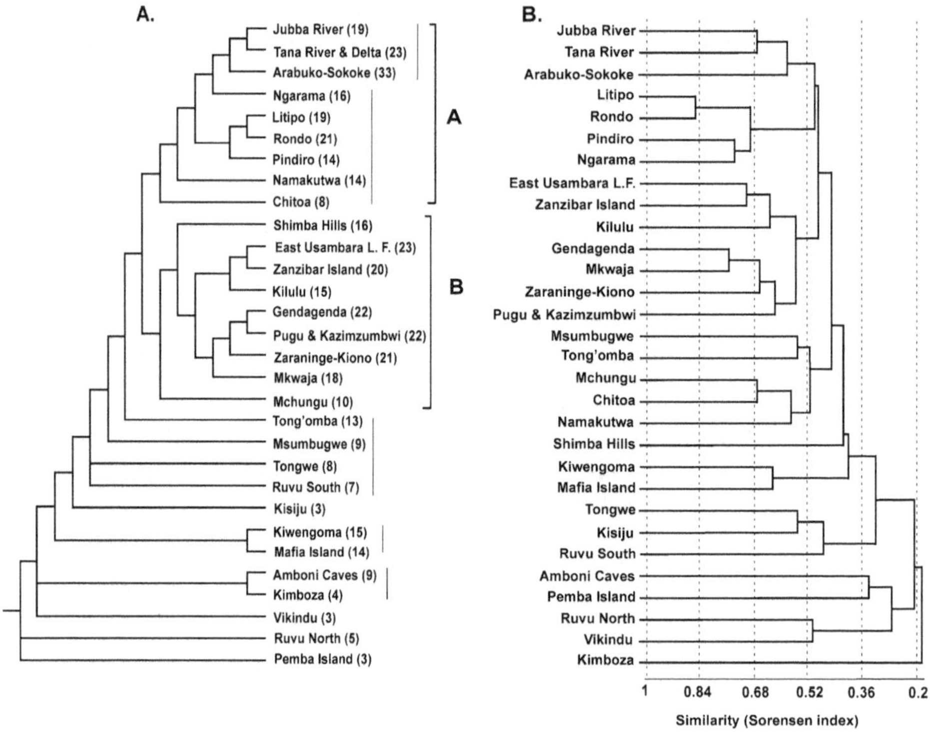

Fig. 3 Biogeographic affinities among 30 East African coastal forests based on the distributions of specialist and generalist mammals. (**A**) Strict consensus of two MPTs (L = 259, CI = 0.33, RI = 0.52) from a parsimony analysis of species distributions; number of species in each forest fragment is given between brackets. (**B**) UPGMA tree from a cluster analysis based on Sorensen's similarity index

faunas. Thus, geographical distance is a less important factor in mammals than in birds, probably because inter-fragment dispersal in mammals is rare over distances of this magnitude (see also the lack of correlation between species richness and isolation in Table 2).

Both the PAD and cluster analyses of reptile species indicated the uniqueness of the reptile fauna in various forest fragments. The PAD analysis resulted in 93 MPTs (L = 62, CI = 0.59, RI = 0.50), whose strict consensus was mostly unresolved (Fig. 4A). The Adams consensus showed only a few two-area clusters, of which only three were geographically defined (Kiwengoma/Tong'omba, Tana/Arabuko-Sokoke and East Usambara/Gendagenda) (Fig. 4A).

This lack of resolution is probably due to the low reptile diversity in each of the forests (Mean ± SD, 3.52 ± 2.68, Table 1; see also Fig. 4A), but also by the fact that at least 50% of reptile species were single-site endemics, a condition common for endemic species in the East African coastal forest (Burgess et al. 1998). The high prevalence of endemism also indicates that inter-forest movement is low in reptiles. This pattern is also reflected in the cluster analysis (Fig. 4B), where the similarity values among clusters (SS = 0.2) were on average lower than in mammals, and much lower than in birds. Thus, there is very little in common between the endemic specialists reptile fauna of the different forest fragments. Moreover, the Mantel test showed that there is no correlation between reptilian compositional similarity and geographic distance among forest fragments (Table 3). Thus, the

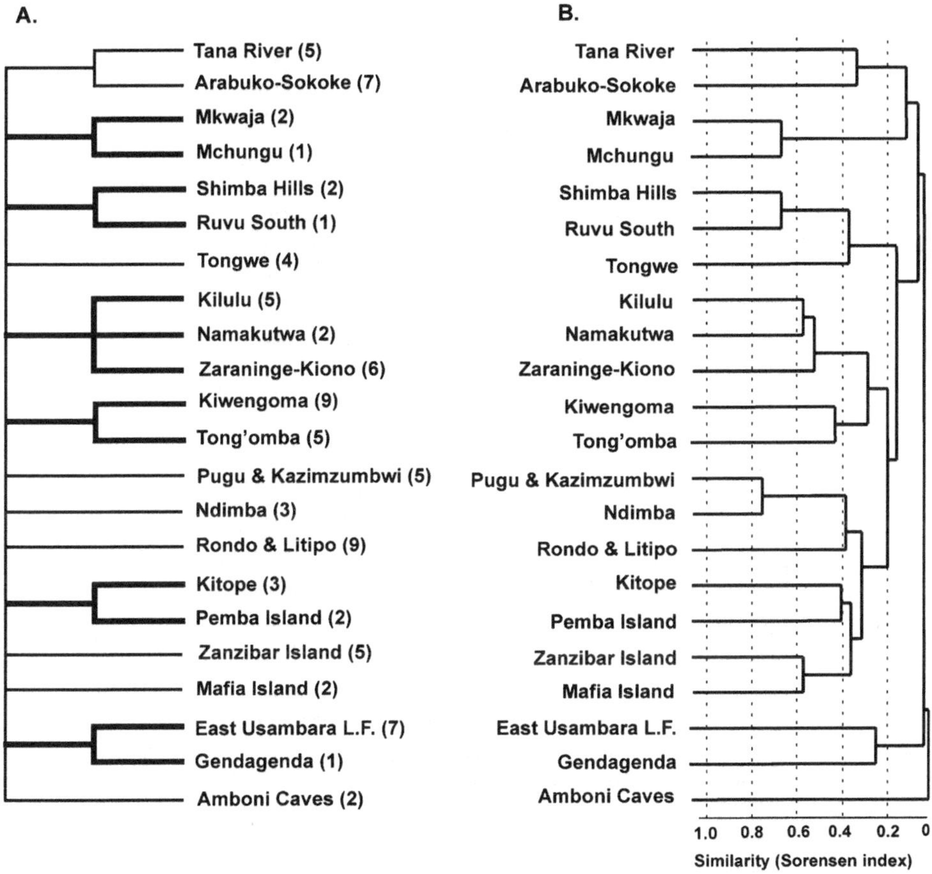

Fig. 4 Biogeographic affinities among 22 East African coastal forests based on the distributions of coastal and forest endemic specialist reptiles. (**A**) Strict and Adams consensus of 93 MPTs (L = 62, CI = 0.59, RI = 0.50) from a parsimony analysis of species distributions; groups further resolved in the Adams consensus are given in bold lines. Number of species in each forest fragment is given between brackets. (**B**) UPGMA tree from a cluster analysis based on Sorensen's similarity index

few endemic species of reptiles with a wide distribution often exhibit a disjunct distribution (Burgess et al. 1998). Such disjunct distributions could be better explained by historical fragmentation dynamics and extinctions of once wider distributions than by contemporary ecological processes such as dispersal.

The PAD analysis for the combined data set (birds + mammals + reptiles) resulted in 35 MPTs (L = 710, CI = 0.33, RI = 0.47). The strict consensus was mostly unresolved. The Adams consensus (Fig. 5) shows similar patterns to those found in mammals and birds. The most distal clades (A–C) are formed by forests in Kenya and northern Tanzania, which roughly correspond to the 'Usambara-Kwale' centre of endemism. Sister-group to these clades, clade (D) includes forests in the south Tanzanian 'Lindi' region, as well as Kiwengoma-Tong'omba, which form the 'Matumbi and Kichi Hills' minor centre of endemism (Burgess 2000). Clade (E) also corresponds to the 'Lindi' region. Finally, the basal clade (F) includes forests that are geographically scattered.

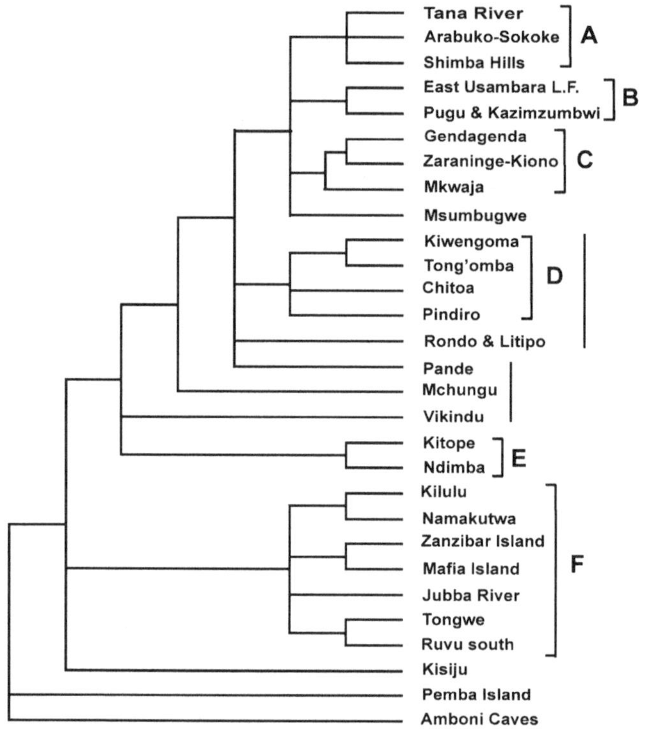

Fig. 5 Biogeographic affinities among 29 East African coastal forests based on the combined distributions of birds, mammals, and reptiles. The tree is the Adams consensus of 35 MPTs (L = 710, CI = 0.33, RI = 0.47) obtained from a parsimony analysis of species distributions

Conclusions

The East African coastal forests have been described as a diverse and heterogeneous forest ecosystem. Both current and historic factors have had a profound effect on the distribution of species (Kingdon and Howell 1993; Hawthorne 1993; Burgess et al. 1998; Lowe and Clarke 2000; Rodgers 2000; de Klerk et al. 2002a, b; Jetz et al. 2004). Like other patchy and fragmented forest systems, current forest attributes, such as size, isolation, median altitude, and elevation range (i.e., habitat diversity), had influenced the species distribution of vertebrates in the coastal forests. However, the relative importance of these ecological factors seem to vary depending on life history characteristics of the taxa studied, such as dispersal ability (birds vs. reptiles) and, within taxa, on the degree of dependence on forest resources (specialists vs. generalists). As expected for vagile species, birds seem to be able to cope with changing ecological conditions, e.g., area, forest type, and diversity, and thus their distribution appear to be driven by extinction and colonization dynamics. Moreover, both bird and mammal forest specialists showed a stronger correlation to variables related to habitat diversity such as altitude range, suggesting that habitat diversity affects species richness of groups with habitat specializations (Ricklefs and Lovette 1999). However, this conclusion requires further investigation using a direct quantification of the number (i.e., quantity) of each habitat type across the forests, as well as empirical data on the specific habitat used by each of the species. Forest generalist mammals and birds showed only a more general species–area relationship. The less vagile endemic reptiles seem to be less correlated to forest area and isolation measure. However, the species richness of the relict coastal endemic reptiles was correlated to altitude range, whereas forest endemic reptiles appear to concentrate in forests with relatively higher altitude. We may conclude that topographical heterogeneity and the relatively moist forests on high altitude may have provided a refugee opportunity (a suitable and stable environment) for coastal and forest endemic reptiles, respectively, both historically and contemporarily during climatic cycles. Thus, the patterns of species richness suggest that past extinctions and dispersal limitations are important factors for the distribution of forest specialist reptiles in the coastal forests. Their (most likely) relict distribution indicates that reptiles could be good candidates for tracing past environmental history, in which case, phylogeny-based (cladistic) biogeography should be the method of choice.

Our study also confirms some of the local centres of endemism identified in previous studies (Burgess 2000; Mlingwa et al. 2000). Regions such as the 'Usambara-Kwale' and the 'Lindi' centre of endemism appear as clusters of areas in mammals and birds, confirming the 'naturalness' of these regions. Moreover, our study shows that other forests should also be included within these two centres of endemism (e.g., Ngarama and Pindiro in the 'Lindi' centre of endemism).

The results from this study also indicate that PAD can be used as an alternative (complementary) method to cluster analysis for finding biogeographic affinities among areas. Unlike cluster analysis, the grouping of areas in PAD is not based on overall similarity but on maximizing congruence between geographic distributions, i.e., the correlation between taxa and areas. The MPT is the simplest explanation to the species distributions given the data: the tree that minimizes the number of changes in the biotic composition of an area required to explain its present biota. Even though species are not subjected to the same constraints as organism traits (i.e., characters are inherited from ancestor to descendant, whereas species can move freely to new areas by dispersal), an explanation that minimizes the number of biotic changes (e.g., colonization/extinction events) is still realistic if communities tend to be maintained. Our results suggest that in fact dispersal can be a strong element in keeping species as part of the community, for example, through the 'rescue effect' (Brown and Kodric-Brown 1977; McDowall 2004). The patterns found by PAD in mammals and reptiles could indicate the existence of a historical component (e.g., habitat continuity over time, fragmentation) that is less influenced by current ecological settings. PAD patterns can reflect both ecological and historical components but how far a given pattern is ecological or historical is unclear, since no phylogenetic information is included in the analysis. On the other hand, cluster methods are more sensitive to methodological artefact, i.e., sample size and index used (Rosen 1988). Moreover, the result from PAD does, in many ways, resemble an analysis of nestedness (Patterson and Atmar 1986; Azeria 2004) with the improvement that it can detect more than one nested subset.

Finally, the variation in biogeographical patterns across taxonomic and habitat specialization elucidates the need for a variety of conservation strategies to protect the vertebrate fauna of the East African coastal forests. Given the high degree of geographical correspondence (and inter-fragment dispersal) found in birds, to design large reserves with good geographical connectivity could be a good strategy for this group. In contrast, the lower affinity and weak geographical correspondence found in forest specialist mammals, and in the rare and widely dispersed (disjunct) endemic reptiles, in particular, indicate that they are less likely to maintain connectivity among forest patches. This fact makes these groups highly vulnerable to disturbances that severe the available resources (e.g. habitat diversity, structure, or stable environment) provided within each of the functionally isolated forest fragments. Thus, these taxa, and the specialist birds, and are likely to persist only when the remnant forests remain intact, in which case a direct 'site-based' protection would be probably the most effective conservation strategy.

Acknowledgements Ermias T. Azeria would like to thank Sida/Sarec for financial support under the auspices of Asmara program, ISP, Uppsala University. Isabel Sanmartín was supported by a European Community Marie Curie Fellowship (MCFI-2000-00794) and by the Swedish Natural Research Council (621-2003-3855). We thank Philip Clarke for his valuable comments on the manuscript, and the authors of the original survey studies for providing us with access to their data. We are especially grateful to Yonas I. Tekle for his assistance in the analysis, and to Christer G. Wiklund and an anonymous referee for their constructive criticisms on the manuscript.

Appendix

Table 4 Comparison between the centres of endemism defined by Burgess et al. (1998) and Burgess (2000) and the groupings of forests (clades A–G, see Figs. 2, 3) obtained from a parsimony analysis of birds and mammals distributions (PAD)

Code	Forest	Endemism centre	Birds	Mammals
JU	Jubba		E	A
BO	Boni		E	–
TR	Tana	TMC	A	A
DK	Dakatcha		E	–
GD	Gede		A	–
AS	Arabuko-Sokoke	AMC	A	A
MT	Mtswakara	UK	F	–
GA	Gandini	UK	E	–
TL	Teleza	UK	F	–
WA	Waa	UK	S (A–E)	–
SH	Shimba Hills	UK	A	B
MU	Muhaka	UK	B	–
DJ	Diani	UK	B	–
KN	Kinondo	UK	S (A–E)	–
DZ	Dzombo	UK	B	–
GN	Gongoni	UK	B	–
BU	Buda	UK	B	–
TM	Timbwa	UK	S (A–E)	–
MR	Mrima	UK	A	–
MA	Marenji	UK	B	–
EU	East Usambaras lowland	UK	C	B
KL	Kilulu	UK	F	B
AC	Amboni Caves	UK	–	C
TW	Tongwe	UK	–	S (A–B)
GG	Gendagenda	UK	S (C-D)	B
MB	Msubugwe	UK	S (C-D)	NS
MJ	Mkwaja		–	B
ZK	Zaraninge-Kiono		S (C-D)	B
RN	Ruvu North		–	C
PA	Pande		S (A-D)	–
PU	Pugu	PM	C	B
KZ	Kazimzumbwi	PM	S (C-D)	B
VI	Vikindu		S (A-D)	C
RS	Ruvu South		–	S (A–B)

Table 4 continued

Code	Forest	Endemism centre	Birds	Mammals
KM	Kimboza		–	–
KI	Kisiju		–?	S (A–B)
MC	Mchungu		S (A-D)	B
NM	Namakutwa		–	A
KW	Kiwengoma	MKH	D	S (A–B)
TO	Tong'omba	MKH	–	S (A–B)
KP	Kitope		D	
NG	Ngarama		D	A
PD	Pindiro		D	A
ND	Ndimba		D	
CH	Chitoa	Lindi	D	A
LI	Litipo	Lindi	D	A
RO	Rondo	Lindi	D	A
NY	Nyangamara		–	–
PI	Pemba island	TZO	S (A–E)	S (A–B)
ZI	Zanzibar Island	TZO	S (A–E)	B
MI	Mafia Island	TZO	S(A–E)	S (A–B)

TMC = Tana Minor center; AMC = Arabuko Minor centre UK = Usambara-Kwale; PM = Pugu minor; MKH = Matumbi and Kichi hills; LN = Lindi; TZO = Tanzania offshore (–) = Data on species distributions not available. S = Forest biota which do not form a clade by themselves, but form basally diverge to clades shown bracket

Table 5 Data matrix of the 83 bird species and 43 forest fragments of the East African coastal forests included in the study. § = forest specialist, the remaining are forest generalist. Forest codes as in Figure 1.

Species code	JR	BO	TR	DK	GD	AS	MT	GA	TL	WA	SH	DJ	KN	TM	MU	GN	BU	MR	DZ	MA	EU	KL	GG	MB	ZK	PA	PU	KZ	VI	KI	MC	KG	KP	CH	NG	PD	ND	LI	RO	NY	PI	ZI	MI
C. f.	1	0	1	1	1	1	1	1	0	0	1	1	0	0	1	1	1	0	0	0	1	1	1	1	1	1	1	1	0	0	0	1	0	0	0	1	0	1	1	0	0	0	0
A. m.	0	0	0	0	1	1	0	0	0	1	1	1	1	1	1	0	0	0	0	1	1	0	0	0	1	0	1	0	0	0	0	0	1	0	0	0	1	0	0	0	1	1	0
A. t.	1	1	1	1	1	1	1	1	1	0	1	1	1	1	1	1	1	1	1	1	1	1	1	1	1	1	1	1	1	1	1	1	1	1	1	1	0	1	1	1	1	1	1
H. a.	0	0	1	0	0	1	0	0	0	0	1	0	0	0	0	0	0	1	0	0	1	0	0	0	0	0	0	0	0	0	0	0	0	0	0	0	0	1	1	1	0	0	0
S. co.§	0	0	1	0	0	1	1	0	1	0	1	1	0	1	1	1	1	1	1	1	1	1	1	0	1	1	1	1	1	0	1	1	1	1	1	1	0	0	1	1	0	0	0
M. a.	0	0	1	0	0	0	0	0	0	0	0	0	0	0	0	0	0	0	0	0	0	0	0	0	0	0	1	0	0	0	0	0	0	0	0	0	0	0	0	0	0	0	0
G. p.	1	1	1	0	0	1	1	1	1	0	1	0	1	1	0	1	1	1	1	1	0	0	1	1	1	1	1	1	1	1	1	1	0	0	1	1	0	1	1	1	0	1	0
S. e.§	0	0	0	0	0	0	0	0	0	0	0	0	0	0	0	0	0	0	0	0	1	0	0	0	0	0	1	?	0	0	0	0	0	0	0	0	0	0	0	0	1	1	0
A. l.§	0	0	0	0	0	0	0	0	0	0	1	0	0	0	0	0	0	0	0	0	1	0	0	0	0	0	1	0	0	0	0	0	0	0	0	0	0	0	1	0	0	0	0
C. d.§	0	0	0	0	0	1	0	0	0	0	1	0	0	0	0	1	0	1	0	0	1	0	0	0	0	0	0	0	0	0	0	0	0	0	0	0	0	0	0	0	1	1	0
T. t.	1	1	1	1	1	1	0	1	1	1	1	1	1	1	1	1	1	1	1	1	1	0	1	1	1	1	1	1	1	0	1	1	1	1	1	1	0	1	1	0	1	1	1
T. c.	0	1	1	1	1	1	0	0	0	1	1	0	0	0	1	1	0	1	0	1	1	0	1	1	1	0	1	1	1	0	1	0	0	0	1	1	0	0	1	0	0	1	1
P. c.	0	0	0	1	0	1	1	1	0	0	1	0	0	0	0	0	0	0	1	1	0	0	0	1	0	0	1	0	1	0	1	1	1	1	1	1	0	1	0	1	1	0	0
T. f.	1	1	1	1	1	1	1	1	1	1	1	1	1	1	1	1	1	1	1	1	1	1	1	1	0	0	0	0	0	0	0	0	0	0	0	0	0	0	0	0	0	0	0
T. l.	0	0	0	0	0	0	0	0	0	0	0	0	0	0	0	0	0	0	0	0	0	0	0	1	1	1	1	1	0	1	1	1	1	1	1	1	1	1	1	0	0	0	0
C. m§	0	0	0	0	0	0	0	0	0	0	0	0	0	0	0	0	0	1	0	0	1	0	1	0	0	0	1	0	0	0	0	1	0	1	1	1	0	1	1	0	0	0	0
C. c.	0	0	1	0	0	0	0	0	0	0	0	0	0	0	0	0	0	0	0	0	0	0	0	0	0	0	0	0	0	0	0	0	0	0	1	0	0	0	0	0	0	0	0
C. a.	0	1	1	1	1	1	1	0	1	1	1	1	1	1	1	1	1	1	1	1	1	0	1	1	1	1	1	1	1	1	1	0	1	0	1	1	0	1	1	0	1	1	1
B. v.§	0	0	0	0	0	0	0	0	0	0	0	0	0	0	0	0	0	0	0	0	1	0	0	0	0	0	0	0	0	0	0	0	0	0	0	0	0	0	0	0	0	0	0
S. w.	1	0	1	1	1	1	0	1	1	0	1	1	1	1	1	1	1	1	1	1	1	1	1	1	1	1	1	1	1	0	1	1	1	1	1	0	1	1	1	1	1	1	0
G. c.	1	1	1	1	1	1	1	1	0	0	1	1	1	0	0	1	0	0	1	0	0	0	0	0	0	0	1	0	0	0	0	0	0	0	0	0	0	0	1	0	0	0	1
O. i.§	0	0	0	0	0	1	0	0	0	0	0	0	0	0	0	0	0	0	0	0	1	0	0	0	0	0	0	0	0	0	0	0	0	0	0	0	0	0	0	0	0	0	0
S. p.	0	0	1	0	0	0	0	0	0	0	0	0	0	0	0	0	0	0	0	0	0	0	0	0	0	0	0	0	0	0	0	0	0	0	0	0	0	0	0	0	0	0	0
C. p.	0	1	1	1	1	1	0	1	0	0	1	0	1	0	0	0	0	0	0	0	0	0	1	1	1	1	1	1	1	0	1	0	0	0	1	1	0	0	1	1	0	0	0
N. b.	0	0	0	0	1	1	1	0	1	0	1	1	0	0	0	0	0	1	1	1	1	1	1	1	1	0	1	1	1	0	1	1	1	0	1	1	0	1	1	0	0	0	1
T. u.	0	0	1	0	1	1	0	1	0	0	1	0	0	0	0	0	0	0	0	0	0	0	0	0	0	0	0	0	0	0	0	0	1	0	1	0	0	0	0	0	0	0	0
A. n.	1	1	1	1	1	1	1	1	1	0	1	1	0	1	1	1	0	1	1	1	1	1	1	1	1	1	1	1	1	0	1	1	1	1	1	1	1	1	1	1	0	1	1
B. br.	0	0	1	0	1	1	1	0	0	0	1	1	1	1	1	1	1	1	1	1	1	0	0	0	0	0	0	0	0	0	0	0	0	0	0	0	0	0	0	0	0	0	0
B. b.	0	0	1	0	1	1	1	1	1	0	1	0	0	0	0	1	1	1	1	1	1	1	1	1	1	1	1	1	0	1	1	1	1	1	1	1	1	1	1	1	0	0	0
S. l.	0	0	0	0	0	0	0	0	0	0	1	1	0	0	0	0	0	0	1	1	1	0	1	0	0	0	1	1	0	0	0	1	1	0	1	1	0	0	0	0	0	0	0
S. o.§	0	0	1	0	1	1	1	1	0	1	1	1	0	0	0	0	1	1	1	1	1	0	1	1	1	1	1	1	1	1	1	1	0	1	1	1	0	1	1	0	1	1	1
P. bi.	0	0	1	0	1	1	0	0	0	1	1	0	1	1	1	1	1	1	1	1	0	1	1	1	1	1	1	1	1	1	1	1	1	1	1	1	1	1	1	1	0	1	1

Table 5 continued

Species code	JR	BO	TR	DK	GD	AS	MT	GA	TL	WA	SH	DJ	KN	TM	MU	GN	BU	MR	DZ	MA	EU	KL	GG	MB	ZK	PA	PU	KZ	VI	KI	MC	KG	KP	CH	NG	PD	ND	LI	RO	NY	PI	ZI	MI
P. s.§	0	0	0	0	1	1	0	1	0	1	1	0	0	1	1	1	1	1	0	1	0	0	1	1	1	1	1	1	0	0	1	1	0	1	1	1	1	1	1	0	0	1	0
C. ab.	1	0	1	1	1	1	0	0	0	0	1	0	0	0	0	0	0	1	0	0	1	0	0	0	1	0	1	0	0	0	0	1	1	1	1	1	1	1	1	1	0	0	0
C. m.	0	1	0	0	0	1	1	1	1	0	1	1	1	0	0	1	1	1	1	0	0	0	0	0	0	0	0	0	0	0	0	0	0	0	0	0	0	0	0	0	0	0	0
S. ca.§	0	0	0	0	0	1	0	0	0	0	1	1	0	0	0	1	1	1	0	1	1	0	1	1	1	1	1	1	1	0	1	1	1	1	1	1	1	1	1	1	0	0	0
P. a.§	0	0	0	0	1	1	0	0	0	0	0	0	1	0	0	0	0	0	0	0	1	0	0	0	0	0	1	0	1	0	0	1	0	1	1	0	0	1	1	0	0	0	0
A. s.§	0	0	0	1	1	1	0	1	0	0	1	0	0	0	0	0	0	0	1	1	0	1	0	0	1	0	1	0	1	0	0	0	0	0	0	0	0	0	0	0	0	0	0
A. mi.§	0	0	0	0	0	0	0	0	0	0	0	0	0	0	0	0	0	0	0	0	1	0	0	0	0	0	1	0	0	0	0	0	0	0	0	0	0	0	0	0	0	0	0
A. v.	0	0	1	0	0	0	0	0	0	0	1	1	1	1	1	1	1	1	0	1	1	0	0	0	0	0	1	1	1	0	1	0	0	0	0	0	0	1	1	0	0	1	1
C. fl.	1	1	1	1	1	1	0	1	0	1	1	1	0	0	0	0	0	1	0	0	1	1	1	1	1	1	1	1	1	1	1	1	1	1	1	1	1	1	1	1	0	0	0
N. g.	1	1	0	1	1	1	1	1	1	0	1	0	0	0	1	1	1	1	1	1	1	1	1	1	1	1	1	1	1	1	1	1	1	1	1	1	1	1	1	1	0	1	0
P. d.§	0	0	0	0	0	1	0	0	0	0	1	1	0	0	1	1	1	1	1	1	1	1	1	1	1	1	1	1	1	1	1	1	1	1	1	1	1	1	1	0	0	0	0
P. f.§	0	1	1	0	1	1	0	0	0	0	1	1	0	0	1	1	1	1	1	1	1	0	1	1	1	1	1	1	1	1	1	1	1	1	1	1	1	1	1	0	0	0	0
P. f.	0	0	0	0	0	0	0	0	0	0	1	0	0	0	0	0	0	0	0	0	1	0	0	0	0	1	1	1	0	0	0	0	1	0	1	0	0	0	1	0	0	0	0
P. t.	0	1	1	0	1	1	0	0	0	1	1	0	1	0	1	1	1	1	0	1	1	0	1	1	0	0	1	0	1	0	0	0	1	0	1	1	0	1	0	0	0	0	0
I. r.§	0	0	0	0	0	0	0	0	0	0	0	0	0	0	0	0	0	0	0	0	1	0	0	0	0	0	1	1	0	0	0	0	0	0	0	0	0	0	0	0	0	0	0
A. f.§	0	0	0	0	0	0	0	0	0	0	0	0	0	0	0	0	0	0	0	0	1	0	0	0	0	0	1	0	0	0	0	0	0	1	0	0	0	0	0	0	0	0	0
C. n.	1	1	1	1	1	1	1	1	1	1	1	1	1	1	1	1	1	1	1	1	1	1	1	1	1	1	1	1	1	1	1	1	0	1	1	1	0	1	1	0	0	1	1
N. r.§	0	1	1	0	1	1	0	0	1	0	1	1	1	1	1	1	1	1	1	1	1	1	1	1	1	1	1	1	1	0	0	1	0	1	1	1	0	1	1	0	0	1	0
P. st.	0	0	0	0	0	0	0	0	0	0	0	0	0	0	0	0	0	1	0	0	1	0	0	1	1	0	1	0	0	0	0	0	0	0	0	0	0	0	0	0	0	0	0
S. g.§	0	0	1	0	0	1	0	0	0	0	1	0	0	0	0	0	0	0	0	0	1	0	0	0	0	0	1	1	0	0	0	0	0	1	0	0	0	1	1	0	0	1	0
S. s.§	0	0	0	0	0	0	0	0	0	0	0	0	0	0	0	0	0	0	0	0	1	0	0	0	0	0	0	0	0	0	0	0	0	0	0	0	0	0	0	0	0	0	0
Z. g.§	0	0	1	0	1	1	0	1	0	1	1	0	0	0	0	0	0	1	0	0	0	0	0	0	0	0	1	1	0	0	0	0	0	0	0	0	0	1	1	0	0	0	0
Z. gu.§	0	0	0	0	0	0	0	0	0	0	0	0	0	0	0	0	0	1	0	0	1	0	0	0	1	0	1	0	0	0	0	0	0	0	0	0	0	0	0	0	0	0	0
M. c.	1	1	1	1	1	1	0	0	0	0	1	0	0	1	1	1	0	1	1	1	1	0	0	0	1	0	1	0	0	0	0	0	0	0	0	0	0	1	0	0	0	0	0
A. ch.§.	0	0	1	0	0	0	0	0	0	0	0	0	0	0	0	0	0	0	0	0	0	0	0	0	0	0	0	0	0	0	0	0	0	0	0	0	0	0	0	0	0	0	0
A. m.§	1	1	1	1	1	1	0	1	1	0	1	1	0	0	1	1	1	1	1	1	1	0	1	1	1	0	1	1	0	0	0	0	0	0	1	0	0	1	1	0	0	0	0
M. k.§	0	0	0	0	0	0	0	0	0	0	0	0	0	0	0	0	0	0	0	0	0	0	1	1	1	0	1	1	0	0	0	1	0	0	0	0	0	0	1	0	0	0	0
E. s.	0	0	0	0	0	1	0	0	0	0	0	0	0	0	0	0	0	0	0	0	0	0	0	0	0	0	0	0	0	0	0	0	0	0	0	0	0	0	0	0	0	0	0
Z. s.§	0	0	0	0	0	0	0	0	0	0	0	0	0	0	0	0	0	0	0	0	1	0	1	0	1	1	1	1	0	0	0	1	1	1	1	1	0	1	1	0	0	0	0
E. h.§	1	0	1	0	1	1	0	0	1	0	0	1	0	0	1	1	1	1	0	1	1	1	1	1	1	1	1	1	0	0	0	0	0	0	0	0	0	0	0	0	0	0	0
T. cy.§	1	1	1	0	1	1	0	1	0	0	1	0	0	0	1	1	1	1	1	1	1	0	1	1	1	1	1	1	1	0	1	1	1	1	1	1	1	1	1	1	0	1	0
B. m.§	0	0	0	0	0	1	0	1	0	0	1	1	1	1	1	1	1	1	1	1	1	0	1	1	1	1	1	1	0	0	1	1	0	0	0	0	0	0	0	0	0	0	0
B. r.§	0	0	0	0	0	0	0	0	0	0	0	0	0	0	0	0	0	0	0	0	0	0	0	0	0	0	0	0	0	0	0	0	1	1	1	1	1	1	1	0	0	0	0

Table 5 continued

Species code	JR	BO	TR	DK	GD	AS	MT	GA	TL	WA	SH	DJ	KN	TM	MU	GN	BU	MR	DZ	MA	EU	KL	GG	MB	ZK	PA	PU	KZ	VI	KI	MC	KG	KP	CH	NG	PD	ND	LI	RO	NY	PI	ZI	MI
B. s.	0	0	1	1	0	1	0	0	0	0	0	0	0	0	0	0	0	0	0	0	0	0	0	0	0	0	0	0	0	0	0	0	0	0	1	1	0	0	1	0	0	1	1
P. p.	1	0	1	0	0	1	0	0	0	1	0	0	0	0	0	0	0	0	0	0	0	0	0	0	0	1	1	1	0	0	1	0	1	0	1	1	0	1	1	0	1	0	1
P. sc.	1	1	1	0	1	1	1	1	0	0	1	0	0	0	0	0	0	1	1	1	0	0	1	1	1	1	1	1	1	0	1	1	1	0	1	1	1	1	1	0	0	0	0
D. c.	1	1	1	1	0	1	0	1	0	1	1	1	0	1	1	0	0	1	0	0	1	0	1	1	1	0	1	1	1	0	1	1	1	1	1	1	1	1	1	1	0	1	1
M. q.	0	0	0	0	1	1	0	0	0	0	0	0	0	0	0	0	0	0	0	0	0	0	1	1	1	0	1	1	1	0	0	1	1	1	1	1	0	1	1	1	0	0	0
D. l.	0	0	1	0	0	0	0	0	0	0	0	0	0	0	0	0	0	0	0	0	1	0	1	1	1	1	1	1	1	0	1	1	1	1	1	1	1	1	1	1	0	0	0
O. c.	0	0	0	0	0	1	0	0	0	0	1	0	0	0	0	0	0	1	0	1	1	0	0	0	0	0	0	0	0	0	0	0	0	0	0	0	0	0	1	0	0	0	0
L. c.	1	1	1	1	1	1	1	1	1	0	1	1	1	1	1	1	1	1	1	1	1	1	1	1	1	1	1	1	1	1	1	1	1	1	1	1	1	1	1	1	1	0	0
A. c.	1	1	1	1	1	1	0	1	0	1	1	1	1	1	1	1	1	1	1	1	1	1	1	1	1	1	1	1	1	0	1	1	1	1	1	1	1	1	1	1	0	1	1
A. ne.	0	0	1	0	1	0	0	0	0	0	0	0	0	0	0	1	1	0	0	0	1	0	1	1	1	0	1	1	0	0	0	1	1	1	1	1	0	1	1	0	0	0	0
A. p.	0	0	0	0	0	1	0	0	0	0	0	0	0	0	0	0	0	0	0	0	1	0	0	0	0	0	0	0	0	0	0	0	0	0	0	0	0	0	0	0	0	0	0
A. r.§	0	0	1	0	1	1	0	1	0	0	1	1	0	0	0	0	0	1	1	1	1	1	1	1	1	1	0	0	0	0	0	0	0	1	1	1	1	1	1	1	0	0	0
A. ru.§	0	0	0	0	0	0	0	0	0	0	0	0	0	0	0	0	0	0	0	0	1	0	0	0	0	0	0	0	0	0	0	0	0	0	1	0	0	0	0	0	0	0	0
N. o.§	1	1	1	0	1	1	1	1	1	1	1	1	1	1	1	1	1	1	1	1	1	1	1	1	1	1	1	1	1	1	1	1	1	1	1	1	1	1	1	0	1	1	1
P. b.	1	1	1	0	1	1	0	0	0	0	1	1	0	0	1	1	1	1	1	1	1	0	1	1	1	1	1	1	1	0	0	1	1	1	1	1	1	1	1	1	0	1	0
P. g.§	0	0	0	1	0	1	0	0	0	0	0	0	0	0	0	0	0	0	0	0	0	0	0	0	0	0	0	0	0	0	0	0	0	0	0	0	0	0	0	0	0	0	0
H. n.	0	1	0	0	1	1	1	1	0	1	1	1	1	1	1	0	0	1	1	1	1	1	1	1	1	1	1	1	1	0	1	1	1	1	1	1	1	1	1	1	0	0	1
M. n.§	0	0	0	0	1	1	0	0	0	0	1	1	1	1	1	1	1	1	1	1	1	0	0	0	0	0	1	0	0	0	0	0	0	0	1	1	1	0	1	0	0	0	0

C. f. = *Circaetus fasciolatus*; *A. m.* = *Accipiter melanoleucus*; *A. t.* = *Accipiter tachiro*; *H. a.* = *Hieraaetus ayresii*; *S. co.* = *Stephanoaetus coronatus*; *M. a.* = *Macheiramphus alcinus*; *G. p.* = *Guttera pucherani*; *S. e.* = *Sarothrura elegans*; *A. l.* = *Aplopelia larvata*; *C. d.* = *Columba delegorguei*; *T. t.* = *Turtur tympanistria*; *T. c.* = *Treron calva*; *P. c.* = *Poicephalus cryptoxanthus*; *T. f.* = *Tauraco fischeri*; *T. l.* = *Tauraco livingstonii*; *C. m.* = *Cercococcyx montanus*; *C. c.* = *Chrysococcyx cupreus*; *C. a.* = *Ceuthmochares aereus*; *B. v.* = *Bubo vosseleri*; *S. w.* = *Strix woodfordii*; *G.c.* = *Glaucidium capens*; *O. i.* = *Otus ireneae*; *S. p.* = *Scotopelia peli*; *C. p.* = *Caprimulgus pectoralis*; *N. b.* = *Neafrapus boehmi*; *T. u.* = *Telecanthura ussheri*; *A. n.* = *Apaloderma narina*; *B. br.* = *Bycanistes (Ceratogymna) brevis*; *B. b.* = *Bycanistes (Ceratogymna) bucinator*; *S. l.* = *Stactolaema leucotis*; *S. o.* = *Stactolaema olivacea*; *P. bi.* = *Pogoniulus bilineatus*; *P. s.* = *Pogoniulus simplex*; *C. ab.* = *Campethera abingoni*; *C. m.* = *Campethera mombassica*; *S. ca.* = *Smithornis capensis*; *P. a.* = *Pitta angolensis*; *A. s.* = *Anthus sokokensis*; *A. mi.* = *Andropadus milanjensis*; *A. v.* = *Andropadus virens*; *C. fl.* = *Chlorocichla flaviventris*; *N. g.* = *Nicator gularis*; *P. d.* = *Phyllastrephus debilis*; *P. f.* = *Phyllastrephus fischeri*; *P. f.* = *Phyllastrephus flavostriatus*; *P. t.* = *Phyllastrephus terrestris*; *I. r.* = *Illadopsis rufipennis*; *A. f.* = *Alethe fuelleborni*; *C. n.* = *Cossypha natalensis*; *N. r.* = *Neocossyphus rufus*; *P. st.* = *Pogonocichla stellata*; *S. g.* = *Sheppardia gunningi*; *S. s.* = *Swynnertonia swynnertoni*; *Z. g.* = *Zoothera guttata*; *Z. gu.* = *Zoothera gurneyi*; *M.c.* = *Muscicapa caerulescens*; *A. ch.* = *Apalis chariessa*; *A. m.* = *Apalas melanocephala*; *M. k.* = *Macrospehnus kretschmeri*; *E. s.* = *Eremomela scotops*; *Z. s.* = *Zosterops senegalensis*; *E. h.* = *Erythrocercus holochlorus*; *T. cy.* = *Trochocercus cyanomelas*; *B. m.* = *Batis mixta*; *B. r.* = *Batis reichenowi*; *B. s.* = *Batis soror*; *P. p.* = *Platysteira peltata*; *P. sc.* = *Prionops scopifrons*; *D. c.* = *Dryoscopus cubla*; *M. q.* = *Malaconotus quadricolor*; *D. l.* = *Dicrurus ludwigii*; *O. c.* = *Oriolus chlorocephalus*; *L. c.* = *Lamprotornis corruscus*; *A. c.* = *Anthreptes collaris*; *A. ne.* = *Anthreptes neglectus*; *A. p.* = *Anthreptes pallidigaster*; *A. r.* = *Anthreptes reichenowi*; *A. ru.* = *Anthreptes rubritorques*; *N. o.* = *Nectarinia olivacea*; *P. b.* = *Ploceus bicolor*; *P. g.* = *Ploceus goland*; *H. n.* = *Hypargos niveoguttatus*; *M. n.* = *Mandingoa nitidula*

Table 6 Data matrix of 66 mammal species and 30 forest fragments of the East African coastal forests included in the study

Sp. code	JR	TR	AS	SH	EU	KL	AC	TW	GG	MB	MJ	ZK	RN	PK	RS	VI	KI	MC	KG	TO	NK	LI	RO	CH	PD	NG	KM	PI	ZI	MI
C. f.	0	0	1	0	1	0	0	0	0	0	0	1	0	1	0	0	0	0	0	0	0	0	0	0	0	0	0	0	1	0
C. hi.	0	0	0	0	0	0	1	0	0	0	0	0	0	0	0	0	0	0	0	0	0	0	0	0	0	0	0	0	0	0
C. j.	0	0	0	0	0	0	0	0	0	0	0	0	0	0	0	0	0	0	1	0	0	0	0	0	0	0	1	0	1	1
C. p.	0	0	1	0	0	0	0	0	0	0	0	0	0	0	0	0	0	0	0	0	0	0	0	0	0	0	0	0	0	0
C. sa.§	0	0	0	0	0	0	0	1	0	0	0	0	0	0	0	0	0	0	0	0	0	0	0	0	0	0	0	0	0	0
C. sb.§	0	0	0	0	0	0	0	0	1	0	0	0	0	0	0	0	0	0	0	0	0	0	0	0	0	0	0	0	0	0
C. sc.§	0	0	0	0	0	0	0	0	0	0	0	1	0	0	0	0	0	0	0	0	0	0	0	0	0	0	0	0	0	0
C. sd.§	0	0	0	0	0	0	0	0	0	0	0	0	0	1	0	0	0	0	0	0	0	0	0	0	0	0	0	0	0	0
C. se.§	0	0	0	0	0	0	0	0	0	0	0	0	0	0	0	0	0	0	0	1	0	0	0	0	0	0	0	0	0	0
G. ro.§	0	0	0	0	0	0	0	0	0	0	0	0	0	1	0	0	0	0	0	0	0	1	1	0	1	0	0	0	0	0
G. z.§	1	1	1	0	1	1	0	1	1	1	1	0	1	1	1	0	0	1	0	1	1	1	1	1	1	1	0	0	1	0
G. co.§	0	0	1	1	0	0	0	0	0	0	0	0	0	0	0	0	0	0	0	0	0	0	0	0	0	0	0	0	0	0
O. c.	1	0	1	0	0	0	1	0	0	1	0	0	0	1	0	0	0	0	0	0	0	1	1	0	0	0	0	0	0	1
O. g.	0	0	1	0	1	1	0	1	1	1	1	1	0	1	0	0	1	0	0	1	1	1	1	0	1	1	0	1	1	0
C. g.§	0	1	0	0	0	0	0	0	0	0	0	0	0	0	0	0	0	0	0	0	0	0	0	0	0	0	0	0	0	0
C. m.§	1	1	1	1	1	1	1	1	1	0	1	1	0	1	1	1	1	1	1	1	1	1	1	1	1	1	1	0	1	1
C. ae.	0	1	1	1	1	1	1	0	1	0	1	0	1	1	0	1	0	0	1	0	0	1	1	0	0	1	0	1	0	1
C. an.§	0	0	0	1	1	1	1	1	1	0	1	1	0	0	0	0	0	1	0	0	0	0	0	0	0	0	1	0	0	0
P. h.	1	1	1	0	0	0	0	0	1	0	1	1	1	1	1	1	0	0	1	1	0	0	0	0	0	1	0	0	0	0
P. pk.§	0	0	0	0	0	0	0	0	0	0	0	0	0	0	0	0	0	0	0	0	0	0	0	0	0	0	0	0	1	0
P. r.§	0	1	0	0	0	0	0	0	0	0	0	0	0	0	0	0	0	0	0	0	0	0	0	0	0	0	0	0	0	0
C. a.	0	1	0	0	0	0	0	0	0	0	0	0	0	0	0	0	0	0	0	0	0	1	1	0	1	1	0	0	0	0
O. m.	0	1	0	0	0	0	0	0	0	0	0	0	0	0	0	0	0	0	0	0	0	0	0	0	0	0	0	0	0	0
C. c.	0	0	1	0	0	0	0	0	0	0	0	0	0	0	0	0	0	0	0	0	0	0	0	0	0	0	0	0	0	0
F. s.	0	0	0	0	0	0	0	0	0	0	0	0	0	0	0	0	0	0	0	0	0	0	0	0	0	1	0	0	0	0
L. s.	1	1	1	0	0	0	0	0	0	0	0	0	0	0	0	0	0	0	0	0	0	0	0	0	0	0	0	0	0	0
P. p.	1	1	1	0	1	0	0	0	1	1	1	1	0	1	0	0	0	1	1	1	1	1	1	1	1	1	0	0	1	0
B. c.	0	0	1	1	1	1	1	0	0	0	0	0	0	0	0	0	0	0	0	0	0	0	0	0	1	0	0	0	1	0
G. sa.	1	0	1	0	0	0	0	0	1	1	0	0	0	1	0	0	0	0	0	0	0	0	0	0	0	1	0	0	1	0
G. sp.	1	1	1	0	1	0	0	0	1	0	0	0	0	1	0	0	0	0	0	0	0	0	1	1	0	0	0	0	1	0
H. i.	0	0	1	0	0	0	0	0	0	0	0	1	0	0	0	0	0	0	1	0	0	0	0	0	0	0	0	0	0	1
I. a.	1	1	1	0	0	1	0	0	0	0	0	0	0	0	0	0	0	0	0	0	0	0	0	0	0	0	0	0	0	0
M. b.§	0	0	0	0	1	0	0	0	1	0	0	0	0	1	0	0	0	0	0	1	0	0	0	0	0	0	0	0	0	0

Table 6 continued

Sp. code	JR	TR	AS	SH	EU	KL	AC	TW	GG	MB	MJ	ZK	RN	PK	RS	VI	KI	MC	KG	TO	NK	LI	RO	CH	PD	NG	KM	PI	ZI	MI
C. cr.	0	1	1	1	0	0	0	0	0	0	1	0	0	1	0	0	0	0	0	0	0	1	0	0	1	1	0	0	0	0
C. ci.	1	1	1	0	1	1	0	0	0	1	1	1	0	1	0	0	0	0	0	0	0	0	1	0	0	0	0	0	1	0
L. o.	1	1	1	1	0	0	0	0	0	1	1	1	0	0	1	0	0	1	1	1	1	1	1	1	1	1	0	0	0	0
D. a.§	0	0	0	1	0	0	0	0	0	0	0	0	0	0	0	0	0	0	0	0	0	0	0	0	0	0	0	0	0	0
D. v.	0	0	0	0	1	0	1	0	0	0	0	0	0	0	0	0	0	0	0	0	0	0	0	0	0	0	0	0	1	1
P. l.	1	1	1	1	1	1	0	0	1	0	1	1	1	1	0	0	0	0	0	1	1	0	0	0	0	0	0	0	0	0
C. a.§	0	0	1	0	0	0	0	0	0	0	0	0	0	0	0	0	0	0	0	0	0	0	0	0	0	0	0	0	1	0
C. h.§	0	1	0	1	0	0	0	0	0	0	0	1	0	0	0	0	0	0	0	0	0	0	0	0	0	0	0	0	0	0
C. m.§	0	0	0	0	1	0	1	0	0	0	0	0	0	0	0	0	0	0	0	0	0	0	0	0	0	0	0	1	1	1
C. n.§	0	0	0	0	0	0	0	0	0	0	0	0	0	0	0	0	0	0	0	0	0	0	0	0	0	0	0	0	0	0
H. n.	0	0	0	1	0	0	0	0	1	0	1	0	0	0	0	0	0	0	1	0	1	0	0	0	0	0	0	0	0	0
N. m.	0	0	1	0	1	1	0	0	0	0	0	0	0	0	0	0	0	0	1	0	0	1	1	0	1	1	0	0	1	1
S. c.	1	1	1	1	0	0	0	0	1	0	1	1	1	0	0	0	0	0	1	1	0	1	1	1	1	1	0	0	0	0
T. s.	1	0	1	0	0	0	0	0	1	0	1	0	0	0	0	0	0	0	0	0	0	1	1	0	1	1	0	0	0	0
H. sp.	1	1	1	1	0	1	0	0	0	1	0	0	0	0	0	0	0	1	1	1	1	1	1	1	0	0	0	0	1	1
P. f.	0	0	0	0	0	0	0	0	0	0	0	0	0	0	0	0	0	0	0	0	1	1	1	0	0	0	0	0	0	0
P. o.	1	1	0	0	0	0	1	0	0	0	0	0	0	0	0	0	0	0	0	0	0	0	0	0	0	0	0	0	0	0
P. p.§	0	1	1	1	1	1	0	0	1	0	1	1	0	0	0	0	0	1	0	0	1	1	1	1	1	1	0	0	1	1
B. h.	0	0	1	1	1	0	0	0	0	0	0	1	0	0	1	0	0	0	0	0	1	1	1	0	0	0	0	0	0	0
C. ga.	0	0	1	0	1	1	0	0	1	0	1	1	0	1	0	0	0	0	0	0	1	0	1	0	0	0	1	0	1	1
G. d.	1	1	0	0	1	0	0	0	1	0	0	1	0	1	0	0	0	1	0	0	1	0	0	0	0	0	0	0	0	0
G. c.	1	0	0	0	0	0	0	0	0	0	0	0	0	0	0	0	0	0	0	0	0	0	0	0	0	0	0	0	0	0
G. co.	0	0	0	0	0	0	0	0	0	0	0	0	0	0	0	0	0	1	0	0	0	0	0	0	0	0	0	0	0	0
M. m.	0	0	0	0	1	0	0	0	0	0	0	0	0	0	0	0	0	0	1	0	0	0	1	0	0	0	0	0	0	0
P. fa.	0	0	0	0	0	0	0	0	0	0	0	0	0	0	0	0	0	0	0	0	0	0	0	0	0	0	0	0	0	0
G. m.	0	0	1	0	0	0	0	0	1	0	0	1	0	0	0	0	0	0	0	0	0	0	0	0	0	0	0	0	0	0
G. p.§	0	0	0	0	0	0	0	0	0	0	0	0	0	1	0	0	0	0	0	0	0	0	0	0	0	0	0	0	0	0
H. a.	0	0	0	0	0	0	0	0	0	0	0	0	0	0	0	0	0	0	0	0	0	0	0	0	1	0	0	0	0	0
H. c.	1	0	1	0	0	0	0	0	0	0	0	0	0	0	0	0	0	0	1	1	0	0	0	0	0	0	0	0	0	1
H. s.	0	1	0	0	1	1	0	1	1	0	0	1	0	1	0	0	0	0	0	0	0	0	0	0	0	0	0	0	0	0
P. t.	0	0	1	0	1	1	0	1	1	0	1	1	0	1	1	0	0	0	1	0	1	1	1	0	0	1	0	0	1	1

Table 6 continued

Sp. code	JR	TR	AS	SH	EU	KL	AC	TW	GG	MB	MJ	ZK	RN	PK	RS	VI	KI	MC	KG	TO	NK	LI	RO	CH	PD	NG	KM	PI	ZI	MI
R. c.§	0	0	1	0	0	0	0	0	0	0	0	0	0	0	0	0	0	0	0	0	0	0	0	0	0	0	0	0	0	0
R. p.§	0	0	0	1	1	0	0	1	1	1	1	1	0	1	1	0	1	1	1	1	0	1	0	0	0	0	0	0	1	1

§ = forest specialist, remaining species are forest generalist. Forest codes as in Fig. 1

Species code: *C. f.* = *Crocidura fuscomurina*; *C. hi.* = *Crocidura hildegardeae*; *C. j.* = *Crocidura jacksoni*; *C. p.* = *Crocidura parvipes*; *C. sa.* = *Crocidura* sp. a; *C. sb.* = *Crocidura* sp. b; *C. sc.* = *Crocidura* sp. c; *C. sd.* = *Crocidura* sp. d; *C. se.* = *Crocidura* sp. e; *G. ro.* = *Galagoides rondoensis*; *G. z.* = *Galagoides zanzibaricus*; *G. co.* = *Galago cocos*; *O. c.* = *Otolemur crassicaudatus*; *O. g.* = *Otolemur garnetti*; *C. g.* = *Cercocebus galeritus*; *C. m.* = *Cercocebus mitis*; *C. ae.* = *Chlorocebus aethiops*; *C. an.* = *Colobus angolensis*; *P. h.* = *Papio hamadryas*; *P. pk.* = *Procolobus pennantii kirkii*; *P. r.* = *Procolobus rufomitratus*; *C. a.* = *Canis adustus*; *O. m.* = *Otocyon megalotis*; *C. c.* = *Caracal caracal*; *F. s.* = *Felis silvestris*; *L. s.* = *Leptailurus serval*; *P. p.* = *Panthera pardus*; *B. c.* = *Bdeogale crassicauda*; *G. sa.* = *Galerell sanguinea*; *G. sp.* = *Genetta*species; *H. i.* = *Herpestes ichneumon*; *I. a.* = *Ichneumia albicauda*; *M. b.* = *Nandinia binotata*; *C. cr.* = *Crocuta crocuta*; *C. ci.* = *Civettictis civetta*; *L. o.* = *Loxodonta africana*; *D. a.* = *Dendrohyrax arboreus*; *D. v.* = *Dendrohyrax validus*; *P. l.* = *Potamochoerus larvatus*; *C. a.* = *Cephalophus adersi*; *C. h.* = *Cephalophus harveyi*; *C. m.* = *Cephalophus monticola*; *C. n.* = *Cephalophus natalensis* A. Smith, 1834; *H. n.* = *Hippotragus niger*; *N. m.* = *Neotragus moschatus*; *S. c.* = *Syncerus caffer*; *T. s.* = *Tragelphus scriptus*; *H. sp.* = *Heliosciurus* spp; *P. f.* = *Paraxerus flavovittis*; *P. o.* = *Paraxerus ocheraceus*; *P. p.* = *Paraxerus palliatus*; *B. h.* = *Beamys hindei*; *C. ga.* = *Cricetomys gambianus*; *G. d.* = *Grammomys dolichurus*; *G. c.* = *Gramnomys caniceps*; *G. co.* = *Grammomys cometes*; *M. m.* = *Mus minutoides*; *P. fa* = *Pelomys fallax*; *G. m.* = *Graphiurus murinus*; *G. p.* = *Graphiurus* n. sp. cf *parvus*; *H. a.* = *Hystix africaeaustralis*; *H. c.* = *Hystix cristata*; *H. s.* = *Hystix* (not identified to spp); *P. t.* = *Petrodromus tetradactylus*; *R. c.* = *Rhynchcyon chrysopygus*; *R. p.* = *Rhynchcyon petersi*

Table 7 Data matrix of 50 reptile species and 25 forest fragments of the East African coastal forests included in the study

Species code	TR	AS	SH	EU	KL	AC	TW	GG	MB	MJ	ZK	PA	PK	RS	KI	MC	KG	NK	KP	TO	ND	RP	PI	ZI	MI
L. u.	0	0	0	0	0	0	1	0	0	0	0	0	0	0	0	0	0	0	0	0	0	0	0	0	0
L. c.§	0	1	0	1	1	0	0	0	0	0	0	0	0	0	0	0	0	0	0	0	0	0	0	0	0
L. b.§	0	0	0	0	1	0	0	0	0	0	1	0	0	0	0	0	1	0	0	0	0	0	0	0	0
L. r.§	0	0	0	0	0	0	0	0	0	0	0	0	0	0	0	0	0	0	0	0	0	0	0	0	0
L. v.§	0	0	0	0	1	0	0	0	0	0	1	0	0	0	0	0	1	1	0	1	0	0	0	1	1
L. w.§	0	0	0	0	0	0	0	0	0	0	0	0	0	0	?	0	0	0	0	0	0	0	0	0	0
L. k.§	0	0	0	0	0	1	0	0	0	0	0	0	0	0	0	0	0	0	0	0	0	0	0	0	0
U. w.	1	0	0	0	0	0	0	0	0	0	0	0	0	0	0	0	0	0	0	0	0	0	0	0	0
C. b.	0	0	0	0	0	1	1	0	0	0	0	0	0	0	0	0	0	0	0	0	0	0	0	0	0
C. u.	0	0	0	0	0	0	0	0	0	0	0	0	0	0	0	0	1	0	0	1	0	0	0	0	0
H. m.§	1	0	0	0	0	0	0	0	0	0	0	0	0	0	0	0	0	0	0	0	0	0	0	0	0
B. t.§	0	0	1	0	0	0	0	0	0	0	0	0	0	0	0	0	0	0	0	0	0	0	0	0	0
B. m.§	0	0	0	0	0	0	0	0	0	0	0	0	0	0	0	0	0	0	0	0	0	0	0	0	0
R. p.	0	0	0	0	0	0	0	0	0	0	0	0	0	0	0	0	0	0	0	0	0	0	0	0	0
R. b.	0	0	1	0	0	0	1	0	0	0	1	0	1	1	0	0	1	0	0	1	0	1	0	1	0
R. br.	0	0	0	0	0	0	0	0	0	0	0	0	0	0	0	0	1	0	0	0	0	0	0	0	0
S. t.§	0	0	0	0	0	0	0	0	0	0	0	0	0	0	0	0	0	0	1	1	0	1	0	0	0
S. d.§	0	0	0	0	0	0	0	0	0	0	0	0	0	0	0	0	0	0	0	0	0	0	0	0	0
S. i.§	0	0	0	0	0	0	0	0	0	0	0	0	0	0	0	0	0	0	0	0	0	0	0	0	0
M. a.	0	0	0	0	0	0	0	0	0	0	0	0	0	0	0	0	0	0	0	0	0	0	0	0	0
M. r.§	0	0	0	0	0	0	0	0	0	0	0	0	0	0	0	0	0	0	0	0	0	1	0	0	0
M. l.	0	0	0	0	0	0	0	0	0	0	0	0	0	0	0	0	1	0	0	0	0	0	0	0	0
S. l.§	0	0	0	0	0	0	0	0	0	0	0	0	0	0	0	0	0	0	0	0	0	1	0	0	0
S. b.§	0	0	0	0	0	0	0	0	0	0	0	0	0	0	0	0	0	0	0	0	0	0	1	0	0
H. l.	0	0	0	0	0	0	0	0	0	0	1	0	0	0	0	0	0	0	0	0	?	?	0	0	0
G. p.§	0	1	0	0	0	0	0	0	0	0	1	0	0	0	0	0	0	0	0	0	0	0	0	0	0
G. v.§	0	0	0	0	0	0	0	0	0	1	0	0	0	0	0	0	0	0	1	0	0	0	0	0	0
C. e.§	0	0	0	0	0	0	0	0	0	0	0	0	0	0	0	0	0	0	0	0	0	?	0	0	0
A. b.§	0	0	0	0	0	0	0	0	0	0	0	0	0	0	0	0	0	0	0	0	0	?	0	0	0
T. o.	0	0	0	0	0	0	0	0	0	0	0	0	0	0	0	0	0	0	0	0	0	0	0	0	0
T. r.§	0	0	0	0	0	0	0	0	0	0	0	0	0	0	0	0	0	0	0	0	0	1	0	0	0
T. p.§	0	0	0	1	0	0	0	0	0	0	0	0	0	0	0	0	0	0	0	0	0	0	0	0	0
R. l.§	0	1	0	0	1	0	0	0	0	0	0	0	0	0	0	0	0	0	0	0	0	0	0	1	0

Table 7 continued

Species code	TR	AS	SH	EU	KL	AC	TW	GG	MB	MJ	ZK	PA	PK	RS	KI	MC	KG	NK	KP	TO	ND	RP	PI	ZI	MI
L. m.§	0	1	0	0	0	0	0	0	0	1	0	0	0	0	0	1	0	0	0	1	0	0	0	0	0
L. spA.§	1	0	0	0	0	0	0	0	0	0	0	0	0	0	0	0	0	0	0	0	0	0	0	0	0
L. spB.§	0	0	0	0	0	0	0	0	0	0	0	0	0	0	0	0	0	0	0	0	0	1	0	0	0
B. g.	0	0	0	0	1	0	1	0	0	0	1	0	1	0	0	0	0	1	0	0	1	1	0	0	0
A. g.	1	1	0	0	0	0	0	0	0	0	0	0	1	0	0	0	0	0	0	0	1	0	0	0	0
A. w.	0	0	0	0	0	0	0	0	0	0	?	0	1	0	0	0	1	0	0	0	0	0	0	0	0
A. t.§	1	1	0	0	0	0	0	0	0	0	0	0	0	0	0	0	0	0	0	0	0	0	0	0	0
D. a.	0	1	0	0	0	0	0	0	0	0	0	0	1	0	0	0	0	0	1	0	1	1	1	1	1
B. v.	0	0	0	1	0	0	0	0	0	0	0	0	0	0	0	0	0	0	0	0	0	0	0	0	0
N. v.§	0	0	0	0	0	0	0	0	0	0	0	0	0	0	0	0	0	0	0	0	0	0	0	0	0
P. j.§	0	0	0	0	0	0	0	0	0	0	0	0	0	0	0	0	0	0	0	0	0	0	0	0	0
P. s.§	0	0	0	1	0	0	0	0	0	0	0	0	0	0	0	0	0	0	0	0	0	0	0	0	0
P. m.§	0	0	0	1	0	0	0	1	0	0	0	0	0	0	0	0	1	0	0	0	0	1	0	1	0
P. n.§	0	0	0	0	0	0	0	0	0	0	0	0	0	0	0	0	0	0	0	0	0	0	0	0	0
T. u.§	0	0	0	1	0	0	0	0	0	0	0	0	0	0	0	0	0	0	0	0	0	0	0	0	0
C. t.	0	0	0	0	0	0	0	0	0	0	0	0	0	0	0	0	1	0	0	0	0	0	0	0	0
D. w.§	0	0	0	1	0	0	0	0	0	0	0	0	0	0	0	0	0	0	0	0	0	0	0	0	0

§ = coastal endemic, others are forest endemic. Note that 13 species were not recoded in the studied forests. Forest codes as in Fig. 1

Species code: *L. u.* = *Lygodactylus uluguruensis*; *L. c.*§ = *Lygodactylus conradti*; *L. b.*§ = *Lygodactylus broadleyi*; *L. r.*§ = *Lygodactylus rex*; *L. v.*§ = *Lygodactylus viscatus*; *L. w.*§ = *Lygodactylus williamsi*; *L. k.*§ = *Lygodactylus kimhowelli*; *U. w.* = *Urocotyledon wolterstorffi*; *C. b.* = *Cnemaspis barbouri*; *C. u.* = *Cnemaspis uzungwae*; *H. m.*§ = *Hemidactylus modestus*; *B. t.*§ = *Bradypodion tenue*; *B. m.*§ = *Bradypodion mlanjense*; *R. p.* = *Rhampholeon platyceps*; *R. b.* = *Rhampholeon brevicaudatus*; *R. br.* = *Rhampholeon brachyurus*; *S. t.*§ = *Sepsina t. tetradactyla*; *S. d.*§ = *Scelotes duttoni*; *S. i.*§ = *Scelotes insularis*; *M. a.* = *Melanoseps ater*; *M. r.*§ = *Melanoseps rondoensis*; *M. l.* = *Melanoseps loveridgei*; *S. l.*§ = *Scolecoseps litipoensis*; *S. b.*§ = *Scolecoseps boulengeri*; *H. l.* = *Holaspis laevis/H. guentheri laevis*; *G. p.*§ = *Gastropholis prasina*; *G. v.*§ = *Gastropholis vittata*; *C. e.*§ = *Chirindia ewerbecki*; *A. b.*§ = *Ancylocranium barkeri*; *T. o.* = *Typhlops obtusus*; *T. r.*§ = *Typhlops rondoensis*; *T. p.*§ = *Typhlops platyrhynchus*; *R. l.*§ = *Rhinotyphlops lumbriciformis*; *L. m.*§ = *Leptotyphlops macrops*; *L. spA.*§ = *Leptotyphlops* sp. nov. A; *L. spB.*§ = *Leptotyphlops* sp. nov. B; *B. g.* = *Bitis g. gabonica*; *A. g.* = *Apparallctus guentheri*; *A. w.* = *Apprallactus werneri*; *A. t.*§ = *Apparallctus turneri*; *D. a.* = *Dendroaspis angusticeps*; *B. v.* = *Buhoma vauerocegae*; *N. v.*§ = *Natriciteres variegata sylvatica*; *P. j.*§ = *Prosymna janii*; *P. s.*§ = *Prosymna semifasciata*; *P. m.*§ = *Philothamnus macrops*; *P. n.*§ = *Philothamnus natalensis*; *T. u.*§ = *Thelotornis usambararicus*; *C. t.* = *Crotaphopeltis tornieri*; *D. w.*§ = *Dipsadoboa werneri*

References

Andrén H (1994) Effects of habitat fragmentation on birds and mammals in landscapes with different proportions of suitable habitat: a review. Oikos 71:355–366

Anonymous (2003) Ecosystem profile: Eastern Arc mountains and coastal forests of Tanzania and Kenya biodiversity hotspot. Critical Ecosystem Partnership Fund (CEPF). Prepared by Conservation International (CI) and International Centre of Insect Physiology and Ecology (ICIPE), Washington, USA

Ås S, Bengtsson J, Ebenhard T (1997) Archipelagoes and theories of insularity. Ecol Bull 46:88–116

Azeria ET (2004) Terrestrial bird community patterns on the coralline islands of the Dahlak Archipelago. Global Ecol Biogeogr 13:177–187

Broadley DG, Howell KM (2000) Reptiles. In: Burgess ND, Clarke GP (eds) The coastal forests of Eastern Africa. IUCN, Gland, Switzerland and Cambridge, UK, pp 191–200

Brooks T, Balmford A, Burgess N, Fjeldså J, Hansen LA, Moore J, Rahbek C, Williams P (2001) Toward a blueprint for conservation in Africa. BioScience 51:613–624

Brooks TM, Mittermeier RA, Mittermeier CG, daFonseca GAB, Rylands AB, Konstant WR, Flick P, Pilgrim J, Oldfield S, Magin G, Hilton-Taylor C (2002) Habitat loss and extinction in the hotspots of biodiversity. Conserv Biol 16:909–923

Brooks DR, Van Veller MGP (2003) Critique of parsimony analysis of endemicity as a method of historical biogeography. J Biogeogr 30:819–825

Brown J, Kodric-Brown A (1977) Turnover rates in insular biogeography: effect of immigration on extinction. Ecology 58:445–449

Burgess ND (2000) Global importance and patterns in the distribution of coastal forest species. In: Burgess ND, Clarke GP (eds) The coastal forests of eastern Africa. IUCN, Switzerland and Cambridge, UK, pp 235–248

Burgess ND, Clarke GP (eds) (2000) The coastal forests of Eastern Africa. IUCN, Gland, Switzerland and Cambridge, UK

Burgess ND, Kock D, Cockle A, FitzGibbon CM, Jenkins P, Honess P (2000) Mammals. In: Burgess ND, Clarke GP (eds) The coastal forests of Eastern Africa. IUCN, Gland, Switzerland and Cambridge, UK, pp 173–190

Burgess ND, Clarke GP, Rodgers WA (1998) Coastal forests of Eastern Africa: status, endemism patterns and their potential causes. Biol J Linn Soc 64:337–367

Burgess N, D'Amico Hales J, Underwood E, Dinerstein E, Olson D, Itoua I, Schipper J, Ricketts T, Newman K (2004) Terrestrial ecoregions of Africa and Madagascar a continental assessment. Island Press, Washington DC, pp 550

Cavieres LH, Arroyo MTK, Posadas P, Marticorena C, Matthei O, Rodríguez R, Squeo FA, Arancio G (2002) Identification of priority areas for conservation in an arid zone: application of parsimony analysis of endemicity in the vascular flora of the Antofagasta region, northern Chile. Biodivers Conserv 11:1301–1311

Clarke GP (2000a) Defining the coastal forests. In: Burgess ND, Clarke GP (eds) The coastal forests of Eastern Africa. IUCN, Gland, Switzerland and Cambridge, UK, pp 9–26

Clarke GP (2000b) Climate and climate history. In: Burgess ND, Clarke GP (eds) The coastal forests of Eastern Africa. IUCN, Gland, Switzerland and Cambridge, UK, pp 47–68

Clarke GP, Karoma NJ (2000) History of anthropic disturbance. In: Burgess ND, Clarke GP (eds) The coastal forests of Eastern Africa. IUCN, Gland, Switzerland and Cambridge, UK, pp 251–262

Cornelius C, Cofré H, Marquet PA (2000) Effects of habitat fragmentation on bird species in a relict temperate forests in semiarid Chile. Conserv Biol 14:534–543

de Klerk HM, Crowe TM, Fjeldså J, Burgess ND (2002a) Patterns of species richness and narrow endemism of terrestrial birds species in the Afrotropical region. J Zool 256:327–342

de Klerk HM, Crowe TM, Fjeldså J, Burgess ND (2002b) Biogeographical patterns of endemic terrestrial Afrotropical birds. Divers Distrib 8:147–162

deMenocal PB (1995) Plio-pleistocene African climate. Science 270:53–59

Diamond AW (1981) The continuum of insularity: the relevance of equilibrium theory to the conservation of ecological islands. An overview of the fourth East African wildlife symposium. Afr J Ecol 19:209–212

Fjeldså J, Lovett JC (1997) Geographical patterns of young and old species in African forest biota: the significance of specific montane areas as evolutionary centres. Biodivers Conserv 6:325–346

Fjeldså J, Ehrlich D, Lambin E, Prins E (1997) Are biodiversity 'hotspots' correlated with current ecoclimatic stability? A pilot study using the NOAA-AVHRR remote sensing data. Biodivers Conserv 6:401–422

Foufopoulos J, Ives AR (1999) Reptile extinctions on land-bridge islands: life-history attributes and vulnerability to extinction. Am Nat 153:1–25

García-Barros E, Gurrea P, Luciáñez MJ, Cano JM, Munguira ML, Moreno JC, Sainz H, Sanz MJ, Simón JC (2002) Parsimony analysis of endemicity and its application to animal and plant geographical distributions in the Ibero-Balearic region (western Mediterranean). J Biogeogr 29:109–124

Gascon C, Lovejoy TE, Bierrgaard RO, Malcolm JR, Stouffer PC, Vasconcellos HL, Laurance WF, Zimmerman B, Toucher M, Borges S (1999) Matrix habitat and species richness in tropical forest remnants. Biol Conserv 91:223–229

Hanski I, Thomas CD (1994) Metapopulation dynamics and conservation: a spatially explicit model applied to butterflies. Biol Conserv 68:167–180

Hawthorne WD (1993) East African coastal forest botany. In: Lovett JC, Wasser SK (eds) Biogeography and ecology of the rain forests of Eastern Africa. Cambridge University Press, Great Britain, Cambridge, pp 57–99

Jetz W, Rahbek C (2002) Geographic range size and determinants of avian species richness. Science 297:1548–1551

Jetz W, Rahbek C, Colwell RK (2004) The coincidence of rarity and richness and the potential signature of history in centers of endemism. Ecol Lett 7:1180–1191

Kearney M (2002) Fragmentary taxa, missing data, and ambiguity: mistaken assumptions and conclusions. Syst Biol 51:369–381

Kingdon J, Howell KM (1993) Mammals in the forests of eastern Africa. In: Lovett JC, Wasser SK (eds) Biogeography and ecology of the rainforests of eastern Africa. Cambridge University Press, Cambridge, UK, pp 338–331

Krebs CJ (1999) Ecological methodology, 2nd edn. Addison-Wesley Educational Publishers, California

Lomolino MV (2000) A call for a new paradigm of island biogeography. Global Ecol Biogeogr 9:1–6

Lomolino MV, Brown JH, Davis R (1989) Island biogeography of montane forest mammals in the American Southwest. Ecology 70: 180–194

Lowe AJ, Clarke GP (2000) Vegetation structure. In: Burgess ND, Clarke GP (eds) The coastal forests of Eastern Africa. IUCN, Gland, Switzerland and Cambridge, UK, pp 103–113

MacArthur RH, Wilson EO (1967) The theory of island biogeography. Princeton University Press, Princeton

McCune B, Mefford MJ (1997) Multivariate analysis of ecological data, version 3.19, MjM Software, Gleneden Beach, Oregon, USA

McDowall RM (2004) What biogeography is: a place for process. J Biogeogr 31:345–351

Mlingwa COF, Waiyaki EM, Bennun LA, Burgess ND (2000) Birds. In: Burgess ND, Clarke GP (eds) The coastal forests of Eastern Africa. IUCN, Gland, Switzerland and Cambridge, UK, pp 149–172

Morand S (2000) Geographic distance and the role of island area and habitat diversity in the species area relationships of four Lesser Antillean faunal groups: a complementary note to Ricklefs and Lovette. J Anim Ecol 69:1117–1119

Moreau RE (1933) Pleistocene climatic changes and the distribution of life in East Africa. J Ecol 21:415–435

Morrone JJ (1994) On the identification of areas of endemism. Syst Zool 43:438–441

Myers N, Mittermeier RA, Mittermeier CG, da Fonseca GAB, Kent J (2000) Biodiversity hotspots for conservation priorities. Nature 403:853–858

Nekola JC, White PS (1999) The distance decay of similarity in biogeography and ecology. J Biogeogr 26:86–878

Newmark WD (2002) Conserving biodiversity in East African forests. Springer, Berlin

Page RDM (1998) GeneTree: comparing gene and species phylogenies using reconciled trees. Bioinformatics 14:819–820

Patterson BD, Atmar W (1986) Nested subsets and the structure of insular mammalian faunas and archipelagos. Biol J Linn Soc 28:65–82

Patterson BD, Brown JH (1991) Regionally nested patterns of species composition in granivorous rodent assemblages. J Biogeogr 18:395–402

Power DM (1975) Similarity among avifaunas of the Galápagos islands. Ecology 56:616–626

Ricklefs RE, Schluter D (1993) Species diversity: regional and historical influences. In: Ricklefs RE, Schluter D (eds) Species diversity in ecological communities. Chicago University Press, Chicago, USA, pp 350–363

Ricklefs RE, Lovette IJ (1999) The roles of island area *per se* and habitat diversity in the species area relationships of four Lesser Antillean faunal groups. J Anim Ecol 68:1142–1160

Rodgers WA (2000) Why a book on coastal forests? In: Burgess ND, Clarke GP (eds) The coastal forests of Eastern Africa. IUCN Gland, Switzerland and Cambridge, UK, pp 3–8

Rosen BR (1988) From fossils to earth history: applied historical biogeography. In: Myers AA, Giller PS (eds) Analytical biogeography: an integrated approach to the study of animal and plant distributions. Chapman and Hall, London, pp 437–481

Rosenzweig ML (1995) Species diversity in space and time. Cambridge University Press, New York
Stuart SN (1981) A comparison of the avifaunas of seven East African forest islands. Afr J Ecol 19:133–152
Swafford DL (2002) PAUP*: phylogenetic analysis using parsimony (*and other methods). Sinauer Associates, Sunderland, MA, USA
Trejo-Torres JC, Ackerman JD (2001) Biogeography of the Antilles based on a parsimony analysis of orchid distributions. J Biogeogr 28:775–794
Trejo-Torres JC, Ackerman JD (2002) Composition patterns of Caribbean limestone forests: are parsimony, classification, and ordination analyses congruent? Biotropica 34:502–515
Van Tongeren OFR (1995) Cluster analysis. In: Jongman RHG, TerBraak CJF, van Tongeren OFR (eds) Data analysis in community and landscape ecology. Pudoc Wageningen, Netherlands, pp 174–212
Watson DM (2002) A conceptual framework for studying species composition in fragments, islands and other patchy ecosystems. J Biogeogr 29:823–834
Welter-Schultes FW, Williams MR (1999) History, island area and habitat availability determine land snail species richness of Aegean islands. J Biogeogr 26:239–249
Whittaker RJ (1998) Island biogeography: ecology, evolution and conservation. Oxford University Press, Oxford
WWF-EARPO (2002) Eastern African coastal forest program. Younge A, Negussie G, Burgess N (Authors). Regional workshop report. Nairobi, 4–7 February 2002, Nairobi, Kenya

Biodivers Conserv (2007) 16:913–930
DOI 10.1007/s10531-006-9087-9

ORIGINAL PAPER

Sooty Shearwater (*Puffinus griseus*) on Guafo Island: the largest seabird colony in the world?

Ronnie Reyes-Arriagada · Paulo Campos-Ellwanger · Roberto P. Schlatter · Cheryl Baduini

Received: 2 November 2005 / Accepted: 5 July 2006 / Published online: 29 November 2006

Abstract Sooty Shearwater (*Puffinus griseus*) is the most common Procellariiform seabird along the south-eastern South American coast. In recent years the wintering population off California has declined noticeably. This decline has been confirmed on the breeding grounds in New Zealand. In Chile, knowledge of the population is limited. Investigations on Isla Guafo were carried out during two seasons (03/04 and 04/05), beginning an ongoing monitoring and evaluation of the breeding population of this species in southern South America. On Isla Guafo we estimate a population of about 4 million birds that nest above 150 m above sea level (m.a.s.l.) under a forest without understory. Sooty Shearwaters on Isla Guafo prefer north and western slopes that we hypothesise protect them from the predominantly southerly winds. Population of the species from New Zealand and Australia are compared with the Isla Guafo population and the state of its conservation is discussed.

Keywords *Puffinus griseus* · Guafo Island · South America · Conservation · Habitat

Introduction

The Sooty Shearwater (*Puffinus griseus*) (Procellaridae) is a medium-sized seabird that breeds in burrows on islands in southern South America, Australia and New Zealand. It is the most common Procellariiform seabird in south-eastern South American and New Zealand. It nests between October and May, and migrates to the northern hemisphere during the austral winter (Watson et al. 1971; Warham et al. 1982; Briggs and Chu 1986; Cooper et al. 1991; Spear and Ainley 1999).

R. Reyes-Arriagada (✉) · P. Campos-Ellwanger · R. P. Schlatter
Instituto de Zoología, Casilla 567 Universidad Austral de Chile, Valdivia, Chile
e-mail: ronniereyes@uach.cl

C. Baduini
Joint Science Department, The Claremont College, Keck Science Center, 925 North Mills Avenue, Claremont, CA 91711, USA

Significant declines in the breeding population of this species have recently been reported (Hamilton et al. 1997; Scofield 2001). Similar declines have been recorded during their migration (Scofield and Christie 2002) and on the wintering grounds (Veit et al. 1996; Oedekoven et al. 2001; Hyrenbach and Veit 2003). This species is abundant in the coastal waters surrounding the Chilean Fjords and offshore along in the Cape Horn current during the breeding season and is found throughout the Chilean-Peruvian (Humboldt) current during the post-nuptial migratory period (March–July) (Spear and Ainley 1999). Nevertheless, the numerical and conservation status of these populations is not yet known, in contrast to what is known about populations in New Zealand (Richdale 1963; Hamilton et al. 1997; Jones 2000; Cruz et al. 2001; Uhlman 2003).

Indeed the state of knowledge about seabird populations in southern Chilean (42°–57° S) is precarious. Zoogeographically it is a complex area that can be divided into 2 zones: (1) the fiords and channels and (2) the outer coastal line (Venegas and Sielfeld 1979). The area is divided by the 10°C mean annual isotherm and lies within the temperate and cold sub-zones in Sub-Antarctic seabird life zones. It is also included in the Magellanic district of the Atlantic Province for the Sub-Antarctic region (Watson et al. 1971).

Some basic survey of seabirds have been carried out in this area (Brown et al. 1975; Devillers and Terschuren 1978) but in general, studies on natural or population biology of seabirds between 42° and 51° S are scarce. Although there are several National Parks that have been established for over three decades and cover large areas of terrestrial habitat this is still rather isolated region. The island of Chiloé is of special zoogeographical interest, and is strongly populated. The development of aquaculture affects coastal waters as well as those in the fiords or estuaries (Schlatter 1984).

The objective of this study is to review the state of knowledge of Sooty Shearwaters world wide and update it with new information for the South American populations based on a study of the large Isla Guafo population, which is only the third documented colony in South American waters (Scofield 2001; Lawton et al. 2006).

Methods

Area of study

We surveyed Isla Guafo between December and March during 2003/2004 and 2004/2005. Isla Guafo is in Chile's region X at 43°61′ S; 74°75′ W, of the south-west corner of Isla Grande de Chiloé at the mouth of Golfo de Corcovado (Fig. 1). The colony begins above 150 m above sea level (m.a.s.l.) and burrows are located along the steep slopes formed by numerous mountain ranges. These slopes predominate in the area and flat land is almost absent. Forest species such as *Drimys winteri* (canelo), *Rhaphithamnus spinosus* (arrayán macho) and *Amomyrtus luma* (luma) dominate, whilst shrubs are scarce and patchy, dominated by *Myrceugenia planipes* (pitrilla). Where shrubs dominate, the vegetation density is high preventing burrow building. The list of plant species and their abundance in colonies is shown in Table 1. Burrows are generally excavated in bare ground (defined here as soil with leave cover, no understory and under a canopy with > 80% tree cover).

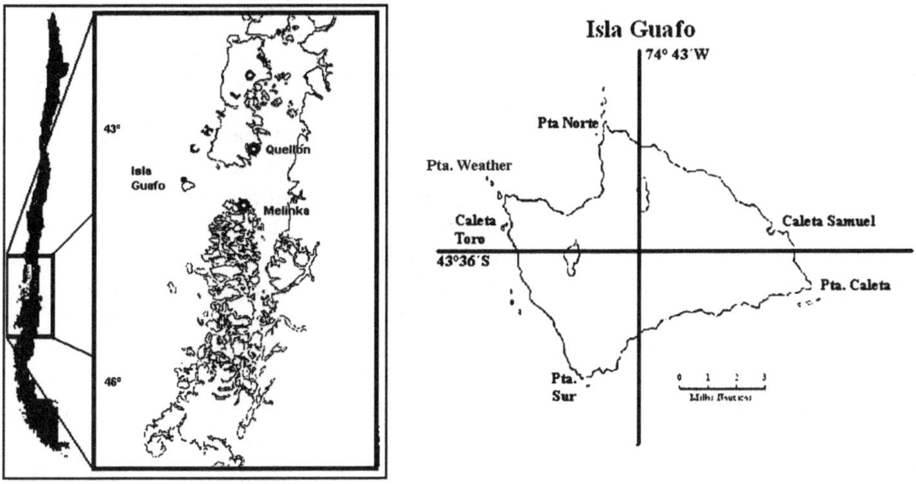

Fig. 1 Map indicating the geographic location of Guafo Island, Chiloé, Chile, highlighting the reference zones of this study

Population

A population estimate was made by calculating the density of occupied nesting burrows in 20,100 m^2 quadrants randomly chosen in the hills between Punta Weather and Caleta Toro (Fig. 3). Not all burrows were active. To discriminate between active and inactive burrows we randomly selected and we marked up to 5

Table 1 List of plant species identified in the working area within the Sooty Shearwater colony, Guafo Island, Chile

Group	Scientific name	Family	Presence
Trees	*Amomyrtus luma*	Mirtaceae	1
	Rhaphithammus spinosus	Verbenaceae	1
	Drymis winteri	Winteraceae	1
	Lomatia ferruginea	Proteaceae	4
Shrubs	*Myrceugenia planipes*	Mirtaceae	1
	Griselina ruscifilia	Cornaceae	3
	Azara lanceolata	Flacourtiaceae	4
Woody climbing plants	*Luzuriaga polliphyla*	Phylesiaceae	2
	Mitraria coccinea	Gesneriaceae	3
	Ercilla syncarpellata	Phytolaccaceae	3
	Hydrangea serratifilia	Hidrangeaceae	3
Epiphytes	*Hymenophyllum secundum*	Hymenopillaceae	1
	H. caudiculatum	Hymenopillaceae	1
	Rigodium implexus	Rigodiaceae	1
	Hypopterygium thouinii	Hypopterygiaceae	2
Ferns	*Megalastrum spectabilis*	Astideaceae	1
	Blechnum magellanicum	Blechnaceae	3
Ground floor	*Uncinia phleoides*	Cyperaceae	3
	Leptostigma arnottianum	Rubiaceae	4

The presence of groups is defined as 1, abundant; 2, upper middle abundance; 3, lower middle abundance; 4, less abundant burrows (if possible) for each of the study quadrants to obtain proportion of quadrant occupation. At the entrance of each burrow, 2–3 toothpicks were loosely placed vertically in the entrance forming a barrier each evening and revisited each morning (Bibby et al. 2000). A burrow was considered occupied if all toothpicks were displaced at least once during the study period of at least 5 consecutive days (the minimum period calculated for shearwaters foraging at sea and returning back to the burrow; Warham 1996). In addition, the number of apparent attempts to build burrows (hereafter referred to as burrow attempts) in each quadrant was recorded. A burrow attempt was defined as an apparently freshly excavated cavity less than 20 cm long generally interrupted by arboreal roots. Rats and feral cats also occur on Guafo. Burrow attempts were attributed to Sooty Shearwaters if they left foot tracks, feathers or guano and if they left large amounts of freshly excavated soil. We believe that burrow attempts are probably made by young or inexperienced males learning to burrow and are loosely indicative of the overall activity of a colony (Warham 1990). It was considered that each burrow attempt was undertaken by a different bird. Approximately 7 ha was surveyed observing presence and burrow attempts in the entire zone, although the quadrate work was limited to 3.2 ha due to the rough island geography. Based on the observations of entrance and exit of birds from the burrows, in the areas of Caleta Toro, Caleta Samuel and NE of Punta Norte (Fig. 1) we estimate the colony is distributed all over the island above 150 m.a.s.l. The total area above the 150 m.a.s.l was calculated by means of the Instituto Geográfico Militar (IGM) chart and AutoCAD™ 2000 software.

Morphometry

Eight morphometric measurements (using vernier callipers) and mass (using Pesola scales) were taken from 113 individual shearwater captured on the surface at night using netting and in burrows. The measurements taken were: mass, bill (from the tip to the beginning of the feathers), total length of the head (from the tip of the beak to the back of the skull), bill depth (at the base of the bill), bill width (at base), tarsus (metatarsus's length), wing (from the edge of the carpal plexus to the tip of the longest primary feather) and body circumference (at the proximal edge of the wing). The weight, wing and body circumference plus the other measurements were taken with an accuracy of 10 g, 1 mm, 0.1 mm, respectively. We compared measurements made during the study with data from the literature using correlation coefficients and mean *t*-tests, respectively.

Use of habitat

Habitat variables were measured using a variety of techniques. Chest height diameter (CHD) of tree trunks was measured using a tree caliper with 1 cm accuracy and trees categorised as thick (> 30 cm), medium (20–30 cm), small (< 20 cm). Site slope was measured using a Sunto™ clinometer with 1° accuracy. Three measurements were made within each quadrant and the average calculated. The aspect (orientation) was measured using a compass to the nearest degrees parallel to the slope. Multiple regression was use to correlate these variables with the number of occupied burrows and burrow attempts allowing us to evaluate the influence of these variables on site use. This also enabled us to estimate overall habitat use. Regression analysis between burrows and burrow attempts enabled us to examine the influence of birds actually occupying burrows to those that were infer are either immature or not currently breeding. The relationship between occupied burrows, burrow attempts was evaluated by means of a contingency table with orientation as a categorical variable using a non-parametric chi square test.

For a random selection of burrow entrances in each quadrate the maximum width, height and diagonal were measured with a tape to 1 mm accuracy. For each of these randomly chosen burrows the direction of orientation was measured to the nearest degree using a Silva^{TM} compass. These four variables were correlated to evaluate degree of protection against adverse weather events such as rain and wind using one way ANOVA with burrow orientation as a predictive variable.

To evaluate the degree of burrow aggregation, the distance between a burrow and the closest neighbour was measured using a measuring tape (1 cm accuracy).

Meteorological variables (rainfall, wind intensity and direction, and temperature) taken the Isla Guafo's weather station in Punta Weather operated by the Armada de Chile were recorded between November 2004 and March 2005. Measurements were carried out every 3 h from midnight. The sequentially sampled No. was 286, 184, 248, 224 and 208, between November and December. The variables considered as dependent were: wind intensity (knots) and temperature (°C). Data were transformed ($\log_{10} x + 1$) to obtain a normal and homocedastic distribution. It was not possible to carry out a variance analysis (ANOVA/MANOVA) of repeated measurements because the expected sphericity (Mauckly's test) was violated in all the variables despite the Greenhouse-Geisser correction (Zar 1999). Consequently an ANOVA/MANOVA for each month was carried out separately. These variables were analysed with the wind direction as a factor (categorised as N, S, W and E) with one way MANOVA if at least two variables correlated significantly and with one way ANOVA when the variables were independent among them (Zar 1999). The a posteriori Tukey's test HSD for unequal N were carried out depending on the case. Wind direction and rainfall (the sum for each month) also categorised as N, S, W and E were compared each month to evaluate their relationships using contingency tables (with chi square).

The statistical analysis was carried out following Zar's methodology (1999) using the STATISTICA 6.0 software. Data from this study was complemented by a literature review of Sooty Shearwater biology in South American waters.

Results

General information

Sooty Shearwaters were first seen near the island at dusk (ca. 18:00 h) flying around the coastline in great flocks. The first individuals arrive at the breeding grounds were after sunset (ca. 21:00 h) with a peak in arrival on the colony between 22:30 h and 23:30 h. The second peak of activity takes place shortly before sunrise (ca. 06:00 h) whilst it is still mostly dark.

The date of egg laying was not ascertained by us, but eggs hatch during the first 10 days of January. Based on an estimate of 56 days incubation for this species (Richdale 1954) it is assumed on Isla Guafo that incubation begins in the first fortnight of November. Most non-breeding adults appear to leave the colony and the surrounding seas before the 20th of March and only chicks were found in burrows examined on 30th March. It is presumed that chicks leave the colony by the end of April and beginning of May as happens in New Zealand (Warham et al. 1982; Sagar and Horning 1998).

Population estimation

The density of burrow attempts and occupation of burrows is shown in Table 2. The burrow density is low and highly variable (0.050 ± 0.049 burrows/m^2). However, the proportion of burrows occupied in Isla Guafo is high (88 ± 30.71%). The burrow attempt density is greater than the burrow density (0.089 ± 0.075 attempts/m^2); and we speculate that the non-breeding and pre-breeding proportion of the population is larger than that of breeding birds. The large variation in the number of mean burrow attempts and mean numbers of occupied burrows is probably due to the low number of quadrants and the clustered distribution of burrow with tight clusters of up to 16 burrows within quadrants (Fig. 2).

The total estimated area above the 150 m.a.s.l. was 84.411 km^2. The population of birds occupying burrows based on the densities in the 20 sampling quadrants was 3,980,000 ± 3,870,000 pairs that were breeding and/or couples starting to reproduce (Fig. 3).

Morphometry

The morphometric measurements for Sooty Shearwaters on Isla Guafo are shown in Table 3. There were no significant differences between the morphometric variables considered among the individuals captured in burrows and those captured on surface therefore, the data was treated as a single unit (t weight = – 0.13, $P = 0.894$; t culmen = 0.519, $P = 0.605$; t head length = 1.675, $P = 0.097$; t peak depth = 0.465, $P = 0.643$; t peak width = 0.502, $P = 0.617$; t circumference = 1.039, $P = 0.302$; t tarsus = 0.328, $P = 0.744$; t wing = 0.259, $P = 0.796$, d.f. = 86). A comparison of means between our results and those reported by Richdale (1963) for Whero Island (46°55′ S), New Zealand was carried out. This is unfortunately the only other work showing standard deviations. *T*-tests of means show significant differences between weight ($t = 63.03$), culmen ($t = 5.73$), peak depth ($t = 213.36$), beak width ($t = 334.94$) and wing ($t = 10.31$); $t_{(2)0.001} = 3.34$, d.f. = 211. All the measures published by Richdale (1963) are greater than the measurement of Isla Guafo Sooty

Table 2 Estimation of density and effective occupation (% and No. of occupied burrows) for burrows of Sooty Shearwater in 100 m^2 quadrants ($n = 20$)

	Mean	SD	Sum
Burrows of Sooty Shearwater	4.95	4.87	99
% occupation	88.00	30.71	
Occupied burrows	4.72	4.59	94.4
Burrows attempts	8.85	7.49	177

The first row shows the mean and SD for number of burrows per quadrant. The second row is the average of % of occupation for the burrows in each quadrant. The results of occupation are extrapolated to the burrow counts carried out in the quadrants, obtaining a third row with the estimation average for the effective occupation of burrows

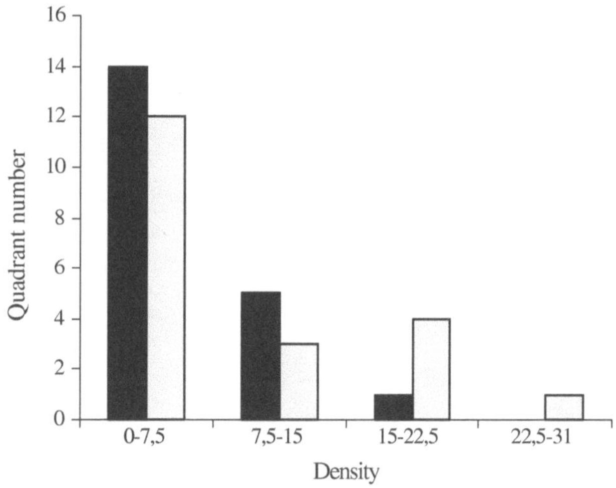

Fig. 2 Distribution of the number of occupied burrows (black bars) and burrow attempts (white bars) per quadrant (n = 20), in four intervals of density

Shearwaters, except the culmen size. Correlation values for the morphometric variables of Sooty Shearwaters from Isla Guafo (Table 4) show weak but significant correlation coefficients among them.

Use of macrohabitat

No significant relationship between the number of occupied burrows and burrow attempts per quadrant were found ($R^2 = 0.103$, $F = 2.056$, d.f. = 1,18, $P = 0.169$)

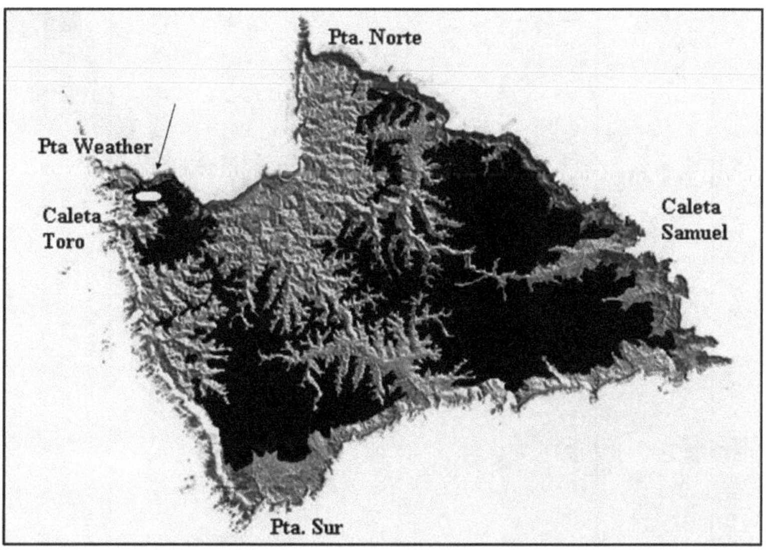

Fig. 3 Nesting Area considered for Sooty Shearwater in Guafo Island, Chile. The black areas represent the total surface above the 150 m.a.s.l. The arrow point out a white ellipse that represents the area with density quadrants used in this study

Table 3 Morphometric measurements for adult of Sooty Shearwater recorded on Guafo Island, Chile, and by observers of different places

Author	Place	Weight (g)	Culmen (mm)	Total head length (mm)	Bill depth (mm)	Bill width (mm)	Tarsus (mm)	Wing (mm)	Wing span (mm)	Circumference (mm)	Tail (mm)	Length (mm)
This work	a	703.23 ± 77.20 (920–520)	42.08 ± 1.72 (49.43–8.80)	98.10 ± 3.15 (111.66–91.14)	12.17 ± 0.71 (14.24–9.72)	8.95 ± 0.95 (11.04–5.05)	56.68 ± 1.52 (60.58–52.53)	29.42 ± 0.59 (30.80–28.30)		31.09 ± 1.21 (34.50–27.30)		
1	b		41 (39–46)				57 (55–66)	293 (270–322)			87 (83–101)	
2	–		42									460
3	c		41 (38–5.6)				55.4 (52.5–59.5)	293 (280–309)			89.4 (84–99.2)	
4	–								940–1040			400–460
5	–								1050			440
6	d								950			450
7	c								1020			450
8	–								1090			400
9	–											400–450
10	e	787 ± 64 (666–978)	41.9 ± 1.6 (38.5–45.5)		14.8 ± 0.6 (13.5–16.25)	13.9 ± 0.6 (12.5–15.25)		30.4 ± 6.9 (287–322)			89.6 ± 2.7 (84–89)	
Chick = 2	a	1110–1320	40.60–43.34	97.85–101.69	11.58–11.68	8.58–8.97	56.27–57.09	23.2–24.4		34.0–35.2		

Measurements done in 2 chicks are also indicated (March 21st). The measurements are shown as mean ± SD (max.–min.). a—Guafo Island, Chile; b—Macquarie island, Australia; c—Chilean coast; d—Argentinean coast; and Whero and Snares Islands, New Zealand. 1—Marchant and Higgins (1990); 2—Falla et al. (1979); 3—Murphy (1936); 4—Harrison (1983); 5—Harrison (1987); 6—Narosky and Yzurieta (2003); 7—Robbins et al. (1966); 8—Araya and Millie (1986); 9—Peterson and Chalyf (1989); 10—Richdale (1963)

Table 4 Correlation coefficients of morphometric measurements for Sooty Shearwater from Guafo Island, Chile, $n = 113$ (*$P < 0.01$)

	Weight	Culmen	Total head length	Bill depth	Bill width	Circumference	Tarsus
Culmen	0.21						
Total head length	0.19	0.47*					
Bill depth	0.12	0.21	0.31*				
Bill width	– 0.00	– 0.02	0.17	0.37*			
Circumference	0.33*	0.19	0.13	– 0.08	– 0.24		
Tarsus	0.08	0.21	0.29*	0.13	0.22	0.03	
Wind	0.06	0.18	0.14	0.07	0.07	– 0.02	0.36*

(Fig. 2). Similarly, no significant relationships was observed between the hill slope inclination and the number of burrows ($R^2 = 0.007$, $F = 0.137$, d.f. = 1,18, $P = 0.715$) and burrow attempts ($R^2 = 0.024$, $F = 0.438$, d.f. = 1,18, $P = 0.516$), although they tend to be located between 19° and 37° of inclination (Fig. 4). The number of occupied burrows and burrow attempts is greater in the areas oriented N and W (Fig. 5). There is no relationship between the chest height diameter of the trees and the number of burrows per quadrant (without tree categories) ($R^2 = 0.04$, $F = 0.749$, d.f. = 1,18, $P = 0.398$) and for each group of dimension (R^2 multiple = 0.173,

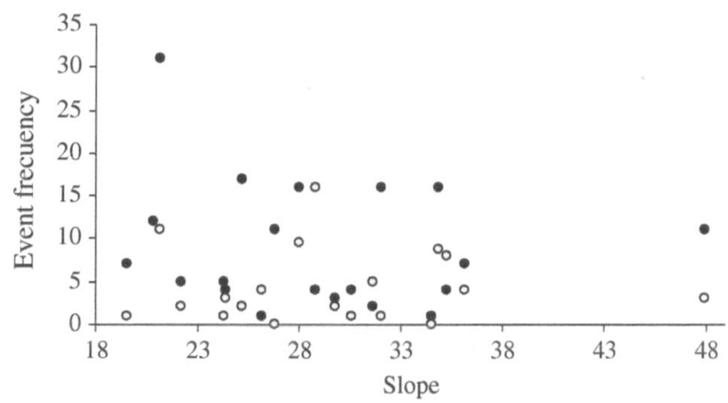

Fig. 4 Relationship between the island's slope inclination and the frequency of occurrence of occupied burrows (white circles) and burrow attempts (black circles)

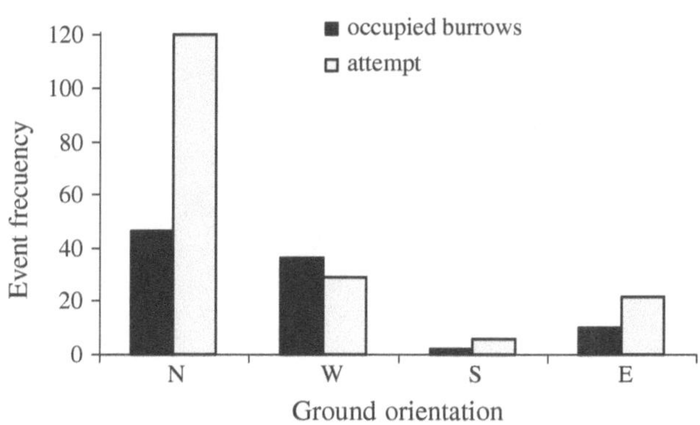

Fig. 5 Occurrence–frequency of occupied burrows and burrow attempts according to exposure of slopes selected as macro-habitat

F = 1.115, d.f. = 3,16, P = 0.372) and neither between the same variables and burrow attempts (R^2 = 0.00, F = 0.086, d.f. = 1,18, P = 0.772 and R^2 multiple = 0.237, F = 1.662, d.f. = 3,16, P = 0.215, respectively). The frequency of occurrence of occupied burrows and attempts are distributed independently of the number of trees in each category (Fig. 6).

Use of microhabitat

The mean height, width and diagonal of burrow entrances was 13.5 ± 4.0, 23.6 ± 7.5 and 22.0 ± 5.8 cm, respectively (n = 78) and there is no correlation between the width and the height (R = 0.086, P = 0.455). However, as expected, we found

Fig. 6 Distribution of the occurrence–frequencies of occupied burrows and attempts, and number of trees for each category at chest height diameter: large (white circle), medium (black circle), small (black square)

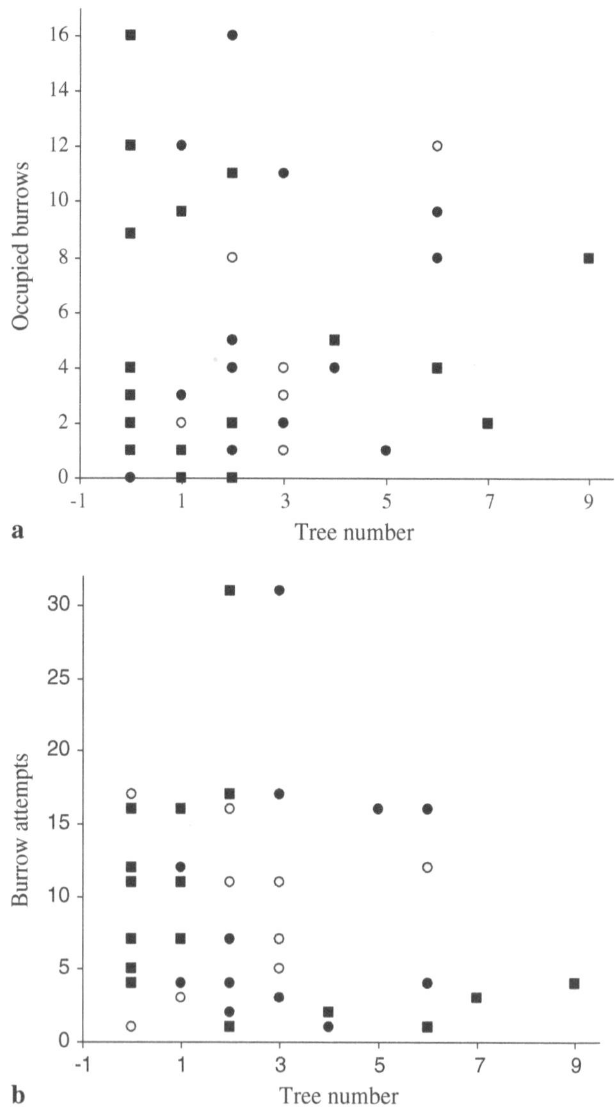

Table 5 One Way ANOVA (model II, 1 tail $P < 0.05$) for the burrow orientation factor with three response variables (height, width, diagonal)

V. S.	SS	d.f.	MS	*F*	*P*
Height	0.006	3	0.002	0.152	0.928
Error	1.038	74	0.014		
Width	0.024	3	0.008	0.512	0.675
Error	1.146	74	0.016		
Diagonal	0.033	3	0.011	0.934	0.429
Error	0.882	74	0.012		

Data were normalised with $\log_{10}(x+1)$

correlation between width, height and the diagonal ($R = 0.242$, $P = 0.03$ for height; $R = 0.689$, $P = 0.000$, for width) because the diagonal is a measurement that is proportional to width and height. There are no significant differences between the burrow exposure and the variables height, width and diagonal (Table 5). There is a tendency of the birds to build their burrows with entrances oriented towards the N and W (65% of the cases) and this conforms to the observed pattern of higher density of occupied burrows with greater slope exposure (Figs. 5 and 7). The burrow aggregations, measured with the distance between burrows show that 38% of the burrows are less than 1.33 m apart and 81% of them were less than 3.9 m apart. This indicates that Sooty Shearwaters show a degree of self-cohesion for the selection and construction of burrows at certain sites (Fig. 8).

Climatic conditions

Monthly rainfalls totalled 53.5, 89.1, 80.9, 27.5, 121.5 mm, between November and March, respectively. Thus rainfall during the study was lowest in November and February and highest in March, mainly during northerly and westerlies winds (Fig. 9). From the total observations for November, January and February almost 50–70% were dominated by southerly winds followed by an alternation of winds from the northerly and westerly directions. During December and March the northerly winds dominated (Fig. 9).

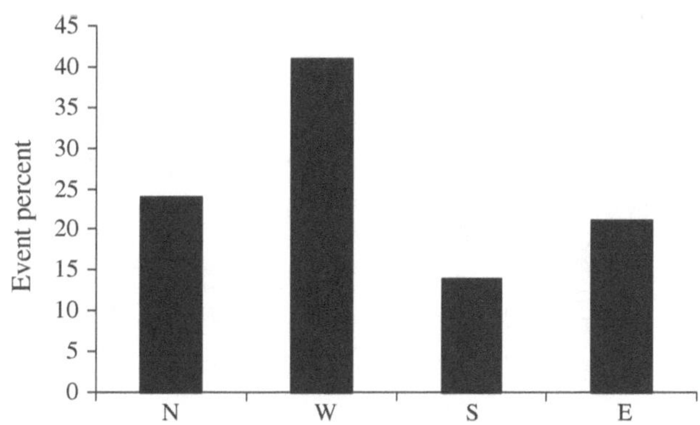

Fig. 7 Percentage of occurrence for the burrow entrances orientation ($n = 78$). *X*-axis indicates the orientation groups: 315–45° (N), 45–135° (E), 135–225° (S), and 225–315° (W)

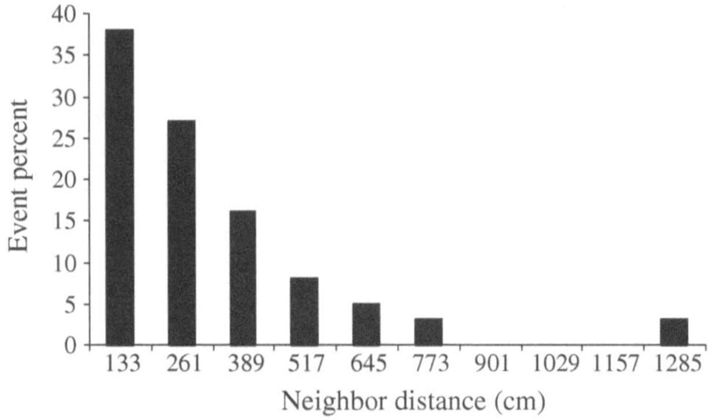

Fig. 8 Distance to the nearest neighbour burrow expressed as % of occurrence (n = 78)

There were significant differences between the variables: wind intensity and temperature related to wind direction during each month (Table 6). There was no significant difference in wind intensity in November and December, but was

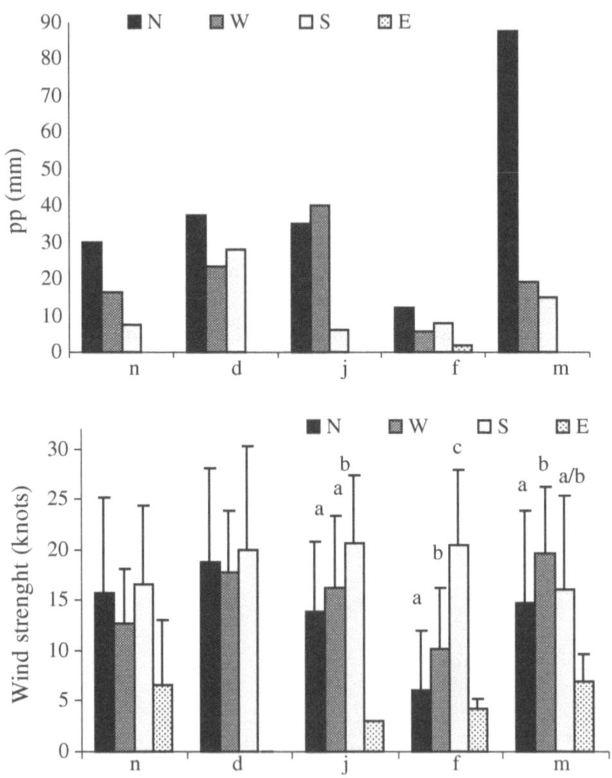

Fig. 9 Rainfall frequencies (mm), winds intensity (knots) and $T°$ (°C) during November and December 2004/2005. The series correspond to N, W, S and E. The series of the other variables correspond to mean and SD. Different letters indicate significant differences ($P < 0.05$) according to Tukey's test HSD for unequal N

Table 6 MANOVA for the wind direction factor (November–March 2004/2005, except December)

Variation source	Pillai's test	d.f. effect	d.f. error	*F*	*P*
O° nov	0.211	4	536	15.83	< 0.00
O° jan	0.237	4	488	16.39	< 0.00
O° feb	0.465	4	434	32.84	< 0.00
O° mar	0.282	4	404	16.58	< 0.00
Variation source	SS	d.f.	MS	*F*	*P*
O° dec (WI)	0.086	2	0.043	0.655	0.521
Error	11.935	181	0.066		
O° dec (*T*°)	92.75	2	46.37	26.18	< 0.00
Error	320.62	181	1.77		

ANOVA was applied for December because the dependent variables were not correlated. Results of Tukey's test HSD a posteriori for N unequal are shown in Fig. 9

significantly different during the three remaining months (Fig. 9). During January and February, the greatest wind intensities came from the south, whereas in March came from the west and south. There were significant differences in the temperature between months and these were lower during southerly winds (Fig. 9).

Discussion

Life cycle

Data indicate that in general the lifecycle of Sooty Shearwater on Isla Guafo was very similar to that on colonies in New Zealand (Richdale 1954, 1963; Marchant and Higgins 1990; Everett and Pitman 1993; Warham 1996; Onley and Bartle 1999; Spear and Ainley 1999), although the date non-breeder depart from the colony in southern Chilean seems be earlier than that on Whero Island (New Zealand; Richdale (1963)). We suggest this may be due to the increase in rainfall from March onwards in this area (Fig. 9).

Population size

Our population estimate for Isla Guafo is significantly greater than that reported by Clark et al. (1984b) of 200,000 individuals. Although the density of burrows in the Isla Guafo colony is quite low (Figs. 2, 3 and 6; Table 2) compared with those in New Zealand (sometime < 2 burrows/m^2; Warham 1990), the population estimate is huge due to the extensive area available for nesting. The heterogeneous nature of the distribution of occupied burrows has caused a large standard deviation and thus wide confidence intervals on our estimate of population size. We found 10% of quadrates had no burrows whilst 70% of the quadrants that had two or more occupied burrows. Furthermore neighbour distance distribution shows that burrows appear to be clustered (Fig. 8).

There is anecdotal evidence that other even larger islands in the southern Chilean Fjords sooty colonies contain populations of Sooty Shearwaters thus the population of this species in South America should be larger than the population reported by Robertson and Bell (1984) for New Zealand. Agreeing with the assumption of Scofield (2001) that Chile may contain the majority of the world's population of

Sooty Shearwaters. Indeed we find that if our estimate is accurate that this is in fact the largest seabird colony in the world. It is rivalled only by the population of Sooty Shearwaters on The Snares (New Zealand) of 2.75 million burrows (Warham and Wilson 1982).

Sooty Shearwaters occur along the coast of New Zealand in a large number of small colonies of between 2 and 2,100 birds on and in large colonies on surrounding islands (Hamilton et al. 1997). It is estimated that there are a total of 20 million pairs in the New Zealand region (Robertson and Bell 1984) Thus it is one of the most abundant seabird species in the Pacific Ocean (Marchant and Higgins 1990). There are also small colonies in southern Australia with approximately 2,000 birds in the largest (Lane and White 1983 and Brothers 1979 cited in Spear and Ainley 1999).

Relationship between populations

Our morphometric data support the finding of Baduini and Warheit (in preparation) that have found that the population of Isla Guafo is genetically different from the populations of New Zealand, Valparaiso (Chile) and Monterrey Bay (Californian Current). We suggest that this difference could explain the Bergman's rule (Scholander 1955) for this species, between populations of Whero Island, New Zealand, and Isla Guafo, Chile.

Factors affected distribution

Activity of the non-breeders (as witnessed by burrowing activity late in the breeding season) appears not be correlated to the density of breeding burrows (Fig. 5) suggesting that non-breeders are not attract by the presence of breeding birds.

Sooty Shearwaters breeding on Guafo prefer steep hill-slopes (Fig. 3). We suggest this may be in response to better drainage due to the heavy fogs and to a lesser extent due to the occasional heavy rainfalls during summer. Landscapes with slopes offer advantages for nesting tubenosed birds (Schramm 1986; Brandt et al. 1995; Catry et al. 2003). The density of birds is not correlated to the presence of a dense forest upperstory. In some cases birds occupy areas with no forest-understory but with a leave covered floor. Alternatively rainfall might have little importance in choosing nesting sites since soils formation is dependent on drainage and soil stability (Warham and Wilson 1982; Jones 2000). Neither collapse nor burrow flooding were observed on the island thus reflecting stability and good drainage of soils.

On Guafo birds tend to locate their burrows on N and W slopes (Fig. 7). During the breeding season winds from the south are frequent and strong bringing dry conditions. During the same period, rainy is more frequent during northerlies and (to a lesser extent) during westerlies (Fig. 9). We suggest therefore that Sooty Shearwaters chose nest site based on there protection against prevalent winds. Exactly why Sooty Shearwaters would make such habitat choices is unclear. Lyver (2002) established that environment in burrows is comparatively uniform due to the isolation from the climatic effects on the surface. Our study suggests that winds influence of burrow location may require further study. The size of burrow entrances did not seem to have been a factor affecting the burrow protection from adverse climatic events as there was no relationship between burrow height width and diagonal in relation to burrow orientation (Table 5).

State of the knowledge for Sooty Shearwater in South America

In South America and particularly in Chile, there is limited information about the populations of this species (Reynolds 1935; Watson et al. 1971; Jehl 1973; Brown et al. 1975; Devillers and Terschuren 1978; Schlatter 1984; Clark et al. 1984a, b, Lawton et al. 2006; Scofield 2001). The first records of nesting without population estimation are from Wollaston and Bayly islands (Reynolds 1935; Watson et al. 1971). Marín (1984) reported 21 burrows with eggs or chicks in the small island Puñihuil, Chiloé, (41°55′ S) located north of Guafo Island. No evidences of nesting has been reported recently there (A. Simeone, personal communication) as a consequence of the increasing tourist activities. Brown et al. (1975) reported that this species is common in Chilean fiords and scarce in the Atlantic Ocean and Drake Passage. It is rarely seen south of the Antarctic Convergence on the South American quadrant as opposed to the New Zealand one (Watson et al. 1971). Jehl (1973) established that Sooty Shearwater is absent in the inner seas during the austral winter. If we compare the abundance distribution of the sooties between the Strait of Magallanes and the central-south zone of Chile (40° S), reported by Jehl (1973) (May), Brown et al. (1975) (March) and Devillers and Terschuren (1978) (January) it is possible to observe an increase in the amount of Sooty Shearwater records towards the north as the summer season moves into autumn-winter. This variation is the result of the migratory process of this species towards the Northern Hemisphere although a small fraction of the population (probably non-breeders) remains in austral waters during winter (Jehl 1973).

Scofield (2001) considers 300,000 birds breed in the Wollaston Islands. Clark et al. (1984b) and Schlatter (personal observations) presume the nesting in Isla Guamblin (45° S) close to and south of Isla Guafo. For the archipelago Diego Ramírez (56°31′ S, 68°44′ W) a population of several thousand pairs (Lawton et al. 2006) have been estimated. Their also occur on Isla Ildefonso (55°50′ S, 69°19′ W) thought no population estimation has been made (Kieran Lawton, personal communication). In the Argentinean Patagonia, there are colonies with about 10,000–20,000 pairs on Las Malvinas (50°59′ S, 61°26′ W) (Leandro Tamini, personal communication). Nesting has been reported on Staten Island east of Tierra del Fuego (54°44′ S, 63°52′ W), by Clark (1986) but more recently discarded by Schiavini et al. (1999). This island needs further survey to confirming Sooty Shearwaters breeding status.

The populations trends of the Sooty Shearwater populations in Chile is unknown. On the Wollaston Islands Scofield (2001) reported a series of introduced predators and sheep that would interact negatively with colonies of shearwaters on these islands. On Guafo Island, the presence of rodents in the colony was evident, during all sample season. Two feral cats was also observed. Most mortality probably mainly by the Crested Caracara *Polyborus plancus*. Thus Sooty Shearwater populations breeding in Chile are under most of the same population pressures than those breeding in New Zealand, where the introduction of predators is considered one of the main causes of the population decreases (Hamilton et al. 1997; Jones 2000; Scofield and Christie 2002; Valenzuela and Grau 2005 for Chile).

Observations suggest Sooty Shearwaters forage mainly inshore and as far as he continental shelf off Guafo. The prey consumed by Sooty in South American waters is still unknown although Cruz et al. (2001) described a wide range of prey for New Zealand sooty populations. The stomach content of two individuals from the Guafo showed fish remains and a small amount of squid.

Uhlman (2003) has summarised incidence of fisheries bycatch of Sooty Shearwater. The interaction of Sooty Shearwaters with fisheries along southern South America is poorly unknown. In Peru there are no records of captures (Jeffrey Mangel, personal communication) and in Argentina, although local fishermen report their capture, data is lacking (Leandro Tamini, personal communication). There are only occasional reports of drowned birds coming ashore south of Concepción (36°30′ S) and near Valdivia (39°40′ S) as a result of the interaction with fisheries using fence nets or artisan fisheries. A very low rate of bycatch (0.005 birds/1,000 hooks) is reported appears with in the *Dissostichus eleginoides* long-line fishery (Moreno et al. 2003).

The Sooty Shearwater has recently been upgraded to "Near Threatened" (IUCN 2006) due to the decrease in the abundance of the species in the California current system (Veit et al. 1996); a decline in populations and colonies in New Zealand (Hamilton et al. 1997). This potentially in the colonies of the austral Chilean archipelagos as a result of the introduction of predators and the interaction with fisheries. In addition, climatic phenomena like El Niño, the global warming, and their relation with the distribution of preys in the ocean, would significantly affect the survival of these birds (Veit et al. 1996; Lyver et al. 1999).

To establish the effect of fisheries on this species in Chile it is essential to evaluate the population size in South America. Some data indicate that trawling fisheries of austral hake (*Merluccius australis*) could be interacting negatively with the populations of this pelagic seabird predator during the reproductive period in Chile (Javier Arata, personal communication). Gillnets have been demonstrated to be dangerous for marine bird species such as the Humboldt penguin (*Spheniscus humboldti*) (Simeone et al. 1999) and pink-footed shearwater (*P. creatopus*) (Schlatter 2004) and thus Sooty Shearwaters may be vulnerable as well. As part of this study we have begun to evaluate the survival and fecundity of this species on Guafo Island as a way to complement further studies on the interaction of Sooty Shearwaters with factors like fisheries and introduced predators.

Acknowledgements The authors wish to thank the Dirección de Extensión, Universidad Austral de Chile (Valdivia, Chile) and The Claremont College (Claremont CA, USA) for partially funding this study and to the Armada de Chile for the logistic support. Thanks are given to Javier Arata, Daniela Guicking, Jeffrey Mangel, Henrik Moller, Kieran Lawton, Philip Lyver, Leandro Tamini, Alejandro Simeone, Paul Scofield and Jorge Valenzuela for providing information. We would also like to thank Paul Scofield for their values comments and reviews on this manuscript. To Cesar Barrales for his help in the field. Ronnie Reyes-Arriagada Ph.D. student of the Ph.D. in Science Programme, on Systematic and Ecology, UACH, is being supported by MECESUP Higher Education Programme (UCO 0214).

References

Araya B, Millie G (1986) Guía de Campo de las Aves de Chile; Séptima Edición. Editorial Universitaria, Santiago de Chile

Bibby CJ, Burgess ND, David AH, Mustoe SH (2000) Bird census techniques, 2nd edn. Academic Press, London, 385 pp

Brandt CA, Parrish JK, Hodges CN (1995) Predictive approaches to habitat quantification: dark-rumped Petrels on Haleakala, Maui. Auk 112:571–579

Briggs KT, Chu EW (1986) Sooty shearwaters off California: distribution, abundance and habitat use. Condor 88:355–364

Brown RGB, Cooke F, Kinnear PK, Mills EL (1975) Summer seabird distributions in Drake Passage, the Chilean fjords and off southern America. Ibis 117:339–356

Catry P, Campos A, Segurado P, Silva M, Strange I (2003) Population census and nesting habitat selection of thin-billed prion *Pachyptila belcheri* on New Island, Falkland Islands. Polar Biol 26:202–207

Clark R (1986) Aves de Tierra del Fuego y Cabo de Hornos, guía de campo. L.O.L.A. Buenos Aires, Argentina, 294 pp

Clark GS, Von Meyer AP, Nelson JW, Watt JN (1984a) Notes on the Sooty shearwater and other avifauna of the Chilean offshore island of Guafo. Notornis 31:225–231

Clark GS, Goodwin AJ, Von Meyer AP (1984b) Extension of known range of some seabirds on the costal of southern Chile. Notornis 31:320–324

Cooper J, Underhill LG, Avery G (1991) Primary moult and transequatorial migration of the sooty shearwater. Condor 93:724–730

Cruz JB, Lalas C, Jillett JB, Kitson JC, Lyver POB, Imber M, Newman JE, Moller H (2001) Prey spectrum of breeding sooty shearwater (*Puffinus griseus*) in New Zealand. New Zeal J Mar Fresh 35:817–829

Devillers P, Terschuren JA (1978) Midsummer seabird distribution in the Chilean fjords. Le Gerfaut 68:577–588

Everett WT, Pitman RL (1993) Status and conservation of shearwater of the North Pacific. In: Vermeer K, Briggs KT, Morgan KH, Siegel-Causey D (eds) The status, ecology, and conservation of marine birds of the North Pacific. Can. Wildl. Serv. Spec. Publ., Ottawa, pp 93–100

Falla RA, Sibson RB, Turbott EG (1979) The new guide to the birds of New Zealand. New Zealand Ornithological Society, Auckland, New Zealand, 247 pp

Hamilton SA, Moller H, Robertson CJR (1997) Distribution of Sooty shearwater (*Puffinus griseus*) breeding colonies along the Otago Coast, New Zealand, with indication of countrywide populations trend. Notornis 44:15–25

Harrison P (1983) Seabirds – an identification guide. Croom Helm, London, Sidney

Harrison P (1987) Seabirds of the world – a photographic guide. Christopher Helm Ltd., London

Hyrenbach KD, Veit RR (2003) Ocean warming and seabird communities of the southern California Current System (1987–98): response at multiple temporal scales. Deep-Sea Res Pt II 50:2537–2565

IUCN (2006) IUCN Red List of threatened species Categories. www.redlist.net

Jehl JR (1973) The distribution of marine birds in Chilean waters in winter. Auk 90:114–135

Jones C (2000) Sooty shearwater (*Puffinus griseus*) breeding colonies on mainland South Island, New Zealand: evidence of decline and predictors of persistence. New Zeal J Zool 27:327–334

Lawton K, Robertson G, Kirkwood R, Valencia J, Schlatter R, Smith D (2006) An estimate of population sizes of burrowing seabirds at the Diego Ramírez archipelago, Chile, using distance sampling and burrow-scoping. Polar Biol 29(3):229–238

Lyver POB (2002) Use of traditional knowledge by Rakiura Maori to guide sooty shearwater harvests. Wildlife Soc B 30:29–40

Lyver POB, Moller H, Thompson C (1999) Changes in the sooty shearwater *Puffinus griseus* chick production and harvest rate precede ENSO events. Mar Ecol Prog Ser 188:237–248

Marchant S, Higgin PJ (eds) (1990) Handbook of Australian, New Zealand, and Antarctic Birds, vol 1. Oxford University Press, Melbourne, 1400 pp

Marín M (1984) Breeding record for the Sooty Shearwater (*Puffinus griseus*) from Chiloé Island, Chile. Auk 101:192

Moreno CA, Hucke-Gaete R, Arata J (2003) Interacción de la pesquería del bacalao de profundidad con mamíferos y aves marinas. Informe final proyecto FIP No. 2001-31. República de Chile, Ministerio de Economía, Fomento y Reconstrucción, Subsecretaría de Pesca, Fondo de Investigación Pesquera,199 pp

Murphy RC (1936) Oceanic birds of South America, vol II. The American Museum of Natural History, New York

Narosky T, Yzurieta D (2003) Aves de Argentina y Uruguay: guía para la identificación: decimoquinta edición. Vazquez Mazzini editores, Buenos Aires Argentina, 364 pp

Oedekoven CS, Ainley DG, Spear LB (2001) Variable responses of seabirds to change in marine climate: California Current, 1985–1994. Mar Ecol Prog Ser 212:265–281

Onley D, Bartle S (1999) Identificación de las aves marinas de los Océanos del Sur. Te Papa Press, Wellington New Zealand, 83 pp

Peterson RT, Chalif EL (1989) Aves de México. Editorial Diana, México, 473 pp

Reynolds PW (1935) Notes of the birds of Cape Horn. Ibis 5:65–101

Richdale LE (1954) Duration of parental attentiveness in the sooty shearwater. Ibis 96:586–600
Richdale LE (1963) Biology of Sooty shearwater *Puffinus griseus*. P Zool Soc Lond 141:1–117
Robbins Ch, Bruun B, Zim H (1966) Birds of North America: a guide to field identification. Goldern Press, New York
Robertson CJR, Bell BD (1984) Seabird status and conservation in the New Zealand region. In: Croxall JP, Evans PGH, Schreiber RW (eds) Status and conservation of the world's seabirds. Inst Counc for Bird Press. Tech. Pub. No. 2, England, pp 573–586
Sagar PM, Horning DS Jr (1998) Mass-related survival of fledging Sooty Shearwaters *Puffinus griseus* at the Snares, New Zealand. Ibis 140:329–339
Schiavini A, Frere E, Yorio P, Parera A (1999) Las aves marinas de la Isla de los Estados, Tierra del Fuego, Argentina: revisión histórica, estado poblacional y problemas de conservación. Ans Inst Pat Punta Arenas (Chile) 27:25–40
Schlatter R (1984) The status and conservation of seabirds in Chile. In: Croxall JP, Evans PGH, Schreiber RW (eds) Status and conservation of the world's seabirds. Inst Counc for Bird Press. Tech. Pub. No. 2, England, pp 261–269
Schlatter R (2004) Fardelas en peligro. Cartas al director Diario El Sur de Concepción, Chile 28/01/04. elsur.cl
Scholander PF (1955) Evolution of climatic adaptation in homeotherms. Evolution 9:15–26
Schramm M (1986) Burrow densities and nest site preferences of petrels (Procellarian) at the Prince Edward Islands. Polar Biol 6:63–70
Scofield P (2001) Status of sooty shearwater, *Puffinus griseus*, in the Wollaston and Hermite groups of southern Chile and concerns for the conservation status of Tierra del Fuego Islands. Unpublished document, 15 pp
Scofield P, Christie D (2002) Beach patrol records indicate a substantial decline in sooty shearwater (*Puffinus griseus*) numbers. Notornis 49:158–165
Simeone A, Bernal M, Meza J (1999) Incidental mortality of Humboldt penguins *Spheniscus humboldti* in gill nets, central Chile. Mar Ornithol 27:157–161
Spear LB, Ainley DG (1999) Migration routes of sooty shearwater in the Pacific Ocean. Condor 101:205–218
Uhlman S (2003) Fisheries bycatch mortalities of sooty shearwaters (*Puffinus griseus*) and short-tailed shearwaters (*P. tenuirostris*). Doc. Science Internal Series 92. Department of Conservation, Wellington, 52 pp
Valenzuela JA, Grau JH (2005) Ocurrence of American mink on the Chonos Archipelago of southern Chile. Oryx 39:15
Veit RR, Pyle P, Mcgowan A (1996) Oceanic warming and long-term change in pelagic bird abundance within the California current system. Mar Ecol Prog Ser 139:11–18
Venegas C, Sielfeld W (1979) Antecedentes para la determinación de un nuevo distrito zoo-geográfico en el litoral exterior de Magallanes. Ans Inst Pat Punta Arenas (Chile) 10:201–208
Warham J (1990) The Petrels, their ecology and breeding systems. Academic Press Limited, United States, 440 pp
Warham J (1996) The behavior, population biology and physiology of the petrels. Academic Press, New York, 611 pp
Warham J, Wilson GJ (1982) The size of the Sooty shearwater population at the Snares islands; New Zealand. Notornis 29:23–30
Warham J, Wilson GJ, Keeley BR (1982) The annual cycle of Sooty shearwater *Puffinus griseus* at the Snares Islands, New Zealand. Notornis 29:269–292
Watson GE, Angle JP, Harper PC, Bridge MA, Schlatter RP, Tickell WLN, Boyd JC, Boyd MM (1971) Birds of the Antarctic and Subantarctic. Antarctic Map Folio Series 14. American geographical Society, NY
Zar JH (1999) Biostatistical analysis. Prentice Hall In., Upper Saddle River, NJ

Biodivers Conserv (2007) 16:931–952
DOI 10.1007/s10531-006-9033-x

ORIGINAL PAPER

Mexico in the international reptile skin trade: a case study

Inés Arroyo-Quiroz · Ramón Pérez-Gil · Nigel Leader-Williams

Received: 27 January 2006 / Accepted: 29 March 2006 / Published online: 12 May 2006

Abstract This paper examines the role of Mexico as importer, manufacturer, producer and distributor centre of reptile skins from non-native and native species, through a combination of documentary research and survey methods. A number of key findings were derived from this study. Although Mexico has adopted the "System for the Conservation, Management and Sustainable Use of Wildlife" (SUMA), the country still relies on reptile skins from non-native species. In contrast, the smaller numbers of skins used from native species mainly derive from captive breeding schemes that although biologically sustainable, provide no incentive for habitat conservation. Sustainable use of reptile skins from native species could positively encourage conservation in Mexico. However, as a megadiverse country with potential to produce wildlife, Mexico will have to implement an appropriate regulatory framework to support local communities to promote the sustainable use of native species.

Keywords Conservation · Mexico · Reptile skin · Sustainable use · Wildlife trade

Introduction

The trade in reptile skins for the leather market is one of the most important aspects of international trade in wildlife. Millions of reptiles are killed, processed and

I. Arroyo-Quiroz · N. Leader-Williams
Durrell Institute of Conservation and Ecology, Department of Anthropology, University of Kent, Canterbury CT2 7NS, UK

I. Arroyo-Quiroz
Centro de Investigaciones en Ecosistemas, UNAM, Morelia, México

R. Pérez-Gil · I. Arroyo-Quiroz (✉)
FAUNAM AC, José Maria Velasco 109 Local 8a, Col. San José Insurgentes, CP 03900 México, DF, México
e-mail: diceines@yahoo.com.mx

manufactured into products every year for the international reptile skin trade (Jenkins and Broad 1994; Hemley 1994; Thorbjarnarson 1999; Hutton and Webb 2003; Broad et al. 2003; TRAFFIC 2006; Zhou and Jiang 2004).

The reptile skin trade flows mainly from tropical or subtropical countries rich in biodiversity to affluent countries or regions which drive the demand, particularly the European Community, the United States (US) and Japan (Jenkins and Broad 1994; Reeve 2002; Roe et al. 2002). Some countries however, such as China, are both consumers and range states (Zhou and Jiang 2004). Among these, the US has been undoubtedly the largest consumer market of reptile skins and reptile skin products in the world (Gaski 1992; Jenkins and Broad 1994; Fleming 1999).

Mexico has been among the major players in the world trade in reptile skins and has played a considerable role as an importer and re-exporter of reptile skins from non-native species and as an exporter of reptile skins from native species (Jenkins and Broad 1994; Pérez-Gil et al. 1996; Fleming 1999). Mexico has had a thriving leather industry that uses native and non-native reptile skins to meet the demand in the US, Europe and Asia for manufactured leather products, at which Mexico has excelled (Pérez-Gil et al. 1996; INE 2000a; Rodríguez 1985a, b; Rose 1991, 1992a, b; Brautigam 1986a, b; Iglesias 1998; Calleja 1994; Bancomext 1994, 1999, 2000, 2001, 2006).

The aim of this paper is to examine the legal use of native and non-native reptiles in the skin trade in Mexico through the analysis of trade data and to investigate how Mexico is promoting sustainable management schemes for reptile species and their contribution to the production and commercialisation of skins.

Data source

The World Conservation Monitoring Centre (WCMC) CITES Trade Database was employed to analyze imports, exports and re-exports of reptile skins and skin products from non-native and native species. The trade records compiled were:

(a) All Mexican imports and re-exports from 1980 to 2001 for the following non-native specific genera and species: *Varanus salvator*, *Tupinambis* spp., *Python reticulatus*, *Caiman* spp., *Varanus niloticus*, *Alligator mississippiensis*, and Crocodylidae.
(b) All Mexican exports from 1980 to 2001 for the following native specific genera and species: *Caiman* spp., *Crocodylus* spp., *Boa constrictor*, *Iguana iguana*, *Chelonia* spp., *Lepidochelys* spp., *Caretta caretta*, and *Eretmochelys imbricata*.

Because Mexico only joined CITES in 1991, net trade figures from the CITES database could not be used. Instead, data on Mexican imports were those compiled from exporting countries as imported to Mexico. Likewise, data on Mexican exports and re-exports were those compiled by importing countries as exported or re-exported from Mexico.

The raw data from the CITES Trade Database was modified in order to facilitate comparisons (see Arroyo-Quiroz (2003) for detailed explanation on data). Trade expressed in terms of pounds were converted to kilograms. A correction factor was also derived in order to express volumes of skin weights as volumes of whole skins

Table 1 Average weight per skin from reptile species used to convert skin weight (kg) into numbers of whole skins

Order	Species	Number of whole skins weighed	Total weight of skins (kg)	Average weight per skin (kg)
Lizards	*Varanus salvator*	654	50.97	0.07
	Tupinambis spp.	92	6.51	0.07
Snakes	*Python reticulatus*	105	28.29	0.26
Crocodilians	*Caiman crocodilus fuscus*	5,440	3,574.40	0.65
	Crocodylidae	131	193.50	1.47

(see also Jenkins and Broad 1994). This correction factor was derived from weighting whole reptile skins from imported shipments arriving at the Mexico City International Airport during June and July, 2003. It was not possible to weigh skins individually, because they were shipped in batches. Therefore the numbers of skins and the total weights for each batch were recorded, and from summing the total numbers of skins from, and the total weights of, all batches, the average weight per skin was calculated (Table 1).

The full range of data available from the CITES Trade Database was too large to represent all fields graphically. Hence, only the most important fields were selected for graphical illustration. The fields most often selected comprised the numbers of whole skins, skin pieces and skin products, because these predominate over all other traded volumes. "Event arrows" were added to mark specific events, such as trade restrictions, that came into effect at given times, in order to help with interpretation of the results.

One key problem with using data from the CITES Trade database occurs if important countries have not reported in certain years (Harris et al. 2003). In order to determine whether there were any problems with incomplete annual reporting for the Mexican datasets presented in this study, the submission of annual reports by the major reptile skin exporting countries was reviewed.

This study also uses data on volumes of various species and genera in trade stored in the LEMIS Trade Database (Law Enforcement Management Information System) held at the US Fish and Wildlife Service. The trade records compiled were for all imports to the US from 1995 to 1999 for the following native specific genera and species: *Caiman* spp., *Crotalus* spp., *Iguana* spp., *Crocodylus* spp., and *Boa constrictor* (see Arroyo-Quiroz (2003) for detailed explanation on data). For graphical illustration the fields selected comprised skin products, because these predominate over all other traded volumes.

Mexican policies on the use of reptile skins from native species were studied through documentary research, survey and semi-structured interviews methods.

Results

CITES annual reporting on trade transactions by the Parties is often complete for any given year (Harris et al. 2003). From 1991–2001, most of the major reptile skin exporting countries that supplied Mexico submitted their annual reports (Table 2). This allows confidence in the following results that seek to interpret trade patterns in reptile skins imported to Mexico.

Table 2 Checklist of annual report submission by the top 11 exporters from 1991–2001 (As of August 2003)

Species	Country	1991	1992	1993	1994	1995	1996	1997	1998	1999	2000	2001
Varanus salvator	Indonesia											
	Malaysia											
Tupinambis spp.	Argentina											
	Paraguay											■
Caiman spp.	Colombia											
Python reticulatus	Indonesia											
	Thailand											
Varanus niloticus	Sudan			■					■		■	■
	Chad							■	■			
A. mississippiensis	USA											
Crocodylidae	Zimbabwe											
	South Africa											

Source: CITES Secretariat (2003)

■ Report Not Submitted

□ Report Submitted

Legal reptile skin imports

Total reptile skin imports 1980–2001

From 1980 to 2001, Mexico imported nearly 9 million reptile skin items (Table 3). These derived from various non-native species of lizard, snake and crocodilian. Numbers of whole skin imports totalled some 5.5 million items (Table 3) and were predominantly of lizards, *Varanus salvator* and *Tupinambis* spp., with lesser numbers of crocodilians, *Caiman* spp., and of snakes, *Python reticulatus* (Fig. 1a). Numbers of skin pieces imports totalled some 3.1 million items (Table 3), and also were predominantly of lizards, *Tupinambis* spp. and *Varanus salvator*, with greater representation of snakes, *Python reticulatus*, and less representation of crocodilians, *Caiman* spp. and *Alligator mississippiensis* (Fig. 1b). Weights of whole skins and skin pieces totalled some 27,000 kg (Table 3). In contrast to records of skin numbers, skin weights were predominantly for crocodilians, *Caiman* spp. and *Alligator mississippiensis* (Fig. 1c).

Table 3 Mexican imports of reptile skins from non-native species 1980–2001

Species	Number of Whole Skins	Number of Skin Pieces	Skin weight (kg)
Varanus salvator	2,181,208	310,617	1,735
Tupinambis spp.	1,760,926	1,591,977	7,236
Caiman spp.	791,701	103,382	12,265
Python reticulatus	580,337	1,003,513	4,793
Varanus niloticus	99,878	–	–
Alligator mississippiensis	59,308	103,013	1,325
Crocodylidae	43,635	4,354	35
Total	5,516,993	3,116,856	27,389

Source: WCMC Trade Database

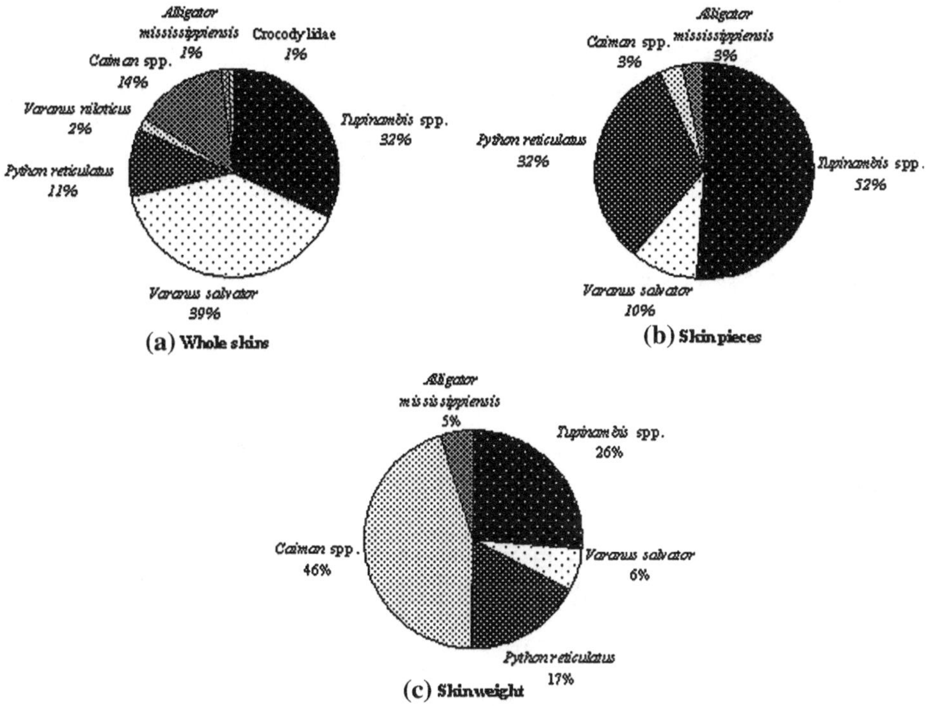

Fig. 1 (**a–c**) Imports of reptile skins by Mexico 1980–2001 (CITES Trade Data)

Total reptile skin imports by year

From 1980 to the early 1990s, Mexico imported increasing quantities of reptile skins (Fig. 2a–c). The numbers of whole reptile skins imported has continued to rise and remained at high levels throughout the 1990s and into the early 2000s (Fig. 2a). In terms of species, *Tupinambis* spp. made up the majority of Mexico's reptile skin imports during the 1980s, whether considered in absolute or proportional terms (Figs. 2 a–c, 3 a–c). From the early 1990s, the numbers of whole skin imports came to be dominated by *Varanus salvator*, whether considered in absolute or proportional terms (Figs.2a, 3a).

In terms of key events in Mexico, imports of reptile skins from non-native species increased steadily after Mexico banned the international trade on native species in 1982 (Fig. 2). Following Mexico's accession to CITES in 1991, imports of whole reptile skins from non-native species continued to rise (Fig. 2a). Even following Mexico's adoption of sustainable resource use policies in 1997, imports of whole reptile skins from non-native species continued to rise (Fig. 2a).

Overall reptile skins by year

From 1980 to 2001, numbers of skin imports for *Varanus salvator*, *Tupinambis* spp., *Caiman* spp., *Python reticulatus* and Crocodylidae considered as whole skins totalled some 5.3 million items (Table 4). In contrast, numbers of skin imports considered as whole skins, skin pieces and skin weights (kg) (overall skins), totalled some 6.8 million

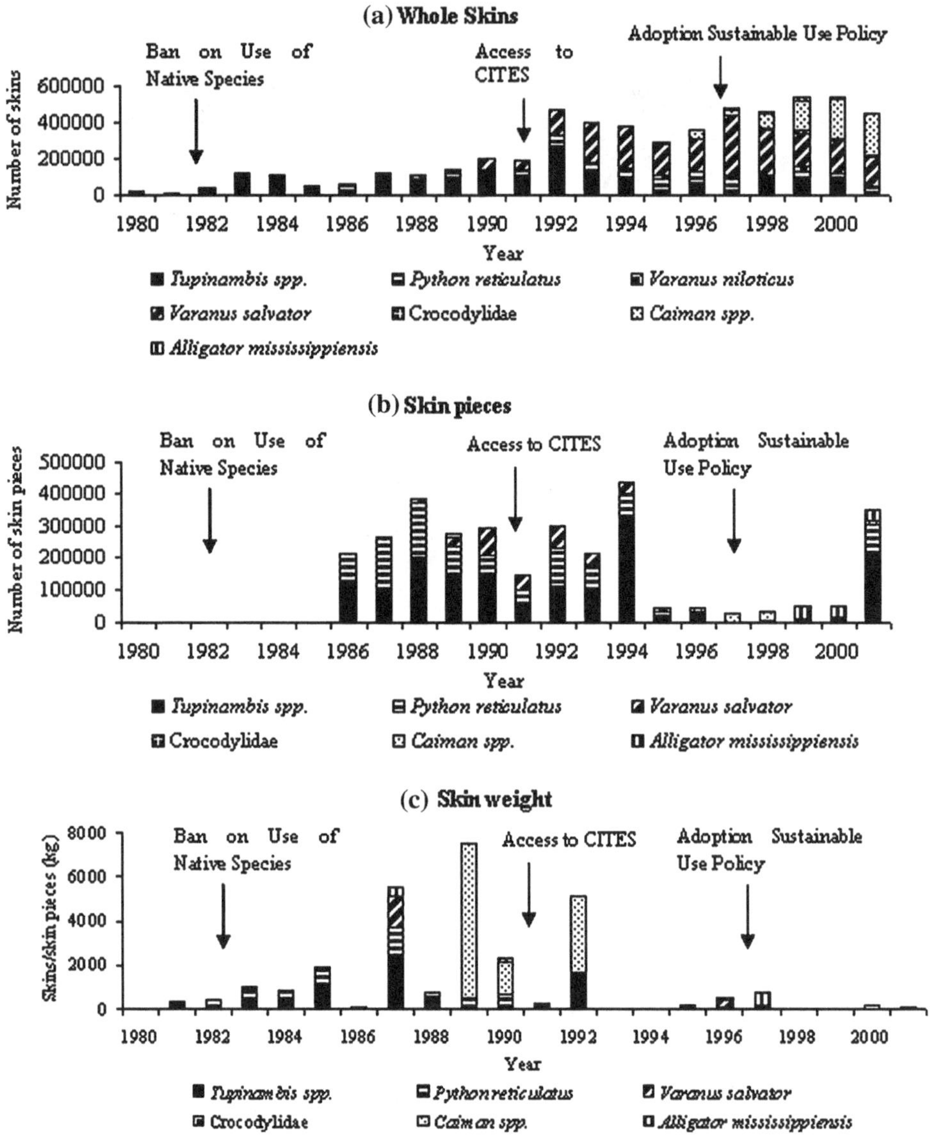

Fig. 2 (**a–c**) Imports of reptile skins by Mexico 1980–2001 (CITES Trade Data)

items during the same period (Table 4). For example, from 1980 to 2001, *Tupinambis* spp. overall skin imports totalled some 2.5 million items (Table 4), were numbers of whole skins totalled some 1.7 million (68%) and numbers of skin pieces and skin weight (kg) equalled nearly 0.8 million whole skins (32%) (Table 4, Fig. 4a). From 1980 to 2001, *Python reticulatus* overall skin imports totalled some 1 million items (Table 4), were numbers of whole skins totalled some 0.5 million (50%) and numbers of skin pieces and skin weight (kg) equalled around 0.5 million whole skins (50%) (Table 4, Fig. 4b).

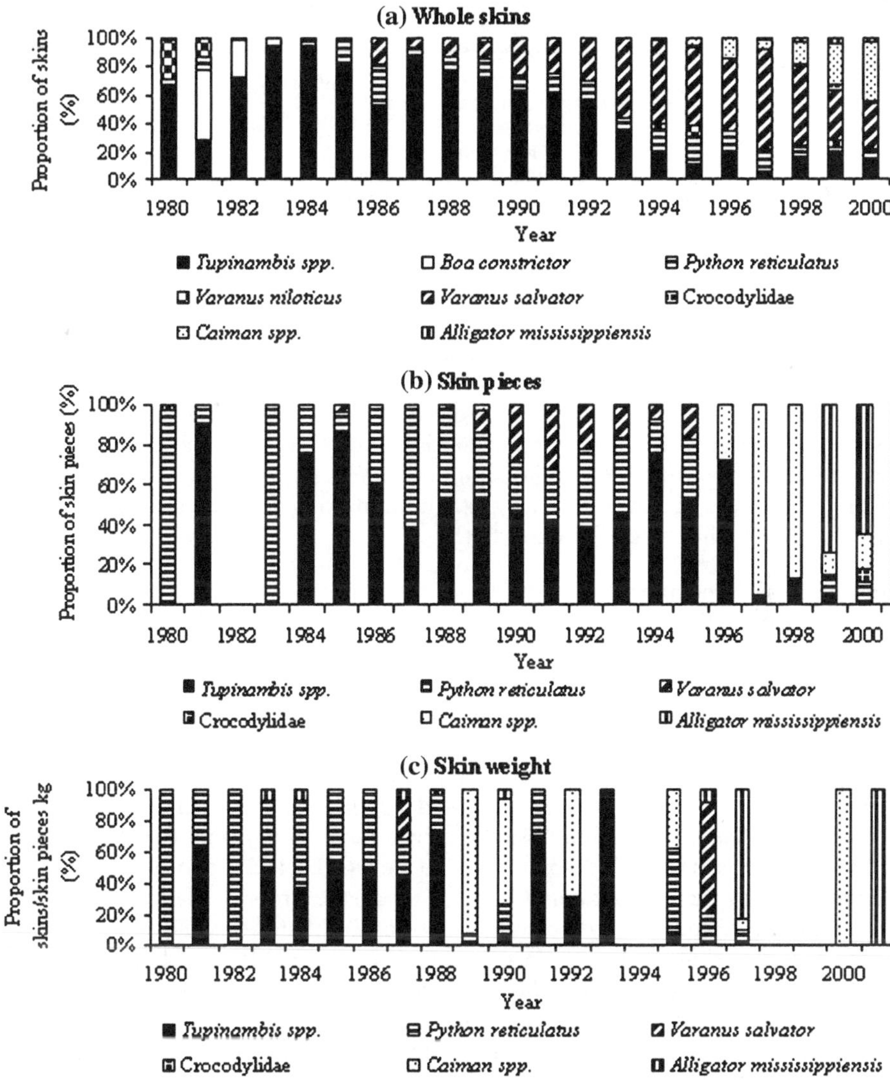

Fig. 3 (a–c) Imports of reptile skins by Mexico 1980–2001 (CITES Trade Data)

Main countries of origin

The main countries of origin for the import of 2.3 million *Varanus salvator* whole skins during 1980–2001 (Table 3) were Indonesia with nearly 1.6 million skins declared as exported, and Malaysia with 0.4 million declared exports (Fig. 5a). The main countries of origin for the import of 2.4 million *Tupinambis* spp. whole skins during 1980–2001 (Table 3) were Argentina, with nearly 2 million skins declared as exported, and Paraguay with 0.3 million declared exports (Fig. 5b). From 1980–2001, Mexico imported around 0.8 million of *Caiman* spp. whole skins (Table 3), of which nearly 0.42 million were reported as unknown origin. The main countries of origin were Colombia, with nearly 0.32 million skins declared as exported, and Brazil with

Table 4 Mexican imports of reptile skins from non-native species 1980–2001

Species	Number of whole skins	Overall skins
Varanus salvator	2,181,208	2,336,638.88
Tupinambis spp.	1,760,926	2,557,421.02
Caiman spp.	791,701	851,364.25
Python reticulatus	580,337	1,083,339.68
Crocodylidae	43,635	45,863.45
Total	5,357,807	6,874,627.28

Source: WCMC Trade Database

0.036 million declared exports (Fig. 5c). The main countries of origin for the import of 1 million *Python reticulatus* whole skins during 1980–2001 (Table 3) were Indonesia with nearly 0.5 million skins declared as exported, Thailand with 0.1 million skins, and Malaysia with 0.1 million declared exports (Fig. 5d). The main countries of origin for the import of 0.1 million *Varanus niloticus* whole skins during 1980–2001 (Table 3) were Sudan with nearly 0.06 million skins declared as exported, and Chad with 0.04 million declared exports (Fig. 5e).

Legal reptile skin reexports

Total reptile skin re-exports 1980–2001

From 1980 to 2001, Mexico re-exported nearly 2 million reptile skin items (Table 5). These derived from various non-native species of lizard, snake and crocodilian. Numbers of whole skin re-exports totalled some 352,500 items (Table 5) and were predominantly of lizards, *Varanus salvator* and *Tupinambis* spp., with lesser numbers of crocodilians, *Caiman* spp., and of snakes, *Python reticulatus* (Fig. 6a). Numbers of skin pieces re-exports totalled some 636,500 items (Table 5), and also were predominantly of lizards, *Tupinambis* spp., with greater representation of snakes, *Python reticulatus*, and lesser representation of crocodilians, *Caiman* spp. (Fig. 6b). Weights of whole skins and skin pieces totalled some 4,500 kg (Table 5), and were predominantly of lizards, *Tupinambis* spp., with greater representation of crocodilians, *Caiman* spp., and lesser representation of snakes, *Python reticulatus* (Fig. 6c). Numbers of skin products re-exported totalled nearly 1 million reptile skin items (Table 5). In contrast to records of skin numbers and skin weights, reptile products were predominantly of crocodilians, *Caiman* spp. and *Alligator mississippiensis* (Fig. 6d).

Total reptile skin re-exports by year

From 1980 to the late 1990s, Mexico re-exported low quantities of whole reptile skins and of reptile skin pieces, whether considered as number or weights, but such re-exports tended to increase from during 1988 onwards (Fig. 7a–c). The numbers of whole reptile skins re-exported remained at constant levels throughout the 1990s but tended to increase into the early 2000s (Fig. 7a). In terms of species, *Tupinambis* spp. made up the majority of Mexico's reptile skin re-exports during the 1980s, whether considered in absolute or proportional terms (Figs. 7 a–c, 8 a-c). In proportional terms, numbers of skin pieces and skin weight (kg) came to be dominated

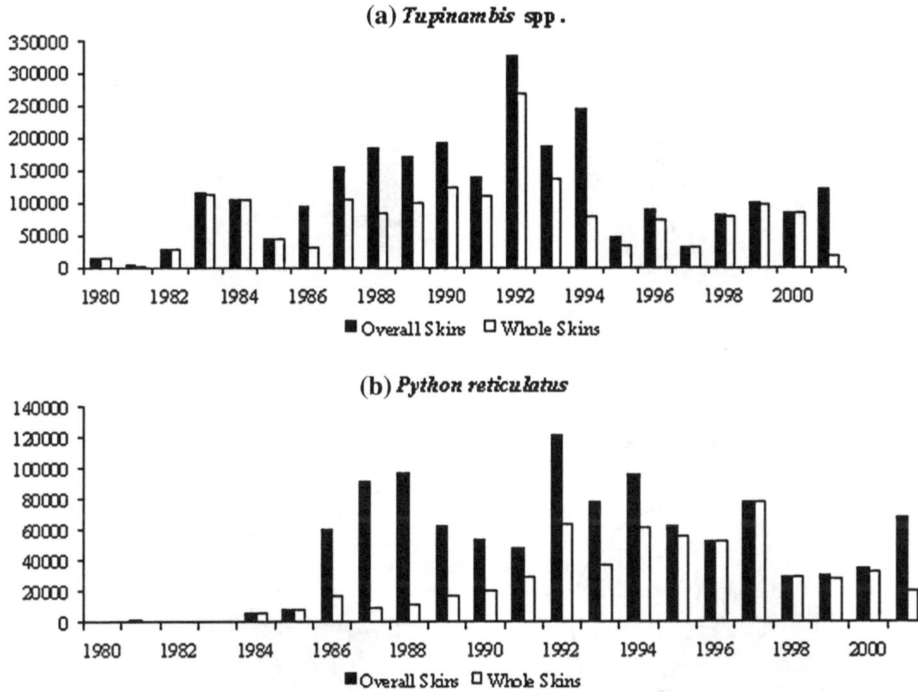

Fig. 4 (**a, b**) Comparison between reptile overall skin and whole skin imports by Mexico 1980–2001 (CITES Trade Data)

Table 5 Mexican Re-exports of reptile skins and products 1980–2001

Species	Number of skins	Number of skin pieces	Skin weight (kg)	Number of skin products
Varanus salvator	167,923	19,235	–	64,214
Tupinambis spp.	86,271	455,703	2,142	76,171
Caiman spp.	52,021	56,918	1,767	530,788
Python reticulatus	31,769	86,689	306	170,997
Varanus niloticus	–	16,446	–	11,819
Alligator mississippiensis	13,993	727	185	79,548
Crocodylidae	513	711	75	31,622
Total	352,490	636,429	4,475	965,159

Source: WCMC Trade Database

by *Python reticulatus* during the 1980s (Fig. 8b, c). From the early 1990s, the numbers of whole skin re-exports came to be dominated by *Varanus salvator*, whether considered in absolute or proportional terms (Figs. 7a; 8a). In contrast, from the early 1990s, the numbers of skin pieces re-exported came to be dominated by *Tupinambis* spp., whether considered in absolute or proportional terms (Figs. 7b, 8b).

In terms of key events in Mexico, re-exports of reptile skins from non-native species increased after Mexico banned the international trade of native species in 1982, whether considered as numbers or weight (Fig. 7a–c). Following Mexico's

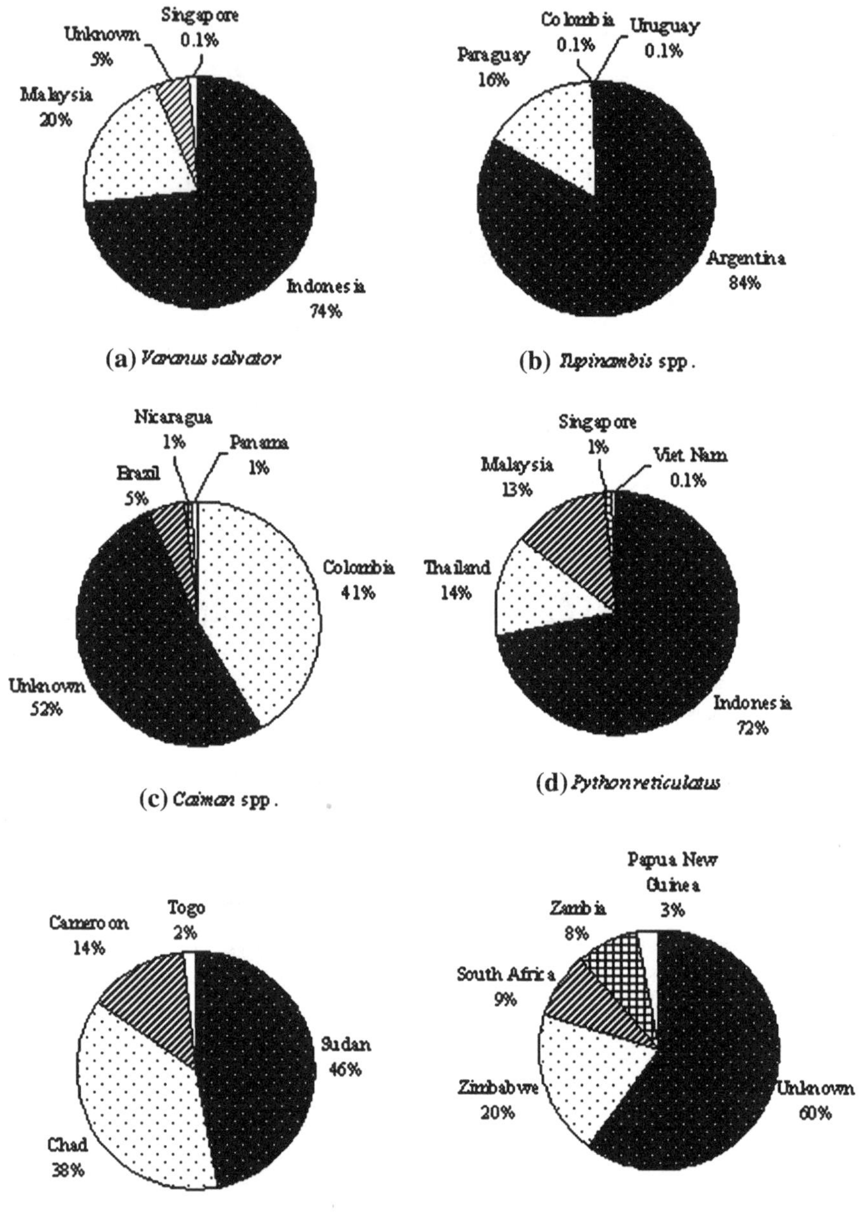

Fig. 5 (**a–f**) Countries of origin for imports of reptile skins by Mexico 1980–2001 (CITES Trade Data)

accession to CITES in 1991, re-exports of whole reptile skins from non-native species remained high (Fig. 7a). Following Mexico's adoption of sustainable resource use policies (1997), re-exports of whole reptile skins from non-native species tended to increase (Fig.7a).

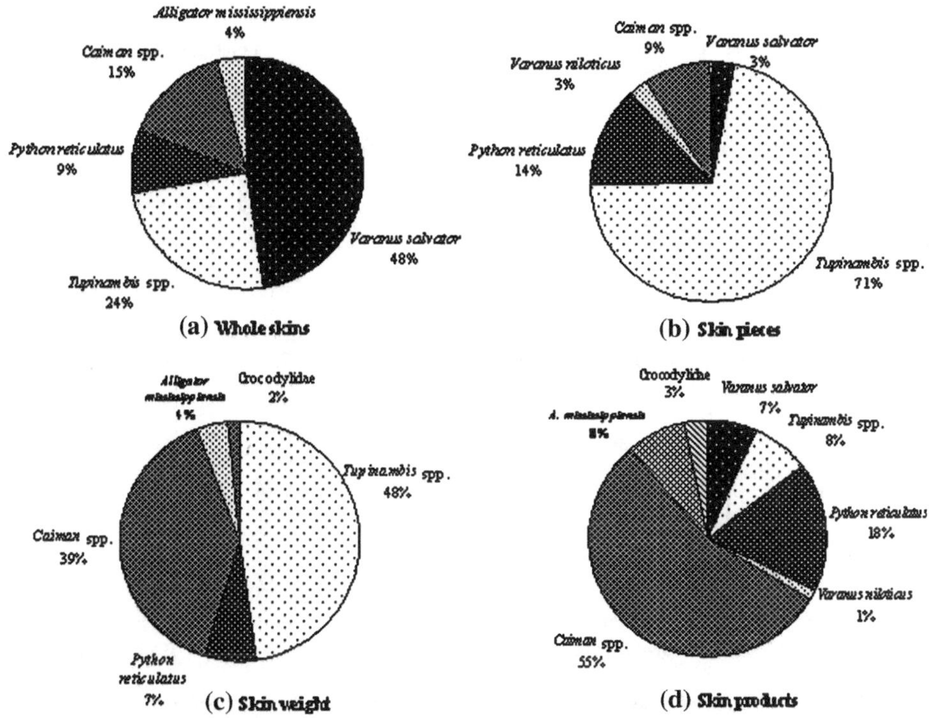

Fig. 6 (**a–d**) Re-exports of reptile skins and products from Mexico 1980–2001 (CITES Trade Data)

Legal reptile skin exports

Total reptile skin exports based on CITES Trade Data

Mexico exports various of its native species of reptile, both as whole skins and as skin products (Table 6). From 1980 to 2001, most exports were from *Chelonia* spp. (95%), followed by *Caiman* spp. (4%) and *Crocodylus* spp. (1%). In terms of skin products, the most exported species were from *Chelonia* spp. (69%), followed by *Lepidochelys* spp. (10%), *Crocodylus* spp. (7%), *Boa constrictor* (6%), *Caiman* spp., and *Iguana iguana* (2%).

Total reptile skin exports by year based on CITES Trade Data

From 1980 to the early 2000s, Mexico exported decreasing quantities of whole reptile skins (Fig. 9a). In contrast, numbers of skin products exported tended to increase from the mid 1980s to peak in 1989, after which exports declined steadily since to reach low quantities in 2001 (Fig. 9b).

In terms of key events in Mexico, exports of whole reptile skins from native species decreased after Mexico banned the international trade on native species in 1982 (Fig. 9a). Even following Mexico's accession to CITES in 1991 and Mexico's adoption of sustainable resource use policies in 1997, exports of whole reptile skins from native species remained at low levels (Fig. 9a). In contrast, exports of skin products from native species increased steadily after Mexico banned the international trade on native species in 1982, but tended to decrease following Mexico's accession to CITES in 1991

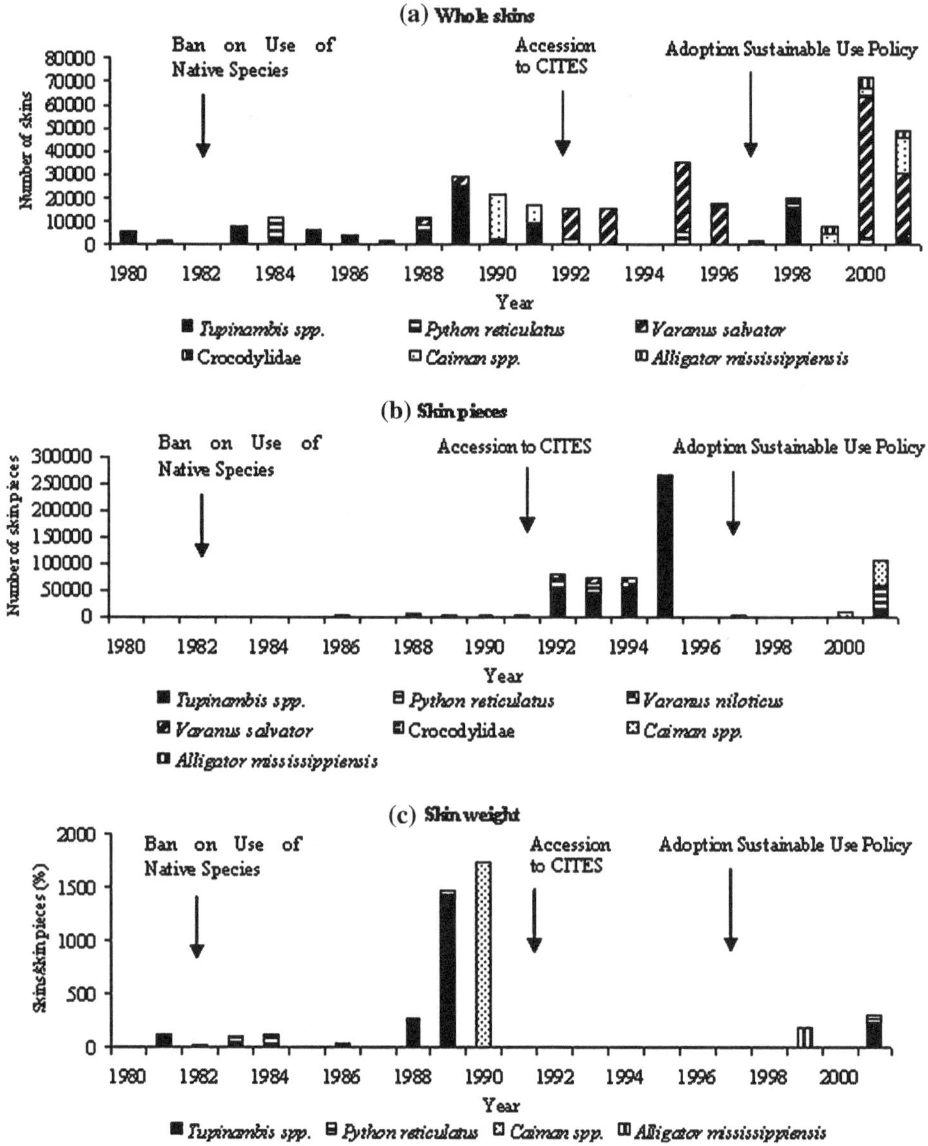

Fig. 7 (**a–c**) Re-exports of reptile skins from Mexico 1980–2001 (CITES Trade Data)

(Fig. 9b). Following Mexico's adoption of sustainable resource use policies in 1997, exports of skin products from native species decreased further and remained at low levels (Fig. 9b).

Main countries of import

From 1980–2001, Mexico exported around 30,000 skins and 35,000 skin products from its various species of native reptile (Table 6). Most of the exports (90–95%) were imported by the US.

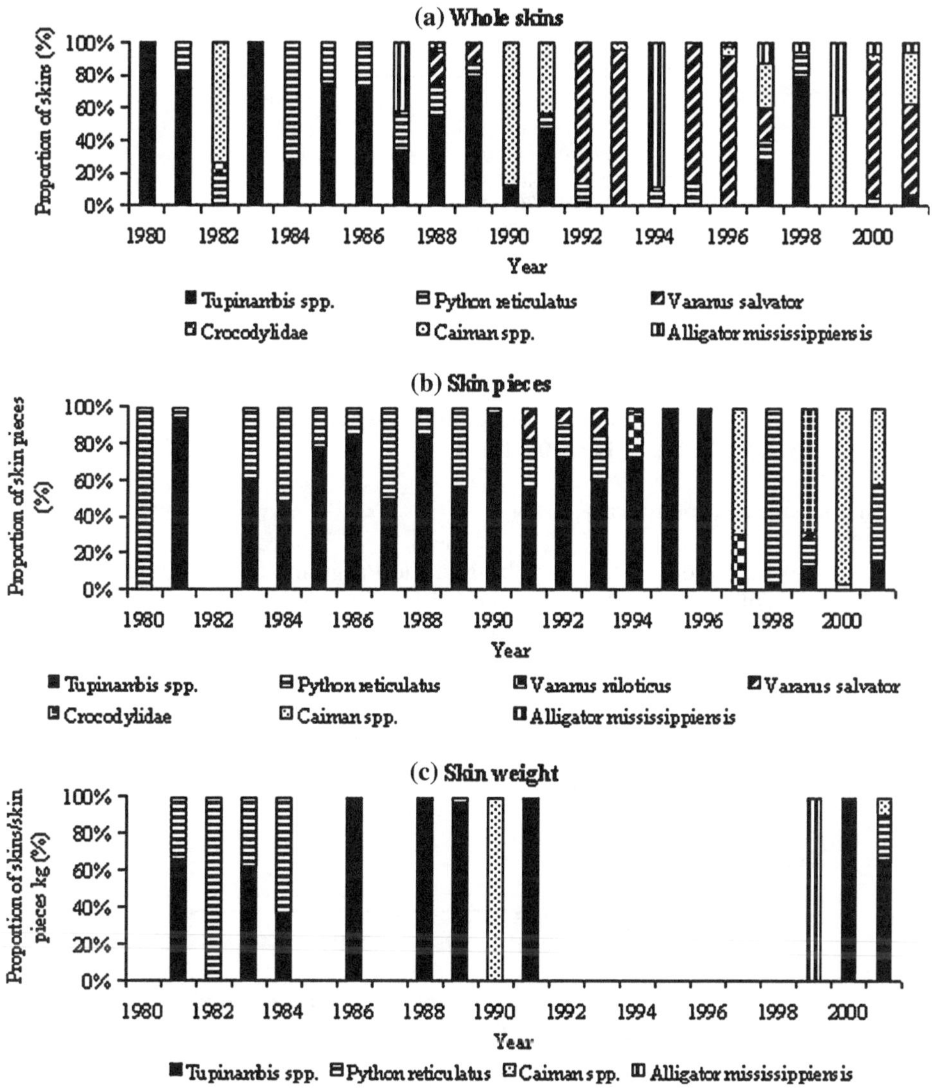

Fig. 8 **(a–c)** Re-exports of reptile skins from Mexico 1980–2001 (CITES Trade Data)

Total reptile skin exports based on LEMIS Trade Data

In terms of quantity of whole skins, the most exported species from Mexico to the US were *Caiman* spp. (97%), followed by *Crocodylus* spp. (2%) and *Crotalus* spp. (1%) (Table 7). In terms of quantity of skin products, the most exported species were *Caiman* spp. (90%), followed by *Crotalus* spp. (6%) and *Iguana iguana* (2%) (Table 7).

During 1995–1999, Mexico exported a total of 2,486 whole skins of *Caiman* spp. to the US (Table 7), and 1,620 of these were in 1999 (Fig. 10a). Mexico also exported a total of 102,691 skin products of *Caiman* spp. to the US (Table 7). The exports of *Caiman* spp. whole skins and skin products increased steadily from 1995–1999, but

Table 6 Mexican exports: reptile whole skins and skin products from native species 1980–2001

Species	Number of skins	Number of skin products
Chelonia spp.	28,405	24,447
Caiman spp.	1,304	1,808
Crocodylus spp.	301	2,331
Iguana iguana	21	754
Lepidochelys spp.	20	3,584
Boa constrictor	12	2,035
Caretta caretta	–	300
Eretmochelys imbricata	–	171
Total	30,063	35,430

Source: WCMC CITES Trade Database

the numbers of skin products were considerably higher than numbers of whole skins during this period (Table 7; Fig. 10a, b).

Sustainable use of reptile skins

The National Institute of Ecology published the Programme of Wildlife Conservation and Diversification of Rural Production in 1997 (INE 1997). The Programme planned to preserve and protect the biodiversity, ecosystems, and wildlife habitat; to

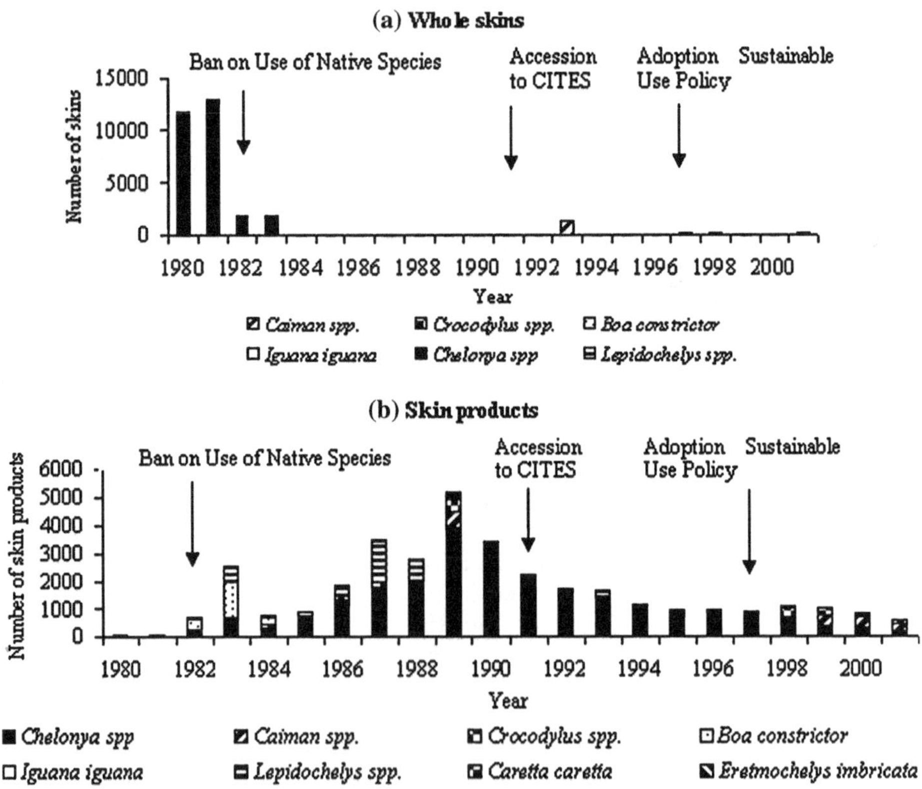

Fig. 9 (**a, b**) Exports of reptile skins by Mexico 1980–2001 (CITES Trade Data)

Table 7 Mexican exports: reptile whole skins and skin products from native species 1995–1999

Species	Number of skins	Number of skin products
Caiman spp.	2,486	102,691
Crotalus spp.	29	6,794
Iguana iguana	2	2,271
Crocodylus spp.	52	1,449
Boa constrictor	2	592
Total	2,571	113,797

Source: LEMIS Trade Data

decrease extinction of species; and to recover species at risk, while also generating opportunities for socio-economic diversification for the rural sector.

The Programme included the launching of the "System for the Conservation, Management and Sustainable Use of Wildlife", or SUMA. The SUMA purports to provide a regulatory framework for the commercial use of wildlife, reconciling socio-economic development through wildlife use and biodiversity protection (DOF 2000; INE 2000b). This is a new approach for Mexico emphasising participation by local people in which the land owners are supposedly convinced of the need to protect and manage the habitat and wildlife, while verifying that conservation is profitable in social, economic, and environmental terms, and that they can accede legally to those benefits when fulfilling the legal procedures established through specific techniques of management. At least in theory, the SUMA programme appears to be a governmental response to overcome the challenges found in previous models of wildlife management.

By allowing the marketing of wildlife under certain conditions, the SUMA is designed to respect CITES and should help improve protection of threatened species. In this regard, the Mexican government recognised in the late 1990s that Mexico was importing large quantities of reptile skins from non-native species instead of looking for alternatives to promote the use of reptile skins from native species (INE 1997; Morales 2002). Reptile species that are included in the SUMA comprise mainly crocodiles (*Crocodylus moreletii*, *C. acutus*), caimans (*Caiman crocodilus*) and iguanas (*Iguana iguana*, *Ctenosaura* spp.). However, the promotion of crocodile, caiman and iguana skin production practices through the SUMA is still incipient in Mexico and the main reptile skin production systems are held by private owners (commercial enterprises) for populations of captive species instead of encouraging the conservation of wild populations and the social and economic

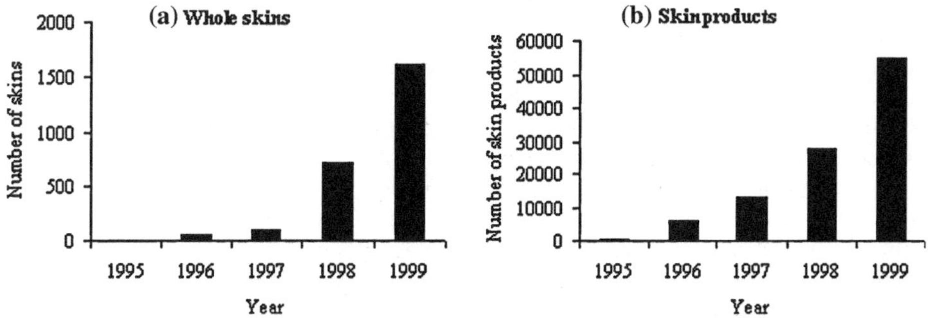

Fig. 10 (**a, b**) Exports of *Caiman* spp. from Mexico to the US 1995–1999 (LEMIS Trade Data)

well-being of local communities (León 1996; León 2001, Pers. comm.; Ross 1995; Villegas 2002, Pers. comm.; Villegas and Vázquez 2002; Morales 2002; Morales 2002, Pers. comm.; Macdonel 2002, Pers. comm.; Díaz 2001; Uriarte 2002, Pers. comm.; Vidal de los Santos 2002, Pers. comm.; Vázquez et al. 2001).

Discussion

Mexico is a key player in the international trade in reptile skins. Mexico functions as a collection and sorting centre of reptile skins and a prime manufacturer of reptile skin products. The US is by far the most important consumer of reptile skins and skin products exported and re-exported from Mexico. Mexico and the US share 1,950 miles of border and a Free Trade Agreement since 1994. The region is an important decision-making centre (Hogenboom 1998), the policies of which can influence world trade patterns and may affect the status of wildlife populations. Mexico, a megadiverse country and biodiversity hot spot (Mittermeier et al. 1998, 1999), has a great potential for contributing to the international discussion on forms of management, techniques of analysis, and regulatory frameworks for ecosystems and biological resources.

The present status of reptile species in Mexico is of serious concern (Hilton-Taylor 2000). Mexico suffers from an active illegal trade in reptile skins and little is known about the legal utilisation of reptile skins by the Mexican leather industry (Arroyo-Quiroz 2003). The use of reptile skins from native species is practically non-existent and the rural communities in Mexico, located in the same areas where wild species are distributed, paradoxically are still immersed in an extremely poor economy.

Legal reptile skin imports

The most numerous reptile skins in reported imports during 1980–2001 were from *Tupinambis* spp., *Varanus salvator* and *Python reticulatus*. Over the period 1980–2001, Mexico's total imports of reptile skins have increased steadily, and the largest numbers of reptile skins were reported during the late 1990s and early 2000s (Fig. 2a). The observed trend was opposite to what was expected i.e. that imports of reptile skins from non-native species would be higher after Mexico adopted the ban on use of native species, whereas imports would tend to decrease after Mexico adopted its policy to sustainably use native species.

Skins of particular species appear to be subject to changing demand in Mexico. For example, during the 1980s, whole skins of *Tupinambis* spp. were the most imported commodities but these were substituted for whole skins of *Varanus salvator* during the 1990s (Fig. 3a). There must be several factors such as market trends and fashion driving Mexico's preferences over time (Freese 1998; Swanson 1992). Unfortunately, consideration of socio-economic aspects of the trade cannot be answered with CITES data alone (Harris et al. 2003).

Levels of trade were high when Mexican imports of whole reptile skins, skin pieces and skin weight were summed separately. From 1980 to 2001, numbers of whole skin imports totalled some 5.3 million items (Table 4). When a conversion factor was used to convert these categories to an approximate number of whole skins, the total number of whole skins animals imported by Mexico increased to nearly 6.9 million (Table 4). Therefore, in terms of the number of animals taken annually, the trade in reptile skins in the Mexican leather market is clearly a very important aspect for the international trade in wildlife. In this regard, the control of reptile skins between the processes of import, manufacture and re-export remains a key challenge for Mexican Authorities. A tracking system for reptile skins and skin products is non-existent. Marking systems are increasingly being used as an additional means to control and track trade, such as the tagging system for crocodile skins (Hutton and Webb 2003) (Fig. 11). However, other reptile skins largely imported by Mexico (e.g. *Python* spp.), as well as all reptile skin products made in Mexico, are unmarked. Thus control and trade recording is dependent on permits and certificates alone.

Argentina (*Tupinambis* spp.) and Indonesia (*Varanus salvator* and *Python reticulatus*) were the major suppliers of reptile skins to Mexico (Fig. 5). However, many (>50%) exports of *Caiman* spp. and *Crocodylus* spp. were recorded as of unknown origin (Fig. 5). As Mexico is also a range State for these taxa, a more detailed investigation of this trade should be undertaken, as much trade in non-ranched or non-farmed crocodilian skins has been knowingly illegal for many years (Ross 1998; Hutton and Webb 2003). Moreover, the long-term sustainability of the trade from the supply side remains an intractable issue in Mexico. To establish a link between the numbers of reptile skins traded in Mexico, and the status of wild populations from which they originate is extremely problematic. Little is known about the populations from which these harvests come or of details of the harvests. As an important consumer of species native to other countries, Mexico, together with the US and Canada, should assist in efforts to study the populations, harvest and trade of some of the main species affected by the North American demand (Fleming 1999).

Legal reptile skin re-exports

The most numerous reptile skins re-exported during 1980–2001 were from *Tupinambis* spp. and *Varanus salvator* (Table 5). In contrast, the most numerous reptile skin products re-exported during 1980–2001 were from *Caiman* spp. and *Varanus salvator* (Table 5). Equally, the prevalence of particular taxa re-exports over time varied depending on the term recorded. For example, re-exports of whole skins during 1980–2001 were dominated by *Varanus salvator*, whereas re-exports of skin pieces were dominated by *Tupinambis* spp. (Table 5).

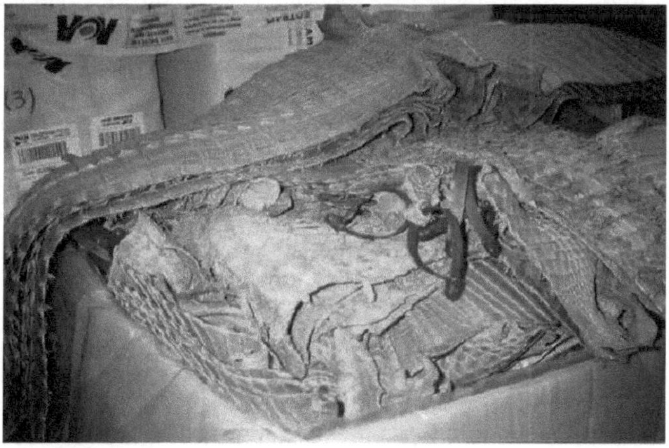

Fig. 11 Imports of finished *Caiman* spp. whole skins by Mexico (Arroyo-Quiroz 2003)

Mexican re-exports in terms of numbers of reptile skins were higher during the 1990s (Fig. 7a, b). It was expected that numbers of reptile skins and skin products re-exported would tend to decrease after Mexico adopted a sustainable use policy. However, such a decrease may be offset by the opening of national markets and it seems very unlikely that the increasing demand in these markets will be satisfied exclusively by the limited domestic production.

Significant levels of re-exports of *Caiman* spp. and Crocodylidae skin products were recorded during the 1990s. Since Mexico is also a range State for these taxa, a more detailed investigation of this trade should be undertaken in order to certify that the skin products re-exported by Mexico indeed have been manufactured with reptile skins from non-native species. This is particularly important, since most of the reptile skin imports by Mexico during 1980–2001 were recorded as being unknown origin (Fig. 5).

Legal reptile skin exports

Over the period 1980–2001, the trade data showed the expected trend for Mexican exports of marine turtle species, since exports decreased after Mexico adopted the ban on use of native species in 1982 (Fig. 9a). However, after adopting the total ban on use of marine turtles and acceding to CITES in 1991, Mexico still exported significant amounts of turtle skin products during the 1990s, essentially of *Chelonia* spp. (Fig.9b). The high volumes of banned *Chelonia* spp. skin products exported by Mexico during the 1990s, compared with the better regulated numbers of *Caiman* spp. and *Crocodylus* spp. skin products exported over the same period is perplexing. There appears no correlation whatsoever between the ban, the sustainable use policy and the reported trade.

Over the period 1995–1999, the most numerous reptile skins and skin products in reported exports to the US were from *Caiman* spp. (Table 7). In terms of numbers of skins and skin products, the observed trend was as expected, since Mexican exports of *Caiman* spp. increased after Mexico adopted a sustainable use policy (Fig. 10). However, it would be better for Mexico in economic terms, if the numbers of *Caiman* spp. whole skins exported decreased, given the value added to skin products. However, Mexico may be exporting *Caiman* spp. whole skins because of its lack of high quality tanneries of wildlife skins (Calleja 1994; Iglesias 1998; León 2001, Pers. comm.).

Over the period 1995–1999, very few whole skins of *Crotalus* spp. were exported to the US (Table 7). This was expected since the country does not have yet a formal skin production scheme for this species. Nevertheless, Mexico exported significant amounts of *Crotalus* spp. skin products to the US during 1995–1999 (Table 7), all specimens taken from the wild. This type of data should encourage CITES Scientific and Management Authorities in Mexico to undertake a thorough investigation on the source of specimens in trade, in order to assess the impact of trade on the survival of this species.

During the 1990s, skins and skin products exported by Mexico from species promoted by the SUMA (e.g. *Crocodylus* spp. and *Iguana iguana*) were low in numbers. Ironically, Mexico exported significant amounts of banned marine turtle skin products from *Chelonia* spp. and *Crotalus* spp. Although Mexico implemented a programme for wildlife conservation and sustainable use, and has the potential to become a significant producer of native reptile skins (e.g. *Crocodylus* spp., *Caiman* spp., *Iguana* spp., and *Crotalus* spp.), Mexico still makes little legal use of skins from native species. This is not the ideal scenario. More reptile skins from native species should be exported, for there is a sizeable market opportunity, as the import, re-export and current export data reveal. With potential to produce wildlife, Mexico should use CITES as a regulatory framework to support local communities in order to promote the sustainable use of species, but also as a means to participate in the global market with sustainable products as successfully as other countries have (Kievit 2000; Hutton and Webb 2003). One might have to consider, though, that even if local communities do gain full proprietorship over wildlife, there is no guarantee that it will be in their interests to conserve wildlife. They might decide to mine the resource and invest the returns elsewhere. It will also depend on factors such as the price they receive for wildlife products and the return they could enjoy from alternative land uses (Dickson 2000).

Also, any future transition to a scenario where native species can fill the market will also require collaboration between professionals and a higher level of mutual respect among different actors - academics, governmental and non-governmental organizations, and industrial and rural producers. The critical issue that remains to be addressed is the actual possibility of transforming the development model Mexico has pursued to a sustainable one. There is still much to accomplish on issues such as: the creation of the necessary human capacity for the technical surveillance of the SUMA at the local level, the fine tuning and enforcement of regulatory legislation of this productive scheme and the consciousness-raising and information disclosure to the citizenship, in order to achieve a higher acceptance of the sustainable use concept and to ensure that sustainable use will in fact lead to improved conservation status of the species.

Sustainable use

The governmental wildlife programme is indeed an instrument with initiatives on the protection and use of wildlife radically different from what prevailed in Mexico for many decades. However, question remains on whether trade in skins from native reptiles under this programme could become an appropriate conservation regime where local communities harvest species on a sustainable basis. Given that currently, in the case of reptile skin production, what still prevails in Mexico is an ongoing use of reptiles from non-native species while the few native species promoted (crocodilians and iguanas) are not harvested but basically subject to privately owned commercial captive breeding schemes, which though some presumably biologically sustainable, do not consider habitat nor species conservation. In the case of such registered and operating reptile production schemes, the great challenge is to ensure that the economic benefits derived from production are also routed toward conserving biodiversity and social and economic benefit for the local communities, as originally envision by the wildlife programme. The future of successful conservation lies in recognizing instances where trade can be beneficial to a species, and creating a mechanism that encourages sustainable use and legal trade, while discouraging unsustainable and illegal exploitation ('t Sas-Rolfes 2000).

Recommendations

Sustainable use could positively encourage the trade of reptile skins in Mexico, which can become a valuable economic and social resource, rather than simply banning the use of such resources. However, before establishing mechanisms to manage markets for sustainability and before designing harvest strategies for sustainable off-take, Mexico should develop market studies to determine which species are at present subject to use and commercialization. Understanding the status of trade in wildlife is very important in formulating management policies for wildlife trade and conservation (Yiming and Dianmo 1998). Mexico needs to characterise the ongoing market of native species, products and by-products (CITES and non CITES) to determine the structure and extent of such markets by differentiating the use of every commodity and their corresponding distribution channels. Discerning these interconnections should contribute to establishing proper guidelines for the commercial use of wild species in Mexico and also work against the difficulty still faced by the country to gather data so as to determine the status of wild species in trade. There is a real need to compile information from across the country on this matter in order to understand the patterns of use, the effects they have on wild populations, and how the sustainable use and intrinsic value of species can better be put to work as conservation tools. Furthermore, how to certify that both sustainable use and incentive-driven conservation (Hutton and Leader-Williams 2003) become core elements of the conservation agenda of Mexico.

Acknowledgements The authors thank TRAFFIC and UNEP-WCMC for the supply of data. Special thanks to Simon Habel, Craig Hoover and John Caldwell for all their support. Gratitude is also extended to all staff of institutions and private individuals who were welcoming and supportive. In particular, the Attorney General for Protection of the Environment (PROFEPA) in Mexico for providing access to custom offices. Special thanks are due to José Bernal, Silvia Philippe, Luis Domenzain, Carlos Contreras, Guadalupe Ávila, Lilia Mondragón, and Uhry Adib.

References

Arroyo-Quiroz, I (2003) Developing countries and the implementation of CITES: a case study of Mexico in the international reptile skin trade. PhD Thesis, University of Kent at Canterbury, UK

Bancomext (1994) Canadá: Perfil de Bota Vaquera. Banco de Comercio Exterior. México. 6 pp

Bancomext (1999) Programa de Actividades Promocionales en el Exterior 1999: Sector Cuero, Calzado y Marroquinería en el Mercado Norteamericano. Banco de Comercio Exterior. México 13 pp

Bancomext (2000) México Exporta: Cuero, Calzado y Marroquinería. Programa de Apoyo Integral Año 2000. Banco de Comercio Exterior. México 51 pp

Bancomext (2001) Bancomext: Importaciones Definitivas. Banco de Comercio Exterior. México

Bancomext (2006) Cuero y Calzado. Banco de Comercio Exterior. México. Available from http://www.bancomext.com (accessed February 2006)

Brautigam A (1986a) Mexican wildlife trade: what do the statistics show? TRAFFIC USA 6(4):4, 9

Brautigam A (1986b) A look at the U.S.-Mexican exotic leather trade. TRAFFIC USA. 6(4):10–11

Broad S et al (2003) The nature and extent of legal and illegal trade in wildlife. In: Oldfield S (ed), The trade in wildilfe. regulation for conservation. Earthscan Publications Ltd., London, UK 210, 3–22 pp

Calleja PM (1994) Micro Industria. Principio y Soporte de la Gran Empresa. La producción de calzado en León, Guanajuato. Universidad de Guadalajara. México. 266 pp

CITES Secretariat (2003) Annual Reports of CITES Parties. Available from http://www.cites.org (accessed August 2003)

Díaz DEP (2001) Proyecto de Conservación, Manejo y Aprovechamiento de las Iguanas en México. Dirección General de Vida Silvestre, SEMARNAP. IV Taller Nacional sobre Manejo de Iguanas en Cautiverio. Puerto Ángel, Oaxaca

Dickson B (2000) Global regulation and communal management. In: Hutton, Dickson (eds), Endangered species threatened convention. the past, present and future of CITES. Earthscan Publications Ltd., London, UK. 202, pp 161–177, 202 pp
DOF (2000) Ley General de Vida Silvestre. Diario Oficial de la Federación 3 de Julio. México
Fleming HE (1999) Exploring Linkages between Trade and Species Conservation in North America. TRAFFIC North America, 161 pp
Freese C (1998) Wild species as commodities. Island Press, Washington DC, 319 pp
Gaski LA (1992) The United States as consumer and regulator in the reptile skin trade. TRAFFIC USA, Washington DC, 46 pp
Harris D et al (2003) CITES Trade Data. An Under-Utilised Wildlife Resource. A Report of the CITES Trade Data Analysis Project. IUCN/SSC Wildlife Trade Programme and TRAFFIC International. RBG Kew/UNEP-WCMC
Hemley G (1994) International wildlife trade. A CITES sourcebook. WWF/Island Press, Washington, DC, 166 pp
Hilton-Taylor C (compiler) (2000) 2000 IUCN Red List of Threatened Species. IUCN, Gland, Switzerland and Cambridge, UK. pp xviii + 61
Hogenboom BB (1998) Mexico and the NAFTA environment debate: the transnational politics of economic integration. International Books, Netherland, 279 pp
Hutton J, Leader-Williams N (2003) Sustainable use and incentive-driven conservation: realigning human and conservation interests. Oryx 37:215–226
Hutton J, Webb G (2003) Crocodiles: legal trade snaps back. In: Oldfield S (ed), The Trade in wildilfe. Regulation for conservation. Earthscan Publications Ltd., London, UK 210, pp 108–120
Iglesias E (1998) Las Industrias del Cuero y del Calzado en México. Instituto de Investigaciones Económicas. UNAM. México, 277 pp
INE (1997) Programa de Conservación de la vida silvestre y diversificación productiva en el sector rural. SEMARNAP. México, 207 pp
INE (2000a) Proyecto para la Conservación, Manejo y Aprovechamiento Sustentable de los Crocodylia en México (COMACROM). SEMARNAP. México, 107 pp
INE (2000b) Estrategia Nacional para la Vida Silvestre. Logros y Retos para el Desarrollo Sustentable. SEMARNAP. México, 212 pp
Jenkins M, Broad S (1994) International Trade in Reptile Skins. A Review and Analysis of the Main Consumer Markets, 1983–91. A TRAFFIC Network Report. TRAFFIC International, WWF Cambridge UK, 68 pp
Kievit H (2000) Conservation of the Nile crocodile: has CITES helped or hindered? In: Hutton, Dickson (eds) Endangered species, threatened convention. The past, present and future of CITES. Earthscan Publications Ltd., London, UK. 202, pp 88–97
León F (1996) La Crianza de Cocodrilos en México: Cocodrilos Mexicanos S.A. de C.V. In: Crocodiles. Proceedings of the 13th working meeting of the Crocodile Specialist Group. UICN. Gland, Switzerland. pp 355–358, 514 pp
Mittermeier RA et al (eds) (1998) Megadiversity: earth's biologically wealthiest nations. CEMEX, Monterrey, Mexico
Mittermeier RA et al (eds) (1999) Hotspots. earth's biologically richest and most threatened ecoregions. CEMEX, Monterrey Mexico, 391 pp
Morales T (2002) Situación Económica de las Iguanas en México. Dirección General de Vida Silvestre, SEMARNAT. V Taller Nacional de Manejo de Iguanas en Cautiverio. Villa Hermosa, Tabasco Mexico
Pérez Gil R et al (1996) Importancia económica de los vertebrados silvestres de México. CONABIO, Pg7 Consultores. México, 170 pp
Reeve R (2002) Policing international trade in endangered species: The CITES treaty and compliance. The Royal Institute of International Affairs/Earthscan Publications Ltd., London, UK, 346 pp
Rodríguez UH (1985a) México, ante la Explotación Internacional de la Fauna Silvestre. In: Ier. Simposio Internacional de Fauna Silvestre Vol.II., SEDUE, México DF, México, pp 849–860
Rodríguez UH (1985b) De la Explotación a la Continuidad de la Fauna Silvestre, Reporte Técnico, Dirección de Flora y Fauna Terrestre, SEDUE. México
Roe D et al (2002) Making a Killing or Making a Living. Wildlife Trade, Trade Controls and Rural Livelihoods. Biodiversity and Livelihoods Issues No. 6. IIED. London UK
Rose D (1991) A North American free trade agreement. The impacts on wildlife trade. TRAFFIC USA Washington DC, 38 pp
Rose D (1992a) Free trade and wildlife trade. Conserv Biol 6:1

Rose D (1992b) The sharks of Mexico: a resource for all seasons. TRAFFIC USA 11(4):4–6
Ross JP (1995) North America: Mexico. CSG Newsletter 14(4):17–18
Ross JP (eds) (1998) Crocodiles: status survey and conservation action plan. SSC Crocodile Specialist Group. IUCN, Gland, Switzerland, 96 pp
Swanson T (1992) The biodiversity convention: a meeting of the minds? In: Global action for biodiversity. Earthscan Publications Ltd., London, UK. pp 79–107, 191 pp
Thorbjarnarson J (1999) Crocodile tears and skins: international trade, economic constraints, and limits to the sustainable use of crocodilians. Conserv Biol 13:465–470
TRAFFIC (2006) What are the main wildlife commodities in trade? TRAFFIC International, Cambridge, UK. Available from http://www.traffic.org/25/wild4_3.htm (accessed February 2006)
't Sas-Rolfes M (2000) Assessing CITES: four case studies. In: Endangered species threatened convention. The Past, Present and Future of CITES. Earthscan Publications Ltd., London, UK. pp 69–87, 202 pp
Vázquez AS et al (2001) Bases Legales para la Colecta Científica y el Aprovechamiento Sustentable de las Iguanas. Unidad de Vida Silvestre de la SEMARNAT, Oaxaca. IV Taller Nacional de Manejo de Iguanas en Cautiverio. Puerto Ángel, Oaxaca. México
Villegas ZF, Vázquez PDB (2002) Problemática, Usos y Forma de Caza de las Iguanas: América Latina-México. Centro de Conservación y Reproducción de Iguanas de la Universidad del Mar. Oaxaca, México
Yiming L, Dianmo L (1998) The dynamics of trade in live wildlife across the Guangxi border between China and Vietnam during 1993–1996 and its control strategies. Biodivers Conserv 7:895–914
Zhou Z, Jiang Z (2004) International trade status and crisis for snake species in China. Conserv Biol 18(5):1386–1394

Biodivers Conserv (2007) 16:953–967
DOI 10.1007/s10531-006-9034-9

ORIGINAL PAPER

Use of forest fragments by blue-winged macaws (*Primolius maracana*) within a fragmented landscape

Maria Flávia Conti Nunes · Mauro Galetti

Received: 29 March 2005 / Accepted: 23 January 2006 / Published online: 27 October 2006

Abstract Parrots are the most threatened group of birds in the world, mainly because of the reduction and fragmentation of their natural habitats. However, few studies have investigated the dynamics of parrot populations in disturbed landscapes on a broad scale. In this paper, we studied the ecological interactions of the vulnerable blue-winged macaw (*Primolius maracana*) in a fragmented landscape surrounding a large protected park in southeastern Brazil. We sampled 36 forest fragments that varied in size, characteristics, degree of isolation and type of surrounding matrix in order to assess the importance of habitat features on the maintenance of these birds. Blue-winged macaws were recorded in 70% of the satellite remnants that were sampled, which included large and small blocks of forest. These areas were used as sites for feeding, nesting or overnight rests, and also provided connectivity for birds' displacements. However, the frequency of macaw visits varied among the remnants, and this was related to habitat features such as patch size, human use of surrounding land, and the proximity to the protected park, to urban areas and to the birds' roosting areas. In general, landscape-scale parameters explained more of the variation in the frequency of visits by macaws than did patch-scale parameters. These results demonstrate the importance of landscape mosaics for the survival of blue-winged macaws.

Keywords Parrots · Threaten species · Atlantic forest · Forest fragmentation · Habitat use · Scale of analyses

---

M. F. C. Nunes (✉)
Centro Nacional de Pesquisa para Conservação das Aves Silvestres (CEMAVE)-IBAMA, BR 230, Km 10, Mata da AMEM, 58310-000 Cabedelo, PB, Brazil
e-mail: maria-flavia.nunes@ibama.gov.br

M. Galetti
Laboratório de Biologia da Conservação, Departamento de Ecologia, Universidade Estadual Paulista (UNESP), CP 199, 13506-900 Rio Claro, SP, Brazil

M. Galetti
Institute for Biological Conservation (IBC), Campinas, SP, Brazil

Introduction

The current extinction crisis has become of great concern in the last decades (Soulé 1987; Wilson 1988; Pimm 1995; Brooks 2000). Recent estimates indicate that one out of eight species of birds has a high risk of becoming extinct in the next 100 years (BirdLife International 2000). These endangered birds included the Psittacidae, the most threatened group, with at least 28% of its species facing some risk of extinction (BirdLife International 2000). The loss and fragmentation of natural habitats are the main causes of the decline of most threatened parrots (BirdLife International 2000).

The blue-winged macaw (*Primolius maracana*) (Vieillot), is a parrot species that is currently considered near-threatened to extinction (BirdLife International 2004). This species was once widespread throughout Brazil, eastern Paraguay, and northern Argentina, but showed a marked retraction in its range in the second half of the 20th century, particularly in the southern portion of its historical distribution (southern Brazil: Belton 1994; Bornschein and Straube 1991; Rosário 1996; Benke 2001; Argentina: Chebez 1996; de la Peña 1999; Paraguay: Lowen et al. 1996; Clay et al. 1998; summarized in Nunes 2003). This reduction in range has been attributed primarily to habitat loss (Olmos 1993) since there is so little forest remaining across most of this species' original distribution (see Fundação SOS Mata Atlântica and INPE 1998).

In São Paulo state, the blue-winged macaw was once widespread in semideciduous forests, but these have been drastically reduced to less than 2% of their original extent (Viana and Tabanez 1996) and the species is now restricted to just a few sites (Nunes 2003). The Caetetus Ecological Station (CES) is one of the largest and most important remnants of semideciduous forest in São Paulo state (Cullen et al. 2000), and it has the largest population of blue-winged macaws in the southern part of this species' original range (Nunes 2000). This population is currently estimated at approximately 150 birds (Nunes 2000). A mosaic of forest remnants, coffee plantations and pasture dominates the landscape around Caetetus, and could be important for maintaining this macaw and other local animal species.

In this study we examined the effects of habitat fragmentation on blue-winged macaws on a broad scale, attempting to assess the interactions of species with the landscape composition and spatial configuration. Such information is especially important for the conservation of parrots, since studies on a patch-level scale have suggested that some parrot species, especially large bodied ones such as macaws, usually do not thrive in small fragments (Aleixo and Vielliard 1995; Willis 1979; Uezu et al. 2005). However, the high mobility of some parrot species allows them to use, or at least to reach, small remnants within fragmented landscapes (Fischer and Lindenmayer 2002; MacNally and Horrocks 2000; Marsden et al. 2000), from which less-mobile species tend to be excluded (Villard and Taylor 1994). Moreover, animal movements tend to vary according to the configuration of the landscape and matrix characteristics (Fischer and Lindenmayer 2002; Roshier 2003).

In this study, we examined the dynamics of a vulnerable parrot species in a fragmented landscape by looking beyond patch boundaries. We examined habitat used by macaws in a series of forest fragments that varied in size, characteristics, degree of isolation, and position within the matrix around the reserve. These data allowed us to assess the importance of different traits in determining the landscape used by the macaws.

Methods

Study area and sampling sites

The study was done in fragments of native vegetation around the Caetetus Ecological Station (Lat 22°24′S, Long 49°42′W), near Gália, in west-central São Paulo state, Brazil (Fig. 1). The reserve, a 2,178 ha forest patch, is the largest block of protected forest in its region. The forest is mainly semideciduous, with a high seasonality and a partial fall of leaves from some trees during the cool, dry season (April–September). The landscape around the reserve is composed of small remnants of native vegetation forest, pastures, and coffee and rubber plantations. The native remnants are mainly semideciduous forest, along with secondary forest and some forest-savanna transitional areas.

We used a 1:250,000 vegetation cover map, produced by Instituto Florestal (Kronka et al. 1994), to select 36 forest patches within a 25 km radius of the CES (Fig. 1). The map was edited to allow corrections and updates, based on the LandsatTM satellite image of April 2001 and using the software Spring 3.5 (Camara et al. 1996). The fragments were randomly selected from the map and stratified by size, with classes of 10–30 ha, 31–50 ha, 51–70 ha, 71–100 ha, 101–200 ha, and > 200 ha. We selected six fragments from each size class for sampling a variety of forest habitats.

Bird recording methods

We searched for blue-winged macaws in the 36 fragments around Caetetus, between August 7th and October 7th 2001, to assess the use of these areas by the species. The frequency of visits by blue-winged macaws was recorded by direct counting at strategic points of observation (Nunes and Betini 2002). In this method, we counted macaws from two points of high visibility around each of these 36 forest fragments. The pairs of point-counts were located outside the remnant in question, but close to its border, and were distant from each other. At each point we delimited the length of forest border where the macaws should be recorded, starting from the area of visibility of the observer. These lengths were estimated in loco or calculated later using Global Positioning System (GPS) and Geographical Information System (GIS) tools, and varied among the point-counts (mean + SD = 544.2 ± 153.8 m).

From both strategic point-counts, we recorded all movements of macaws flying out of or into the forest fragment, over a period of 7.5 h, during one morning (6:00–10:30 h) and one afternoon (15:00–18:00 h). For each visual contact with macaws, we recorded the route of the flocks (into or out of the fragment), the number of individuals, the time, the direction of flight and any relevant observations. A blue-winged macaw visitation index was calculated for each sampled fragment as the total number of records of individuals flying into the fragment divided by the sum of the border length sampled by the two observers (records of macaws/km).

Habitat data

We related the blue-winged macaw visitation index to features of the fragments and their surrounding landscapes. Based on the edited Caetetus map, we calculated landscape metrics and indices using GIS ArcView 3.2 with the extension Spatial

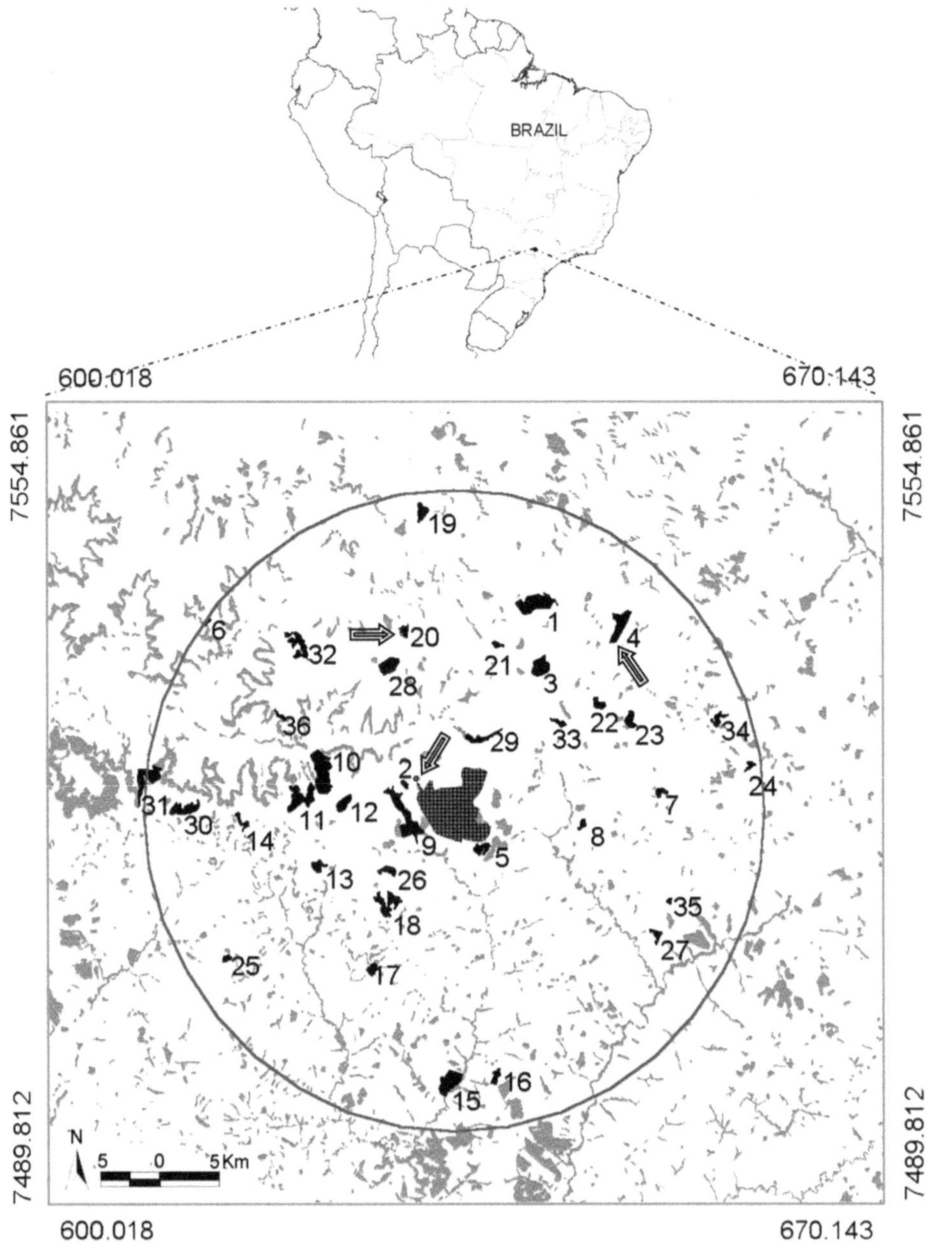

Fig. 1 Location of Caetetus Ecological Station in south-east Brazil and the study area in a 25 km radius (UTM co-ordinate system, zone 22). The reserve is indicated by the hatch pattern patch, the sampled fragments by the black patches, the non-sampled fragments by the gray patches, and the matrix by the white background. The arrows indicate the location of blue-winged macaw comunal roosts

Analist 1.1 (ESRI 1996a,b) and the software Fragstats 3 (McGarigal et al. 2002). Three features of fragments were considered: their vegetation category (C) (old-growth forest or secondary forest), size (S) and shape (SI) (shape index; see

Table 1 Landscape indices used to access habitat use by blue-winged macaw around Caetetus Ecological Station, São Paulo, Brazil

Index	Formula	Description	Units	Range
Shape index of the patch (SI)	$SI = P/P_{min}$	P = perimeter of the patch P_{min} = minimum perimeter possible for a maximally compact patch of the corresponding patch area	Non-dimensional	1—no limit
Isolation index (II)	$II = \sum[NPA/(DNP)^2]$	NPA = neighboring patch area in a 3 km radius from the focal patch DNP = nearest distance between the focal patch and the neighboring patch	Non-dimensional	0—no limit

Table 1). Five features of the surrounding landscapes were also considered: isolation (II) (isolation index; see Table 1), the main land use of the surroundings (LU) (pasture, coffee, citric or *Eucalyptus* plantations), distance to the protected area, Caetetus Ecological Station (DC), distance to the closest city (DCC) and distance to a known macaw sleeping roost (DR).

The variables were tested for normality and, when necessary, transformed to achieve normality. The following transformations were used for the indicated variables: square root for the blue-winged macaw visitation index and for the distance to a known macaw roost (srtDR), and natural logarithms for fragment size (log S), fragment shape index (log SI) and fragment isolation index (log II). The variables distance to Caetetus Ecological Station and distance to the closest city were not transformed.

The possible correlations between each pair of explanatory variables were also examined since such a relationship could create multicollinearity problems in multiple models. Univariate analyses were run between explanatory variables and the blue-winged macaw visitation index. Pearson's product–moment correlations (between quantitative variables) and one-way ANOVA (among categorical and quantitative variables) were used for these statistical analyses.

The multiple linear regression model was used to quantify the relationships among groups of explanatory variables and the blue-winged macaw visitation index. Construction of the multiple regression model and selection of the variables were done by evaluating the best fit using statistical procedures for diagnosis that included the inspection of assumptions, significances, coefficients of correlation, multicollinearity problems and leverages. All statistical analyses were done with Statistica 6.0 software (StatSoft 2001).

Table 2 Remnants of native vegetation in a 25 km radius around the Caetetus Ecological Station (196343.75 ha), south-east Brazil

Category	Total area of cover (ha)	Number of fragments	Mean size (ha)	SD of size (ha)
Forest	16915.59	821	20.60	86.87
Secondary forest	16326.9	974	16.76	37.88
Savanna	4510.71	125	36.08	62.10
Forested savanna	798.21	51	15.65	15.90
Swamp	1810.71	53	34.16	44.13
Total	40362.12	2,024	123.25	246.88

Results

The native vegetation around the protected area covers approximately 25% of the study area (25 km radius), and is greatly reduced and highly fragmented, with marked variation in shape and size among remnants (Table 2, Fig. 1). Within this landscape, we observed flocks of blue-winged macaws and other species of parrots, including scaly-headed parrots (*Pionus maximiliani*), white-eyed parakeets (*Aratinga leucophthalmus*), yellow-chevroned parakeets (*Brotogeris chiriri*), reddish-bellied parakeets (*Pyrrhura frontalis*) and blue-winged parrotlets (*Forpus xanthopterygius*).

The blue-winged macaws occurred throughout the landscape around the protected area and were extremely mobile, usually crossing the landscape at mid-high altitude and landing in patches of native vegetation, isolated trees in the matrix, and agricultural fields. We observed macaws feeding on coffee bushes (*Coffeea arabica*, Rubiaceae), one of the main crops in the region, and other exotic fruits such as chinaberry tree (*Melia azedarach*, Meliaceae).

We located three macaw sleeping roosts within the study area (roost 1: Lat 49°44′S, Long 22°16′W; roost 2: Lat 49°43′S, Long 22°23′W; and roost 3: Lat 49°34′S, Long 22°17′W). Two of these (roost 1 and 2) were in isolated guapuruvus trees (*Schizolobium parahyba*, Leguminosae) on pasture, an canopy species which usually shows many hollows of fallen branches, and one roost (roost 3) was in trees at a fragment edge. We counted the maximum number of blue-winged macaws during 1 day in each roost and we recorded: 23 macaws at roost 1, 38 macaws at roost 2, and 30 macaws at roost 3. Such roosts do not hold all the population of blue-winged macaws of the region, because there are some macaws that roost in small flocks in the Caetetus Ecological Station and other forests fragments, and we cannot assure that all comunal roosts were found in this study.

We recorded macaws, either visually or aurally, in 29 of the 36 sampled fragments. However, aural-only records and those of individuals flying out of patches were not included in the visitation indices, which left 25 patches for which the species was recorded. We obtained 173 records of macaws flying to satellite patches around the protected park, usually in pairs (56,6%) but also in groups of up to eight individuals; larger groups were seen on a few occasions.

The species was easily detected in all of 29 fragments, mainly within the first 3 h of sampling, suggesting we detected all fragments visited by blue-winged macaws (Fig. 2). Macaws flew into patches mainly during the morning (8–10 h) and left the

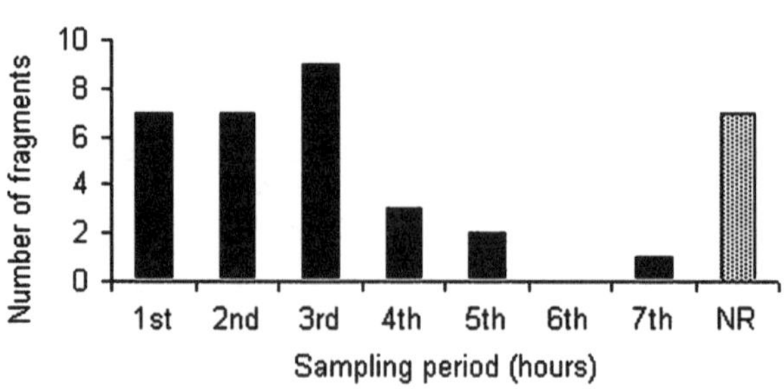

Fig. 2 Sampling period (hours) during which blue-winged macaws were first recorded in each fragment around the Caetetus Ecological Station, SP, Brazil. NR—fragments where macaws were never recorded

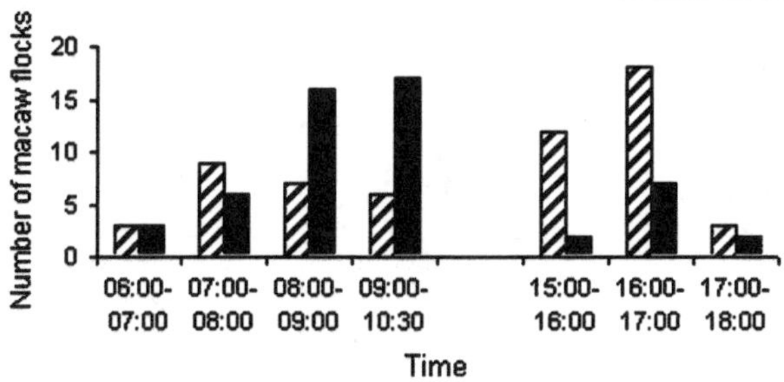

Fig. 3 Variation in the routes of blue-winged macaw flocks during the day in satellite forest fragments around the Caetetus Ecological Station, SP, Brazil. The bars represent the direction of flight of the flocks: black bars—towards the forest patch, hatched bars—away from the forest patch

patches mainly during the afternoon (16–17 h) (Fig. 3). In general, there was no significant difference between the number of records of macaws arriving and leaving the fragments [Wilcoxon matched pairs test, $Z = 0.042$, $n = 36$, $P > 0.05$], which suggested that most of these records probably corresponded to the same individual on two-way flights. Hence, the use of records of only one-way flights in the visitation index avoided the double counting of individuals.

Significant associations were found between pairs of landscape metrics and indices, despite the fact that each of them measured a peculiar feature of the landscape: shape was positively correlated with size [Pearson's product–moment $r = 0.433$, $P < 0.025$], distance to Caetetus was positively correlated with distance to a known macaw roost [$r = 0.597$, $P < 0.001$], and vegetation categories were not equally distributed over different distances to a known macaw roost [ANOVA, $F_{(2,33)} = 3.286$, $P > 0.05$].

The frequency of blue-winged macaw visits was associated with certain features of the patch and landscape (Appendix 1). The visitation index was positively correlated with patch size [$r = 0.414$, $P < 0.025$] and negatively correlated with distances to a known macaw roost [$r = -0.686$, $P < 0.001$] and with distances to the Caetetus Ecological Station [$r = -0.493$, $P < 0.001$] (Fig. 4). The main type of land use around the patches (LU) also affected the blue-winged macaw visitation index [$F_{(3,32)} = 4.511$, $P > 0.025$] since the index mean was lower in patches surrounded by pasture than by other land uses [Tukey HSD test, $P > 0.05$] (Fig. 5).

A dichotomous variable "pasture versus non-pasture" was derived from "the main land use of the surroundings" variable and showed that patches surrounded by pastures had a lower visitation index [t test, $t_{(34)} = 3.595$, $P < 0.001$], and were smaller in size [$t_{(34)} = 2.151$, $P < 0.05$], farther from Caetetus Ecological Station [$t_{(34)} = 14.474$, $P < 0.025$] and farther from a known macaw roost [$t(34) = 3.554$, $P < 0.025$].

In the best fit multiple regression model, the frequency of visits of macaws was explained by size and distance to a known macaw sleeping roost, as well as by distance to the closest city, a non-significant variable in univariate analyses [$r = 0.055$, $P > 0.05$] which now showed a significant positive association with the visitation index. Hence, larger fragments, which are closer to blue-winged macaw roosts and farther from urban localities, are likely to be chosen by macaws during their movements. This multivariate model was highly significant [$F_{(3,32)} = 15.843$, $R^2 = 0.597$, $P < 0.001$] and explained almost 60% of the variation in the visitation index.

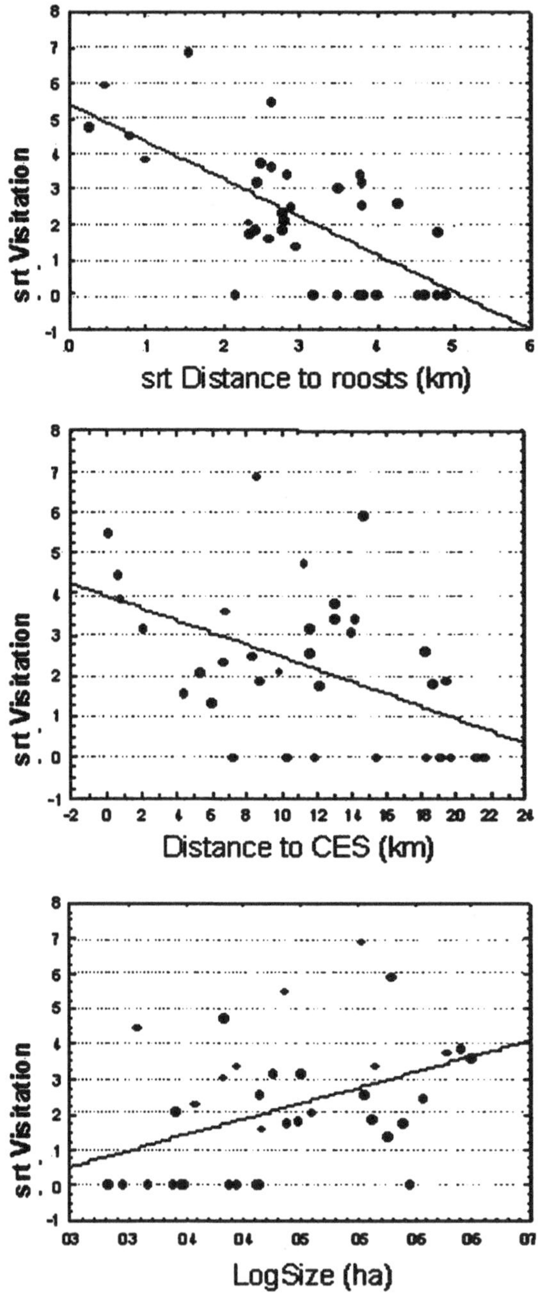

Fig. 4 Relationships between the visitation index for blue-winged macaws and habitat features that were significant in Pearson's product–moment correlations in the fragments around Caetetus Ecological Station, SP, Brazil

The variable "distance to Caetetus Ecological Station" was not included in the multiple model, despite its significant correlation with the visitation index in univariate analyses. Distance to Caetetus Ecological Station was highly correlated to distance to a known macaw sleeping roost, which resulted in redundant information for these variables. This correlation probably reflected the fact that one of the three

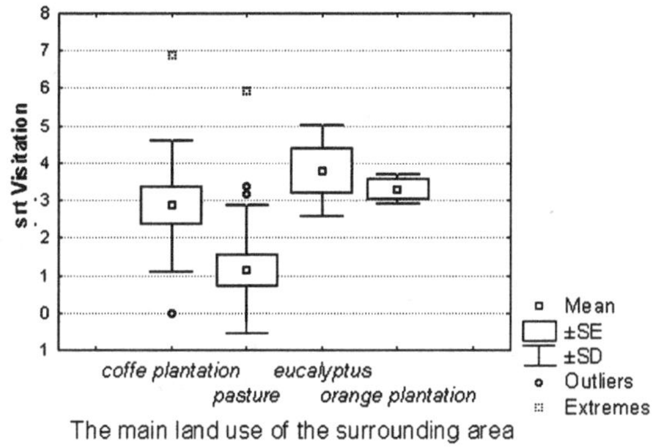

Fig. 5 Means visitation indices for blue-winged macaws in satellite forest fragments in relation to the land use of the surrounding area around the Caetetus Ecological Station, SP, Brazil

roosts was located less than 1 km from the edge of the reserve. Despite this strong correlation, when the variable "distance to Caetetus Ecological Station" replaced "distance to a known macaw sleeping roost" in the multiple model, the explanatory power of the model was reduced [$F_{(3,32)} = 6.588$, $R^2 = 0.381$, $P < 0.025$].

The relationship between the blue-winged macaw visitation index and landscape features was unaffected by the main land use of the surroundings or the dichotomous variable "pasture versus non-pasture" in a multivariate model. Since neither of the latter two variables improved the model they were not included in the multiple model.

Discussion

Habitat selection

Our results revealed the ability of the vulnerable blue-winged macaw to thrive in a mosaic of landscape containing only 25% of the original native forest, even so we do not have information if this population is stable or in decline. The presence of macaws was not restricted to large blocks of forest and the species was managed to spread into the fragmented landscape and adapted their foraging behavior to include exotic species. Such tolerance and adaptability to habitat modification have also been reported for the other two species of *Primolius*, the blue-headed macaw (*P. couloni*) and the yellow-collared macaw (*P. auricolis*) (del Hoyo et al. 1997; Juniper and Parr 1998).

During the study, we frequently recorded blue-winged macaws flying to, and landing in, anthropogenic landscapes and forest fragments around the Caetetus Ecological Station. Such observations suggested that blue-winged macaws included these areas in their home-range. A similar pattern was also observed in Sooretama Biological Reserve (Espirito Santo state, Brazil), where the species left the reserve in the morning to travel to orange plantations and small satellite forest fragments and then returned in the afternoon (Marsden et al. 2000).

The blue-winged macaw was recorded in nearly all of the fragments sampled in the Caetetus region, but the frequency of visits varied according to characteristics of the fragments and their surroundings. The landscape features around the fragments (such as distance to the Caetetus Ecological Station, distance to the closest city, distance to a known macaw sleeping roost and the main land use of the surroundings) explained more of the variation in the visitation index than actual measured features of the fragments (such as shape index and vegetation category). This pattern has been observed for generalist and highly mobile species (Naugle et al. 1999; Price et al. 1999; Graham and Blake 2001; Suarez-Seoane and Baudry 2002) because most of them can cross open areas and use small fragments of non-specific quality (Villard and Taylor 1994; Andrén 1994).

Among the landscape-scale parameters, distance to a known macaw sleeping roost and distance to the closest city showed the strongest associations with the frequency of visitations. Distance to Caetetus Ecological Station was also important, but its high correlation with the distance to a known macaw sleeping roost hindered the interpretation of the relative effects of both variables: distance to a known macaw sleeping roost could cause the distance to Caetetus Ecological Station to show a spurious correlation with visitation rate or, perhaps, both features were simultaneously important for patch selection by the birds, what would be reflecting the dispersion of flocks of macaws that sleep in the roosts and others that sleep in trees inside the reserve.

The tendency to gather at sleeping roosts is a common behavior among parrots (Chapman et al. 1989), particularly in fragmented environments where they can find isolated trees that provide greater visibility for security (Gilardi and Munn 1998). Such roosts apparently serve as information centers where parrots can gather at the end of the afternoon to spend the night and leave at dawn to feed (Chapman et al. 1989). The short distance between some blue-winged macaw roosts and an adjacent fragment could result in a high frequency of visits since it is energetically more economical to visit close fragments than distant ones (Eckert and Randall 1988).

The correlation between the visitation index and the distance to Caetetus Ecological Station could be explained by the "mainland" effect (Wiens 1997; Hanski 1998). The reserve, which is the largest patch of forest in the region, would correspond to a core area that serves as a source of macaws that disperse to other patches. Hence, the forest remnants closest to the Caetetus Ecological Station would receive more macaws than more distant fragments. Conversely, the correlation between the visitation index and the distance to the closest city may reflect the influence of disturbance factors that operate in areas close to cities, where there is considerable human activity.

The presence of pasture as the main type of landscape surrounding the fragments negatively affected the frequency of blue-winged macaw visits, perhaps because of the smaller amount of native vegetation left on cattle ranches. There is extensive cattle ranching in the region around the Caetetus Ecological Station, and this has produced large areas of deforestation. Brazilian law compels landowners to preserve at least 20% of the Atlantic forest on their property, but this law is rarely if at all enforced, and most farms do not have native vegetation on their land. Indeed, cattle ranches usually keep only a few isolated trees in their pastures to provide shade for the cattle. Consequently, this landscape can markedly influence habitat connectedness and the distance between patches.

We had expected that isolation would be highly related to a lower frequency of blue-winged macaw visits because it is generally an important factor for mobile organisms in highly fragmented landscapes (Andrén 1994). However, there was no significant correlation between these two parameters, perhaps because the distance involved in these analyses (3 km) was not large enough to detect such relationship, particularly considering the high mobility of macaws. Among the patch-scale parameters, only fragment size was significantly associated with the frequency of macaw visits. The size of a fragment is usually an important factor in explaining the number of species that inhabit an area (MacArthur and Wilson 1967; Price et al. 1999; Graham and Blake 2001; Miller and Cale 2000), especially for sedentary and restricted animals (Blake and Karr 1987; Naugle et al. 1999; Boulinier et al. 2001; Lee et al. 2002). The occurrence of very mobile frugivores is also affected by patch size because of interactions with the abundance and diversity of fruit resources (Price et al. 1999). However, these species are often influenced by a combination of fragment and landscape features (Graham and Blake 2001; Price et al. 1999).

The visitation index was not related to fragment shape, which means that the species was unaffected by the proportion of edges in the patches. Indeed, our observations indicated that blue-winged macaws were not restricted to mature forest since we regularly observed them perched at the edges of fragments. In addition, there was no significant difference between the visitation indices for old-growth and secondary forest patches. This may reflect the plasticity of blue-winged macaws to explore different types of forest. In contrast, the vegetation category apparently affected the choice of roosting sites, which tended to be closer to fragments of old-growth forest.

Unexplained variation in the visitation index in the multiple regression model may be due to some differences in patch-level factors among fragments, such as quality and availability of resources, and vegetation structure. The inclusion of data with finer spatial and temporal resolution should improve the performance of our model, however our model with coarser resolution data was managed to explain a great part of variation of the visit index.

Implications and recommendations for conservation

The presence of a large block of semideciduous forest (Caetetus Ecological Station) may have precluded the local extinction of the blue-winged macaw in the region, at moment. The Caetetus Ecological Station represents a core area for this macaw, because it is unknown whether landscapes containing only small fragments could sustain macaw populations. Price et al. (1999) suggested that a net of fragments can maintain highly mobile frugivores, but that some species can be lost when the fragment size falls below a certain threshold. In our study, patch size was an important predictor of blue-winged macaw sightings, which means that preservation of Caetetus Ecological Station is key to uphold the blue-winged macaw population at the region of our study.

However, small fragments may also be important for the species' survival, and the ability of blue-winged macaws to use a range of different forest fragments indicates the importance of a mosaic of habitats to contribute to the maintenance of this species in such fragmented region. Small fragments provide landscape connectivity for blue-winged macaw in Caetetus region, by offering shelter and rest for the birds during movement, and work as extra sites for feeding, overnight roosts, and may be for nesting. Since this study was done during the dry season, usually a period of food scarcity for parrots (Galetti 1993), the relevance of these areas for the conservation of blue-winged macaws is even greater. These findings are important because they show the necessity of up scaling conservation strategies to protect these parrots.

On the region of our study, it is needed to look at beyond the Caetetus boundaries. Hence, the survival of the blue-winged macaw also depends on the co-operation of land owners, since most of the extra-reserve fragments are privately owned. There is a need to encourage the conservation of remnants of native vegetation on surrounding private land, particularly the large fragments near roosts and the Caetetus Ecological Station. The presence of patches of trees in the matrix may be important too, because they tend to increase the connectivity of the landscape (Bolger et al. 2001; Fischer and Lindenmayer 2002; Graham and Blake 2001).

Efforts must be concentrated on conserving the population of *P. maracana* in the region of the Caetetus Ecological Station because this is the largest remaining population of this species in the southern part of its distribution (Nunes 2003). This population could serve as a source of blue-winged macaws for the colonization of other areas from which the species has been extinguished. The conservation of "source populations", at the extreme of geographic distribution of species which respond factors at landscape scale, tends to be vital for maintaining these species' regional occurrences (see Wilson et al. 2002). Therefore we expect that a fortuitous extinction of the population at Caetetus could have a wider negative impact on the species' geographic distribution as a whole.

Acknowledgments We thank the Instituto Florestal for allowing us to work in Caetetus, and S. J. Marsden, M.A. Pizo, C. A. Vettorazzi, D. Eaton, A. Keuroghlian, A. Townsend Peterson and N. Collar for suggestions and comments on the manuscript. This work was supported by the New England Zoo Society, Fundação O Boticário de Proteção à Natureza, FAPESP and the American Bird Conservancy. M.G. is supported by a research fellowship from CNPq.

Appendix

Appendix 1 The fragments sampled, their features and the visitation indices for blue-winged macaws

Fragment	C	LU	S (ha)	SI	II	DCC (km)	DC (km)	DR (km)	Bwm visit index
1	Forest	Coffee plantations	321.21	1.75	1.01	1.34	13.01	6.22	14.00
2	Forest	Coffee plantations	21.78	1.28	88.25	6.19	0.66	0.59	20.00
3	Forest	Coffee plantations	164.97	1.23	1.41	6.19	8.76	5.79	3.39
4	Forest	Pasture	200.61	1.72	0.68	1.37	14.70	0.21	35.00
5	Secondary forest	Eucalyptus	78.75	1.98	1759.46	8.32	0.09	6.84	30.00
6	Forest	Coffee plantations	19.17	1.40	1392.96	1.73	21.26	15.95	0.00
7	Secondary forest	Orange plantations	45.99	1.41	4.14	2.90	14.03	12.17	9.23
8	Forest	Pasture	31.41	1.34	2.05	6.70	7.20	13.99	0.00
9	Forest	Eucalyptus	363.96	2.61	1274.88	3.08	0.70	0.97	14.71
10	Forest	Orange plantations	394.47	1.71	418.54	2.29	6.83	6.87	12.90
11	Forest	Coffee plantations	260.46	2.21	543.56	0.92	8.28	8.28	6.06
12	Forest	Coffee plantations	99.09	1.36	5.83	2.41	5.34	5.37	4.17
13	Forest	Pasture	70.38	1.52	3.73	7.74	11.61	14.44	10.00
14	Secondary forest	Pasture	51.03	2.31	26.61	4.12	14.25	14.20	11.36

Appendix 1 continued

Fragment	C	LU	S (ha)	SI	II	DCC (km)	DC (km)	DR (km)	Bwm visit index
15	Forest	Pasture	218.88	1.45	722.60	7.49	18.58	23.01	3.13
16	Secondary forest	Pasture	51.57	1.60	27.68	6.76	18.35	23.69	0.00
17	Forest	Eucalyptus	63.00	1.28	115.10	7.74	11.65	14.48	6.45
18	Forest	Coffee plantations	191.79	2.68	3.04	2.06	5.98	8.59	1.82
19	Forest	Coffee plantations	87.93	1.41	1.15	4.08	19.42	7.59	3.33
20	Forest	Coffee plantations	46.62	1.17	1.74	7.50	11.27	0.06	22.45
21	Secondary forest	Pasture	30.69	1.54	1.41	5.29	9.87	7.71	4.35
22	Forest	Pasture	60.57	1.37	0.85	3.63	10.33	4.58	0.00
23	Forest	Pasture	78.84	1.45	21.48	2.90	12.18	5.47	2.99
24	Secondary forest	Pasture	32.40	1.71	3.33	5.37	21.57	14.57	0.00
25	Forest	Pasture	29.52	1.22	11.19	8.75	19.74	20.53	0.00
26	Secondary forest	Pasture	64.08	1.80	2.54	0.53	4.38	6.63	2.50
27	Secondary forest	Pasture	47.79	1.77	4.53	6.21	15.45	22.73	0.00
28	Forest	Coffee plantations	151.74	1.27	3.52	9.32	8.55	2.38	47.37
29	Forest	Coffee plantations	90.90	2.56	121.91	11.58	2.09	5.92	10.00
30	Forest	Coffee plantations	159.03	1.86	24.75	2.70	18.18	18.17	6.67
31	Forest	Pasture	230.76	2.31	6688.39	4.40	21.30	21.32	0.00
32	Secondary forest	Eucalyptus	172.89	3.14	1.62	4.87	12.99	7.99	11.43
33	Secondary forest	Coffee plantations	35.73	1.73	2.77	7.11	6.67	7.61	5.26
34	Forest	Pasture	63.18	2.34	12.01	3.81	19.19	10.09	0.00
35	Secondary forest	Pasture	16.83	1.18	1.64	4.71	15.46	22.72	0.00
36	Forest	Pasture	23.67	1.85	2380.85	8.02	11.90	12.11	0.00

Abbreviations: C, category; LU, main land use in the surroundings; S, size; SI, shape index; II, isolation index; DCC, distance to the closest city; DC, distance to the Caetetus Ecological Station; DR, distance to the closest macaw roost; Bwm, blue-winged macaw

References

Aleixo A, Vielliard JM (1995) Composição e dinâmica da floresta Santa Genebra, Campinas, São Paulo, Brasil. Rev Bras Zool 12:493–511

Andrén H (1994) Effects of habitat fragmentation on birds and mammals in landscapes with different proportions of suitable habitat – a review. Oikos 71:355–366

Belton W (1994) Aves do Rio Grande do Sul, distribuição e biologia. Universidade do Vale do Rio dos Sinos, São Leopoldo, Brazil

Benke GA (2001) Lista de referência das aves do Rio Grande do Sul. Fundação Zoobotânica do Rio Grande do Sul, Porto Alegre, Brazil

BirdLife International (2000) The world list of threatened birds. Lynx Editions and Bird Life International, Barcelona, Spain and Cambridge, England

BirdLife International (2004) Threatened birds of the world 2004 CD-ROM. BirdLife International, Cambridge, England

Blake JG, Karr JR (1987) Breeding birds of isolated woodlots: area and habitat effects. Ecology 68:1724–1734

Bolger DT, Scott TA, Rotenberry JT (2001) Use of corridor-like landscape structures by bird and small mammal species. Biol Conserv 102:213–224

Bornschein M, Straube FC (1991) Sobre o status de três espécies de aves no Estado do Paraná: *Crax faciolata, Ara maracana* e *Psarocolis decumanos*. In: Resumos do I Congresso Brasileiro de Ornitologia, Belém, Brazil, 21–26 July 1991, pp 33. Ed. Super Cores, Belém, Brazil pp 33
Boulinier T, Nichols JD, Hines JE, Sauer JR, Flather CH, Pollock KH (2001) Forest fragmentation and bird community dynamics: inference at regional scales. Ecology 82:1159–1169
Brooks T (2000) Conservation biology, living on the edge. Nature 403:26
Camara G, Souza RCM, Freitas UM, Garrido J (1996) SPRING: integrating remote sensing and GIS by object-oriented data modeling. Comput Graph 20:395–403
Chapman CA, Chapman LJ, Lefebvre L (1989) Variability in parrot flock size – possible functions of communal roosts. Condor 91:842–847
Chebez JC (1996) Fauna misionera. L.O.L.A. Press, Buenos Aires, Argentina
Clay RP, Capper DR, Barnett JM, Burfield IJ, Esquivel EZ, Fariña R, Kennedy CP, Perrens M, Pople RG (1998) White-winged nightjars and cerrado conservation: the key findings of Project Aguará Ñu 1997:1. Cotinga 9:52
Cullen L, Bodmer RE, Padua CV (2000) Effects of hunting in habitat fragments of the Atlantic forests, Brazil. Biol Conserv 95:49–56
del Hoyo J, Elliott A, Sargatal J (1997) Handbook of the birds of the world, Sandgrouse to Cuckoos. Lynx Editions, Barcelona, Spain
de la Peña MR (1999) Aves argentinas, lista y distribución. L.O.L.A. Press, Buenos Aires, Argentina
Eckert R, Randall D (1988) Animal physiology: mechanisms and adaptations. W. H. Freeman, New York, New York, USA
ESRI (1996a) Environmental Systems Research Institute. Using ArcView GIS. ESRI Press, Redlands, California, USA
ESRI (1996b) Environmental Systems Research Institute. Usin ArcView Spatial Analyst. ESRI Press, Redlands, California, USA
Fischer J, Lindenmayer DB (2002) The conservation value of paddock trees for birds in a variegated landscape in southern New South Wales. 2. Paddock trees as stepping stones. Biodivers Conserv 11:833–849
Fundação SOS Mata Atlântica and INPE (1998) Atlas da evolução dos remanescentes florestais e ecossistemas associados do domínio da Mata Atlântica no período de 1991–1995. Fundação S.O.S. Mata Atlântica, São Paulo, Brazil
Galetti M (1993) Diet of the scaly-headed parrot (*Pionus maximilliani*) in a semidesciduous forest in southeastern of Brazil. Biotropica 25:491–425
Gilardi JD, Munn CA (1998) Patterns of activity, flocking, and habitat use in parrots of the Peruvian Amazon. Condor 100:641–653
Graham CH, Blake JG (2001) Influence of patch- and landscape-level factors on bird assemblages in a fragmented tropical landscape. Ecol Appl 11:1709–1721
Hanski I (1998) Metapopulation dynamics. Nature 396:41–49
Juniper T, Parr M (1998) Parrots: a guide to the parrots of the world. Pica Press, London, England
Kronka FJN, Matsukuma CK, Nalon MA, Del Cali IH, Rossi M, Mattos IFA, Shin-Ike MS, Pontinhas AS (1994) Inventário florestal do Estado de São Paulo. Instituto Florestal de São Paulo, São Paulo, Brazil
Lee M, Fahrig L, Freemark K, Currie DJ (2002) Importance of patch scale vs. landscape scale on selected forest birds. Oikos 96:110–118
Lowen JC, Bartrina L, Clay RP, Tobias JA (1996) Biological surveys and conservation priorities in eastern Paraguay: the final reports of projects CANOPY '92 and YACUTINGA '95. CBS Conservation Publications, Cambridge, England
MacArthur RH, Wilson EO (1967) The theory of island biogeography. Princeton University Press, Princeton, NJ, USA
MacNally RM, Horrocks G (2000) Landscape-scale conservation of an endangered migrant: the swift parrot (*Lathamus dicolor*) in its winter range. Biol Conserv 92:335–343
Marsden SJ, Whiffin M, Sadgrove L, Guimarães PR (2000) Parrot populations and habitat use in and around two lowland Atlantic forest reserves, Brazil. Biol Conserv 96:209–217
McGarigal K, Cushman SA, Neel MC, Ene E (2002) FRAGSTATS: spatial pattern analysis program for categorical maps. Computer software program produced at the University of Massachusetts, Amherst, Massachusetts, USA
Miller JR, Cale P (2000) Behavioral mechanisms and habitat use by birds in a fragmented agricultural landscape. Ecol Appl 10:1732–1748
Naugle DE, Higgins KF, Nusser SM, Johnson WC (1999) Scale-dependent habitat use in three species of prairie wetland birds. Landsc Ecol 14:267–276

Nunes MF (2000) Ecologia e conservação do maracanã-verdadeiro *Primolius maracana* (Psittacidae) em um fragmento florestal do sudeste do Brasil. Bachelor Monograph. Universidade Estadual Paulista, Rio Claro, São Paulo, Brazil

Nunes MF (2003) Distribuição do maracanã-verdadeiro *Primolius maracana* (Psittacidae): modelo de habitat e fatores que influenciam na persistência de suas populações remanescentes. MSc thesis. ESALQ, Universidade de São Paulo, Piracicaba, SP, Brazil

Nunes MF, Betini GS (2002) Estimativas de tamanhos populacionais de psitacídeos. In: Galetti M, Pizo MA (eds) Ecologia e conservação de psitacídeos no Brasil. Melopsittacus Publicações Científicas, Belo Horizonte, Brazil, pp 123–140

Olmos F (1993) Birds of the Sierra da Capivara National Park in the "caatinga" on north-eastern Brasil. Bird Conserv Int 3:21–36

Pimm SL (1995) The future of biodiversity. Science 269:347–350

Price OF, Woinarski JCZ, Robinson D (1999) Very large area requirements for frugivorous birds in monsoon rainforests of the Northern Territory, Australia. Biol Conserv 91:169–180

Rosário LA (1996) As aves em Santa Catarina: distribuição geográfica e meio ambiente. Fundação Amparo and Tecnologia and Meio Ambiente, Florianópolis, Santa Catarina, Brazil

Roshier DA (2003) On animal distributions in dynamic landscapes. Ecography 26:539–544

Soulé ME (1987) Viable populations for conservation. Cambridge University Press, Cambridge, England

StatSoft, Inc. (2001) STATISTICA, data analysis software system, version 6. http://www.statsoft.com

Suarez-Seoane S, Baudry J (2002) Scale dependence of spatial patterns and cartography on the detection of landscape change: relationships with species' perception. Ecography 25:499–511

Uezu A, Metzger JP, Vielliard JME (2005) Effects of structural and functional connectivity and patch size on the abundance of seven Atlantic Forest bird species. Biol Conserv 123:507–519

Viana VM, Tabanez AA (1996) Biology and conservation of forest fragments in the Brazilian Atlantic moist forest. In: Schellas J, Greenberg R (eds), Forest patches in tropical landscapes. Island Press, Washington D.C., USA, pp 151–167

Villard M, Taylor PD (1994) Tolerance to habitat fragmentation influences the colonization of new habitat by forest birds. Oecologia 98:393–401

Wiens JA (1997) Metapopulation dynamics and landscape ecology. In: Hanski IA, Gilpin ME (eds) Metapopulation biology: ecology, genetics, and evolution. Academic Press, Cambridge, England, pp 43–52

Willis EO (1979) The composition of avian communities in remanescent woodlots in southern Brazil. Papeis Avulsos Zool 33:1–25

Wilson EO (1988) Biodiversity. National Academy Press, Washington D.C., USA

Wilson RJ, Ellis S, Baker JS, Lineham ME, Whitehead RW, Thomas CD (2002) Large-scale patterns of distribution and persistence at the range margins of a butterfly. Ecology 83:3357–3368

Biodivers Conserv (2007) 16:969–985
DOI 10.1007/s10531-006-9036-7

ORIGINAL PAPER

Commercialization and use of snakes in North and Northeastern Brazil: implications for conservation and management

Rômulo Romeu da Nóbrega Alves ·
Gentil Alves Pereira Filho

Received: 24 January 2006 / Accepted: 7 April 2006 / Published online: 16 May 2006

Abstract Snakes are sold in many markets and religious article stores in Brazil. Besides their use as food, snakes are exploited in a variety of ways, such as pets, or for use in traditional medicine and magic/religious rituals (especially in Afro-Brazilian religions). Despite widespread commercialization, there is a general lack of information about this snake trade, which makes it difficult to evaluate its magnitude and its impact on reptile populations. This work documents the commercialization and use of snakes in five cities in Northeastern (São Luís, Teresina, João Pessoa and Campina Grande) and Northern (Belém) Brazil, through interviews with 119 merchants of biological products in outdoor markets and religious articles stores. The data was gathered through the use of semi-structured questionnaires, complemented by semi-directed interviews. The products derived from 11 snake species were being commercialized for medicinal or magical/religious purposes. *Boa constrictor*, *Crotalus durissus* and *Eunectes murinus* were the species most commonly sold. The economic importance of snakes as sources of medicines and religious products demonstrates the need for the development of sustainable use programs for these species.

Keywords Biodiversity · Ethnozoology · Traditional medicine · Zootherapy · Wildlife trade

Introduction

Brazil has the richest fauna and flora of all of South and Central America. Approximately 650 reptile species are currently known to occur in Brazil: 610

R. R. N. Alves (✉)
Departamento de Biologia, Universidade Estadual da Paraíba, Av. das Baraúnas, 351/Campus Universitário, Bodocongó, 58109-753, Campina Grande, PB, Brasil
e-mail: romulo_nobrega@yahoo.com.br

G. A. P. Filho · R. R. N. Alves
Programa de Pós-Graduação em Ciências Biológicas (Zoologia), Departamento de Sistemática e Ecologia, Universidade Federal da Paraíba, 58051-900, João Pessoa -PB, Brasil

Squamata (330 snakes, 230 lizards, 50 amphisbaenids), 6 caimans, and 35 turtles (Rodrigues 2005). A significant part of this fauna had been used by traditional human populations, and some are still being used by modern societies.

Reptile populations are being seriously reduced throughout the world. Factors responsible for the observed declines are thought to include the alteration, destruction, or fragmentation of habitat; climate change; disease; as well as impacts from non-indigenous species, ultraviolet radiation, and xenobiotic chemicals (Gibbons et al. 2000). The collection of wild animals for subsistence or commercialization has also been invoked as a factor contributing to the declines seen in certain species (Gibbons et al. 2000).

Besides their use as food, reptiles have been exploited for various purposes, such as commercialization as pets and use in traditional medicine and in magical–religious rituals (Franke and Telecky 2001; Fitzgerald et al. 2004; Zhou and Jiang 2004). Unfortunately some species are not being used in a sustainable manner and are being commercialized illegally, generating a good deal of concern in relation to the impact of this use on natural populations. Some of these snakes species are listed as rare or threatened species, and over exploitation is one of the causes of this problem.

Snakes are commercialized in outdoor markets (Almeida and Albuquerque 2002; Freire 1996; Costa-Neto 1999) and religious article stores for medicinal and magic–religious purposes in Brazil. Traditional outdoor produce markets, where herbal merchants are also installed, represent important sources of information concerning the use of the native flora and fauna of a given region. Albuquerque (1997) pointed out the role of these public markets in terms of their social and symbolic functions related to the medicinal and magical/religious use of biological products. For Jain (2000), this is a rich field that is commonly neglected.

Despite the intensive use and commercialization of snakes in Brazil, there is a general lack of information about this trade, which makes it difficult to evaluate its magnitude and its impact on these reptiles. Although some information concerning the use of wildlife for medicinal purposes is available from the published ethnobiological literature, in most cases little is know regarding the harvesting and commercialization of the species involved, either the market dynamics or the conservation impact of these activities.

Given Brazil's large cultural and biological diversity, the country may serve as a useful case study in or quest for knowledge about faunal resources used for medicinal and religious purposes. This work documents the commercialization and use of snakes in five cities in Northeastern (São Luís, Teresina, João Pessoa and Campina Grande) and Northern Brazil (Belém), and had two specific purposes: (1) to catalog the snake species being sold for medicinal or magic/religious purposes and, (2) to identify the main locations where these animals are sold. Considering that information about this theme is rare, it is hoped that the data presented here may be useful in developing strategies for the conservation and sustainable use of these animals.

Methods

Study areas—The localities surveyed are shown in Fig. 1, and are briefly described as follows:

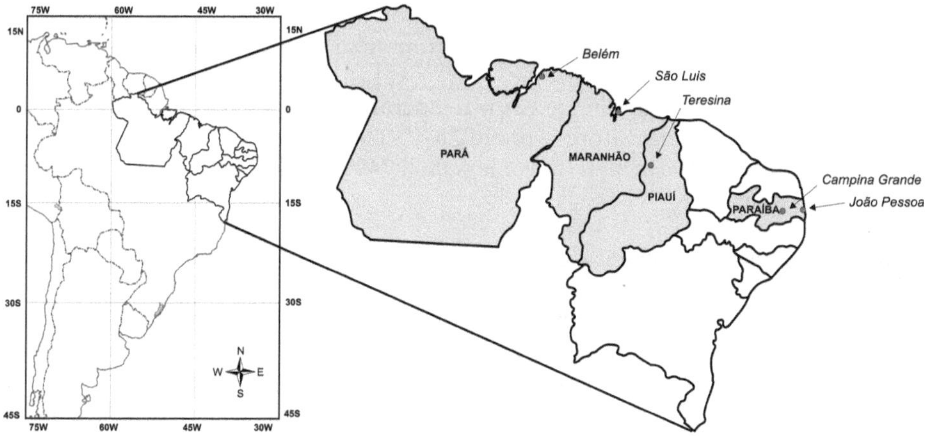

Fig. 1 Map showing the surveyed cities and states in Brazil

São Luís—Maranhão State

São Luís (02°31′47″ S and 44°18′10″ W) is the capital of Maranhão State and is located on São Luís Island on the banks of the São Marcos and São José bays, on an island in the deltas of the Pindaré and Itapecuru Rivers, bordering the Atlantic Ocean, in NE Brazil. The population of the metropolitan area is 1,227,659 (the 16th largest city in Brazil). The climate is tropical/equatorial (hot and semi-humid) with a 4–5 month dry period. The average annual temperature is approximately 26°C, the relative humidity varies from 70% to 80%, and the average annual rainfall totals 2,340 mm. A good part of the island has suffered antropic action, but some areas still remain native vegetation, especially the herbaceous and arboreal species typical of coastal *restinga* vegetation and the mangrove areas (Ahid et al. 1999).

According to the Brazilian Institute of Geography and Statistics (IBGE), the city had 870.028 inhabitants in 2000, with 837.584 living in the urban zone, and 32.444 in rural areas. However, recent data indicates that the population has grown to 978.824 inhabitants (http://www.ibge.gov.br).

Teresina—Piauí State

Teresina is the capital and largest city of the State of Piauí, and is the only inland capital in the northeastern region of Brazil. The municipality of Teresina is located in the central northern part of the State (05°05′21″ S and 42°48′07″ W) and covers 1,809 km^2. The central part of the city is situated between the Parnaíba and Poti rivers. The city of Teresina has over 752,000 inhabitants, while the greater metropolitan area has approximately 996,000 inhabitants. Teresina is located near the equator, and temperatures only vary between 26°C and 38°C, with the lowest temperatures occurring in the first 5 months of the year.

João Pessoa—Paraíba State

João Pessoa is the capital of Paraíba State (07°06′54″ S and 34°51′47″ W), occupies 210 km^2, and is located at the easternmost point of the Americas. It is the state's largest and most important city, with a population of about 660,000 (about one sixth of the total population of the state). The metropolitan area encompasses three other satellite cities with ca. 270,000 people.

João Pessoa has a mild climate, with temperatures between 20°C and 31°C degrees, with average temperatures around 26°C. The city is 47.5 m above sea level, and the humidity is always high (never less than 74%), with cooler weather between March and August.

Campina Grande—Paraíba State

Campina Grande is located in the central part of the State of Paraíba (7°13′11″ S and 35°52′31″ W) and is the second most populated city in the state (behind the capital, João Pessoa, located about 120 km away). Aguiar (2005) noted that it is the largest and most important city in the interior of Paraíba and one of the major Northeastern population centers, with significant farming and cattle ranching, as well as industry and commerce. The city of Campina Grande covers 970 km^2, and all but 411 km^2 correspond to urban area. According to IBGE the estimated population for the year of 2005 was 376.132.

Belém—Pará State

Belém is the capital and also the largest city of the state of Pará (01°27′21″ S and 48°30′16″ W). Its population is over 1.3 million, making it the 10th largest city in Brazil. Its metropolitan area has approximately 2.01 million inhabitants. It and Manaus are the most important cities in the Brazilian Amazon region.

Belém has hundreds of outdoor markets and shops selling agricultural commodities, fish, and a wide range of the Amazonian flora and fauna. One of the regions' most famous outdoor markets is the "Ver-O-Peso" in the city of Belém, where fruit, fish, meat, medicinal and other herbs, as well as handicrafts are sold. Products reach the market by both boat and truck. Additional riverside ports host wholesale fruit and vegetable markets while scores of smaller, open-air retail markets are dispersed throughout the city (Shanley et al. 2002).

Procedures

The present study was carried out between January, 2004 and November, 2005. Visits were made to outdoor markets and religious articles stores where products derived from snakes are commonly sold. Information on the use and commercialization of snakes for medicinal and religious purposes was collected through interviews with 114 merchants. Of these, 40 were owners of religious articles stores, and 74 sold herbs or roots for medicinal purposes in open markets. The merchants interviewed were distributed among the following towns: Belém (37), São Luís (37), Teresina (26), João Pessoa (10), and Campina Grande (4).

The sampling method was non-random, and the interviewees were pre-defined (Albuquerque and Paiva 2004). Attempts were made to interview all of the snake merchants, however interviews were cancelled in some situations, or failed to provide much information, because interviewees were reluctant to answer our questions. The data was gathered employing the user/researcher technique, following Freire (1996). The first phase consists of acting as a costumer (user) interested in buying medicinal products and asking numerous questions about which parts of the animal are more appropriate for different treatments, how to prepare the material, etc. In the second (researcher) phase, the interviewer explains the objectives of the research to the interviewee, pointing out the importance of his/her participation. This initial contact is fundamental to information gathering, as the people involved in these activities tend to omit information as they know that some of the species are protected by environmental laws.

The information obtained through semi-structured interviews (Bernard 1994) was complemented by free interviews (Huntington 2000). Information was sought concerning the snake species used for medicinal and magic/religious purposes, as well as which parts are used in each situation. This information was checked repeatedly so that the interviewer could and add or modify information about any reptile mentioned. Conversations were recorded and later transcribed (when permitted by the interviewees).

The names of the snakes were noted exactly as used by the merchants. The animals mentioned were identified by direct inspection during the interviews or from photographs of the snakes (or their parts) that were taken during interviews.

Results

Snake species commercialized

Eleven snake species are commercialized for medicinal or magic/religious purposes: *Crotalus durissus* (rattlesnake), *Bothrops* sp. (lancehead), and *Lachesis muta* (bushmaster) of the family Viperidae; *Boa constrictor* (boa), *Epicrates cenchria* (rainbow boa), *Eunectes murinus* (anaconda) and *Corallus caninus* (emerald tree boa) of the family Boidae; *Micrurus* sp. (true coral snake) of the family Elapidae; *Spilotes pullatus* (tiger snake), *Oxyrhopus trigeminus* (false coral snake) and *Leptophis ahaetula* (parrot snake) of the family Colubridae.

Additional information concerning snakes being sold in the localities surveyed:

Jibóia (Boa)—Boa constrictor (Linnaeus, 1758)

Medium sized snakes reaching 3.5 m. They feed on lizards, birds and small mammals. The Boas have arboreal habits but can also be found on the ground; they demonstrate both nocturnal and diurnal activity. Their skin colors and patterns are striking, but aid in camouflage under natural conditions (Freitas 1999). They are found throughout Brazil in different biomes (Peters and Orejas-Miranda 1970).

Sucuri (Anaconda)—Eunectes murinus (Linnaeus, 1758)

The largest Brazilian snake can reach 9.0 m and has aquatic habits. They kill their prey by constriction (wrapping their bodies around them and squeezing), and feed on birds, alligators and large mammals. These snakes also prey on domesticated animals such as cows and dogs, and demonstrate both nocturnal and diurnal activity (Freitas 1999). They are found in the Amazon Region, the Pantanal and near the São Francisco River.

Cascavel (Rattlesnake)—Crotalus durissus (Wagler, 1824)

This venomous snake is responsible for many accidents in Brazil. Rattlesnakes can be easily identified by the rattle at the end of their tails. They are a terrestrial species with nocturnal habits, can reach 1.8 m, and feed on marsupials, small mammals and rodents. This species lives primarily in open areas, but can be found in forested regions (Marques et al. 2001) and is well distributed throughout Brazil (Peters and Orejas-Miranda 1970).

Jararaca (Lancehead)—Bothrops Genus (Wagler, 1824)

Snakes of the genus Bothrops are responsible for most accidents involving snake bites in Brazil and in many South America countries (Campbell and Lamar 2004). These pit vipers vary in length from 0.70 m to 2.0 m. They are nocturnal, terrestrial, and feed on marsupials and rodents which they kill with their venom. They are well distributed throughout Brazil, including some islands along the coast.

Surucucu (Bushmaster)—Lachesis muta (Linnaeus, 1766)

This is a large venomous snake reaching up to 4.0 m. They have brown and grayish markings, and responsible for few attacks on humans. Although these animals are fairly large, they are rarely seen, spending most of their time sheltered in rodents' burrows. They feed on mammals and are active nocturnally. They are found throughout the Amazon and the Atlantic Coastal Forest (Campbell and Lamar 2004).

Salamanta (Rainbow Boa)—Epicrates cenchria (Linnaeus, 1758)

This species reaches 1.80 m, and has nocturnal and terrestrial habits. It feeds on lizards, mammals and birds. The Rainbow Boa has a very characteristic color, making it easy to identify (Freitas 1999). It can be found in the Amazon, the Atlantic Coastal Forest, as well as the *Caatinga* drylands.

*Azulão bóia (Parrot snake)—*Leptophis *ahaetula (Linnaeus, 1758)*

This species is diurnal and arboreal. It feeds principally on frogs, but also takes arboreal lizards, grasshoppers, birds, and birds' eggs. These reptiles can reach up to 1.7 m, and have a green or blue color. They are well distributed throughout the Amazon, the Atlantic Coastal Forest, and the *Caatinga* drylands.

*Cobra coral verdadeira (True Coral Snake)—*Micrurus *Genus (Wagler, 1824)*

This genus is represented by terrestrial fossorial snakes that can be active either during the day or night. They can reach till 1.2 m, and feed on worm lizards and other species of snakes, using their venom to kill. They are well distributed throughout the Amazon and Atlantic Coastal Forests and the *Caatinga* drylands (Peters and Orejas-Miranda 1970).

Periquitambóia (Emerald tree boa)—Corallus caninus *(Linnaeus, 1758)*

This is a large snake that can reach 2 m in length. It has nocturnal habits and feeds mostly on small mammals, which are killed by constriction. These serpents are endemic to the Amazon forest.

Caninana (Tiger snake)—Spilotes pullatus *(Linnaeus, 1758)*

This snake can attain 3 m, and is extremely aggressive. It feeds on mammals and birds, which are swallowed while still alive. Its activity is strictly diurnal, and it can be found widely throughout Brazil.

Falsa coral (False Coral Snake)—Oxyrhopus trigeminus (Duméril, Bibron & Duméril, 1854)

These snakes mimic the true coral snake; the principal visible difference is that their color rings are incomplete on the lower belly. *Oxyrhopus trigeminus* has nocturnal habits and feed on lizards and mammals. It can reach 1 m long and is found widely throughout Brazil.

Commercialization

Snakes were commonly sold in outdoor markets and stores that sell religious articles in the cities surveyed. None of the shops visited sold snakes exclusively, but usually also commercialized plants and other animals for different types of medications. They are occasionally sold whole snakes, but most of the time they were already cut into pieces (flesh, skin, tail, eyes, head, cloaca, fat, rattle) with a single snake providing many different kinds of raw materials. For example, the skin, tail, cloaca, eyes, head, excrement, fat, and teeth of *Boa constrictor* are sold separately. In one shop in the city of Belém, a Boa was kept alive (in a plastic recipient) (Fig. 2) and, according to the vendor, its feces were periodically collected and sold, with most sales being for magic and religious purposes.

The snakes (or their parts) are usually stored in plastic or glass jars. The snake fat, one of the most popular products, is stored in 1 or 2 l jars and later divided into smaller flasks (20 ml) (Fig. 3). Whole snakes are generally stored in PET (Polyethylene Terephthalate) bottles with alcohol as the conservative (Fig. 4).

Whole snakes are not usually displayed directly to the public, but are kept in closed rooms behind the stores and presented to customers only when solicited. In Belém, however, whole preserved snakes were often displayed to potential costumers. Snake parts, such as skin or fat, are often displayed together with other medicinal products derived from plants or other animals. One shop in São Luís had a wooden sign openly advertising snakes and their organs (Fig. 5).

The merchants interviewed revealed that they obtained the snakes (or parts) periodically from middlemen, or directly from people from the rural areas that capture snakes and other animals to be sold. Additionally, the merchants may also buy them in large central markets where wildlife products are sold. Belém is well known as a center for these products and many people from other states buy animals there. This is not surprising, as Belém is located in the fauna rich Amazon region.

Fig. 2 A live *Boa constrictor* in the São Braz market, Belém, Pará State, Brazil

Fig. 3 Examples of products derived from snakes and used for medicinal and magical–religious purposes. Head and body fat of *Boa constrictor* (Boa) and the rattle and body fat of *Crotalus durissus* (rattlesnake)

Prices for whole snakes vary depending on the species, availability, demand, and the body part desired, but vary from about U$ 4.5 to U$ 24. The merchant will usually then divide the animals in parts that are sold individually. Snake fat is bought at wholesale prices ranging from U$ 11.2 to U$ 46.7 per litter, and is subsequently sold for U$ 0.90 in 20 ml bottles. Just the head of a Boa may be sold for U$ 4.5. Information concerning

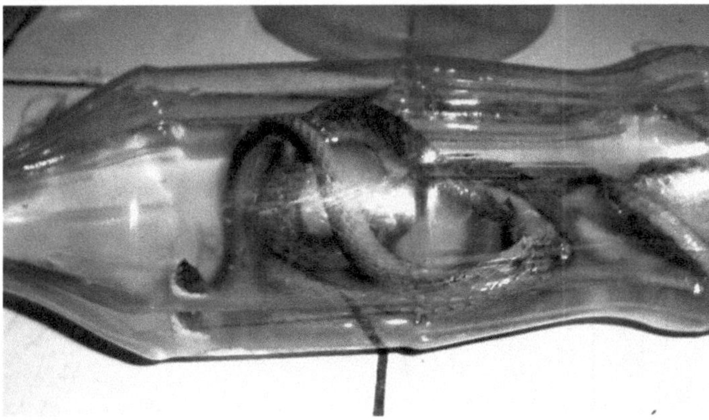

Fig. 4 *Leptophis ahaetula* conserved in alcohol in the open market of São Luís, MA

the volume of sales is extremely hard to come across as there are no official records and the individual merchants do not keep track of their sales. The three most commonly offered species were *Crotalus durissus*, *Boa constrictor*, and *Eunectes murinus*. The first two species have the most diverse uses (Table 1).

Uses

According to the interviewees, snake products can be used to treat 21 different diseases (or illness). Some snake species have multiple uses, and provide a large

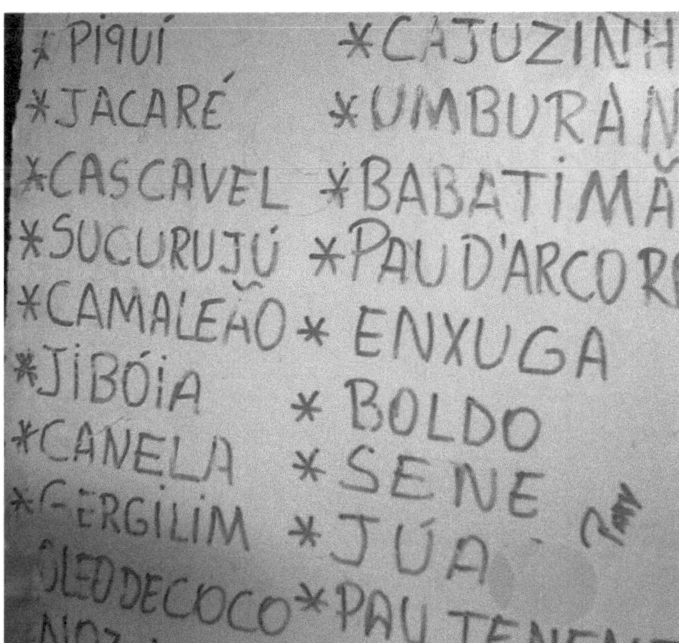

Fig. 5 Sign announcing the sale of snakes (cascavel = *Crotalus durissus*, sucurujú = *Eunectes murinus* and jibóia (Boa) = *Boa constrictor*) in a store in the in São Luís public market, MA

Table 1 Species of snakes encountered at the localities surveyed that are used in popular medicine and magic/religious ceremonies

Family/Species/local name	Medicinal use (MU)		Magic–religious use (MRU)		Number of mentions (ME—MRU)					
Boidae	Parts	Disease (or illness)	Parts	Indications	BE	SL	TE	JP	CG	Total
Boa constrictor (Linnaeus, 1758)/Boa, "jibóia"	Fat	Asthma, ulcer, stomach ache, infection, erysipelas, inflammation, rheumatism, luxation, diabetes, kidney disease, heart disease, leprosy	Skin, tail, vagina, eye, head, faeces, fat, teeth	Afro-Brazilian rituals, magic spells, trade (attract costumers), attract sexual partners, eliminate animals illness, amulet to protect against "evil eye"	15–13	11–15	0–4	0–4	1–0	59
Eunectes murinus (Linnaeus, 1758)/anaconda, "sucurujú," "sucuri"	Fat	Rheumatism, infection, erysipelas, inflammation, asthma, thrombosis		Afro-Brazilian rituals	13–1	8–0	8–0	0–0	0–0	30
Epicrates cenchria (Linnaeus, 1758)/Rainbow boa, "salamanta"	Fat	Rheumatism, sore throat			–	–	–	–	1–0	1
Corallus caninus (Linnaeus 1758)/Emerald tree boa, "cobra papagaio"	Whole	Pain relief in injuries caused by sting of the insect and snake bites				1–0	–	–	–	1

Table 1 continued

Family/Species/local name	Medicinal use (MU)		Magic–religious use (MRU)		Number of mentions (ME—MRU)					
Boidae	Parts	Disease (or illness)	Parts	Indications	BE	SL	TE	JP	CG	Total
Viperidae										
Crotalus durissus (Linnaeus, 1758)/Neotropical rattlesnake, "cascavel"	Fat, skin, rattle, head, eye	Gastritis, rheumatism, spine, kidney disease, swelling, asthma, cancer, osteoporosis, boils, thrombosis	Skin, tail, vagina, head, fat	Magic spells, protect trade, attract sexual partners, trade, religious rituals, repel evil spells	9–2	16–5	20–0	2–0	2–0	56
Bothrops sp./lancehead, "jararaca"	Fat	Rheumatism, pain relief in injuries caused by sting of the insect and snake bites			–	2–0	–	–	–	2
Colubridae										
Lachesis muta (Wied, 1825)/Bushmaster, surucucu	Fat	Rheumatism		Afro-Brazilian rituals	3–1	–	–	–	–	4
Leptophis ahaetula—Parrot snake	Whole	Pain relief in injuries caused by sting of the insect and snake bites			–	1–0	–	–	–	1
Spilotes pullatus (Linnaeus, 1758)—Tiger snake, caninana	Whole	Pain relief in injuries caused by sting of the insect and snake bites			1–0	–	–	–	–	1
Elapidae										
Micrurus sp.—coral snake	Whole, fat	Rheumatism, stings of the snakes and insects			–	1–0	–	–	1	2
Oxyrhopus trigeminus (Duméril, Bibron & Duméril, 1854)	Whole	Stings of the snakes and insects			–	–	–	–	1–0	1

Fig. 6 Head of *B. constrictor* used to prepare extracts commercialized in religious articles stores in São Luís, MA

range of raw material from which many medications are prepared and different diseases are treated. For example, the Boa is used to treat infirmities such as asthma, ulcers, stomach aches, infections, skin infections, inflammations, rheumatism, diabetes, renal dysfunction, heart disease and leprosy (Table 1).

Zootherapeutic products derived from snakes are used in a variety of manners. The hard organs of these animals (such as the teeth and skin) are dried in the sun and then crushed or ground. The powder is used to prepare teas or is mixed with food. The fat is ingested directly or used as an ointment

Four species, *Boa constrictor, Crotalus durissus, Eunectes murinus*, and *Lachesis muta*, have magical/religious attributes beyond their utilization as medicines. The first two species provide a large range of products used for many magic/religious purposes, mainly in Afro-Brazilian rituals (Table 1). Additionally, parts of these animals are also used in the preparation of purifying "baths" which protect people from the negative thoughts of others. For the "baths," is used the "water" of Boa. This water is prepared using parts of snakes (head or body) immersed in alcohol or *patchouly* oil (*Pogostemon* sp.) (Figs. 6, 7). These extracts are mixed with water during the baths or applied as perfumes afterwards. According to some of the interviewees, this procedure will assure the user success in love and/or financial matters. Only the species Boa is utilized for preparing these extracts. Some parts are burned as incense.

Another very popular product derived from snakes is the "*patuá*," a kind of amulet that is hung around the neck, glued on a piece of clothe, or kept in ones

Fig. 7 *B. constrictor* (Jibóia, or Boa) used to produce extracts commercialized in religious articles stores in Belém, PA

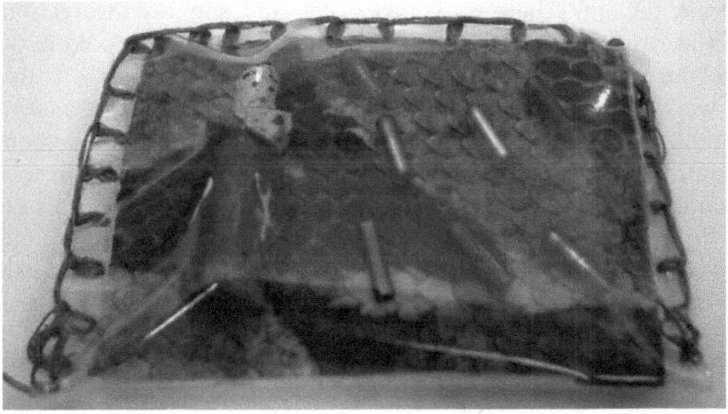

Fig. 8 Examples of amulets (*patuás*) that contain snake parts (*Boa constrictor*)

pocket or wallet. They are square or rounded objects, usually made of leather or some plastic material, containing animal parts (such as pieces of snake skin or a dolphin eye). *Patuás* can be produced using one or more animal part derived from one or more species of animal (Fig. 8). For example, the same *patuá* can have pieces of snake skin and a dolphin eye. According to the shop owners where this material is sold, these amulets are very popular among customers seeking good luck, love, and financial success.

There are some industrialized goods encountered in religious articles stores that are allegedly made from snake parts, such as the "attractive Boa powder" that is used to attract sexual partners. The merchants affirm that they buy this product directly from salesmen from Bahia State, but they do not have any direct knowledge of the details of its production.

Discussion and conclusions

Although the Brazilian legislation forbids the commercialization of the wildlife, products and sub products of snakes (and other animals) were encountered being sold in all the localities surveyed. Similar situations occur in many other Brazilian cities, and wild animals are commonly found with herbal merchants (Freire 1996; Costa-Neto 1999; Almeida and Albuquerque 2002; Silva et al. 2004). These animal parts are not generally displayed to the public, however, as are the herbs and other products. This kind of behavior indicates that the merchants are aware of the illegality of this activity and the possibility of inspection by environmental officials.

The number of snake species found being commercialized in this study is superior to the number reported from others Brazilian cities. Freire (1996) reported the commercialization of 8 snakes species in the public markets of Maceió in Alagoas state; Almeida and Albuquerque (2002) found the 2 snake species being sold by merchants of medicinal herbs in Caruaru, Pernambuco State; Silva et al. (2004) encountered one snake species being sold in Recife, the capital of Pernambuco State; and Costa-Neto (1999) identified one snake species commercialized in Feira de Santana, Bahia State. These results suggest that there is a geographic continuum of commercialization of snake species in the country, and reinforces the need to further address the links between snake sales and snake conservation and their management in Brazil. There are probably other species being sold in the localities surveyed, as the merchants interviewed mentioned the names of other kinds of snakes that we were not able to identify (by their verbal descriptions).

Among the snakes encountered, *C. durrisus* and *B. constrictor* demonstrated the greatest versatility of uses (Table 1). These two species (in addition to *E. cenchria* and *E. murinus*) have known medicinal use in other Brazilian states (Branch and Silva 1983; Freire 1996; Costa-Neto 2000; Almeida and Albuquerque 2002). This demonstrates that the sale and use of snakes to produce folk medicines is widespread throughout Brazil, and that these activities are expanding to urban areas in open markets and fairs in major cities.

In addition to their medicinal uses, snake parts also have magical/religious uses. Researchers have previously noted the use of plants in Afro-Brazilian religious rituals (Albuquerque 1997; Voeks 1997), although few studies make reference to the important role of animals. Some investigators of zootherapy have made references to the animal species used in Afro-Brazilian religious rituals. Fitzgerald et al. (2004) noted a similar situation in Mexico (at the Plateros market, located in front of the Sanctuary for the Miraculous Child of Atocha) where religious objects, images, candles, altars, etc., are sold, together with natural and traditional remedies derived from reptiles, such as dried rattlesnakes, rattlesnake pills, and rattles. Stuffed toads and turtles, and boots made from reptile skins were also commercialized. In Mexico, snakes (and other reptiles) are used in magical–religious ceremonies by many sectors of the population (SEMARNAP-PROFEPA 1998). The role of sacred serpents may be related to their traditional symbolic relationship with health, continuity, and eternity (Metazá 1823; Bruno and Maugeri 1990; Angeletti et al. 1992) in many cultures throughout the world.

The term Afro-Brazilian religion embraces a variety of religious manifestations that exist in Brazil today. Some descended directly from traditional African religions, while others, only a few decades old, have as their main characteristics the cult of African spiritual entities, trances, and the integration of elements from Catholicism (Ferreti 2002). Animals play a very important role in the Afro-Brazilian ritual practices, and are frequently used as fundamental ingredients in the healing processes of physical and spiritual problems, as well as offers to the divinities (*orixás*), or as thanks for blessings achieved. Lody (1987) supports this view of the importance of the ritual use of animals in Afro-Brazilian cults when he observed that "...the flesh and blood in ceremonial offerings establishes the necessary links for the perpetuation of the African divinities' properties in the *terreiros* (where the rituals occur)." A whole animal (or its pieces) can have multiple uses in a ritualistic context (Santos 1993; Souza 1994) and clearly reflects an African influence. In the dry regions of Nigeria, animal products are used in cultural ceremonies (e.g.; in funerals or when leaders take office), in traditional rites (e.g., to invoke or reconcile with the Gods), and have a very significant role in the traditional pharmacopoeia (Adeola 1992). In the present work, although the merchants interviewed did not reveal details about Afro-Brazilian religious rituals, they did confirm that snakes (and others animals) are commonly used. These many folk uses of snakes underscore their importance within the context of biodiversity conservation in Brazil.

Conservation implications

The commercialization of snakes raises concerns in relation to the conservation of these reptiles, especially for the highly exploited species. In Brazil, in spite of the existence of intense commercial dealings with snakes, no official statistics on the use of snakes for medicinal and magic/religious purposes are available, probably because of the clandestine nature of the this trade. Although the species identified in this study are not included in lists of endangered animals, the uses should be made aware of the importance of the sustainable use of these animals.

This present work indicates that the species *B. constrictor* and *Crotalus durissus* are the two most commercialized species, and are thus more susceptible to overexploitation. The merchants themselves admit a decrease in availability of these two species, which may be taken as an indication of decreasing natural populations. Species with high commercial values may benefit from conservation and management to prevent overexploitation. Collection for commercialization may effect these populations, although the magnitude of this impact has not been thoroughly investigated and, as such, is poorly understood.

The demand for live snakes and snake parts for use in traditional medicine appears to have led to significant population reductions in some areas (Fitzgerald et al. 2004). It must be emphasized though, that many factors may affect reptile populations in Brazil. Rodrigues (2005) points out that habitat destruction is the principal threat to natural populations of reptiles in that country. As such, studies focusing on the use and commercialization of snakes for medicinal or magical/religious purposes are necessary to evaluate the magnitude of their effects on the natural populations in

Brazil. As pointed Alves and Rosa (2005), the use of animals for medicinal purposes is part of a body of traditional knowledge that is becoming increasingly more relevant to discussions on conservation biology and the sustainable management of natural resources.

References

Adeola MO (1992) Importance of wild animals and their parts in the culture, religious festivals, and traditional medicine, of Nigeria. Environ Conserv 19(2):125–134

Aguiar AVC (2005) A transferência de informação tecnológica entre a Universidade Federal da Paraíba e as empresas de base tecnológica do pólo tecnológico de Campina. MSc Dissertation, Universidade Federal da Paraíba

Ahid SMM, Lourenço-de-Oliveira R, Saraiva LQ (1999) Canine heartworm on São Luís Island, Northeastern Brazil: a potential zoonosis. Cad Saúde Pública 15(2):405–412

Albuquerque UP (1997) Plantas medicinais e mágicas comercializadas nos mercados públicos de Recife-Pernambuco. Ciência e Trópico 25:7–15

Albuquerque UP, Paiva RF (2004) Métodos e técnicas na pesquisa etnobotânica. Editora Livro Rápido/NUPEEA, Recife

Almeida CFCBR, Albuquerque UP (2002) Uso de plantas e animais medicinais no estado de Pernambuco (Nordeste do Brasil): Um estudo de caso. Interciencia 27(6):276–284

Anageletti LR, Agrimi U, Cúria C, French D, Mariani-Costantini R (1992) Healing rituals and sacred serpents. Lancet 340:223–225

Alves RRN, Rosa IL (2005) Why study the use of animal products in traditional medicines? J Ethnobiol Ethnomed 30:1–5

Bernard R (1994) Research methods in anthropology: qualitative and quantitative approaches. Sage Publications, Thousand Oaks

Branch L, Silva MF (1983) Folk medicine in Alter do Chão, Pará, Brasil. Acta Amazônica 13(5–6): 737–797

Bruno S, Maugeri S (1990) Serpenti d'Italia e d'Europe. Editoriale Giorgio Mondadori, Milan

Campbell JA, Lamar WW (2004) The Venomous Reptiles of the Western Hemisphere. Cornell University Press, Ithaca

Costa-Neto EM (1999) Healing with animals in Feira de Santana city, Bahia. J Ethnopharmacol 65:225–230

Costa-Neto EM (2000) Conhecimento e usos tradicionais de recursos faunísticos por uma comunidade Afro-Brasileira. Resultados Preliminares. Interciencia 25(9):423–431

Ferreti M (2002) As religiões Afro-Brasileiras no Maranhão. Boletim da comissão Maranhense de Folclore 22:4–7

Fitzgerald LA, Painter CW, Reuter A, Hoover C (2004) Collection, trade, and regulation of reptiles and amphibians of the Chihuahuan Desert ecoregion. TRAFFIC North America. World Wildlife Fund, Washington DC

Franke MS, Telecky TM (2001) Reptiles as pets: an examination of the trade in live reptiles in the United States. The Humane Society of the United States

Freire FC (1996) Répteis utilizados na medicina popular no Estado de Alagoas. Monograph. Universidade Federal de Alagoas

Freitas MA (1999) Serpentes da Bahia e do Brasil: suas características e hábitos. Editora Dall

Gibbons JW, Scott DE, Ryan TJ, Buhlmann KA, Tuberville TD, Metts BS, Greene JL, Mills T, Leiden Y, Poppy S, Winne C (2000) The global decline of reptiles, déjà vu amphibians. BioScience 50:653–666

Huntington HP (2000) Using traditional ecological knowledge in science: methods and applications. Ecol Appl 10(5):1270–1274

Jain SK (2000) Human aspects of plant diversity. Econ Bot 54:459–470

Lody R (1987) Candomblé. Religião e resistência cultural Ed Ática. SP, São Paulo

Marques OAV, Eterovic A, Sazima I (2001) Serpentes da Mata Atlântica: Guia ilustrado para a Serra do Mar. Ed. Holos, Ribeirão Preto

Metazá L (1823) Monografia dé serpenti di Roma e suoi contorni. de Romanis, Roma

Peters JA, Orejas-Miranda B (1970) Catalogue of the neotropical squamata, part I. Snakes. Bull US Nat Mus 297:viii + 347

Rodrigues MT (2005) The conservation of Brazilian reptiles: chellenges for megadiverse country. Conserv Biol 19(3):659–664

Santos JE (1993) Os nagô e a morte. 6ª edição. Petrópolis, Editora Vozes

SEMARNAP-PROFEPA (1998) Memorias del Curso Taller de Identificación de Productos y Subproductos de Fauna Silvestre de México. Diaproy. CREDES. Mazatlán, Sinaloa (4–8 de agosto) 266 pp

Shanley P, Luz L, Swingland IR (2002) The faint promise of a distant market: a survey of Belém's trade in non-timber forest products. Biodivers Conserv 11(4):615–636

Silva MLV, Alves AGC, Almeida AV (2004) A zooterapia no Recife (Pernambuco): uma articulação entre as práticas e a história. Biotemas 17(1):95–116

Souza LM (1994) O diabo e a terra de Santa Cruz: feitiçaria e religiosidade popular no Brasil colonial. Companhia das letras, São Paulo

Voeks RA (1997) Sacred leaves of Candomble: African magic, medicine, and religion in Brazil. University of Texas Press, Austin

Zhou Z, Jiang Z (2004) International trade status and crisis for snake species in China. Conserv Biol 18(5):1386–1394

Biodivers Conserv (2007) 16:987–995
DOI 10.1007/s10531-006-9019-8

ORIGINAL PAPER

Philopatry, dispersal patterns and nest-site reuse in Lesser Grey Shrikes (*Lanius minor*)

A. Krištín · H. Hoi · F. Valera · C. Hoi

Received: 27 July 2005 / Accepted: 27 February 2006 / Published online: 9 July 2006

Abstract To nest in the same breeding area, territory or even nest-site in successive years may provide a possibility to look at mechanisms involved in breeding habitat selection and could also be an important tool for conservation, management and restoration attempts. In this study we examine site fidelity towards the breeding area as well as the nesting site in a dense and stable population of the Lesser Grey Shrike *Lanius minor*, a long-distance migrant and highly endangered passerine species, at its present northern border of its breeding range. Overall we recovered 48 out of 146 (32.8%) adults between 1996 and 2000. Recovery rate is significantly higher for males (31 of 77, 40.25%) than for females (17 of 69, 24.6%). Recovery rate of nestlings is much lower since only 51 of 790 (6.5%) were recovered and there is no significant sex difference. Furthermore, our results from 1989 to 2000 revealed that more than 30% (97/319) of the nests were built in the same nest tree in successive years and more than half (183/319 = 57.4%) of the nests in the same or neighboring trees (up to 20 m), but very seldom by the same individuals. The fact that nest reuse in successive years is almost exclusively done by different individuals suggests habitat copying and public information of individual birds. Due to optimal ecological breeding conditions other parameters like inbreeding avoidance or increased genetic variability could be important factors in nest-site selection strategies and consequently result in a ''disperse over stay strategy''.

Keywords Nest-site selection · Philopatry · Habitat copying · Lesser Grey Shrike · Territory fidelity · Nest-site tradition

A. Krištín
Institute of Forest Ecology of SAS, Štúrova 2, SK-960 53 Zvolen, Slovakia

H. Hoi (✉) · C. Hoi
Konrad Lorenz Institute for Ethology, Austrian Academy of Sciences, Savoyenstraße 1a, A-1160 Vienna, Austria
e-mail: h.hoi@klivv.oeaw.ac.at

F. Valera
Estación Experimental de Zonas Áridas (CSIC), General Segura 1, E-04001 Almería, Spain

Introduction

Many migratory birds return precisely to the same site for breeding attempts in successive breeding seasons which suggest fitness benefits over dispersal (Greenwood 1980; Gowaty and Plissner 1997). In fact, it has been shown that individuals experiencing a higher reproductive success in 1 year are more likely to return to the same territory in the following breeding season (Gavin and Bollinger 1988; Paton and Edwards 1996; Haas 1997, 1998), whereas individuals suffering a low reproductive success may more likely disperse (Robinson 1985). Site fidelity, however, may not only affect survival and reproductive success of individuals but also demography, dynamics (Brown and Kodric-Brown 1977; Freemark et al. 1995; Schmidt 2004) and genetic variability of a population (Lande and Barrowclough 1987). In small, isolated or fragmented populations female-biased dispersal can lead to male-biased sex ratios (Yosef 1992). Recent research has shown that females often have a limited ability to search for mates and may therefore effectively be lost from the breeding population if they disperse into areas empty of conspecifics (Dale 2001). Finally, the level of site fidelity could be a determinant for habitat quality and be used to (i) compare the state of different populations or (ii) evaluate the suitability of an area for a species. Thus site fidelity is an important argument for conservation, management and restitution attempts as well (Saunders et al. 1991; Fahrig and Merriam 1994). Dispersal patterns are still poorly known for most especially long distance migrant passerines (Holmes et al. 1996). In this study we focus on different parameters of site fidelity in the highly endangered Lesser Grey Shrike *Lanius minor*. We examined the return rates of (i) juveniles to the natal area, (ii) adults to the breeding area, and (iii) adults to the same territory. Furthermore, we analyzed sex specific differences (Lemon et al. 1996; Konczey et al. 1997) and the role of mate fidelity (Harvey et al. 1979; Payne and Payne 1993) for site fidelity. We studied the frequency of nest-site reuse in several years and addressed the question whether this is due to the same or different individuals. We also investigated nest-site tenacity to the same tree, the core territory within 20 m and the territory within 100 m of adult birds.

The population density of Lesser Grey Shrikes (*Lanius minor*) has declined over large parts of its range and in some places there are only small isolated populations left (Lefranc and Worfolk 1997; Knysh and Pertsov 2003; Giralt and Valera 2003). This species performs one of the longest migratory movements among passerines, winters in South Africa and travels about 10,000 km (Lefranc and Worfolk 1997), Thus the Lesser Grey Shrike provides a possibility to study philopatry and fidelity to nest-sites not only for its own but also for conservation and management aspects.

A preliminary study on a dense breeding population indicates a high return rate of adults and a high nest tree tradition (Kristin et al. 1999). But otherwise there are only scarce and old data on philopatry, breeding site and mate fidelity of this species (Hantge 1957; Warncke 1958; Cramp and Perrins 1993; Glutz von Blotzheim and Bauer 1993). Our results suggest a constant high rate of breeding site fidelity which is highest for males and lowest for juveniles but, in contrast to many other shrike species, this does not result in nest-site (territory) fidelity. In fact, the reuse of the same nest-site or territory is mainly due to different individuals.

Methods

We conducted the study between May and July 1989–2000 (most intensively between 1996 and 2000) in Central Slovakia (40°35–38 N, 19°18–22 E). The study sites comprise 20 km^2 (450–850 m a. s. l.) of traditionally cultivated area characterized by high diversity of habitats (for detailed information and habitat description see Kristin et al. 2000; Wirtitsch et al. 2001).

A total of 146 adults and 790 nestlings were ringed in the breeding seasons of 1989–1999, most of them (143 adults and 612 nestlings) between 1996 and 1999. The number of investigated breeding pairs per season varied between 63 and 75 (1996–2000). The breeding population has been found to be stable for actually a whole decade (Kristin et al. 2000). Adult birds were trapped and re-trapped by means of bowl-traps with a cricket inserted as a lure. During the early hatchling period we used mist nets located close to the nests as well. Adults were provided with one or two color aluminum rings combined with standardized numbered alu-rings (Ringing Centre of National Museum Prague, ring series Z), which enabled individual identification also by visual observation. In a few cases rings and color combinations of recovered adult birds could only be identified with a telescope. We provided nestlings with a standardized alu-ring on the right leg when they were 6–10 days old. Identification was only possible by re-trapping.

Data on philopatry and reuse of nests were collected by checking all individuals, nests and territories used in previous breeding seasons in five successive years (1996–2000). For investigation of crucial factors influencing philopatry two distinct categories of adults were tested separately. We distinguished between individuals faithful to the breeding area (all those birds which were recovered in the study area in the following years) and individuals which were ringed in the study area, but could not be recovered by 2000.

Inter-nest distances were taken from a local map of a scale 1:25,000. For calculation of nest-site tenacity and territory fidelity we distinguished between: (i) nest-site faithful birds which returned to the same nest tree the following year, (ii) nest core territory faithful birds which returned to an area within a radius of 20 m around the original nest and (iii) territory faithful birds which returned to the same territory—an area within a radius of 100 m around the original nest. For the purposes of comparison we used territories with a radius of 200 m around the nest as well (for territory size see Wirtitsch et al. 2001).

Results

Fidelity towards the breeding area

In total, we recovered 48 out of 146 (32.8%) adults between 1996 and 2000. Recovery rate was higher for males (31 of 77, 40.3%) than for females (17 of 69, 24.6%) (χ^2-test: $\chi^2 = 6.9$, $P = 0.008$). Six males (7.8%) returned twice in successive years and one female (1.5%) three times in successive years (1997–2000). Year to year philopatry varied from 11.7% to 28.9% among males, and from 5.8% to 15.8% among females. Twenty seven out of 31 males (87.1%) and 11 out of 17 females (64.7%) were already recorded in the first year after ringing, four males (12.9%) and six females (35.2%) were located later on (second to fifth year). Recovery rate of nestlings is much lower since only 51 out of 790 (6.5%) nestlings were recovered and there is no significant sex difference. About 24 nestlings were recovered as males and 27 as females, which does not deviate from by chance assuming an even sex ratio for nestlings (χ^2-test: $\chi^2 = 0.16$, $P > 0.6$, $n = 790$). One female and one male ringed as nestling were recovered two times.

Fidelity towards nest-site and territory

Our results show that more than 30% (97/319) of all nests were built in the same nest tree in successive years and more than half (57.4%, 183/319) of the nests in the same or a neighboring tree (within 20 m) (Fig. 1). In fact, 11.0% ($n = 319$) of the nest-sites (including a radius of 20 m around the original nest) were used in five successive years (1996–2000), 12.5% in four and 11.3% in three successive years. Including the period from 1989 to 1995 there are two nest trees, which were used for seven years. However, the high overall fidelity to the nesting site is not due to nest-site fidelity of individual birds. Males, in general, tend to return closer to the previous nest-site than females (Fig. 2), but male nest tree tenacity as well as nest core territory fidelity (including 20 m or even 100 m around the former nest tree) is very low when using the proportion of all males ringed. It is also low when using only the proportion of males which we actually recovered (Fig. 1). In fact, only two out of 31 (6.4%) returned males nested in the same nest tree twice. Three other males returned to the same territory within 20 m around the original nest, two others within 100 m (Fig. 1) and two within 200 m. In total, we recovered 9 out of 31 males (29%) within 200 m around the original nest. According to recoveries the maximum age of males was five years and of females six years. Both individuals were ringed as nestlings and found after five, respectively six years, 350 and 2800 m away from the natal site.

Female Lesser Grey Shrikes, however, (n = 17 females recovered) have never been found to breed in the same nest tree or its surrounding (within 100 m) again. The female breeding closest to the previous nesting place settled 130 m away. Other females were found more than 450 m away from previous nests. Nestlings (n = 18) never returned to the natal site or territory (within 100 or 200 m).

This, in general, low specific nest-site tenacity and its variation in relation to sex and age is reflected in the mean dispersal distance (Fig. 2). A two-way ANOVA (age and sex as independent factors) revealed that the distance from the previous nest tree, or natal site is significantly larger for nestlings than for adults (mean±SE: 2739.2±372.5 m, n = 18, vs. 1651.3±279.8 m, n = 51, respectively, F = 4.9, P = 0.03, df = 69). Neither differences among sexes are significant (F = 2.16, P = 0.14, df = 69) nor the interaction between age

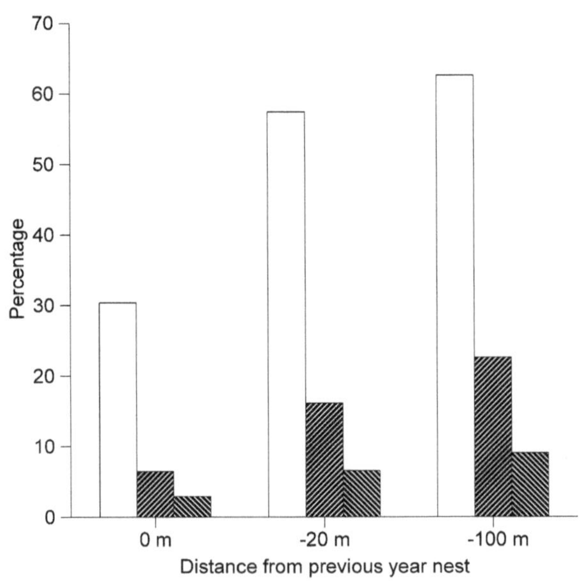

Fig. 1 Reuse of the same nest tree (0 m), the same nest core territory within 20 m around the original nests and the same nest territory within 100 m around the original nest by male Lesser Grey Shrikes in successive years (open bars, n = 319 nests), and reuse expressed as percentage of returned males (right hatched bars, n = 31) and percentage of all ringed males (left hatched bars, n = 77)

Fig. 2 Mean distances (±SE) between nests of recovered birds ringed as adult males (M), adult females (F) and nestlings (PM, PF) and their previous nest. Number of recovered individuals is given in parenthesis

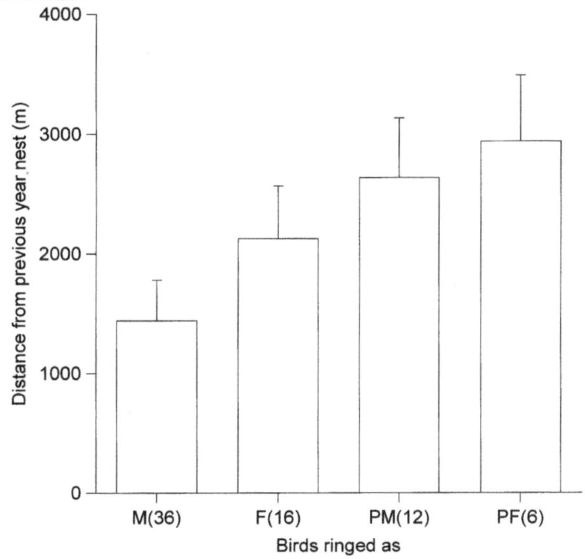

and sex (F = 1.00, P = 0.3, df = 69), Males, however, tended to return nearer to the previous nest-site (Fig. 2). The most distant recovery for nestlings was made 5850 m (for males) and 4980 m (for females) off the natal site. For adults the most distant recovery was 5100 m (females) and 4950 m (male) away from a previous breeding site. However, these results are limited by the size of our study area. Birds which dispersed more than 9.000 m could not be detected because they were outside our study area.

Mate fidelity

The low nest-site fidelity of individuals is accompanied by their low mate fidelity. In the course of our study we ringed 64 complete pairs. However, in all cases, where we could re-sight both partners (n = 5), pairs divorced and had new mates and in two cases females paired with neighbors of the previous year.

Discussion

Our results demonstrate that adult Lesser Grey Shrikes have a comparatively high and significantly male biased return rate to the breeding area, which is in line with many other migratory passerines (Greenwood 1980; Clarke et al. 1997). About 40% of all male and 25% of all female Lesser Grey Shrikes returned to the breeding area. Similar results can be found in other closely related shrike species, i.e. in Woodchat Shrikes *Lanius senator* (Ullrich 1987), Red-backed Shrikes *L. collurio* (Jakober and Stauber 1987, 1989; Šimek 2001), Great Grey Shrikes *L. excubitor* (Yosef 1992), Loggerhead Shrikes *Lanius ludovicianus* (Collister and Smet 1997), or Brown Shrikes *L. cristatus* and Bull headed Shrikes *L. bucephalus* (Takagi 2003) but also in genetically more distant passerines (i.e. Barn swallow *Hirundo rustica*, Tree swallow *Iridoprogne bicolor*, Nightingale *Luscinia megarhynchos*, Willow warbler *Phylloscopus trochilus*, see review in Sokolov 1991). Studies on the Lesser Grey Shrike are rare and none of them mentioned any evidence of a high return rate in this species (e.g. Hantge 1957; Warncke 1958; Lovászi et al. 2000; Knysh and Pertsov 2003; Giralt personal communication). If past reproductive success of individuals increases breeding site fidelity (see Gavin and Bollinger 1988; Paton and Edwards 1996; Switzer 1997; Haas 1997, 1998; Hoover 2003; Sedgwick 2004), our results may indicate a healthy LGS population. This is, in fact, supported by the average high breeding success (see Kristin et al. 2000; Hoi et al. 2004) compared to other populations (Cramp and Perrins 1993; Lovászi et al. 2000; Giralt and Valera 2004).

Secondly, we found that nest-site tradition was high in this species. About one third of all nest trees was used repeatedly and almost two thirds of the nests could be found in the same or a neighboring tree in successive years. In two cases we found a nest in the same nest tree in seven successive years and in some years even exactly at the same place at the tree. Such a long nest-site tradition is common in several long-lived, mainly non-passerine bird species (e.g. storks and raptors) and seems to be unusual in passerines (but see Yosef 1992). He found that some bushes were used as nest-site for about nine consecutive breeding seasons in the closely related Northern Shrike (*Lanius excubitor*). One would expect a frequent nest-site reuse due to high nest-site fidelity of individuals or conspecifics (see in Haas and Sloane 1989). But, in fact, the opposite is true in the Lesser Grey Shrike. The third and most surprising result is that returning males usually disperse less far than females (Fig. 2) but only 6% of all returning males reused the same nest tree and only 16% of the returning males used the same or a neighboring tree. This result suggests that most of this ''nest-site tradition'' is due to different individuals. Together with the finding that Lesser Grey Shrikes always switched their mates in consecutive years (see results) mate switching could also be considered as an important factor responsible for the low nest site fidelity of individuals (see also Haas and Sloane 1989).

The question of how nest-site tradition can work across individuals remains. Conspecific attraction, for instance, could be one mechanism explaining cross individual nest-site tradition (Stamps 1988, 1991, Muller et al. 1997; Ward and Schlossberg 2004; Parejo et al. 2005). ''Habitat copying'' which means that individuals use the reproductive performance of conspecifics to assess habitat suitability and choose their future breeding site might be another possible mechanism (Boulinier and Danchin 1997; Danchin et al. 1998; Doligez et al. 1999; Doligez et al. 2003). Furthermore, an innate preference for specific habitat features also would be sufficient to explain cross individual nest-site tradition, especially for our study population which lives in a very stable and predictable environment (Wirtitsch et al. 2001). However, at the moment we can only speculate about the exact mechanism behind nest-site choice and this topic has to be addressed in future investigations.

A second question related to cross individual nest-site tradition is why most individuals change nest-sites and territories between successive years. It is known that familiarity with the territory provides an advantage over intruders which might be greater for males than for females (Paton and Edwards 1996; Schjorring et al. 2000). Familiarity with foraging and nesting sites, for instance, may allow a quicker start of breeding or may, in general, improve the competitive abilities for breeding opportunities (Lozano and Lemon 1999). So why is changing of the breeding site the rule? In search of a better site, dispersal may be a consequence of breeding in a low quality habitat, or due to nest predation (Robinson 1985). In the Lesser Grey Shrike this explanation is contradicted by the fact that other individuals immediately replace the former, which will be unlikely if the habitat is poor and nest predation is in general very low (Kristin et al. 2000). Alternatively, one can assume that changing to a different breeding site must be beneficial or at least not more costly than reusing the same site. This would consequently mean that the benefits of familiarity with a territory may not be very important for individuals of our population. However, as already mentioned, males do not disperse very far from the original site, so they may be familiar with the surrounding territories as well. When selecting a future breeding site, individuals might use 'public information', which means the local reproductive success of all conspecifics in a breeding patch. Patch reproductive success may even better integrate the effect of all environmental factors on breeding success (Boulinier and Danchin 1997; Danchin et al. 1998; Doligez et al. 1999, 2003). The open habitat in our study area (Wirtitsch et al. 2001) may simplify to gather information for future breeding attempts. Furthermore, habitat quality seems to be optimal for most sites of the study area (Wirtitsch et al. 2001). Breeding success is very high all over the study site (Kristin et al. 2000; Hoi et al. 2004). We could not identify habitat parameters influencing territory choice neither is there a settlement order across different years (Wirtitsch et al. 2001). All these arguments suggest, that habitat quality and resource availability might be very important for individuals to return to the breeding area (breeding site philopatry) but is probably of minor importance for territory choice. Due to the optimal breeding conditions other parameters like inbreeding avoidance or increasing of genetic variability (Pusey 1987) could turn into more important factors in nest-site selection strategies and consequently result in a ''disperse over stay strategy''.

To conclude, our results revealed that on the one hand high nest-site tradition across individuals, which is probably due to habitat choice based on public information and the tendency to breed aggregated and on the other hand between year dispersal of individuals for reasons related to genetic variability (Lande and Barrowclough 1987; Pusey 1987; Weatherhead and Forbes 1994) may be important features when dealing with management concepts of the Lesser Grey Shrike (Ward and Schlossberg 2004). An ongoing study examining the mechanisms of territory choice may additionally shed light on habitat choice of this highly endangered bird species.

Acknowledgements This study was supported by the bilateral agreement between the Austrian and Slovak Academy of Sciences. Anton Kristin was partially supported by the grant of the Slovak Grant agency (No: 2/6007/25). Francisco Valera was supported by the Programa de Ayudas para el Retorno de Investigadores de la Consejería de Educación y Ciencia (Junta de Andalucía).

References

Boulinier EK, Danchin R (1997) The use of conspecific reproductive success for breeding patch selection in territorial migratory species. Evol Ecol 11:505–517

Brown JH, Kodric-Brown A (1977) Turnover rates in insular biogeography: effect of immigration on extinction. Ecology 58:445–449

Clarke AL, Saether BE, Roskaft E (1997) Sex biases in avian dispersal: a reappraisal. Oikos 79(3):429–438

Collister DM, Smet De K (1997). Breeding and natal dispersal in the Loggerhead Shrike. J Field Ornithol 68:273–282

Cramp S, Perrins CM (eds) (1993) The birds of the Western Palearctic. vol 7. Oxford University Press, Oxford, New York

Dale S (2001) Female-biased dispersal, low female recruitment, unpaired males, and the extinction of small and isolated bird populations. Oikos 92(2):344–356

Danchin E, Boulinier T, Massot M (1998) Conspecific reproductive success and breeding habitat selection: implications for study of coloniality. Ecology 78:2415–2428

Doligez B, Danchin E, Clobert J, Gustafsson L (1999) The use of conspecific reproductive success for breeding habitat selection in a non-colonial, hole-nesting species, the collared flycatcher. J Anim Ecol 68:1193–1206

Doligez B, Cadet C, Danchin E, Boulinier T (2003) When to use public information for breeding habitat selection? The role of environmental predictability and density dependence. Anim Behav 66:973–988

Fahrig L, Merriam G (1994) Conservation of fragmented populations. Conserv Biol 8:50–59

Freemark KE, Dunning JB, Hejl SJ, Probst JR (1995) A landscape ecology perspective for research, conservation and management. In: Martin TE, Finch DM (eds) Ecology and management of neotropical migratory birds. Oxford University Press, Oxford, NewYork, pp 381–427

Gavin TA, Bollinger EK (1988) Reproductive correlates of breeding-site fidelity in Bobolinks (*Dolichonyx oryzivorus*). Ecology 69:96–103

Giralt D, Valera F (2003). Decline of the spanish population of lesser grey shrike: is it a consequence of living off limits?. Die Vogelwarte 42:37

Glutz von Blotzheim U, Bauer K (1993) Handbuch der Vögel Mitteleuropas, Band 13. Aula Verlag, Wiesbaden

Gowaty PA, Plissner JH (1997) Breeding dispersal of eastern bluebirds depends on nesting success but not on removal of old nests: an experimental study. J Field Ornithol 68(3):323–330

Greenwood PJ (1980) Mating systems, philopatry and dispersal in birds and mammals. Anim Behav 28:1140–1162

Haas CA (1997) What characteristics of shelterbelts are important to breeding success and return rate of birds? Am Midland Nat 137:225–238

Haas CA (1998) Effects of prior nestling success on site fidelity and breeding dispersal: an experimental approach. Auk 115:929–936

Haas CA, Sloane SA (1989) Low return rates of migratory Loggerhead Shrikes: winter mortality or low site fidelity? Wilson Bull 101:458–460

Hantge E (1957) Zur Brutbiologie des Schwarzstirnwürgers (*Lanius minor*). Vogelwelt 8:137–147

Harvey PH, Greenwood PJ, Perrins CM (1979) Breeding area fidelity of Great Tits (*Parus major*). J Anim Ecol 48:305–313

Hoi H, Kristin A, Valera F, Hoi C (2004) Clutch enlargement in Lesser Gray Shrikes (*Lanius minor*) in Slovakia when food is superabundant: a maladaptive response? Auk 121:557–564

Holmes RT, Marra PP, Sherry TW (1996) Habitat-specific demography of breeding black-throated blue warblers (*Dendroica caerulescens*): implications for population dynamics. J Anim Ecol 65:183–195

Hoover JP (2003) Decision rules for site fidelity in a migratory bird, the Prothonotary warbler. Ecology 84:416–430

Jakober H, Stauber W (1987) Dispersionsprozesse in einer Neuntöter-Population. Beih. Veröffentlichungen Naturschutz Landschaftspflege Baden-Württemberg 48:119–130

Jakober H, Stauber W (1989) Beeinflussen Bruterfolg und Alter die Ortstreue des Neuntöters (*Lanius collurio*)? Die Vogelwarte 35:32–36

Knysh N, Pertsov D (2003) Lesser grey shrike (*Lanius minor*) in the North-East of Ukrajine. Die Vogelwarte 42:37–38

Konczey R, Toth L, Törek J (1997) Site fidelity of great- and blue tits in the Pilis-Visegrad Mountains. Opuscula Zoologica (Budapest) 29–30:103–111

Kristin A, Hoi H, Valera F, Hoi C (1999) Nest site fidelity and breeding biology in the Lesser Grey Shrike (*Lanius minor*) in Central Slovakia. Ring 21:74

Kristin A, Hoi H, Valera F, Hoi C (2000) Breeding biology and breeding success of the Lesser Grey Shrike (*Lanius minor*). Ibis 142:305–311

Lande R, Barrowclough GF (1987) Effective population size, genetic variation and their use in population management. In: Soulé M (eds) Viable population s for conservation. Cambridge University Press, Cambridge, UK, pp 87–123

Lefranc N, Worfolk T (1997) Shrikes. A guide to the shrikes of the world. Pica Press, Sussex pp 192

Lemon RE, Perreault L, Lozano GA (1996) Breeding dispersions and site fidelity of American redstarts (*Setopaga ruticilla*). Can J Zool 74:2238–2247

Lovászi P, Bártol I, Moskát C (2000) Nest-site selection and breeding success of the lesser Grey Shrike (*Lanius minor*) in Hungary. Ring 22:157–164

Lozano GA, Lemon RE (1999) Effects of prior residence and age on breeding performance in Yellow Warblers. Wilson Bull 111:381–388

Muller KL, Stamps JA, Krishnan VV, Willits NH (1997) The effects of conspecific attraction and habitat quality on habitat selection in territorial birds (*Troglodytes aedon*). Am Nat 150:650–661

Parejo D, Danchin E, Aviles JM (2005) The heterospecific habitat copying hypothesis: can competitors indicate habitat quality? Behav Ecol 16(1):96–105

Paton PWC, Edwards TC Jr (1996) Factors affecting interannual movements of Snowy Plovers. Auk 113:534–543

Payne RB, Payne LL (1993) Breeding dispersal in Indigo Buntings: circumstances and consequences of breeding success and population structure. Condor 95:1–24

Pusey AE (1987) Sex biased dispersal and inbreeding avoidance in birds and mammals. Trends Ecol Evol 2:295–299

Robinson SK (1985) Coloniality in the yellow-rumped Cacique as a defense against nest predators. Auk 102:506–519

Saunders DA, Hobbs RJ, Margules CR (1991) Biological consequences of ecosystem fragmentation: a review. Conserv Biol 5:18–32

Schjorring S, Gregersen J, Bregnballe T (2000) Sex difference in criteria determining fidelity towards breeding sites in the great cormorant. J Anim Ecol 69:214–223

Schmidt KA (2004) Site fidelity in temporally correlated environments enhances population persistence. Ecol Lett 7(3):176–184

Sedgwick JA (2004) Site fidelity, territory fidelity and natal philopatry in willow flycatchers (*Empidonax traillii*). Auk 121:1103–1121

Šimek J (2001). Patterns of breeding fidelity in the Red-backed Shrike (Lanius collurio). Ornis Fennica 78(2):61–71

Sokolov LV (1991) Philopatry and dispersal of birds. Institute of Zoology, St. Petersburg, p 232

Stamps JA (1988). Conspecific attraction and aggregation in territorial species. Am Nat 131:329–347

Stamps JA (1991) The effect of conspecifics on habitat selection in territorial species. Behav Ecol Sociobiol 28:29–36

Switzer PV (1997) Past reproductive success affects future habitat selection. Behav Ecol Sociobiol 40:307–312

Takagi M (2003) Philopatry and habitat selection in Bull-headed and Brown shrikes. J Field Ornithol 74:45–52

Ullrich B (1987) Beringungsergebnisse aus einer Brutpopulation des Rotkopfwürgers (*Lanius senator*) im mittleren Albvorland, Kreis Göppingen und Esslingen. Orn. Jh. Baden-Württ 3:107–112

Ward MP, Schlossberg S (2004) Conspecific attraction and the conservation of territorial songbirds. Conserv Biol 18:519–525

Warncke K (1958) Zur Brutbiologie des Schwarzstirnwürgers. Vogelwelt 79:177–181

Weatherhead PJ, Forbes MRL (1994) Natal philopatry in passerine birds genetic or ecological influences. Behav Ecol 5:426–433

Wirtitsch M, Hoi H, Valera F, Kristin A (2001) Habitat composition and use in the lesser grey shrike (*Lanius minor*). Folic Zoologica 50:137–150

Yosef R (1992) Male-biased breeding site fidelity in a population of Northern Shrikes. Condor 94:1025–1027

Biodivers Conserv (2007) 16:997–1008
DOI 10.1007/s10531-006-9010-4

ORIGINAL PAPER

Conservation biogeography of anurans in Brazilian Cerrado

José Alexandre Felizola Diniz-Filho · Luis Mauricio Bini · Míriam Plaza Pinto · Thiago Fernando L. V. B. Rangel · Priscilla Carvalho · Sibelius Lellis Vieira · Rogério Pereira Bastos

Received: 30 January 2006 / Accepted: 27 February 2006 / Published online: 29 April 2006

Abstract The increasing rates of declines in anuran populations worldwide are creating demands for urgent strategies to maximize conservation efforts. This may be critical in regions for which few detailed data on diversity, abundance and distribution are available, such as in the Cerrado of Central Brazil. In this paper, we used a macroecological approach based on the extent of occurrence of 131 species of Anura (Amphibia) in the Cerrado region to design a regional network of potential areas that preserves all anuran species. The final network, obtained using a simulation annealing algorithm based on complementarity, has a total of 17 cells, widely distributed throughout the biome. Minimum costs solutions were obtained in respect to total human population size, soybean production and bovine density, because these are the factors associated with human occupation that historically are more likely to cause broad scale habitat losses. The macro-scale approach used here can provide overall guidelines for conservation and define the focus for more local and effective conservation efforts.

Keywords Anurans · Cerrado · Macroecology · Optimization · Reserve network

J. A. F. Diniz-Filho · L. M. Bini · R. P. Bastos
Departamento de Biologia Geral, ICB, Universidade Federal de Goiás, Cx. P. 131, CEP 74001-970 Goiânia, GO, Brasil

J. A. F. Diniz-Filho (✉)
Departamento de Biologia/MCAS, Universidade Católica de Goiás (UCG), Goiânia, Goiás, Brasil
e-mail: diniz@icb.ufg.br
Fax: +55-062-5211480

M. P. Pinto · T. Fernando L. V. B. Rangel
Graduate program in Ecology & Evolution, ICB, Universidade Federal de Goiás, Goiânia, Brasil

P. Carvalho
Graduate program in Ecology of Continental Aquatic Ecosystems, NUPELIA, Universidade Estadual de Maringá, Paraná, Brasil

S. L. Vieira
Departamento de Ciência da Computação/MCAS, Universidade Católica de Goiás (UCG), Goiânia, Goiás, Brasil

Introduction

Reservation is the main strategy adopted by governments to preserve biodiversity (Margules and Pressey 2000; Aaron et al. 2001). Although this strategy could be effective to diminish habitat loss if the reserves were selected and managed adequately, scientific criterions are hardly taken into account when a reserve is to be established (Possingham et al. 2000; Aaron et al. 2001). Unfortunately, political and economic interests are usually more important than scientific criterions when establishing reserve networks, and many reserves encompass areas of unsuitable habitat for the maintenance of native species, or are defined by cultural or scenic reasons only.

The most important criterion for locating and designing reserve systems should be to achieve maximum representation (or persistence) of biodiversity with the smallest possible cost (Pressey et al. 1997; Margules and Pressey 2000). This optimization involves many different aspects, including spatial distribution of reserves, their connections, overall area, shape and percentage of suitable habitats ensuring species persistence (Possingham et al. 2000; Cabeza and Moilanen 2001; Briers 2002; Lawler et al. 2003; Williams et al. 2004). Also, socio-economic factors associated with the development of human populations at local and regional scales, including population size, growth rate and land use, should be taken into account in the optimization models (Abbitt et al. 2000). This is potentially important because many recent papers found broad-scale correlations between species richness and human population density (Balmford et al. 2001; Araújo 2003; Chown et al. 2003; Luck et al. 2004; Gaston and Evans 2004). These correlations have been interpreted as an indicative that processes driving species richness, mainly related to high ecological productivity and occupation of more suitable habitats, also drive human populations, under the 'more-individuals' energy hypothesis (Balmford et al. 2001). More importantly, because of this positive correlation, a conflict between biodiversity conservation and human development might occur, both because of direct impacts on the environment (i.e., habitat conversion) or because of higher land prices and increase in other associated costs to conserve biodiversity (Luck et al. 2004; but see Faith 2001a, b; Huston 2001; Diniz-Filho et al. 2006).

There is a growing concern about the decline in amphibian populations worldwide (Stuart et al. 2004), creating demands for urgent strategies to maximize conservation efforts for these populations, especially in regions in which few detailed data on diversity, abundance and distribution are available (Young et al. 2000). This is exactly the case of Brazilian Cerrado, one of the global biodiversity hotspots, in which rates of habitat conversion are very high (due to a recent expansion of soybean cultures and cattle ranching (Klink and Machado 2005) and may imply in a quick loss of many endemic and rare species (Sala et al. 2000; Myers et al. 2000; Stuart et al. 2004). Previous attempts to establish conservation priorities in the Cerrado region were usually based on subjective criteria (but see Cavalcanti and Joly 2002). However, lack of detailed data on species distribution and abundance for most groups of organisms constrained many of these previous initiatives. It is also important to note that anurans have the highest level of endemism among vertebrates in Cerrado region (Klink and Machado 2005; Silvano and Segalla 2005), so they may be an important indicator group to establish a network with a relatively high efficiency in terms of preserving more of the biome biodiversity.

In this paper we used macroecological data of geographic distribution (extents of occurrence, following Gaston 1994, 2003) to evaluate spatial patterns in species richness and endemism of anurans in Cerrado. More importantly, we evaluated how these patterns can be optimally represented using complementarity-based and irrepleaceability procedures, defining which regions of the biome are more important to represent total species richness of anurans. We also found which networks represent all Anuran species but, simultaneously, have the minimum amount of human activities (measured as human population and estimates of intensity of soybean cultures and cattle ranching). Although broad-scale approaches are usually considered coarse to establish reserve networks (e.g., Gaston and Rodrigues 2003), they allow an overview of diversity patterns and, thus, can furnish overall guidelines for downscaled conservation strategies and help defining the focus for more local and effective conservation efforts, within the new framework of conservation biogeography (Whittaker et al. 2005). This hierarchical approach may be particularly useful in poorly known and threatened regions of the world, which demands urgent actions due to a combination of high rates of habitat loss and fast human occupation, for instance as the case of Brazilian's Cerrado.

Methods

Geographic distributions, measured as extents of occurrence based on minimum convex polygons (see Gaston 2003), for the 131 species of anurans that can be found in Brazilian Cerrado (e.g., Colli et al. 2002) were mapped with a spatial resolution of 1° grid cell, using as a basis a grid with 181 cells covering the Cerrado Biome (Fig. 1). These 131 species are distributed in 6 families and 29 genera, out of which 47 are endemics to Cerrado region (Table 1). A detailed species list and references are available from the authors upon request (see also Diniz-Filho et al. 2004a, b, 2005, 2006, for a discussions of this dataset).

A binary matrix was constructed by recording the species whose geographic ranges overlap each cell, and species richness was calculated by summing the species present in cells. Total species richness, richness for endemic species and corrected weighted endemism (CWE - given by the average of the inverse of geographic ranges of each species in the Cerrado, for each cell—see Bickford et al. 2004), were also mapped.

Based on the occurrence of the 131 species in the 181 cells of the Cerrado biome, we used an optimization procedure to select the minimum number of cells necessary to represent all species at least once (Church et al. 1996; Possingham et al. 2000; Polasky et al. 2000, 2001; Cabeza and Moilanen 2001). Simulated annealing procedure on Site Selection Mode (SSM) routine of SITES software (Andelman et al. 1999) was used to find these combinations of cells (a network), by performing one hundred runs with 10,000,000 iterations. A relatively high value of penalty for loosing a species was set, so all solutions tended to represent all species with a minimum number of cells.

There are frequently multiple ways (i.e., combination of cells) that satisfy this representation goal, and the solutions were combined to generate a map that gives the relative importance of each cell in these multiple minimum networks, by considering the frequency in which it occurs in the representative combinations of cells or alternative networks. This is an estimate of the irreplaceability of the cell (see Meir et al. 2004), ranging from 0.0 (minimum irreplaceability) to 1.0 (maximum irreplaceability), measuring the likelihood of a given cell to ensure achievement of a set of conservation targets (see also Ferrier et al. 2000).

We also added to SSM a cost for each cell, estimated by different variables expressing human occupation of the Cerrado, and minimized this cost, while representing all species at least once. Some previous studies minimized directly only the total number of people

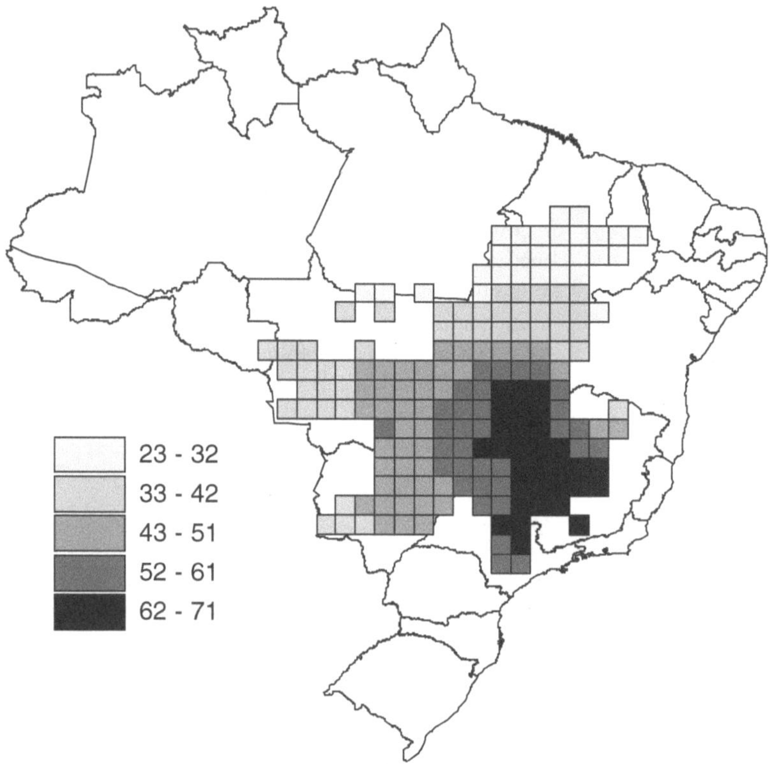

Fig. 1 Spatial patterns of species richness of anurans in the Cerrado region

within the networks (i.e., Chown et al. 2003). However, the recent mode of human occupation in Cerrado is based on a fast expansion of highly technological agriculture and extensive cattle ranching practices (Klink and Machado 2005), which are, in turn, usually weakly correlated with human population density. For example, the Pearson's r coefficients of correlation between human population density and soybean productivity and between human population density and bovine herd density were 0.11 and 0.29, respectively. Thus, due to this particularity, we also generate networks with a minimum amount of soybean productivity and density of bovine, as described below.

Data for human population for each cell in the Cerrado region was obtained from the official census of Brazilian population for the year 2000, done by the Brazilian Agency of Geography and Statistics (IBGE) (see www.ibge.gov.br). For each cell covering the Cerrado biome, human population was obtained by summing urban or rural population from 1054 municipalities whose geopolitical limits are within the Cerrado borders. The same procedure was used to obtain the average soybean productivity and bovine density for each cell, based on data of official Brazilian Agricultural Census, from 1996, also from IBGE. These two variables were used as surrogates of occupation by technological agriculture and extensive cattle ranching (see Rangel et al. in press, for a multivariate analysis of socio-economic factors of Cerrado occupation). Thus, amongst many possible solutions that represent all species, we also found a combination of cells in which there is smallest total human occupation, and so may be useful to minimize potential conservation conflicts (Balmford et al. 2001; Diniz-Filho et al. 2006).

Table 1 Families and genera of Anura from Brazilian Cerrado used in the conservation biogeography analyses, with number of species and number of endemic species in each genus

Family	Genus	Number of species	Number of endemic species
Bufonidae	*Bufo*	8	0
Dendrobatidae	*Colostethus*	1	1
	Epipedobates	3	1
Hylidae	*Aplastodiscus*	1	0
	Bokermannohyla	8	4
	Corythomanthis	1	0
	Dendropsophus	13	5
	Hypsiboas	15	7
	Lysapus	2	0
	Phasmahyla	1	1
	Phyllomedusa	4	2
	Pseudis	3	1
	Scinax	13	4
	Trachycephalus	2	0
Leptodactylidae	*Adenomera*	3	1
	Barycholos	1	1
	Crossodactylus	2	0
	Eleutherodactylus	5	2
	Hylodes	1	0
	Leptodactylus	18	7
	Odontophrynus	4	2
	Physalaemus	7	3
	Proceratophrys	2	2
	Pseudopaludicola	5	1
	Thoropa	1	0
Microhylidae	*Chiasmocleis*	3	1
	Dermatonotus	1	0
	Elachistocleis	2	1
Ranidae	*Rana*	1	0

Results

The maximum value for anuran species richness was found in the central-southern region of the Cerrado Biome, decreasing towards northeastern region (Fig. 1). Patterns of richness for endemic species and CWE are also similar, with restricted species more concentrated in the southeastern part of the biome (Fig. 2). This result, as indicated below, will be important to understand the spatial configuration of the reserve network.

The simulated annealing procedure indicated that 17 regions (i.e., cells) must be considered in order to represent all species in the Biome at least once. The combination of solutions with 17 cells provided by SSM (Fig. 3) reveals that the regions must be widely distributed across the entire Cerrado Biome and encompass the states of Goiás, Minas Gerais, Tocantins, Bahia, Maranhão, Mato Grosso, and Mato Grosso do Sul, and cells with maximum irreplaceability (i.e., cells that appear in all 100 solutions) are concentrated in the southeastern part of the Biome. Other cells with high irreplaceability are found close to Pantanal, in the southwestern part of the Cerrado, and a group of cells with moderate irreplaceability is found in the northwest part of the biome. Also, it is important to note that because of the strong spatial autocorrelation in richness and species ranges (see Diniz-Filho et al. 2003, 2004a), these solutions are very similar and congruent across geographic space.

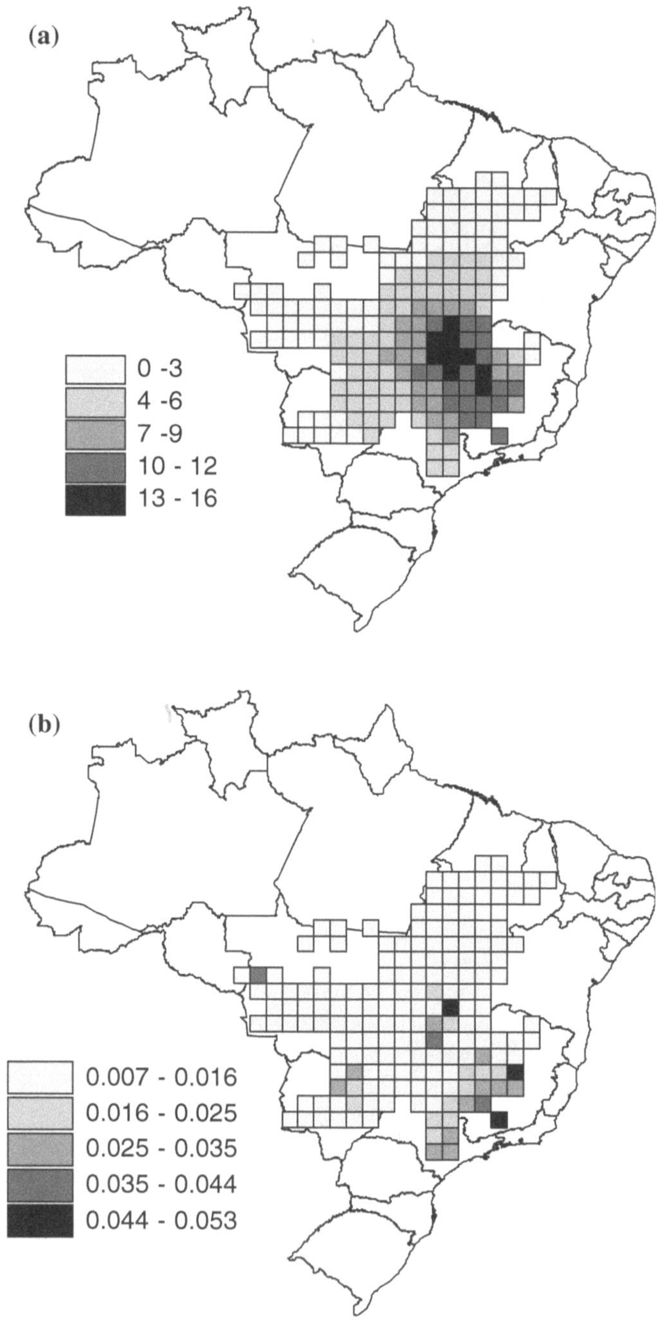

Fig. 2 Spatial patterns of richness for endemic species (**a**) and of endemism (CWE) (**b**)

SSM was also used to represent all species while minimizing the total amount of human population, soybean productivity and bovine density. These three solutions (Fig. 4) are similar, with cells allocated preferentially in the south part of the biome and with a single

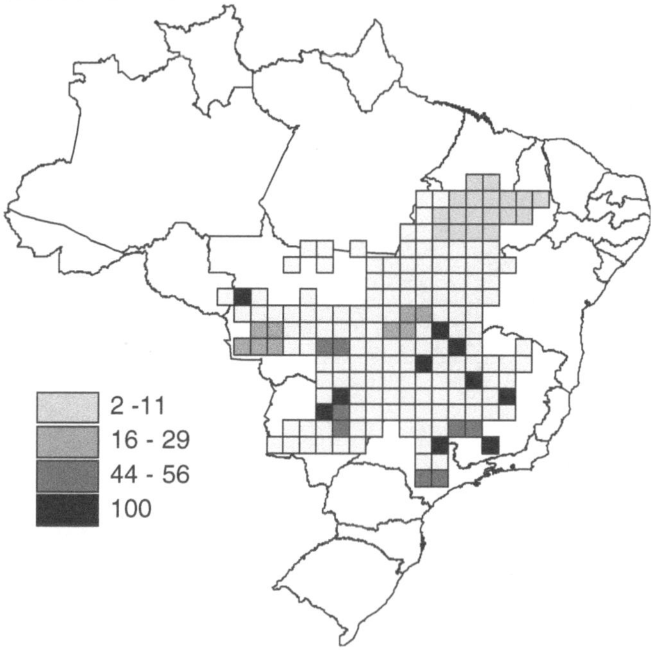

Fig. 3 Sum of the 100 SSM near-optimal solutions with 17 cells (irreplaceability)

cell in the northeastern. For human population size and bovine herd density, SSM found a single minimum solution (Fig. 4a, b), whereas six equivalent minimum solutions were found for soybeans (Fig. 4c). Irreplaceability calculated by combining these six solutions is also higher the southeastern region of the biome, and in a single cell in the extreme west of the Biome.

Discussion

Reserve networks defined by optimal complementarity solutions, based on regional biodiversity analyses, have been successfully implemented or proposed for different parts of the world (Csuti et al. 1997; Araújo 1999; Chown et al. 2003). The analyses performed in this paper showed that conservation efforts for the anurans in Cerrado biome should be concentrated in at least 17 regions (cells) of Central Brazil, covering seven different states (Goiás, Minas Gerais, Tocantins, Bahia, Maranhão, Mato Grosso, and Mato Grosso do Sul). Due to type of data used (i.e., extents of occurrence) and patterns of beta diversity, multiple solutions to represent all species are available, showing a certain level of flexibility in the system, although many areas with maximum irreplaceability are concentrated in the southeastern region of the biome. This is expected since SSM solutions are constrained, in geographical terms, by a number of small-ranged species whose distributions are concentrated in the southeastern of the Biome. Despite this, by considering the Cerrado biome as a whole, it is clearly necessary to establish a national geopolitical coordination in conservation planning to minimize the loss of overall efficiency (see Rodrigues and Gaston 2002).

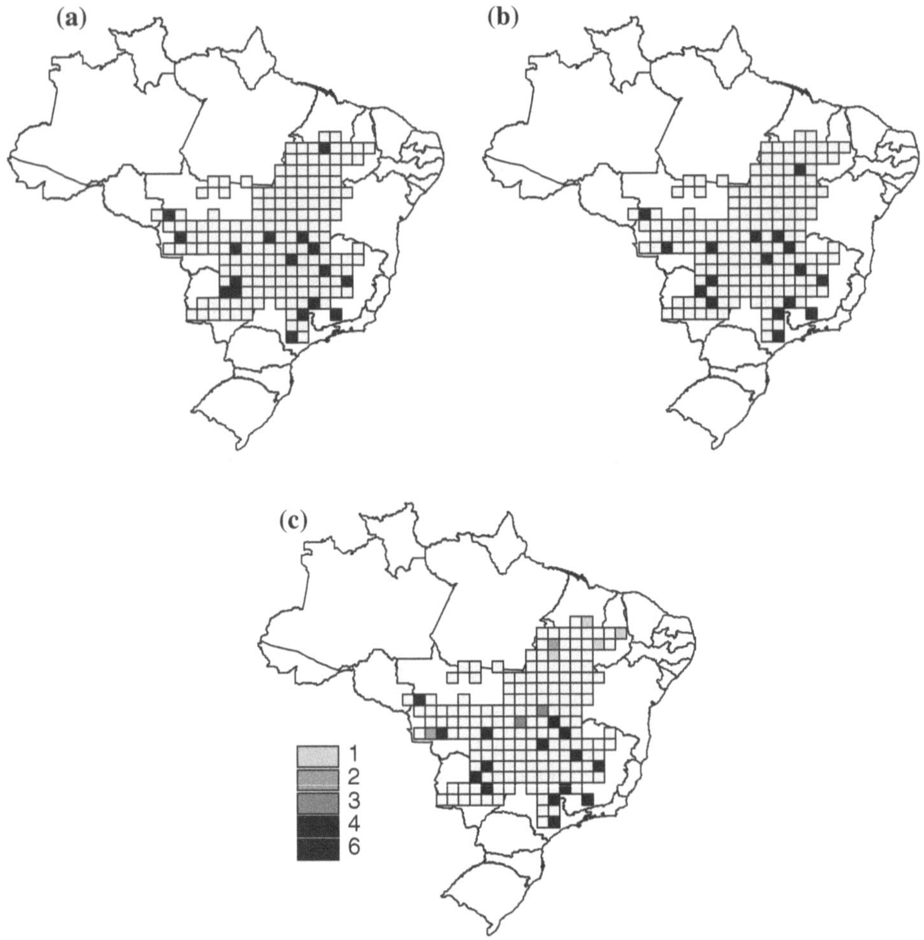

Fig. 4 Minimum solutions of SSM in respect to human population size (**a**) and bovine density (**b**). In (**c**) we show the sum of six optimum solutions minimizing costs related to soybean productivity

A network with 17 select cells represents about 10% of the grid system used to map species‘ ranges (cells with 1° of latitude and longitude), and is similar, both in size and spatial configuration, to a system previously established using a subset of the species used here, a coarsely defined grid and based on a simple heuristic and sequential algorithm of reserve selection (Diniz-Filho et al. 2004b). The 17 cells in the multiple solutions obtained by the simulated annealing algorithm are widely distributed in the space, as expected if beta-diversity increases with increasing geographic distance (see Maurer 1994), and thus encompasses a great amount of the environmental heterogeneity at regional scales, such as different vegetation types (Ratter and Dargie 1992; Ratter et al. 1996; Bridgewater et al. 2004). Also, and more importantly, it is possible to generate networks that contain a smaller amount of human occupation (defined by three different surrogates of these patterns), revealing a possibility to minimize conservation-human development conflicts (sensu Balmford et al. 2001; see also Chown et al. 2003; Diniz-Filho et al. 2006). For modern agricultural spatial patterns, based on soybean cultures, six minimum cost solutions were found, revealing even more flexibility (compared with the single solution provided by minimizing human population and bovine density) to establish a broad-scale conservation system targeted to represent all anuran species. However, most irreplaceable cells are still situated in the south-eastern and central regions of the biome, as a function of endemism of many species, including some recently discovered (Diniz-Filho et al. 2005). The problem is that this region is the one most densely occupied by human populations and their activities, so establishing protected areas becomes both difficult and expensive.

Since there is a strong autocorrelation pattern for grid richness (Diniz-Filho et al. 2003), alternative reserve systems could also work if based on cells close to the ones chosen here, as showed by the multiple solutions obtained by simulated annealing in SSM. This occurs because short distance autocorrelation in species richness is usually a function of range overlap, reflecting low species turnover (Diniz-Filho et al. 2003). These alternative systems would be also important when considering the previous efforts in defining reserves in the Cerrado Biome (e.g. Emas and Chapada dos Veadeiros National Parks, in Goiás State, that are adjacent to some of the cells selected in this paper. Thus, in a context of gap analyses at broad scale (e.g., Rodrigues et al. 2004), these previously established reserves, after a more detailed evaluation of local parameters (see below), could be used as starting points for defining priorities in the allocation of conservation efforts along the entire system. Of course, a better understanding of how the current system of conservation units preserves anuran diversity requires a more detailed study, due to differences in scale focused for each analysis.

Despite the growth of macroecology research program worldwide (see Blackburn and Gaston 2003), macro-scale approaches are obviously considered coarse to establish reserve networks. Although they can provide overall guidelines for conservation and define the focus for more local and effective conservation efforts in Neotropical regions, it is also important to be aware of the limitations of this approach. A first general problem with macro-scale approaches is the definition of the extent of occurrence, based on biogeographical data, which is by definition overestimated (Gaston 2003). Although each region in our study has ca. 12,000 km^2, and probably has at least one population of the species listed (assuming continuous ranges), we cannot ensure that viable population for all species will be found within these regions. More studies are necessary to evaluate these parameters at local scale because, in general, regional patterns of species richness and abundance for Cerrado species are poorly described, with inventories restricted to a few regions in the central and south-eastern part of the biome (Cavalcanti and Joly 2002). In this case, the hierarchical approach suggested here (i.e., defining regions using biogeographic data and only then analysing local areas within regions by local sampling), would be in principle more effective than select reserves based only on spatially restricted and detailed local datasets. Also, this approach can be improved in a near future if a few additional parameters linking species‘ persistence with patterns of regional occurrence and habitat suitability are obtained. This can improve the system by adopting a 'filtering' strategy based on extents of occurrence (Araújo and Williams 2000; Williams and Araújo 2000; Araújo et al. 2002). Another possibility is to downscaling data to a finer resolution, based on modeled species distributions, but this procedure still requires relatively high density of local records of occurrence to minimize uncertainness in modeling process (Araújo et al. 2005).

Spatial variation in total species richness and especially in the level of endemism (which possess an important role in establish the networks) may be biased since most sampling efforts have been historically concentrated into the southern and central regions of the Cerrado biome (Diniz-Filho et al. 2005). Suppose, for example, the discovery of new endemic species in the northern part of the biome (e.g. in the Tocantins State), a plausible assumption due to the paucity of faunal inventories in this region. This would increase the relative importance (i.e., the irreplaceability) of those northern regions and, consequently, increase the length of the entire reserve network. Increasing knowledge could also show that geographic distribution of some currently known species could also expand towards these northern regions, and this would counteract this effect of increasing network size due to endemisms. On the other hand, updating data on ranges of known species would even invert the current pattern of the reserves, since species that are today considered endemic to southern would be also preserved in the northern.

Edge effects observed in the network would partially counteract the increase in the number of regions necessary for reservation caused by adding more endemics species in the northern regions, and also change network length in the future (see Diniz-Filho et al. 2005). In a broader scale approach (i.e., analysing the entire country), non-endemic species found only at the margins of the biome would have been previously included in richer regions outside the Cerrado, decreasing the importance of some regions selected in the edge of the biome. Thus, the selected regions in the centre of the biome would be more stable to changes if a national scale planning were implemented (Rodrigues and Gaston 2002). In general, these results indicate how difficult is to predict changes in the network patterns due to the increase in the comprehensiveness in biodiversity data in the entire biome (Diniz-Filho et al. 2005). Anyway, the irreplaceability patterns and reserve networks described here are the best possible conservation bioegeography design based on current knowledge of the anuran species distribution in the Cerrado.

Despite these problems, our analyses revealed general patterns of anuran species richness and endemism in Cerrado biome, which may be important for conservation purposes. The regional system presented here can furnish guidelines for future conservation and research programs, taking into account both patterns of species richness, endemism, human development and land use to define priority regions for conservation. In this context, the next step in this research program is to add habitat suitability dimensions to filter regional occurrence and incorporate issues of increasing species persistence into reserve design.

In addition, at local scales, we suggest an increasing in sampling efforts within the regions identified above with the aim of identify suitable habitats and to estimate population and meta-population parameters. Data gathered in this way will serve to evaluate the validity of areas to maintain viable populations and also to increase our knowledge about patterns of richness and endemism by updating geographic distribution of known and describing new species.

Acknowledgements Financial support for this study came from a PRONEX program of CNPq and SECTEC-GO (proc. 23234156). Work by JAFDF, LMB and RPB were also partially supported by other CNPq projects (procs. ns. 300762/94-1, 300367/96-1; 400381-97.4). Our research program in macroecology and biodiversity has also been continuously supported by CAPES and FUNAPE-UFG.

References

Aaron GB, Gullison RE, Rice RE, Fonseca GAB (2001) Effectiveness of parks in protecting tropical biodiversity. Science 291:125–128

Abbitt RJF, Scott JM, Wilcove DS (2000) The geography of vulnerability: incorporating species geography and human development patterns into conservation planning. Biol Conserv 96:169–175

Andelman S, Ball I, Davis F, Stoms D (1999) SITES v. 1.0, An analytical toolbox for designing ecoregional conservation portfolios. Technical report, The Nature Conservancy. http://www.biogeog.ucsb.edu/projects/tnc/toolbox.html

Araújo MB (1999) Distribution patterns of biodiversity and the design of a representative reserve network in Portugal. Div Distribut 5:151–163

Araújo MB (2003) The coincidence of people and biodiversity in Europe. Global Ecol Biogeogr 12:5–12

Araújo MB, Williams PH (2000) Selecting areas for species persistence using occurrence data. Biol Conserv 96:331–345

Araujo MB, Williams PH, Turner A (2002) A sequential approach to minimise threars within selected conservation areas. Biodiv Conserv 11:1011–1024

Araújo MB, Thuiller W, Williams PH, Reginster I (2005) Downscaling European species atlas distributions to a finer resolution: implications for conservation planning. Global Ecol Biogeogr 14:17–30

Balmford A, Moore JL, Brooks T, Burgess N, Hansen LA, Williams P, Rahbek C (2001) Conservation conflicts across Africa. Science 291:2616–2619

Bickford SA, Laffan SW, de Kok RPJ, Orthia LA (2004) Spatial analysis of taxonomic and genetic patterns and their potential for understanding evolutionary histories. J Biogeogr 31:1715–1733

Blackburn TM, Gaston KJ (2003) Macroecology: concepts and consequences. Blackwell Science, Oxford

Bridgewater S, Ratter JA, Ribeiro JF (2004) Biogeographic patterns, β-diversity and dominance in the cerrado biome of Brazil. Biodiv Conserv 13:2295–2318

Briers RA (2002) Incorporating connectivity into reserve selection procedures. Biol Conserv 103:77–83

Cabeza M, Moilanen A (2001) Design of reserve network and the persistence of biodiversity. Trends Ecol Evol 16:242–248

Cavalcanti RB, Joly CA (2002) Biodiversity and conservation priorities in the cerrado region. In: Oliveira PS, Marquis RJ (eds) The cerrados of Brazil: ecology and natural history of a neotropical savanna, pp 351–367

Chown SL, van Rensburg BJ, Gaston KJ, Rodrigues ASL, van Jaarsveld AS (2003) Energy, species richness, and human population size: conservation implications at a national scale. Ecol Appl 13:1233–1241

Church RL, Stoms DM, Davis FW (1996) Reserve selection as a maximal covering location problem. Biol Conserv 76:105–112

Colli GR, Bastos RP, Araújo AFB (2002) The character and dynamics of the Cerrado herpetofauna. In: Oliveira PS, Marques RJ (eds) The Cerrado of Brazil: ecology and natural history of a neotropical savanna. Columbia University Press, Columbia, pp 223–241

Csuti B, Polasky S, Williams PH, Pressey RL, Camm JD, Kershaw M, Kiester AR, Downs B, Hamilton R, Huso M and Sahr K (1997) A comparison of reserve selection algorithms using data on terrestrial vertebrates in Oregon. Biol Conserv 80:83–97

Diniz-Filho JAF, Bini LM, Hawkins BA (2003) Spatial autocorrelation and red herrings in geographical ecology. Global Ecol Biogeogr 12:53–64

Diniz-Filho JAF, Bini LM, Bastos RP, Vieira CM, Souza MC, Motta JAO, Pombal JP Jr, Peixoto JC (2004a) Anurans from a local assemblage in Central Brazil: linking local processes with macroecological patterns. Brazilian J Biol 64:41–52

Diniz-Filho JAF, Bini LM, Vieira CM, Souza MC, Bastos RP, Brandão D, Oliveira LG (2004b) Spatial patterns in species richness and priority areas for conservation of anurans in the Cerrado region, Central Brazil. Amphibia-Reptilia 25:63–75

Diniz-Filho JAF, Bastos RP, Rangel TFLVB, Bini LM, Carvalho P, Silva RJ (2005) Macroecological correlates and spatial patterns of anurans description dates in Brazilian Cerrado. Global Ecol Biogeogr 14:469–477

Diniz-Filho JAF, Bini LM, Pinto MP, Rangel TFLVB, Carvalho P, Bastos RP (2006) Anuran species richness, complementarity and conservation conflicts in Brazilian Cerrado. Acta Oecologica 29:9–15.

Faith DP (2001a) Overlap of species richness and development-opportunity does not imply conflict. Science 293:1591–1592

Faith DP (2001b) Those complementarity analyses do not reveal extent of conservation conflict in Africa. Science 293:1591–1592

Ferrier S, Pressey RL, Barrett TW (2000) A new predictor of the irreplaceability of areas for achieving a conservation goal, its application to real-world planning, and a research agenda for further refinement. Biol Conserv 93:303–325

Gaston KJ (1994) Rarity. Chapman & Hall, London

Gaston KJ (2003) The structure and dynamics of geographic ranges. Oxford University Press, Oxford

Gaston KJ, Rodrigues ASL (2003) Reserve selection in regions with poor biological data. Conserv Biol 17:188–195

Gaston KJ, Evans KJ (2004) Birds and people in Europe. Proc R Soc London Ser B 271:1649–1655

Huston MA (2001) People and biodiversity in Africa. Science 293:1591–1592

Klink CA, Machado RB (2005) Conservation of the Brazilian cerrado. Conserv Biol 19:707–713

Lawler JJ, White D, Master LL (2003) Integrating representation and vulnerability: two approaches for prioritizing areas for conservation. Ecol Appl 13:1762–1772

Luck GW, Ricketts TH, Daily GC, Imhoff M (2004) Alleviating spatial conflicts between people and biodiversity. Proc Nat Acad Sci – USA 101:182–186

Margules CR, Pressey RL (2000) Systematic conservation planning. Nature 405:243–253

Maurer BA (1994) Geographical population analysis: tools for the analysis of biodiversity. Cambridge University Press, Cambridge

Meir E, Andelman S, Possingham HP (2004) Does conservation planning matter in a dynamic and uncertain world? Ecol Lett 7:615–622

Myers N, Mittermeier RA, Mittermeier CG, Fonseca GAB, Kent J (2000) Biodiversity hotspots for conservation priorities. Nature 403:853–858

Polasky S, Camm JD, Solow AR, Csuti B, White D, Ding R (2000) Choosing reserve network with incomplete species information. Biol Conserv 94:1–10

Polasky S, Csuti B, Vossler CA, Meyers SM (2001) A comparison of taxonomic distinctness versus richness as criteria for setting conservation priorities for North American birds. Biol Conserv 97:99–105

Possingham H, Ball I, Andelman S (2000) Mathematical methods for identifying representative reserve networks. In: Ferson S, Burgman M (eds) Quantitative methods for conservation biology. Springer-Verlag, New York, pp 291–306

Pressey RL, Possingham HP, Day JR (1997) Effectiveness of alternative heuristic algorithms for identifying indicative minimum requirements for conservation reserves. Biol Conserv 80:207–219

Rangel TFLVB, Bini LM, Diniz-Filho JAF, Pinto MP, Carvalho P, Bastos RP (in press). Human development and biodiversity conservation in Brazilian Cerrado. Appl Geogr

Ratter JA, Dargie TCD (1992) An analysis of the floristic composition of 26 Cerrado areas in Brazil. Edin J Bot 49:235–250

Ratter JA, Bridgewater S, Atkinson R, Ribeiro JF (1996) Analysis of the floristic composition of the Brazilian Cerrado vegetation II: comparison of the woody vegetation of 98 areas. Edinb J Bot 53:153–180

Rodrigues ASL, Gaston KJ (2002) Rarity and conservation planning across geopolitical units. Conserv Biol. 16:674–682

Rodrigues ASL, Andelman SJ, Bakarr MI, Boitani L, Brooks TM, Cowling RM, Fishpool LDC, Fonseca GAB, Gaston KJ, Hoffmann M, Long JS, Marquet PA, Pilgrim JD, Abbitt RJB, Pressey RL, Schipper J, Sechrest W, Stuart SN, Underhill LG, Waller RW, Watts MEJ, Yan X (2004) Effectiveness of global protected area network in representing species diversity. Nature 428:640–643

Sala OE, Chapin FS, Armesto JJ, Berlow E, Bloomfield J, Dirzo R, Huber-Sanwald E, Huenneke LF, Jackson RB, Kinzig A, Leemans R, Lodge DM, Mooney HA, Oesterheld M, Poff NL, Sykes MT, Walker BH, Walker M, Wall DH (2000) Global biodiversity scenarios for the year 2100. Science 287:1770–1774

Silvano DL, Sagallo MV (2005) Conservation of Brazilian amphibians. Conserv Biol 19:653–658

Stuart SN, Chanson JS, Cox NA, Young BE, Rodrigues ASL, Fischman DL, Waller RW (2004) Status and trends of amphibian declines and extinctions wordwide. Science 306:1783–1786

Young BE, Lips KR, Reaser JK, Ibanez R, Salas AW, Cedeno JR, Coloma LA, Ron S, La Marca E, Meyer JR, Munoz A, Bolanos F, Chaves G, Romo D (2000) Population declines and priorities for Amphibian conservation in Latin America. Conserv Biol 15:1213–1223

Whittaker RJ, Araujo MB, Paul J, Ladle RJ, Watson JEM, Willis KJ (2005) Conservation Biogeography: assessment and prospect. Div Distribut 11:3–23

Williams PH, Araújo MB (2000) Using probability of persistence to identify important areas for biodiversity. Proc R Soc London B 267:1959–66

Williams JC, ReVelle CS, Levin SA (2004) Using mathematical optimization models to design nature reserves. Front Ecol Environ 2(2):98–105

Biodivers Conserv (2007) 16:1009–1026
DOI 10.1007/s10531-006-9040-y

ORIGINAL PAPER

Avian responses to tourism in the biogeographically isolated high Córdoba Mountains, Argentina

Lisandro Heil · Esteban Fernández-Juricic · Daniel Renison · Ana M. Cingolani · Daniel T. Blumstein

Received: 23 June 2005 / Accepted: 25 March 2006 / Published online: 9 July 2006

Abstract Species do not respond identically to the presence of humans, and this may have consequences at higher-levels of ecological organization. We established bird transects on and off recreational trails in the high Córdoba Mountains of Argentina, a biogeographic island characterized by high levels of endemism, to examine the effect of human visitation at three different levels: (a) community (avian species richness and diversity), (b) guild (relative density of carnivores, granivores, insectivores, and omnivores), and (c) population (relative density of individual bird species). Human presence in the high Córdoba Mountains decreased avian species richness and diversity, and reduced insectivorous relative density, but we did not detect significant effects on granivores, omnivores, and carnivores. At the population level, 6 of 28 species were negatively affected by human visitation; four of these species are of conservation concern. Our results show negative responses to recreationists at multiple levels (e.g., reductions in density, displacement of species from highly visited areas), which may be related to spatial and temporal access to suitable resources, physical disturbance or species-specific tolerance thresholds. Our study area had lower levels of human visitation relative to other protected areas in

L. Heil · D. Renison
Cátedra de Ecología General (FCEFyN - UNC), Av. Vélez Sarsfield 299, 5000 Córdoba, Argentina

E. Fernández-Juricic (✉)
Department of Biological Sciences, California State University Long Beach, 1250 Bellflower Blvd. (Mailstop 3702), Long Beach, CA 90840, USA
e-mail: efernand@csulb.edu

A. M. Cingolani
Instituto Multidisciplinario de Biología Vegetal (CONICET-UNC), CC 495, Vélez Sársfield 299, 5000 Córdoba, Argentina

D. T. Blumstein
Department of Ecology and Evolutionary Biology, University of California, 621 Young Drive South, Los Angeles, CA 90095-1606, USA

the Northern Hemisphere, which raises the issue of whether this kind of biogeographically isolated habitat may be too fragile to sustain increasing levels of tourism.

Keywords Endemic species · Guild · Human disturbance · Mountainous ecosystem · Recreational activities · South America

Introduction

Conventional wisdom suggests that the impact of tourism is localized because tourists often prefer to visit areas close to established facilities (Priskin 2003), and consequently wildlife further away may be minimally affected. This is an important management paradigm that might be challenged in the future as the number of tourists visiting natural areas worldwide is expected to double by 2020 (Christ et al. 2003), which raises concerns about the large scale environmental impacts of recreational activities (e.g., development of infrastructure, disease transmission, invasion of non-native species, reduced habitat quality, etc.). Importantly, tourism has become profitable for developing countries, energizing local economies and, in some countries, generating revenues for the conservation of protected areas (Sekercioglu 2002; Christ et al. 2003; but see Kiss 2004). For instance, tourism is contributing between 3% and 25% of the Gross Domestic Product of developing nations worldwide (Diaz Benevides and Perez-Ducy 2001). Therefore, managers face the challenge of finding strategies that would promote coexistence between humans and wildlife. Ultimately, the goal is to increase the chances of wildlife viewing without eroding ecological integrity in protected areas (Blumstein et al. 2005).

Previous studies have shown that the effects of tourists on wildlife may trigger short-term behavioral responses (Frid and Dill 2002), modify predation risk assessment (Webb and Blumstein 2005), reduce breeding performance (Beale and Monaghan 2004a; Müllner et al. 2004), influence habitat selection (Gill et al. 1996; Gutzwiller and Anderson 1999) and population abundance (van der Zande et al. 1984; Miller et al. 1998), and modify community structure (Skagen et al. 1991; Gutzwiller 1995; Fernández-Juricic 2002). Most of these studies have been conducted in the Northern Hemisphere, which has experienced a higher volume of recreationists than areas in the Southern Hemisphere. However, tourism has increased sharply in the Southern Hemisphere recently, particularly within biodiversity hotspots (Christ et al. 2003). Thus, to increase our ability to predict the large-scale and long-term effects of tourism and to devise proper management strategies, it is necessary to assess population and community responses to recreationists in threatened ecosystems in parts of the world without a long history of human visitation.

We studied the effects of human visitation on birds inhabiting a biogeographically isolated area in South America (high Córdoba Mountains, Argentina), with many plant and animal endemic species (Cabido et al. 1998; 2003). Our goal was to assess the effects of recreationists in areas with low (off-trails) and high (on-trails) levels of visitation at different ecological levels: (a) community (avian species richness and diversity), (b) guild (relative density of carnivores, granivores, insectivores, and omnivores), and (c) population (relative density of individual bird species).

Some of the mechanisms proposed to explain wildlife responses to recreationists predict that higher frequencies of human visitation reduce the spatial and temporal access to foraging and breeding resources, which could eventually lead to reductions in species occurrence and density if disturbance is prevalent in space and time (Gill et al. 1996; Fernández-Juricic 2002; Frid and Dill 2002; Fernández-Juricic et al. 2003). Therefore, we predicted a decrease in species richness, species diversity, relative density of guilds and individual species in highly visited areas controlling for differences between study sites, habitat structure, and altitude.

Methods

Study area

The high Córdoba Mountains (1,500 to 2,800 m) are almost 1,000 m higher than the surrounding mountain systems in central Argentina, and their altitude and weather make their ecological conditions different from surrounding habitats (Luti et al. 1979; Cabido et al. 1998). Vegetation consists of a mosaic of tussock grasslands, lawns, granite outcrops, eroded areas with exposed rock surfaces and low densities of *Polylepis australis* woodlands and shrublands (Cingolani et al. 2004). Mean precipitation is 854 mm, which is concentrated between October and March (Renison et al. 2002). The climate is temperate with cold winters, and the mean annual temperature is 8°C (Cabido 1985). Forty one endemic plant and animal taxa are found in this area (Cabido et al. 2003), including 12 endemic sub-species of birds (Nores 1995; Miatello et al. 1999).

In 1997, part of the high Córdoba Mountains (26,000 ha above 1,500 m.a.s.l.) was expropriated to create the Quebrada del Condorito National Park, while the private lands surrounding the Park were declared National (12,000 ha) and Provincial (117,000 ha) Water Reserves. Because the high Córdoba Mountains are relatively close to several big cities, and because hiking and climbing began increasing in Argentina in the 1990's, tourists visit the area year round, and generally hike and use tents or mountain lodges to overnight.

We selected three sites which are the most widely visited in the high Córdoba Mountains: (a) Champaquí mountain (S 31° 59′; W 64° 49′) with an un-maintained trail of 14.0 km that usually varies in width from 0.8 m to 2 m, but may reach a width of 30–40 m in areas where there is no clear trail; (b) Quebrada del Condorito National Park (S 31° 37′; W 64° 42′) with a trail of 7.8 km which is well maintained and usually varies in width from 0.8 to 1.5 m; and (c) Los Gigantes (S 31° 24′; W 64° 47′) with an un-maintained trail of 10.8 km with similar characteristics to the Champaquí mountain trail (National Parks Administration, pers. comm; personal observations). For the sake of simplicity, hereafter we will call these areas Champaquí, Condorito, and Gigantes, respectively. Visitor rates to the Champaquí trail vary between 6,000 and 12,000 visitors per year, mostly trekking groups that may use horses for cargo and occasionally motorcycles. Eight local families also use the trail (mainly on horseback). Visitors to the Condorito trail vary between 3,000 and 4,000 per year, mostly trekking and some bicycles. No horses or motorcycles are allowed in the trail and there is no use by local inhabitants. In the Gigantes trail, visitation rates vary between 3,000 and 6,000 per year, mostly trekking. In a few areas, there is also some occasional use by motorcycles and vehicles, but the steep terrain precludes motorized vehicles in the rest of trail. Around five local families use the trail (mainly on horseback), but there is no horse use by visitors. Hunting is illegal in all the Córdoba Mountains.

General sampling procedures

Using a Geographic Information System (GIS) with vegetation, erosion, topography, human settlements, and trails layers at a pixel size of 30 × 30 m (Cingolani et al. 2004), we selected 21 pairs of locations for transects. The number of locations for transects in each study site was determined according to the availability of straight trail sections and the possibility of obtaining appropriate off-trail transects (see below). As a result, the pairs of transects differed among study sites (Champaquí, n=10; Condorito, n=6; Gigantes, n=5). No more transects could be established in the study areas without relaxing our transect selection criteria.

From each pair, one transect was placed on a trail used by tourists (on-trail) and the other transect was placed on an area with similar topography, vegetation, natural rock outcrops, and rock exposed by erosion but without tourist use (off-trail). The criteria for determining habitat similarity for each pair of transects was based on the GIS thematic layers. After locating sectors with trails that were fairly straight for at least 600 m and that were at least 300 m from each other (average distance between the closest ends of transects was 1,145 ± 1,129 m), we chose off-trail transects that were more than 200 m away from on-trail transects but with similar proportion of habitat composition (see also Results), altitude, slope orientation, and aspect, and as a result were presumably influenced by similar potential confounding factors. All transects were selected in areas with no human settlements or camping areas near them. Livestock is the main determinant of vegetation types in the High Córdoba Mountains (Cingolani et al. 2003). As our vegetation types were similar on and off trails, long term grazing pressure can be assumed to be approximately equal.

In a preliminary study, we estimated that in the habitat with the least visibility (sparse *Polylepis* woodland and shrubland), detection probabilities started dropping substantially 30 m away from the centre of the transect. Thus, we established transects that were 60 m wide and 600 m long (3.6 ha).

Bird surveys were conducted between December 2002 and March 2003, which encompassed the breeding and post-breeding seasons, because bird species richness (Ordano 1996) and the frequency of human visitation (National Parks Administration, pers. comm.) are at the highest levels during this period, which would increase the chances of human–wildlife interactions. We surveyed birds between 0800 and 1200 and 1530 and 1930 because those were the periods with the highest bird activity. Adult and fledgling individuals were counted during surveys. Pairs of on- and off-trail transects were surveyed consecutively on a given day (but each pair visited in random order between surveys). We visited each transect three times to obtain better estimates of species presence and abundance between areas with low and high visitation. All surveys were conducted by only one person (LH) during favorable weather conditions to avoid inter-observer bias.

In each transect, we recorded the identity and abundance of all individuals seen or heard within 30 m at each side of the center of the transect. We included in the analyses individuals that were only using the transects (e.g., discarded observations with individuals flying high). The observer was trained to visually estimate the 30 m with <10% error prior to beginning the surveys. Vegetation is mainly composed of short grassing lawns, many flat rocky areas, and tussock grasslands where bird detection was relatively easy. Potential small biases in detection probability between habitats should not affect our on and off-trail comparisons due to the similarity in habitat composition between pairs of transects. Since four species of the genus

Anthus (*A. correndera*, *A. furcatus*, *A. hellmayri* and *A. lutescens*) were difficult to distinguish, we lumped all individuals in one category (*Anthus sp.*).

We could not get density estimates following distance sampling (Buckland et al. 2001), because the studied species had relatively low abundances (Ordano 1996), and the assumptions of this method could not be met. Consequently, we developed a relative index of density by averaging the number of individuals detected on each transect over the three visits and dividing that by the area of each transect (3.6 ha). This index can only be used for comparisons between the conditions studied, but not to make inferences about the absolute density of the species. The relative density index was expressed as number of individuals per 10 ha.

We used the vegetation map in the GIS (Cingolani et al. 2004) to determine habitat composition in each transect. In a 90 × 600 m area including a transect, we counted the number of pixels of each of seven distinct habitat types (*Polylepis* woodland and shrubland, thick tussock grassland with hydromorphic lawn, thin tussock grassland, lawn, outcrop with tussock grassland, natural rock outcrop mixed with exposed rock, and rock exposed by erosion, Appendix 1, as described by Cingolani et al. 2004). We also calculated the average altitude of each transect. These variables were included in the statistical analyses to control for variations in habitat composition between transects.

Statistical analyses

We used a nested ANOVA to check whether habitat composition would vary between on- and off-trail transects *within* and among study sites.

From our surveys, we calculated the following dependent variables in each transect: species richness (mean number of species detected per transect), species diversity (Shannon diversity index), relative density of carnivores, granivores, insectivores, and omnivores, and relative density of individual species. In the statistical analyses, we only included species present in at least five transects (following Fernández-Juricic 2002). We analyzed these dependent variables with General Linear Models (GLM), including the following categorical factors: visitation frequency (2 levels, on- and off-trail transects), study area (three levels, Champaquí, Condorito and Gigantes), and the interaction between visitation frequency and study area. We also included the following covariates (all continuous): transect altitude and habitat composition (number of pixels of seven different habitat types, see Appendix 1). The results of step-wise selection procedures could be misleading with unbalanced designs (in our case, different number of transects per study site), but such effect could be minimized by running full models. We included all variables in the analyses, but for the sake of clarity in the presentation we reported only the main factors (frequency of visitation and study area), their interaction, and the significant covariates. All other factors not reported were non-significant ($P > 0.05$). We also ran the same models for species richness, species diversity, and guild relative density but excluding the species that were significantly affected by visitation frequency (see Results) to determine if the community and guild level patterns were consistent or relied on the effects of individual species.

The effects of human disturbance may become noticeable only at certain densities. For instance, Reijnen and Foppen (1995) found that the negative effects of road noise were intensified when the density of breeding birds was lower. Using a logistic regression, we assessed whether the probabilities of finding a significant effect of human visitation could be affected by the relative densities of each species.

We checked the normality and homoscedasticity of our data before and after running the analyses. We log-transformed some variables (see Results) to meet assumptions of linear models. Throughout, we reported untransformed means ± SE.

Results

We recorded a total of 45 bird species in a total of 151.2 ha (considering the four species from the genus *Anthus* as one species, Appendix 2). The mean relative density of all recorded species was 22.55 ± 3.14 individuals per 10 ha.

Habitat composition varied among some study sites (*Polylepis* woodland and shrubland, $F_{2, 36}$=5.47, P=0.008; thick tussock grassland with hydromorphic lawn, $F_{2, 36}$=14.23, P<0.001; outcrop with exposed rock, $F_{2,36}$=7.02, P=0.002; the other relationships were non-significant, $F_{2, 36} > 3.25$, $P > 0.05$). However, there were no differences in habitat composition between on- and off-trail transects within each study site ($F_{3,36}$ varied between 0.04 and 0.87, $P > 0.05$).

Community level

Species richness was lower in on-trail transects compared to off-trail transects (Fig. 1a). Species richness also varied among study sites (Fig. 1a). The interaction between study area and visitation frequency was not significant (Table 1). Only three covariates were significant (Table 1): species richness was positively affected by altitude and the amount of lawn, and negatively by the amount of rock exposed by erosion. A similar result was found without considering the species that were affected by human disturbance (see Population level section): species richness was lower in on-trail than in off-trail transects, with variations among study sites and similar effects of the covariates (Table 1).

Species diversity (Shannon index) was lower in on-trail relative to off-trail transects, but no variation was found among sites (Table 1, Fig. 1b). The study site vs. visitation frequency interactions were not significant; however, species diversity was positively associated with the amount of lawn and with altitude (Table 1). Species diversity calculated without considering the species affected by visitation frequency (see Population level section) was also lower in on-trail relative to off-trail transects, but it varied among study sites (Table 1), and was positively associated with *Polylepis* woodland and shrubland and altitude, and negatively with rock exposed by erosion (Table 1).

Guild level

At the guild level, we found lower carnivorous relative density in on-trail relative to off-trail transects (Table 2, Fig. 2a). Study area and study area vs. visitation frequency interaction did not affect carnivorous relative density (Table 2), nor did any of the other covariates ($P > 0.05$). However, when we ran another model with carnivorous relative density excluding the two carnivorous species that were affected by human visitation (see Population level section), we found no significant effect of

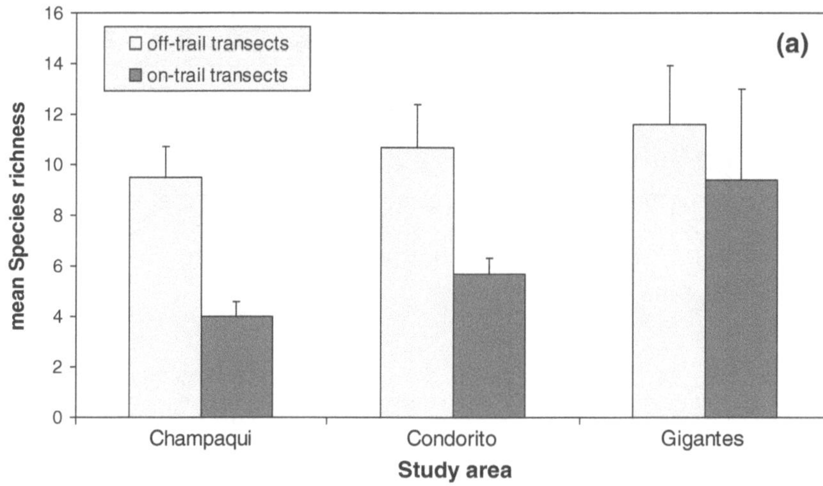

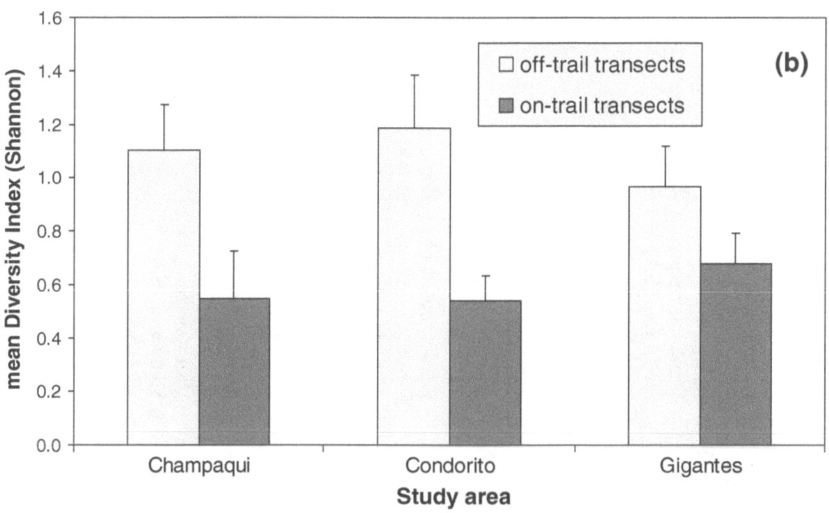

Fig. 1 Species richness (**a**) and species diversity (**b**) in areas with low (off-trail) and high (on-trail) levels of human visitation in the high Córdoba Mountains

the frequency of visitation (off-trail transects = 1.808 ± 0.431 individuals/10 ha, on-trail transects = 0.749 ± 0.244 individuals/10 ha) and the other factors (Table 2).

Granivorous relative density did not differ between visitation frequencies (off-trail transects = 5.776 ± 1.735 individuals/10 ha, on-trail transects = 6.041 ± 2.917 individuals/10 ha), but it did vary among study sites (Table 2, Champaqui = 3.702 ± 1.508 individuals/10 ha, Condorito = 2.701 ± 0.907 individuals/10 ha, Gigantes = 14.167 ± 5.746 individuals/10 ha). Granivorous relative density was also positively affected by altitude, and negatively by the amount of rock exposed by erosion (Table 2).

Insectivorous relative density was lower in on-trail transects relative to off-trail transects and varied among study sites (Table 2, Fig. 2b). The study site vs. visitation frequency interaction was not significant (Table 2), but insectivorous relative density

Table 1 Results from general linear models explaining variation in species richness and diversity as a function of human visitation, study area, and habitat-related characteristics

Dependent factor	Independent factors	*F*	d.f.	*P*
Species richness $R^2 = 0.71$	Intercept	0.05	1, 28	0.758
	Visitation frequency (VF)	13.33	1, 28	**0.001**
	Study area (SA)	6.41	2, 28	**0.005**
	VF × SA	0.40	2, 28	0.675
	Rock exposed by erosion (−)	9.53	1, 28	**0.005**
	Lawn (+)	4.25	1, 28	**0.048**
	Altitude (+)	14.91	1, 28	**<0.001**
Species richness without species affected by VF $R^2 = 0.69$	Intercept	2.68	1, 28	0.551
	Visitation frequency (VF)	5.09	1, 28	**0.032**
	Study area (SA)	8.40	2, 28	**0.001**
	VF × SA	0.28	2, 28	0.757
	Rock exposed by erosion (−)	8.82	1, 28	**0.006**
	Lawn (+)	5.38	1, 28	**0.027**
	Altitude (+)	18.04	1, 28	**<0.001**
Species diversity $R^2 = 0.51$	Intercept	0.58	1, 28	0.451
	Visitation frequency (VF)	8.18	1, 28	**0.007**
	Study area (SA)	0.28	2, 28	0.756
	VF × SA	0.89	2, 28	0.423
	Lawn (+)	5.77	1, 28	**0.023**
	Altitude (+)	4.21	1, 28	**0.049**
Species diversity without species affected by VF $R^2 = 0.61$	Intercept	0.01	1, 28	0.952
	Visitation frequency (VF)	4.33	1, 28	**0.046**
	Study area (SA)	8.27	2, 28	**0.002**
	VF × SA	0.34	2, 28	0.711
	Rock exposed by erosion (−)	7.53	1, 28	**0.010**
	Polylepis woodland and shrubland (+)	4.89	1, 28	**0.035**
	Altitude (+)	12.99	1, 28	**0.001**

Shown are covariates with $P<0.05$. Significant P-values are marked in bold

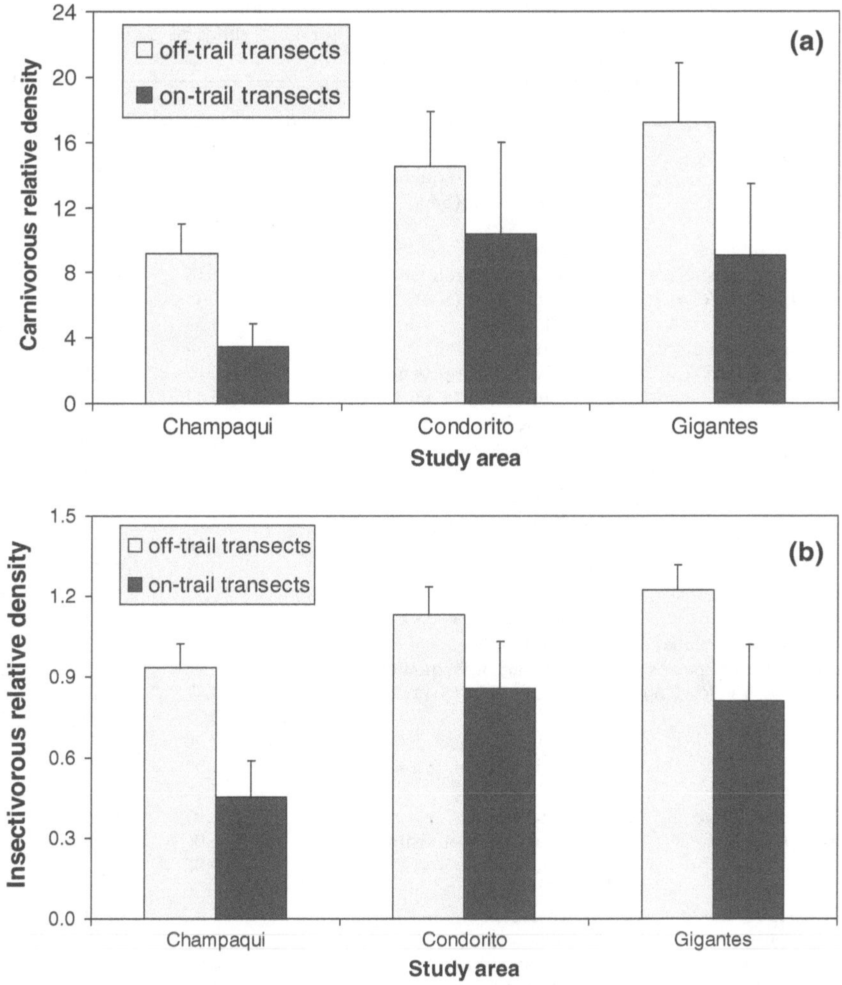

Fig. 2 Relative density (individuals/10 ha) of (**a**) carnivores and (**b**) insectivores in areas with low (off-trail) and high (on-trail) levels of human visitation in the high Córdoba Mountains

was positively associated with altitude and the amount of lawn (Table 2). Insectivorous relative density without considering the species affected by human visitation (see Population level section) was also lower in on-trail (9.082 ± 1.658 individuals/10 ha) transects compared to off-trail transects (5.996 ± 2.013 individuals/10 ha), and differed among study sites (Champaqui =3.889 ± 1.219 individuals/10 ha, Condorito =10.802 ± 3.126 individuals/10 ha, Gigantes = 10.925 ± 2.559 individuals/10 ha) (Table 2). Three covariates were positively associated with this insectivorous relative density: altitude, and the amount of lawn and thin tussock grassland (Table 2).

Omnivorous relative density did not differ between on-trail (5.996 ± 1.576 individuals/10 ha) and off-trail (4.629 ± 1.643 individuals/10 ha) transects, and was not affected by any of the other studied factors (Table 2). Similar results were found with omnivorous relative density without considering the species affected by human

Table 2 Results from general linear models explaining variation in the relative density of carnivorous, granivorous, insectivorous, and omnivorous foraging guilds as a function of human visitation, study area, and habitat-related characteristics

Dependent factor	Independent factors	*F*	d.f.	*P*
Carnivorous relative density R^2 = 0.46	Intercept	0.36	1, 28	0.552
	Visitation frequency (VF)	5.56	1, 28	**0.025**
	Study area (SA)	1.30	2, 28	0.287
	VF × SA	1.69	2, 28	0.201
Carnivorous relative density without species affected by VF R^2 = 0.03	Intercept	0.10	1, 28	0.749
	Visitation frequency (VF)	2.18	1, 28	0.151
	Study area (SA)	1.37	2, 28	0.269
	VF × SA	0.53	2, 28	0.589
Granivorous relative density R^2 = 0.65	Intercept	0.54	1, 28	0.467
	Visitation frequency (VF)	0.17	1, 28	0.682
	Study area (SA)	6.47	2, 28	**0.005**
	VF × SA	1.29	2, 28	0.289
	Rock exposed by erosion (−)	5.59	1, 28	**0.025**
	Altitude (+)	11.07	1, 28	**0.002**
log Insectivorous relative density R^2=0.65	Intercept	0.48	1, 28	**0.490**
	Visitation frequency (VF)	14.21	1, 28	**<0.001**
	Study area (SA)	4.10	2, 28	**0.027**
	VF × SA	0.01	2, 28	**0.996**
	Lawn (+)	11.72	1, 28	**0.002**
	Altitude (+)	4.36	1, 28	**0.045**
log Insectivorous relative density without species affected by VF R^2=0.66	Intercept	2.03	1, 28	**0.164**
	Visitation frequency (VF)	5.96	1, 28	**0.021**
	Study area (SA)	8.71	2, 28	**0.001**
	VF × SA	0.05	2, 28	**0.952**
	Lawn (+)	10.29	1, 28	**0.003**
	Thin tussock grassland (+)	4.97	1, 28	**0.034**
	Altitude (+)	7.54	1, 28	**0.010**
Omnivorous relative density R^2=0.04	Intercept	1.97	1, 28	**0.171**
	Frequency of visitation (FV)	0.06	1, 28	**0.801**
	Study area (SA)	0.92	2, 28	**0.413**
	FV × SA	1.86	2, 28	**0.174**
Omnivorous relative density without species affected by VF R^2=0.02	Intercept	1.42	1, 28	**0.244**
	Visitation frequency (VF)	0.28	1, 28	**0.603**
	Study area (SA)	0.62	2, 28	**0.547**
	VF × SA	1.85	2, 28	**0.174**

Shown are covariates with $P < 0.05$. Significant *P*-values are marked in bold

visitation (see Population level section): no differences between on-trail (4.629 ± 1.571 individuals/10 ha) and off-trail (3.791 ± 1.617 individuals/10 ha) transects, and no effect of the other studied factors (Table 2).

Population level

We modeled the effects of visitation frequency on 28 of the 45 species recorded, and found that the relative densities were lower in on-trail transects in 6 species (three insectivores: *Anthus* sp., *Asthenes modesta cordobae*, *Asthenes sclateri sclateri*; two carnivores: *Buteo albicaudatus*, *Falco sparverius*; and one omnivore: *Turdus chiguanco*, Table 3, Fig. 3). Two of these species are endemic (*Asthenes modesta cordobae*, *Asthenes sclateri sclateri*, Appendix 2), and two are declining (*Buteo albicaudatus*, *Turdus chiguanco*, Appendix 2).

Table 3 Results from general linear models explaining variation in the relative density of individual species in relation to human visitation frequency, altitude, and habitat-related characteristics

Species	Model R^2	VF	SA	FV × SS	Altitude	PWS	RP	OER	OTG
AEAN	0.47	0.44	4.94**	0.80					
AGMO	0.06	0.08	0.36	0.01					
ANTHSP	0.51	7.20*	0.07	1.66					
ATMO	0.56	4.33*	0.63	0.75	(+) 6.26*				
ASSC	0.51	11.16**	0.57	0.57					
BOAY	0.09	0.05	0.51	0.53					
BUAL	0.58	5.23*	0.16	3.20		(+) 5.43*			
BUPO	0.52	0.01	2.01	0.02		(−) 5.72*			
CAMA	0.63	0.05	1.20	5.92*	(+) 5.11*	(−) 4.24*			
CAIN	0.54	0.23	1.46	1.34	(+) 5.55*		(−) 6.37*		
CAAU	0.02	0.23	1.74	1.08					
CICO	0.15	0.02	0.52	2.11					
CIFU	0.56	1.06	1.06	2.77	(+) 15.89**		(−) 4.29*		
CIOR	0.55	1.15	6.26**	0.91	(+) 5.82*				
CIPL	0.60	0.05	0.37	1.09				(−) 5.09*	
FASP	0.24	4.51*	0.13	0.06					
MURU	0.50	1.11	0.68	0.80	(+) 11.89**				
NOCY	0.43	0.28	0.91	1.35					
NOMA	0.35	0.26	0.24	1.76	(−) 6.17**				
PHUN	0.60	0.25	4.34*	0.54	(+) 11.72***		(+) 6.28*		
POPL	0.09	2.09	0.09	0.52					
SASP	0.52	0.01	2.39	0.46	(+) 4.56*				
STLO	0.43	0.04	3.39*	1.08					
TRAE	0.02	0.38	0.18	1.91					
TUCH	0.61	4.19*	4.29*	0.45					
VACH	0.01	0.64	0.47	0.64					
VUGR	0.04	0.53	1.60	0.18					
ZOCA	0.63	1.71	9.05**	0.09	(+) 4.88*	(−) 9.55**	(−) 13.61***		(+) 4.87*

Shown for each factor are the direction of the relationship (+, positive; −, negative), the *F*-values and *P*-values when significant. Study area and its interaction with visitation frequency had 2, 28 degrees of freedom; the remainder of the factors had 1, 28 degrees of freedom. All densities were log-transformed, but MURU, TRAE, and STLO. Shown are covariates with $P<0.05$

*, $P<0.05$; **, $P<0.01$; ***, $P<0.001$; VF, visitation frequency; SA, study area; PWS, *Polylepis* woodland/shrubland; RP, rock exposed by erosion; outcrop with exposed rock, OER; outcrop with tussock grassland, OTG; AEAN, *Aeronautes andecolus*; AGMO, *Agriornis montana fumosus*; ANTHSP, *Anthus* sp.; ATMO, *Asthenes modesta Córdobae*; ASSC, *Asthenes sclateri sclateri*; BOAY, Bolborhynchus aymara; BUAL, *Buteo albicaudatus*; BUPO, *Buteo polyosoma*; CAMA, *Carduelis magellanica*; CAIN, *Catamenia inornata*; CAAU, *Cathartes aura*; CICO, *Cinclodes comechingonus*; CIFU, Cinclodes fuscus; CIOR, *Cinclodes olrogi*; CIPL, Cistothorus platensis; FASP, *Falco sparverius*; MURU, *Muscisaxicola rufivertex achalensis*; NOCY, *Nothiochelidon cyanoleuca*; NOMA, *Nothura maculosa*; PHUN, *Phrygilus unicolor cyaneus*; POPL, *Polyborus plancus*; SASP, *Sappho sparganura*; STLO, *Sturnella loica obscura*; TRAE, *Troglodytes aedon*; TUCH, *Turdus chiguanco*; VACH, *Vanellus chilensis*; VUGR, *Vultur gryphus*; ZOCA, *Zonotrichia capensis*

There was no relationship between the probability of finding a significant visitation frequency effect and the relative density of the studied species (logistic regression, X^2=0.005, d.f. = 1, P=0.938).

We found a significant interaction between visitation frequency and study area for *Carduelis magellanica* (Table 3). This granivore was not affected by visitation frequency in Champaqui (off-trail transects = 0.012 ± 0.056 individuals/10 ha, on-trail transects = 0.024 ± 0.056 individuals/10 ha), but its relative density was lower in on-trail transects in Condorito (off-trail transects = 0.261 ± 0.091 individuals/10 ha,

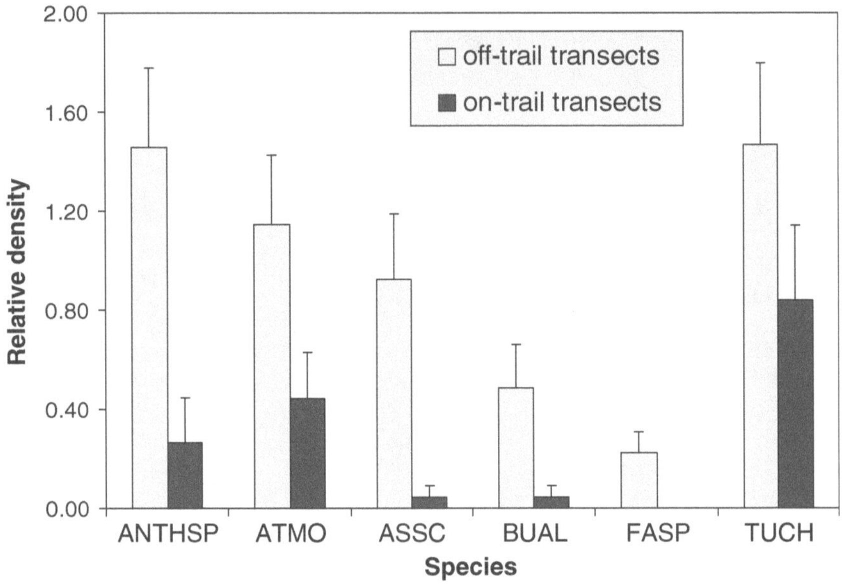

Fig. 3 Relative densities (individuals/10 ha) of *Anthus* sp. (ANTHSP), *Asthenes modesta cordobae* (ATMO), *Asthenes sclateri* (ASSC), *Buteo albicaudatus* (BUAL), *Falco sparverius* (FASP); *Turdus chiguanco* (TUCH) in areas with low (off-trail) and high (on-trail) human visitation in the high Córdoba Mountains

on-trail transects = 0.021 ± 0.081 individuals/10 ha), and higher in on-trail transects in Gigantes (off-trail transects = 0.018 ± 0.072 individuals/10 ha, on-trail transects = 0.214 ± 0.083 individuals/10 ha).

We also found that the relative density of six species (two insectivores, two granivores, two omnivores) varied among study sites (Table 3). Altitude significantly affected ten species (four insectivores, four granivores, two omnivores, Table 3), which in all cases but one (*Nothura maculosa*) increased their relative densities at higher altitudes. The amount of different habitat types affected several species in different ways (Table 3): *Polylepis* woodland and shrubland, four species (two granivores, one granivore, one carnivore), rock exposed by erosion, four species (three granivores, one insectivore); outcrop with exposed rock, one insectivore species; and outcrop with tussock grassland, one granivore species.

Discussion

Our findings suggest that human disturbance may be involved in the responses found at different levels (community, guild, population). Six species, four of them of conservation concern, had lower densities in areas with relatively high levels of human visitation in the high Córdoba Mountains. Two of the four guilds studied had lower densities in areas with high visitation levels (carnivores and insectivores); although, the effects on carnivores were apparently caused by two abundant species negatively affected by visitors. Finally, species richness was lower in highly visited areas, even when the six species sensitive to disturbance were not considered. We compare these results with previous studies and suggest some mechanisms involved in these responses.

Like ours, some studies found a decrease in the occurrence of species associated with recreational activities in different habitat types (Hammitt and Cole 1987; Miller et al. 1998; Riffell et al. 1996; Fernández-Juricic 2002). The displacement of species from highly visited areas has been suggested to be the result of reductions in habitat quality (e.g., Gill et al. 1996; Fernández-Juricic 2002; Frid and Dill 2002; Fernández-Juricic et al. 2003). Physical disturbance may decrease habitat structural complexity (Miller et al. 1998; Laiolo 2003) and/or affect the spatial and temporal patterns of resource exploitation due to the presence of pedestrians (Soulé et al. 1992; Fernández-Juricic 2000). We found that habitat composition was not significantly different between off- and on-trail areas; this suggests that a reduction in the proportion of suitable habitat available to birds may be underlying human–wildlife interactions in this habitat. However, future studies should rule out physical effects of disturbance (namely, soil compaction, soil erosion, and increase in trail width and depth; Deluca et al. 1998; McDougall and Wright 2004; Roovers et al. 2004) and explore this mechanism experimentally (Fernández-Juricic et al. 2003) before reaching definite conclusions.

Taking aside the effects of recreational activities on single species, only the insectivorous guild consistently had lower relative densities in highly visited areas. Similar effects on insectivores were also found in other studies conducted in urbanized landscapes (Chuan Lim and Sodhi 2004; but see Fernández-Juricic 2002). The decrease in the density of insectivores with higher human disturbance has been usually associated with food limitation (Canaday 1996; Laiolo 2003; Chuan Lim and Sodhi 2004) rather than low tolerance levels (Fernández-Juricic 2002). The composition of insect communities can indeed change at habitat edges associated with human disturbance (Webb et al. 1984; Shure and Phillips 1991; Suarez et al. 1998; Bolger et al. 2000; Kitahara et al. 2000), in some cases with invasive species being more abundant near human-induced edges and native species farther from those edges (Webb 1989). If bird insectivores rely mainly on native insects, variations in insect community composition with recreationists might potentially affect food availability; a possibility that deserves further study in the high Córdoba Mountains.

We found that 21.4% of the species whose abundance could be modeled decreased their densities in highly visited areas. However, the proportion of species affected by recreational activities was lower than in other studies, which reported negative effects in 53–92% of species (van der Zande and Vos 1984; Fernández-Juricic 2002; van der Zande et al. 1984; Miller et al. 1998). The low proportion of species affected by recreationists may have to do with the reduced levels of human visitation to the high Córdoba Mountains (3,000–12,000 visitors per year) relative to other areas (60,000 visitors per year in Laiolo 2003; 3,500,000 visitors per year in Miller et al. 1998; > 5,000,000 visitors per year in Fernández-Juricic 2002) or with species-specific differences in tolerance to disturbance (Blumstein et al. 2003). Alternatively, the low proportion of species affected could be related to behavioral adjustments to disturbance (Frid and Dill 2002). For instance, Forsman et al. (1998) found in boreal forests that territories tended to be more clumped when animals were exposed to an artificial predator during the breeding season. Territorial species in the High Córdoba Mountains could reduce the perceived risk imposed by the presence of visitors by reducing inter-specific neighbor distances; a strategy that might not be profitable at the intra-specific level due to competition effects (Pulliam and Caraco 1984). If behavioral responses like this are enough to reduce human disturbance, population level effects might not be noticeable (see also, Gill et al. 2001; Beale and Monaghan 2004b).

Conservation implications

Understanding human–wildlife interactions in this kind of environment provides an opportunity to assess how the spread of recreational activities might affect species with limited geographic distribution, which has become a top-priority for conservation biology (Soulé et al. 2001). Despite the low number of species negatively affected by recreationists, two of them are endemic and two are in decline. Furthermore, our relatively low number of transects suggest that more species could be decreasing due to tourism, but could not be detected with our samples. In threatened ecosystems, species of high conservation concern could be the first ones to be influenced by tourism, and raises the issue of whether or not biogeographic islands may be too fragile to support tourism. Given the geographic isolation of the high Córdoba Mountains, alternative habitats may not be available for these species, which could affect their persistence. Besides ecological loss, the decreased abundance of these species might also reduce wildlife-viewing opportunities and the economic benefits for local communities.

What strategies could then promote coexistence? First, small-scale manipulative studies to determine sustainable levels of human visitation must be conducted (e.g., Rodriguez-Prieto and Fernández-Juricic 2005). Second, visitation should be limited to few trails rather than expanding the area visitors can access, because the mere presence of visitors could reduce the access to suitable habitat for some species and guilds. Third, in areas with high visitation levels, increasing protective cover (e.g., *Polylepis* woodland) might also reduce negative impacts of human disturbance.

Acknowledgements We thank the British Embassy in Buenos Aires for funding the field work, the "Comisión Nacional de Actividades Espaciales" for providing satellite images, "Club Andino Córdoba and Club Andino Villa Carlos Paz" for providing lodging during part of the field work, Argentinean National Parks for providing permits, and Javier Heredia, Luis Volkmann, and Manuel Nores for their useful comments and suggestions. EFJ was supported by the College of Natural Sciences and Mathematics, California State University Long Beach. DR and AMC are researchers of CONICET (Argentina).

Appendix 1

Table 4 Brief description of the different habitat types identified in the study areas in the high Córdoba Mountains (based on Cingolani et al. 2004)

Habitat types	Description
Polylepis woodland and shrubland	Dominated by *Polylepis* woodland/shrubland, with shrubby grassland and rock outcrop, but low total rock cover. Generally occurred below 2000 m on steep slopes in mid to low topographic positions, but was also found on flat sites in ravine bottoms or in gentle slopes. Erosion and grazing pressure were present

Table 4 continued

Habitat types	Description
Thick tussock grassland with hydromorphic lawn	Generally located in low, flat, sometimes flooded positions. Some stands lack the hydromorphic lawn type, being mostly covered by *Poa* grassland. Proportion of active erosion edges was relatively high, but restricted to the margins of water courses. Grazing pressure was variable
Thin tussock grassland	Dominated by *Deyeuxia/Festuca* tussocks. Generally found on gentle slopes and flat summits at all altitudes, although the dominant species shifted with altitude. Below 1,900 m *Festuca* dominated at all topographic positions, being gradually replaced by *Deyeuxia* as altitude increased, up to 2,300 m where this species dominated at all topographic positions. Erosion activity was low, and livestock pressure intermediate
Lawn	Largely dominated by *Alchemilla-Carex* lawn, with some patches of other types. Rock exposed by erosion was generally found at the bottom of concavities which get flooded in the rainy season. Located at sites with less than a 10% slope, usually at high altitudes. Erosion was more active, and grazing pressure more intense, than in tussock grasslands
Outcrop with tussock grassland	A mixture of natural outcrops, exceptionally reaching 120 m tall, and tussock grasslands, together with small patches of other types, including *Polylepis* woodlands or shrublands. Mainly located on mid and upper steep slopes. Erosion activity was intermediate, and grazing pressure low
Outcrop with exposed rock	Dominated by rock, with small vegetation patches. Found at similar topographic positions than outcrop with tussock grassland, although at somewhat higher altitudes in more exposed and northerly sites. Activity of erosion and grazing pressure were high
Rock exposed by erosion	Its bare rock cover was higher than 80%, most of which was exposed due to erosion. Generally located in flat sites with high erosion activity. The grazing pressure on the few remaining vegetation patches was very intense

Appendix 2

Table 5 List of all recorded species, emphasizing their conservation status and degree of endemism (based on Miatello et al. 1999)

Common name	Scientific name	Guild	Conservation status (degree of endemism)
Andean swift	*Aeronautes andecolus*	Insectivorous	Unknown
Black-bellied shrike-Tyrant	*Agriornis montana fumosus*	Insectivorous	Unknown (subendemic)
Tufted tit-Tyrant	*Anairetes parulus*	Insectivorous	Unknown
"*Anthus sp.*"*		Insectivorous	Unknown
Cordilleran Canastero	*Asthenes modesta cordobae*	Insectivorous	Unknown (subendemic)
Córdoba Canastero	*Asthenes sclateri sclateri*	Insectivorous	Unknown (subendemic)
Gray-hooded Parakeet	*Bolborhynchus aymara*	Granivorous	Declining/ CITES II
White-tailed Hawk	*Buteo albicaudatus*	Carnivorous	Declining
Red-backed Hawk	*Buteo polyosoma*	Carnivorous	Declining
Swainson's Hawk	*Buteo swainsoni*	Carnivorous	Unknown
Hooded Siskin	*Carduelis magellanica*	Granivorous	Unknown
Band-tailed Seedeater	*Catamenia analis*	Granivorous	Declining

Table 5 continued

Common name	Scientific name	Guild	Conservation status (degree of endemism)
Plain-colored Seedeater	*Catamenia inornata cordobensis*	Granivorous	Declining (subendemic)
Turkey Vulture	*Cathartes aura*	Carnivorous	Unknown
White-winged Cinclodes	*Cinclodes atacamensis*	Insectivorous	Unknown (subendemic)
Chestnut-winged Cinclodes	*Cinclodes comechingonus*	Insectivorous	Abundant
Bar-winged Cinclodes	*Cinclodes fuscus*	Insectivorous	Abundant
Olrogs's Cinclodes	*Cinclodes olrogi*	Insectivorous	Abundant (endemic)
Cinereous Harrier	*Circus cinereus*	Carnivorous	Unknown
Grass Wren	*Cistothorus platensis*	Insectivorous	Abundant
Field Flicker	*Colaptes campestris*	Insectivorous	Abundant
Golden-breasted Woodpeacker	*Colaptes melanolaimus*	Insectivorous	Abundant
Black Vulture	*Coragyps atratus*	Carnivorous	Unknown
Peregrine Falcon	*Falco peregrinus*	Carnivorous	Declining
American Kestrel	*Falco sparverius*	Carnivorous	Unknown
Common Snipe	*Gallinago gallinago*	Insectivorous	Declining
Common Miner	*Geositta cunnicularia contrerasi*	Insectivorous	Unknown (subendemic)
Rufous-banded Miner	*Geositta rufipennis otowi*	Insectivorous	Unknown (subendemic)
Black-chested Buzzard-Eagle	*Geranoaetus melanoleucus*	Carnivorous	Declining
Spectacled Tyrant	*Hymenops perspicillata*	Insectivorous	Unknown
Brown-capped Tit-Spinetail	*Leptasthenura fuliginiceps*	Insectivorous	Declining
Rufous-naped Ground-Tyrant	*Muscisaxicola rufivertex achalensis*	Insectivorous	Declining (subendemic)
Blue-and-White Swallow	*Nothiochelidon cyanoleuca*	Insectivorous	Abundant
Spotted Tinamou	*Nothura maculosa*	Omnivorous	Abundant
Band-tailed Sierra-Finch	*Phrygilus alaudinus*	Granivorous	Unknown
Ash-breasted Sierra-Finch	*Phrygilus plebejus naroskyi*	Granivorous	Unknown (subendemic)
Plumbeous Sierra-Finch	*Phrygilus unicolor cyaneus*	Granivorous	Abundant (subendemic)
Crested Caracara	*Polyborus plancus*	Carnivorous	Abundant
Red-tailed Comet	*Sappho sparganura*	Omnivorous	Declining/ CITES II
Long-tailed Meadowlark	*Sturnella loica obscura*	Omnivorous	Abundant (subendemic)
House Wren	*Troglodytes aedon*	Insectivorous	Abundant
Chiguanco Thrush	*Turdus chiguanco*	Omnivorous	declining
Southern Lapwing	*Vanellus chilensis*	Insectivorous	Abundant
Andean Condor	*Vultur gryphus*	Carnivorous	Declining/ CITES I
Rufous-collared Sparrow	*Zonotrichia capensis*	Granivorous	Abundant

**Anthus sp.*" includes the following species: *A. correndera*, *A. furcatus*, *A. hellmayri* and *A. lutescens*

References

Beale CM, Monaghan P (2004a) Human disturbance: people as predation-free predators? J Appl Ecol 41:335–343

Beale CM, Monaghan P (2004b) Behavioural responses to human disturbance: a matter of choice? Anim Behav 68:1065–1069

Blumstein DT, Anthony LL, Harcourt RG, Ross G (2003) Testing a key assumption of wildlife buffer zones: is flight initiation distance a species-specific trait? Biol Conserv 110:97–100

Blumstein DT, Fernandez-Juricic E, Zollner PA, Garity SC (2005) Inter-specific variation in avian responses to human disturbance. J Appl Ecol 42:943–953

Bolger DT, Suarez AV, Crooks KR, Morrison SA, Case TJ (2000) Arthropods in urban habitat fragments in southern California: area, age, and edge effects. Ecol Appl 10:1230–1248

Buckland ST, Anderson DR, Burnham KP, Laake JL, Borchers DL, Thomas L (2001) Introduction to distance sampling: estimating abundance of biological populations. Oxford Univ. Press, Oxford

Cabido M (1985) Las comunidades vegetales de la Pampa de Achala, Sierras de Córdoba, Argentina. Docum Phytosociol 9:431–443

Cabido M, Funes G, Pucheta E, Vendramini F, Días S (1998) A chorological analysis of the mountains from Central Argentina. Is all what we call Sierra Chaco really Chaco? Contribution to the study of the flora and vegetation of the Chaco. XII. Candollea 53:321–331

Cabido M, Antón A, Cabrera M, Cingolani A, Di Tada I, Enrico L, Funes G, Haro G, Polop J, Renison D, Rodríguez V, Roqué Garzón J, Rosacher C, Zak M (2003) Línea de Base y Programa de Monitoreo de la biodiversidad del Parque Nacional Quebrada del Condorito y la Reserva Hídrica Provincial Pampa de Achala. Administración de Parques Nacionales Report, Córdoba

Canaday C (1996) Loss of insectivorous birds along a gradient of human impact in Amazonia. Biol Conserv 77:63–77

Christ C, Hillel O, Matus S, Sweeting J (2003) Tourism and biodiversity: mapping tourism's global footprint. Conservation International, Washington DC

Chuan Lim H, Sodhi NS (2004) Responses of avian guilds to urbanisation in a tropical city. Landscape Urban Plan 66:199–215

Cingolani AM, Cabido MR, Renison D, Solís Neffa V (2003) Combined effects of environment and grazing on vegetation structure in Argentine granite grasslands. J Veg Sci 14:223–232

Cingolani AM, Renison D, Zak MR, Cabido MR (2004) Mapping vegetation in a heterogeneous mountain rangeland using Landsat data: an alternative method to define and classify land-cover units. Remote Sens Environ 92:84–97

Deluca TH, Patterson WA, Freimund WA, Cole DN (1998) Influence of llamas, horses, and hikers on soil erosion from established recreation trails in western Montana, USA. Environ Manage 22:255–262

Diaz Benevides, Perez-Ducy E (2001) Tourism in the least developed countries. Madrid, WTO and UNCTAD

Fernández-Juricic E (2000) Local and regional effects of pedestrians on forest birds in a fragmented landscape. Condor 102:247–255

Fernández-Juricic E (2002) Can human disturbance promote nestedness? A case study with birds in an urban fragmented landscape. Oecologia 131:269–278

Fernández-Juricic E, Sallent A, Sanz R, Rodríguez-Prieto I (2003) Testing the risk-disturbance hypothesis in a fragmented landscape: nonlinear responses of House Sparrows to humans. Condor 105:316–326

Forsman JT, Mönkkönen M, Inkeröinen J, Reunanen P (1998) Aggregate dispersion of birds after encountering a predator: experimental evidence. J Avian Biol 29:44–48

Frid A, Dill LM (2002) Human-caused disturbance stimuli as a form of predation risk. Conserv Ecol 6(1):11. Online URL: http://www.consecol.org/vol6/iss1/art11

Gill JA, Sutherland WJ, Watkinson AR (1996) A method to quantify the effects of human disturbance on animal populations. J Appl Ecol 33:786–792

Gill JA, Norris K, Sutherland WJ (2001) Why behavioural responses may not reflect the population consequences of human disturbance. Biol Conserv 97:265–268

Gutzwiller KJ (1995) Recreational disturbance and wildlife communities. In: Knight RL, Gutzwiller KJ (eds) Wildlife and recreationists: coexistence through management and research. Island Press, Washington DC, pp 169–181

Gutzwiller KJ, Anderson SH (1999) Spatial extent of human-intrusion effects on subalpine bird distributions. Condor 101:378–389

Hammitt WE, Cole DN (1987) Wildland recreation: ecology and management. John Wiley and Sons, New York

Kiss A (2004) Is community-based ecotourism a good use of biodiversity conservation funds? Trends Ecol Evol 19:232–237

Kitahara M, Sei K, Fujii K (2000) Patterns in the structure of grassland butterfly communities along a gradient of human distrubance: further analysis based on the generalist/specialist concept. Res Popul Ecol 42:135–144
Laiolo P (2003) Diversity and structure of the bird community overwintering in the Himalayan subalpine zone: is conservation compatible with tourism? Biol Conserv 115:251–262
Luti R, Solis M, Galera F, Ferreira N, Berzal M, Nores M, Herrera M, Barrera J (1979) Vegetación. In: Vázquez JB, Miatello RA, Roqué ME (eds) Geografía Física de la Provincia de Córdoba. Ed. Bolett, Buenos Aires. pp 297–368
McDougall KL, Wright GT (2004) The impact of trampling on feldmark vegetation in Kosciuszko National Park, Australia. Aust J Bot 52:315–320
Miatello R, Baldo J, Ordano M, Rosacher C, Biancucci L (1999) Avifauna del Parque Nacional Quebrada del Condorito y Reserva Hídrica Provincial de Achala, Córdoba, Argentina. Una lista comentada. Talleres Gráficos de Euder S.R.L., Córdoba
Miller SG, Knight RL, Miller CK (1998) Influence of recreational trails on breeding bird communities. Ecol Appl 8:162–169
Müllner A, Linsenmair KE, Wikelski M (2004) Exposure to ecotourism reduces survival and affects stress response in hoatzin chicks (*Opisthocomus hoazin*). Biol Conserv 118:549–558
Nores M (1995) Insular biogeography of birds on mountain-tops in north western Argentina. J Biogeogr 22:61–70
Ordano M (1996) Estudio de una comunidad de aves altoserrana (Córdoba, Argentina) durante un ciclo anual. Revista de la Asociación de Ciencias Naturales del Litoral 27:83–94
Priskin J (2003) Tourist perceptions of degradation caused by coastal nature-based recreation. Environ Manage 32:189–204
Pulliam HR, Caraco T (1984) Living in groups: is there an optimal group size? In: Krebs JR, Davies NB (eds) Behavioural ecology, an evolutionary approach, 2nd edn. Sinauer, Sunderland, MA, pp 122–147
Reijnen R, Foppen R (1995) The EFFECTS of car traffic on breeding bird populations in woodland. IV. Influence of population-size on the reduction of density close to a highway. J Appl Ecol 32:481–491
Renison D, Cingolani AM, Schinner D (2002) Optimizing restoration of *Polylepis australis* woodlands: when, where and how to transplant seedlings to the mountains? Ecotropica 8:219–224
Riffell SK, Gutzwiller KJ, Anderson SH (1996) Does repeated human intrusion cause cumulative declines in avian richness and abundance? Ecol Applic 6:492–505
Rodríguez-Prieto I, Fernández-Juricic E (2005) Effects of direct human disturbance on the endemic Iberian frog (*Rana iberica*) at individual and population levels. Biol Conserv 123:1–9
Roovers P, Verheyen K, Hermy M, Gulinck H (2004) Experimental trampling and vegetation recovery in some forest and heathland communities. Appl Veg Sci 7:111–118
Sekercioglu CH (2002) Impacts of birdwatching on human and avian communities. Environ Conserv 29:282–289
Shure DJ, Phillips DL (1991) Patch size of forest openings and arthropod populations. Oecologia 86:325–334
Skagen SK, Knight RL, Orians GH (1991) Human disturbance of an avian scavenging guild. Ecol Appl 1:215–225
Soulé ME, Alberts AC, Bolger DT (1992) The effects of habitat fragmentation on chaparral plants and vertebrates. Oikos 63:39–47
Soulé ME, Orians GH, Boersma DP (2001) Conservation biology: research priorities for the next decade. Island Press, Covelo, CA
Suarez A, Bolger D, Case T (1998) Effects of fragmentation and invasion on native ant communities in coastal Southern California. Ecology 79:2041–2056
van der Zande AN, Vos P (1984) Impact of a semi-experimental increase in recreational intensity on the densities of birds in groves and hedges on a Lake Shore in the Netherlands. Biol Conserv 30:237–259
van der Zande AN, Berkhuizen JC, van Latesteijn HC, ter Keurs WJ, Poppelaars AJ (1984) Impact of outdoor recreational activities on the density of a number of breeding bird species in woods adjacent to urban residential areas. Biol Conserv 30:1–39
Webb NR, Clarke RT, Nicholas JT (1984) Invertebrate diversity on fragmented Calluna-heathland: effects of surrounding vegetation. J Biogeogr 11:41–46
Webb NR (1989) Studies on the invertebrate fauna of fragmented heathland in Dorset, UK, and the implications for conservation. Biol Conserv 47:153–165
Webb NV, Blumstein DT (2005) Variation in human disturbance differentially affects predation risk assessment in western gulls. Condor 107:178–181

Biodivers Conserv (2007) 16:1027–1037
DOI 10.1007/s10531-006-9049-2

ORIGINALPAPER

Population decline of loggerhead turtles: two potential scenarios for Fethiye beach, Turkey

Ç. Ilgaz · O. Türkozan · A. Özdemir · Y. Kaska
M. Stachowitsch

Received: 12 April 2005 / Accepted: 10 April 2006 / Published online: 16 August 2006

Abstract Based on nesting data over a 12-year period (1993–2004), this study points to a negative population trend of the loggerhead turtle population at Fethiye beach, Turkey. The number of nests fluctuated from a maximum of 186 in 1995 to a minimum of 58 in 2004. Successively smaller peaks at 3-year intervals were followed by successively smaller troughs. Two analyses—one representing a dampened oscillation, the other retaining the period and the amplitudes of the nesting cycles—predict that nest number will drop to about 40–50 by 2015, i.e. to about 22–27% of its highest value. This drop at Fethiye does not correspond with a visible increase at neighboring beaches, leading to the interpretation that the number of nesting turtles here is declining. Moreover, the carapace size of emerging adult females is apparently decreasing, as are clutch sizes. Such a potential negative trend at a key Turkish nesting beach is cause for concern, an incentive for continued study, and a call for more coordinated and effective conservation programs in this region of the Mediterranean.

Ç. Ilgaz
Fauna and Flora Research and Application Center, Dokuz Eylül University, 35150 Buca-İzmir, Turkey

O. Türkozan (✉)
Faculty of Science and Arts, Department of Biology, Adnan Menderes University, 09010 Aydın, Turkey
e-mail: turkozan@adu.edu.tr

A. Özdemir
Faculty of Education, Department of Science Education, Adnan Menderes University, 09010 Aydın, Turkey

Y. Kaska
Faculty of Science and Arts, Department of Biology, Pamukkale University, Denizli, Turkey

M. Stachowitsch
Department of Marine Biology, University of Vienna, Althanstrasse 14, A-1090 Vienna, Austria

Keywords *Caretta caretta* · Conservation · Mediterranean Sea · Population decline · Sea turtles · Trend analysis

Introduction

In the Mediterranean, the major nesting grounds for loggerhead turtles, *Caretta caretta*, are Greece and Turkey (Baran and Kasparek 1989; Margaritoulis 2000), with smaller numbers recorded in Cyprus (Broderick and Godley 1996), Egypt (Kasparek 1993; Clarke et al. 2000), Libya (Laurent et al. 1995), Tunisia (Laurent et al. 1990), Israel (Kuller 1999), Syria (Kasparek 1995) and Lebanon (Newburry et al. 2002).

Fethiye beach (Fig. 1) is one of 13 key Turkish nesting sites for loggerhead turtles (Margaritoulis et al. 2003). This area is also important because it is used as a feeding ground by juvenile green turtles (Türkozan and Durmus 2000). The continuity of the loggerhead turtle population studies at Fethiye beach, which have been carried since 1993 (Türkozan and Baran 1996; Baran and Türkozan 1996; Türkozan 2000), underlines the status of this site and makes it a case study for Turkish beaches and conservation efforts.

"Assessing the population size of sea turtles is a difficult task, particularly in those species that occupy different habitats during their life cycle and migrate large distances" (Broderick et al. 2002). The strong site fidelity of female turtles to specific

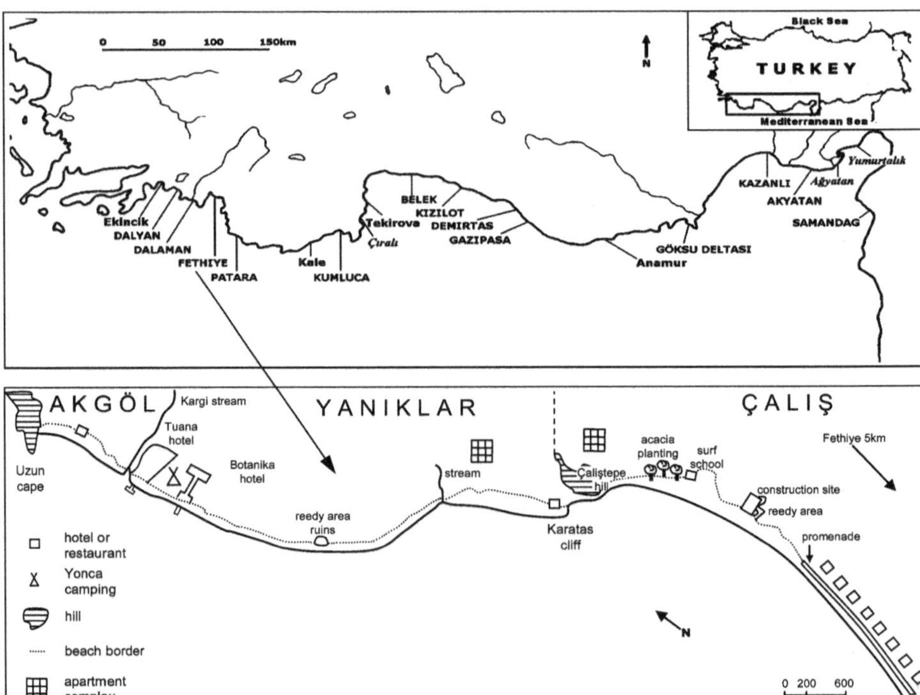

Fig. 1 Top: Localities with loggerhead nesting beaches in Turkey (upper case indicates main nesting beaches; lower case, secondary nesting beaches; italics, additional unsubstantiated records). Bottom: General map of Fethiye beach showing the three investigated beach subsections

nesting beaches provides a basis for population assessment provided that the study is sufficiently long and methods are standardized (Chaupka 2001). One approach to evaluating the status of a sea turtle species is to examine long-term changes or trends in the number of nesting females or nests; nest number is a useful index for the number of nesting females, especially in cases where full beach censuses of adults are impractical on a multi-annual basis (National Research Council 1990). This indirect estimate of population size requires the fewest assumptions about the biology of a particular sea turtle species. Evaluating trends in sea turtle nesting populations, however, requires many years of data because of the high annual variation in nesting numbers (Bjorndal et al. 1999). According to the National Research Council (1990), a decade or more, depending on longevity, might be required to measure a real change in a population. "Surveys longer than a decade become increasingly valuable for management purposes because they can transcend the short-term fluctuations that obscure long-term trends" (National Research Council 1990). In addition to annual nesting data, knowing the interval between breeding seasons and the clutch frequency per season of females provides further information for assessing population trends (Broderick et al. 2002). Such trends have, for example, been assessed for green turtles nesting at Tortuguero, Costa Rica (Bjorndal et al. 1999), green turtles in Southeast Asia (Chaupka 2001), and olive ridley turtles in Orissia, India (Shanker et al. 2003), using time series analysis for interpretations and predictions. In another approach, Hatase et al. (2002) used the annual recruitment dynamics and body size fluctuations of loggerhead turtle populations at Senri Beach, Japan, to show a population decline.

The aim of the present study is to identify potential trends in nesting numbers of the loggerhead turtle population at one key Turkish beach over recent years. This is the first population assessment study for the Mediterranean basin using trend analysis.

Materials and methods

Twelve nesting seasons were evaluated with respect to the number of nests and the size of nesting females, seasonal distribution of nesting, clutch size, and the structure of Fethiye beach. The first, 5-year published data set (Türkozan and Baran 1996; Baran and Türkozan 1996; Türkozan 2000) stems from 1993 to 1997 ("previous study"), the second set from 1998 to 2004 ("present study"). The two were combined because the surveys used standardized methodologies and were carried out by observers who had undergone the same training. For the long-term trend analysis, two non-linear regressions were calculated for these data using SPSS 10, one reflecting a dampened oscillation, the other retaining the period and the amplitudes of the cycles in the empirical data (both based on an exponentially decreasing median value). These two final solutions were obtained by checking different periodic regression methods.

The current investigation was carried out from June to September 1998–2004, encompassing 75.8% of the nesting events (range: 66.9–87.9). The nesting records of May and early June were provided by a local volunteer and also reconstructed based on later hatching events.

We followed standard beach survey procedures: Teams of 2–3 persons conducted night patrols (ca. 21:00–2:00) and morning patrols (ca. 6:00–8:00). During night patrols, each encountered turtle—following oviposition or while returning to the water—was measured and tagged with monel tags on the right front flipper. Carapace lengths and widths (straight) from nuchal notch to caudal tip and widest point of carapace were measured in centimeters using wooden calipers. For those turtles recaptured in one season, the mean of the measurements was used. During morning patrols, the shape and pattern of tracks were noted and those tracks that resulted in nests were marked. The nest locations were confirmed with probes and then marked. Tracks with no nests were counted as non-nesting emergences.

Study site

The beach is located in Fethiye Bay, Muğla Province, Turkey, and is approximately 8 km long (Fig. 1, top). Three subsections were distinguished based on their features and on practical fieldwork considerations. The Akgöl beach subsection extends from Uzun Cape in the north to the mouth of Kargı stream in the south (Fig. 1, bottom). It is approximately 1 km long and 50 m and more wide. The front of this beach consists of pebbles up to 2 cm in diameter. Behind this zone, the beach becomes much steeper and is composed of sand mixed with pebbles: sand is the dominant substrate at some places. Except for short stretches at both ends, this subsection is not suitable for nesting because a length of about 300–400 m is covered with pebbles. The hinterland here consists of farmland.

The second subsection (Yanıklar) extends from the mouth of the Kargı stream to the hill called Çalıstepe (Fig. 1, bottom). This beach is approximately 4.5 km long and its width varies between 50 m and 80 m. The first few meters of the beach gently slope up from the sea and consist of pebbles. Behind this zone, sand becomes the dominant substrate. The hinterland here is a large wetland, mostly covered by a forest which is partly inundated until June. Several small creeks enter the sea along this beach subsection.

The third subsection (Çalıs) extends from Çalıstepe to the end of the promenade of Çalıs town (intersection between the main road and the beach) (Fig. 1, bottom). It is about 2.5 km long and 17–19 m wide, delimited over about half its length by a concrete wall topped by a broad promenade. The promenade is lined by tourist infrastructures (hotels, restaurants, bars, etc.). The beach is completely sandy and flat along the promenade stretch (this is the turtle nesting area) but is steeper and consists of pebbles elsewhere.

Results and discussion

In Fethiye, the 12-year average was 112 ± 34.9 nests (range: 58–186) per season with an average density of 14 ± 4.4 nests/km (range: 7.3–23.9). At all three beach subsections, the nests tended to be concentrated along certain stretches. There were strong annual fluctuations in the number of nests, ranging from a minimum of 58 (in 2004) to a maximum of 186 (in 1995), a 221% difference. This difference is very similar to the 224% documented by Margaritoulis and Rees (2001) between the minimum and maximum nest counts in Greece (Kyparissia Bay) over the course of 17 seasons; interestingly, their peak value was also recorded in 1995. Broderick and Godley (1996), for example, recorded a 112% increase between the lowest and highest nest numbers in northern Cyprus in three seasons, and even smaller fluctuations (ca. 54%) are known for loggerhead nests in Florida (Heppel et al. 2003).

In Fethiye, the major peak in 1995 was followed by a considerable drop in 1996 (Fig. 2). Subsequent peaks were recorded in 1998 (119), 2001 (114) and 2003 (106), each being followed by a renewed decline (Fig. 2). Of these nests, 7.1% were produced in May, 61.2% in June, 31% in July and 0.7% in August. Note that the peaks for overall nest numbers per year are reflected in the curves for June and to a lesser extent July (Fig. 3).

The first three peaks are positioned in 3-year intervals, and the fourth peak represents a 2-year interval. Such a fluctuation would be consistent with the generally accepted 3-year cycle of nesting by adult female loggerhead turtles in populations around the world (Miller 1997). Broderick et al. (2002) estimated the median interval for loggerhead turtles as 2 years in the Mediterranean.

Interestingly, each peak between 1993 and 2004 is successively lower than the preceding one. Moreover, with the exception of 1997, the minima are also successively smaller (Fig. 2). This development indicates a negative trend. The nesting data from 1993 to 2004 were therefore analyzed in greater detail. The trend in the annual number of nests cannot be optimally described by a linear function ($r^2 = 0.29$; $P > 0.05$). Two non-linear approaches were therefore used to examine the 12-year data set (Fig. 2). Both confirm that the number of nests is decreasing. The first assumes an oscillation (decreasing dampened oscillation) around an exponentially decreasing median value

$$f(y) = -64.51e^{-0.216(y-1992)}\sin[1.66(y-1992)+0.29] + 164.87 \cdot 0.9403^{(y-1992)}.$$

Biologically, such a dampened oscillation would mean a reduction not only in the mean number of nests, but that the fluctuation in nest number (i.e. the amplitudes) is also reduced, presumably due to ever fewer adult females. This means weaker peak years and less abrupt drops after the maxima, while retaining the period (ca. 3 years)

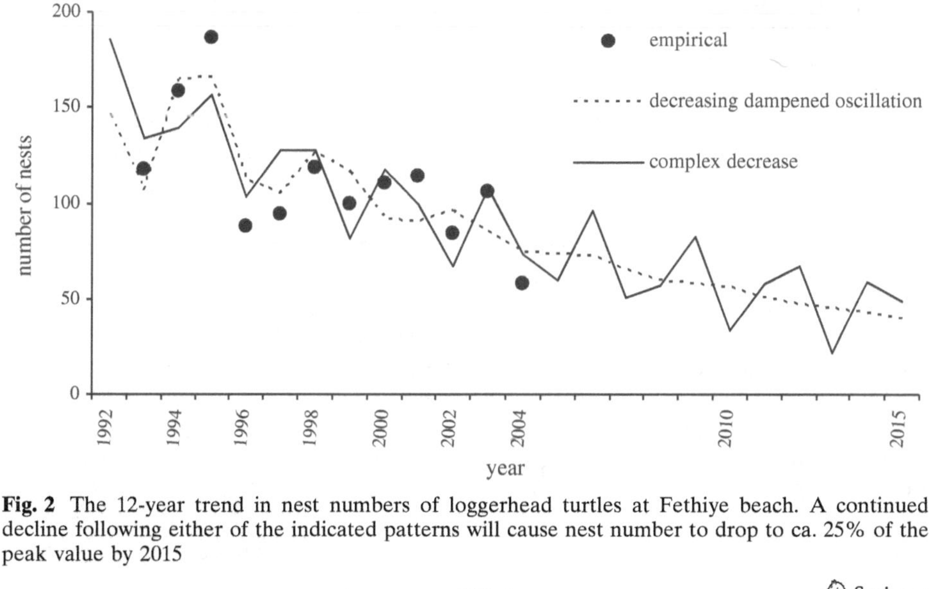

Fig. 2 The 12-year trend in nest numbers of loggerhead turtles at Fethiye beach. A continued decline following either of the indicated patterns will cause nest number to drop to ca. 25% of the peak value by 2015

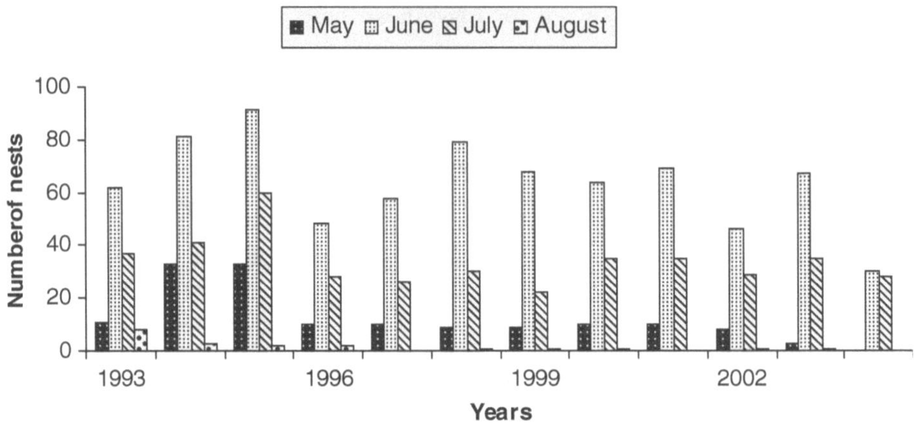

Fig. 3 Distribution of numbers of nests with respect to the months of the nesting seasons

of the oscillations ($r^2 = 0.7314$; $P < 0.001$). This function, and this interpretation, was originally applied to the first 10 years of data (1993–2002), where it appeared as though the amplitudes were decreasing. It fits the initial empirical values well, even though the major drop in 2004 signals a strong amplitude, and a complete flattening of the curve (as indicated after 2010) is unlikely.

Based on the relatively large difference between 2003 and 2004, a second approach, depicting a complex decrease, was calculated

$$f(y) = 160.59 \cdot 0.943^{(y-1992)} - 25.4\cos[2\pi(y-1992)/2.81 - 3.04].$$

Here, the values also decrease around an exponentially decreasing mean, but both the period and the amplitudes of the cycles in the empirical data are retained ($r^2 = 0.6776$; $P < 0.001$). Biologically, this would also reflect a decrease in the mean nest number, but one that continues to show strong cycles of nesting.

The fluctuations in Fethiye are clearly occurring in the framework of a considerable decreasing trend. This stands in contrast to the fluctuating but increasing number of loggerhead nests recorded in the Index Nesting Beach Survey Program in Florida (Fig. 11.1b in Heppel et al. 2003). In fact, the drop more closely parallels the decreasing trend in leatherback females in Playa Grande, Costa Rica (Fig. 11.1e in Heppel et al. 2003), reflecting a decline in Pacific leatherbacks that the authors describe as "dangerous".

Accordingly, our interpretation is that the number of nesting adult females is declining along Fethiye beaches. Does the decline in Fethiye follow the dampened oscillation or the complex model? Measurements of tagged turtles, clutch sizes, and the most recent nest number (from 2005) each provide further evidence for the overall decline and tend to support the complex model.

One-hundred-four females were tagged and measured from 1998 to 2004 (this study), 87 for the first time, along with 17 recaptures. The dimensions of these nesting females are presented in Fig. 4. The mean straight carapace lengths (SCL) during this 7-year period ranged from 69.1 ± 4.6 (range: 63–78) to 72.6 ± 5.8 (range: 58–85) cm with a mean of 70.6 ± 5.2 cm (range: 55–85) ($N = 100$) (Fig. 4). In both turtle carapace parameters (SCL and SCW), size generally tended to decrease over the 12-year period (highest SCL 73.2 and highest SCW 54.4 in 1994) (Fig. 4). Spearman rank correlation supported this negative correlation between the SCL and years ($N = 191$, Spearman $R = -0.19$, $t = -2.7$, $P < 0.01$). In particular, the drop in 2000 is evident and, with few exceptions (e.g. SCW in 2004), the values of both measures remained clearly below those recorded in the 1990s. A decrease in mean

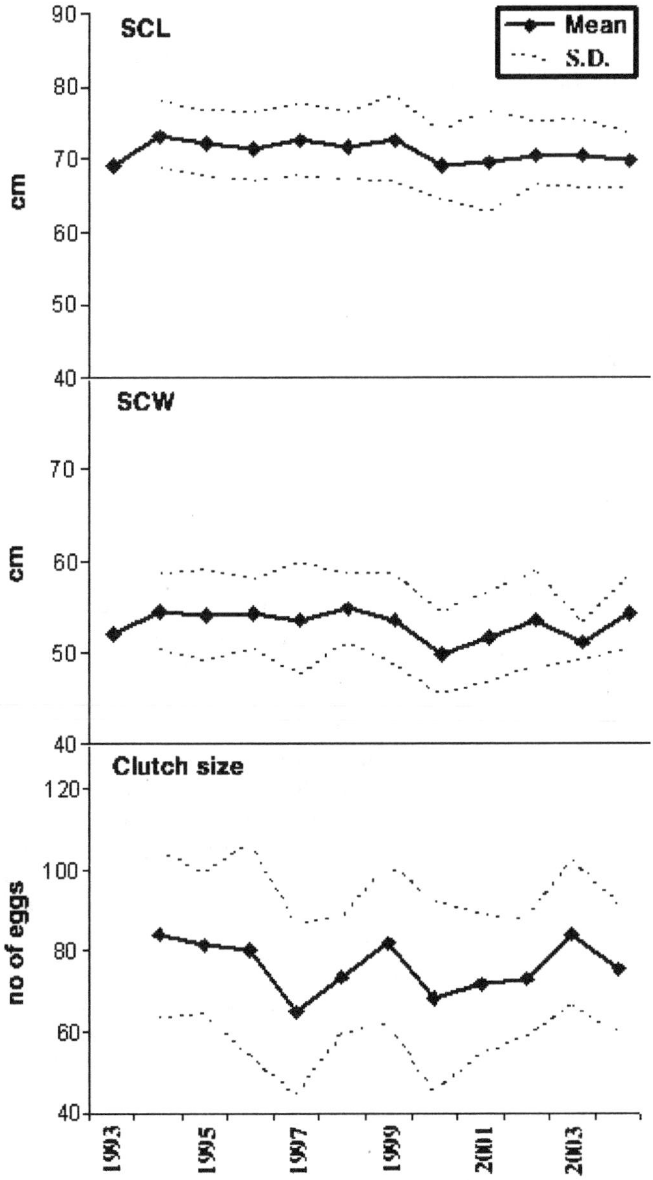

Fig. 4 Dimensions of SCL (straight carapace length), SCW (straight carapace width) and clutch sizes of nesting loggerhead turtle females tagged on Fethiye beach (1993 values from Türkozan and Baran 1996; 1994 values from Baran and Türkozan 1996; 1995–1997 values from Türkozan, 2000).

size would argue for an increase in the relative proportion of recruiting versus remigrating turtles (Hatase et al. 2002).

Based on the assumption that smaller turtles produce smaller clutches, the decreasing size of emerging turtles in Fethiye is also supported by the clutch size data: again, the values dropped in 2000 and generally (exceptions 1997, 2003) remained below those of the 1990s ($N = 101$, $r = -0.285$, $t = -2.96$, $P < 0.001$). Whether the decrease in carapace size reflects increased recruitment of young adults (a positive sign of population recovery) or the selective loss or shift to other beaches of larger females (a negative development) will require further investigation.

Of the 172 individuals tagged over the 12 consecutive seasons, 25 were remigrants. The number of recorded remigrants ranged from 2 to 6 per season, with a mean of 3.4 individuals. In general, the interval between successive nesting years (the remigration interval) reflects conditions encountered on the feeding grounds, with good feeding years reducing the remigration interval and vice versa (Hays 2000). In green turtles, higher sea surface temperatures also positively affected the remigration interval (Solow et al. 2002). The remigration interval in Fethiye was 1 to 7 years with a mean of 3.35 years. Broderick et al. (2002) reported the median interval for the Mediterranean loggerhead turtles as 2 years. Remigration intervals of >5 years are probably artifacts caused by tagged turtles being missed by beach patrols when they return in the interim (Mortimer and Carr 1987). This no doubt holds true for Fethiye beach as well: many turtles were missed, particularly in the early part of the season before arrival of the research teams.

Any decline in the number of nesting females or nests at a particular beach should also be examined in the framework of beach fidelity. Four of the tagged Fethiye females, for example, originated from other Turkish nesting beaches; they must have been tagged in the 1988 marine turtle general survey (Baran and Kasparek 1989) because we have other records after that year. Conversely, one turtle that emerged on Dalaman beach in 2002 had been previously tagged at Fethiye in 1993 (Kaska 2004). Similar shifts were recorded from Senri beach, Japan, where the loggerhead turtle population declined (Sato et al. 1997). According to the authors, this decline was due to extensive resort developments rather than individual movements.

This perspective is also helpful in discussing the two decline models (Fig. 2). Accordingly, the decrease that both models show may either reflect an actual decrease in the population, or the adult females may be seeking other beaches to lay their eggs, for example due to beach deterioration (e.g. loss of sandy stretches, increased tourism pressure) in Fethiye. The data from adjoining beaches along this otherwise rocky coast (Fig. 1, top), however, such as Patara Beach 50 km and Çıralı 125 km to the east, or Dalaman Beach 40 km and the famed Dalyan Beach 50 km to the west, do not indicate a corresponding increase in nest numbers (Fig. 5). We are therefore inclined to accept the interpretation involving an actual decline in the loggerhead turtle population here. This is further supported by the decreasing turtle and clutch sizes: a shift to another beach would not necessarily be reflected by changes in these parameters, unless the shift involved a disproportionate number of older, larger females.

What proportion of the Mediterranean population is affected by the Fethiye scenario? During the 1998–2004 nesting seasons, a total of 1712 emergences were recorded, with 691 (40.4%) resulting in nests. In this 7-year period, the peak nest number in one season was 119 in 1998; in the overall 12-year period, the maximum value was 186 (in 1995). Assuming that each female nests an average of 3 times per

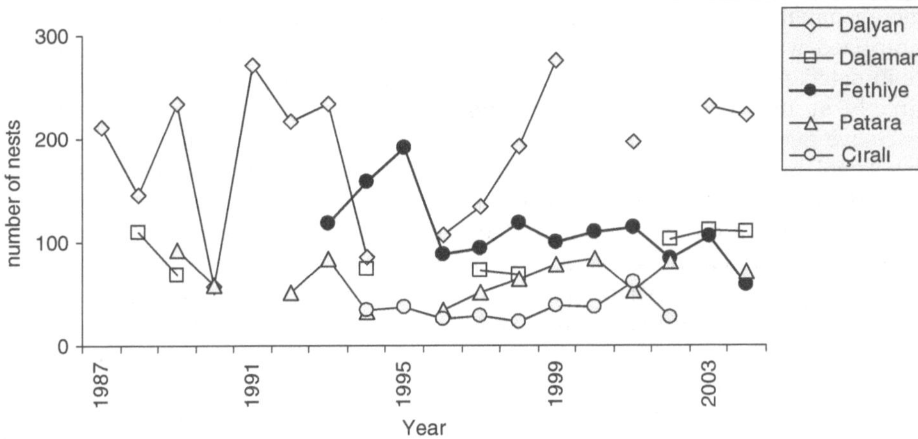

Fig. 5 The number of nests at neighboring nesting beaches. As opposed to Fethiye, no decline is apparent elsewhere. (Data compiled from Türkozan et al. 2003; Oruç et al. 2003; Kaska 2004).

season every 2–3 years (Groombridge 1990), approximately 64 (1/3 of the max. number of nest recorded) loggerheads visit Fethiye beach. Groombridge (1990) estimated 2000 *C. caretta* females nesting annually in the Mediterranean, and Broderick et al. (2002) used four different methods to update this number to 2280–2787. Accordingly, ca. 2.3–2.8% of the *C. caretta* population of the Mediterranean nest on this beach. In reviewing loggerhead turtle nesting in Turkey, Türkozan et al. (2003) concluded that approximately 1267 (range 663–1991) nests were laid annually on the 20 beaches studied. The average number of nests (112) for 12 consecutive years makes this beach one of the most important nesting sites in Turkey (8.8% of the nests laid annually in that country). Similarly, Canbolat (2004) reviewed both loggerhead and green turtle nesting in Turkey and concluded that 500–800 loggerheads nest there annually. His calculation of the annual nesting potential of Fethiye (5.4% of all nests laid annually on Turkish beaches), however, did not take the highest value (186 nests in 1995) into account.

Regardless of whether the decline at Fethiye is dampened (model 1) or not (model 2), the analysis predicts that nest number will drop to about 40–50 nests by 2015, i.e. to about 22–27% of its highest value in 1995. The latest nest data (from 2005; not included in Fig. 2) provides a final clue as to the type of decrease: After the all-time low in 2004 (58), the value increased to 80 nests in 2005. This lies on the curve of model 1 (dampened oscillation), below the preceding peak in 2003 and even below the trough of 2002. An only slight increase or a renewed drop in 2006 would strengthen the dampened oscillation scenario, whereas a larger increase in 2006 would support model 2. Ultimately, should female numbers decline further, the oscillations may well become more dampened.

Regardless of which scenario better describes the decline in Fethiye, the overall development should be a wake-up call for sea turtle conservation efforts in this region. Both anthropogenic and natural causes can be cited as potential reasons behind the observed negative trends at this particular beach. Recent severe winter storms, for example, have altered certain beach sections, replacing sand with pebbles and cobbles. Fethiye is also subject to many of the threats facing beaches worldwide, including sand removals. At the Yaniklar subsection, this includes increasing encroachment by two major hotel complexes (Fig. 1, bottom), with their persistent motorized water sport activities during the day, and beach discos, fireworks and illuminated piers at night. Although a large student camp with 40 wooden huts directly at the southern tip of the beach (2001–2003) has been removed, the extensive restaurant/bar complex and improved road access remains; road access to the Akgöl subsection has also been substantially enlarged. In the Çalis subsection, more than 800 densely spaced acacia trees were planted in 14 rows over a stretch of 200 m directly on the beach in 2000/2001; the first row begins in the sand immediately after the initial cobble band, and their roots now form an impenetrable barrier for nesting loggerheads. Major apartment/vacation home complexes (Fig. 1, bottom) have been built behind the beaches, and all three sections are experiencing an ever increasing number of beach bars and watersports rental stands along with their attendant dense rows of beach chairs, water vehicles and lighting. These developments, combined with the ongoing destruction of immediately adjoining wetlands for major upcoming construction projects, are incompatible with the Special Protected Area status of Fethiye beach. The possible decline of loggerhead turtles on a beach as important as Fethiye is a cause for concern, an incentive for continued study, and a call for more coordinated and effective conservation programs in this region of the Mediterranean.

Acknowledgements We thank all the volunteers from Dokuz Eylül, Adnan Menderes, Pamukkale and Vienna Universities for their help during the fieldwork. M.S. thanks his past and present field coordinators Nadja Ziegler and Christine Fellhofer for their crucial support and acknowledges sponsorship by the 'Society of the Friends of Schoenbrunn Zoo' and Gulet Touropa Touristik. We are very grateful to Prof. Dr. Johann Hohenegger of the Department of Paleontology in Vienna for calculating and representing the long-term nesting trend in Fig. 2. Fritz Sobotka drew the final version of Fig. 1.

References

Baran İ, Kasparek M (1989) Marine turtles Turkey. Status survey 1988 and recommendations for conservation and management. Heidelberg, Germany

Baran İ, Türkozan O (1996) Nesting activity of the loggerhead turtle, *Caretta caretta*, on Fethiye Beach, Turkey in 1994. Chelonian Conserv Biol 2(1):93–96

Bjorndal AK, Wetherall JA, Bolten AB, Mortimer JA (1999) Twenty-six years of green turtle nesting at Tortuguero, Costa Rica: an encouraging trend. Conser Biol 13(1): 126–134

Broderick AC, Godley BJ (1996) Population and nesting ecology of the Green Turtle, *Chelonia mydas*, and the Loggerhead Turtle, *Caretta caretta*, in northern Cyprus. Zool Middle East 13:27–46

Broderick AC, Glen F, Godley BJ, Hays GC (2002) Estimating the number of Green and Loggerhead Turtles nesting annually in the Mediterranean. Oryx 36(3):1–9

Canbolat AF (2004) A review of sea turtle nesting activity along the Mediterranean coast of Turkey. Biol Conser 116:81–91

Chaupka M (2001) Historical trends, seasonality and spatial synchrony in green sea turtle egg production. Biol Conserv 101:263–279

Clarke M, Campbell AC, Hameid WS, Ghoneim S (2000) Preliminary report on the status of marine turtle nesting populations on the Mediterranean coast of Egypt. Biol Conserv 94:363–371

Groombridge B (1990) Marine turtles in the Mediterranean: distribution, population status, conservation. Council of Europe Nature and Environment Series, Strasbourg

Hatase H, Goto K, Sato K, Bando T, Matsuzawa Y, Sakamoto W (2002) Using annual body size fluctuations to explore potential causes for the decline in nesting population of the loggerhead turtle *Caretta caretta* at Senri Beach, Japan. Mar Ecol Prog Ser 245:299–304

Hays GC (2000) The implications of variable remigration intervals for the assessment population size in marine turtles. J Therm Biol 206:221–227

Heppel SS, Snover ML, Crowder LB (2003) Sea turtle population ecology. In: Lutz P, Musick JA, Wyneken J (eds) Biology of sea turtles vol II, Florida, USA, pp 275–299

Kaska Y (2004) Deniz kaplumbağalarını koruma planı projesi. Muğla-Dalaman Bölgesi'ndeki kıyı ve sulak alan ekosistemlerinin arastırılması II. alt projesi. T.C. Çevre ve Orman Bakanlığı, Ankara

Kasparek M (1993) Survey of the Mediterranean coast between Alexandria and El-Salum, Egypt. Mar Turtle Newsl 63:8–9

Kasparek M (1995) The nesting of marine turtles on the coast of Syria. Zool Middle East 11:51–62

Kuller Z (1999) Current status and conservation of marine turtles on the Mediterranean coast for Israel. Mar Turtle Newsl 86:3–5

Laurent L, Navira S, Grissac DJ, Bradai MN (1990) Les tortues marines des Tunisie: premiers données. Bull Soc Herp Fr 53:1–17

Laurent L, Bradai MN, Hadoud DA, Gomati HM (1995) Marine turtle nesting activity assessment on Libyan coast. Phase I: survey of the coast between Egyptian border and Sirte- RAC/SPA (MAP-UNEP), Tunis

Margaritoulis D (2000) An estimation of the overall nesting activity of the loggerhead turtle in Greece. In: Abreu-Grobois FA, Briseño-Dueñas R, Márquez-Millán R, Sarti-Mártinez L (eds) Proceedings of the 18th International Sea Turtle Symposium, NOAA Technical Memorandum NMFS-SEFSC-436 Mazatlan, Mexico, pp 48–50

Margaritoulis D, Rees AF (2001) The loggerhead turtle, *Caretta caretta*, population nesting in Kyparissia Bay, Peloponnesus, Greece: results of beach surveys over seventeen seasons and determination of the core nesting habitat. Zool Middle East 24:75–90

Margaritoulis D, Argano R, Baran İ, Bentivegna F, Bradai MN, Camiñas JA, Casale P, De Metrio G, Demetropoulos A, Gerosa G, Godley BJ, Haddound DA, Houghton J, Laurent L, Lazar B (2003) Loggerhead turtles in the Mediterranean Sea: present knowledge and conservation perspectives. In: Bolten AB, Witherington BE (eds) Loggerhead sea turtles, Washington, USA, pp 175–198

Miller JD (1997) Reproduction in sea turtles. In: Lutz P, Musick JA (eds) Biology of sea turtles, Florida, USA, pp 51–81

Mortimer JA, Carr A (1987) Reproduction and migrations of the Ascension Island green turtle (*Chelonia mydas*). Copeia 1987:103–113

National Research Council (1990) Decline of sea turtles. Causes and prevention. National Academy Press, Washington, p 259

Newburry N, Khalil M, Venizelos L (2002) Population status and conservation of marine turtles at El-Mansuri, Lebanon. Zool Middle East 27:47–60

Oruç A, Türkozan O, Durmus SH (2003) Deniz kaplumbağalarının izinde deniz kaplumbağası yuvalama kumsalları değerlendirme raporu 2003. WWF-Turkey, İstanbul, Turkey

Sato K, Bando T, Matsuzawa Y, Tanaka H, Sakamoto W, Minamikawa S, Goto K (1997) Decline of the loggerhead turtle, *Caretta caretta*, nesting on Senri beach in Minabe, Wakayama, Japan. Chelonian Conserv Biol 2(4):600–603

Shanker K, Pandav B, Choudhury BC (2003) An assessment of the olive ridley (*Lepidochelys olivacea*) nesting population in Orissa, India. Biol Conserv 115:149–160

Solow RA, Bjorndal KA, Bolten AB (2002) Annual variation in nesting numbers of marine turtles: the effect of sea turtle surface temperature on remigration intervals. Ecol Lett 5:742–746

Türkozan O, Baran İ (1996) Research on the loggerhead turtle, *Caretta caretta*, of Fethiye Beach. Turk J Zool 20:183–188

Türkozan O (2000) Reproductive ecology of the loggerhead turtle, *Caretta caretta*, on Fethiye and Kızılot beaches, Turkey. Chelonian Conserv Biol 3(4):686–692

Türkozan O, Durmus SH (2000) A feeding ground for juvenile green turtles, *Chelonia mydas*, on the western coast of Turkey. Brit Herp Soc Bull 71:1–5

Türkozan O, Taskavak E, Ilgaz Ç (2003) A review of the biology of the loggerhead turtle, *Caretta caretta*, at five major nesting beaches on the Southwestern Mediterranean Coast of Turkey. Herpet J 13:27–33

Biodivers Conserv (2007) 16:1039–1053
DOI 10.1007/s10531-006-9059-0

ORIGINAL PAPER

Habitat use, roost selection and conservation of bats in Tsingy de Bemaraha National Park, Madagascar

Amyot Kofoky · Daudet Andriafidison · Fanja Ratrimomanarivo · H. Julie Razafimanahaka · Daniel Rakotondravony · Paul A. Racey · Richard K. B. Jenkins

Received: 7 February 2005 / Accepted: 24 April 2006 / Published online: 9 July 2006

Abstract Although the land mammals of Madagascar have been the subject of many studies, the island's bats have yet to feature prominently on the research or conservation agenda. In this study we used mist nets, acoustic sampling and cave surveys to assess habitat use, seasonality and roost selection. Four microchiropteran species (*Triaenops rufus*, *T. furculus*, *Miniopterus manavi* and *Myotis goudoti*) appeared to be strongly associated with the forest interior based on trapping, but analysis of time-expanded echolocation recordings revealed that *T. rufus* and *M. manavi* were frequently recorded in forest edges and clearings. Bat activity was significantly lower inside the forest than at the interface between agricultural land and forest. The caves visited most often by tourists were low in bat abundance and species richness. Anjohikinakina Cave, which was visited infrequently by people, was used by five species and contained between 54% (winter) and 99% (summer) of bats counted in 16 caves and is a site of national importance for bat conservation. *Hipposideros commersoni* was only netted in our study area during October and may be a migrant to the site or present but inactive during the austral winter. The forest

A. Kofoky · D. Andriafidison · F. Ratrimomanarivo · H. J. Razafimanahaka · R. K. B. Jenkins
Madagasikara Voakajy, B.P. 5181, Antananarivo 101, Madagascar

D. Rakotondravony · D. Andriafidison
Département de Biologie Animale, Faculté des Sciences, Université d'Antananarivo, B.P. 906, Antananarivo 101, Madagascar

H. J. Razafimanahaka
Département des Eaux et Forêt, Ecole Supérieure des Sciences Agronomiques, Université d'Antananarivo, Antananarivo 101, Madagascar

P. A. Racey · R. K. B. Jenkins (✉)
School of Biological Sciences, University of Aberdeen, Aberdeen AB24 2TZ, UK
e-mail: jenkins@wanadoo.mg

Present Address:
F. Ratrimomanarivo
WWF, B.P. 738, Antananarivo 101, Madagascar

surrounding the caves is therefore important because it provides cover for emerging bats and a potential source of invertebrate prey whilst the forest edge is important to foraging bats.

Keywords Acoustic sampling · Caves · Chiroptera · Forest dependency · Karst · Roost

Introduction

Bats make a significant contribution to mammalian species richness and biomass in the tropics, but despite the extensive amount of previous research on Madagascar's vertebrates (Goodman and Benstead 2003) there have been few ecological studies on Malagasy Chiroptera (Eger and Mitchell 2003). Research biologists have traditionally focused on the island's endemic land mammals with lemurs, rodents and tenrecs being popular topics of study (e.g. Stephenson et al. 1994; Ramanamanjato and Ganzhorn 2001; Ganzhorn et al. 2003; Kappeler and Rasoloarison 2003).

Until recently the taxonomy of Malagasy bats was based mainly on a series of collections made in the 1970s (Peterson et al. 1995) and ecological information was almost completely lacking. Over the last few years, new information has helped to fill some of the gaps on species distribution (Goodman et al. 2005b) and have led to revisions of the existing bat taxonomy and the description of new species (e.g. Goodman and Cardiff 2004; Goodman and Ranivo 2004). Progress has also been made by other workers in describing the echolocation calls of Malagasy microchiropterans (Russ et al. 2003) and in assessing the conservation status (Eger and Mitchell 2003; MacKinnon et al. 2003) and diet of fruit bats (Bollen and Van Elsacker 2002; Hutcheon 2003). However, there has been little attempt to describe habitat requirements for bats in Madagascar with the result that the impact of large-scale habitat alteration or small-scale variation in habitat structure remains unknown.

Bats provide a unique set of challenges for conservation biologists because many species of conservation concern occur outside forests, in degraded or open habitats. Furthermore, as many species aggregate in conspicuous and accessible roosts that are often accessible to people, successful conservation measures should include the protection of roost sites, as well as foraging habitats (Hutson et al. 2001). Agricultural areas between forest fragments are used by some bat species and farmland is sometimes an important linking habitat between isolated forest fragments (Estrada et al. 1993). Likewise, other studies from South America have shown that non-forest habitats (e.g. farmland, savanna) constitute important habitats for bats (Bernard and Fenton 2003) and that certain forest-dependent species are reliable indicators of habitat disruption (Fenton et al. 1992). One of the first steps to address the conservation priorities of bats in Madagascar should therefore be to assess forest dependency and habitat requirements. Using presence and absence data from sites in the dry forests of Madagascar, Goodman et al. (2005b) report that few of the 25 bats species recorded from the region appear to be dependent on expanses of intact natural forest and they suggest that conservation of roosting sites, especially caves, is the most important goal for chiropteran conservation.

Here we aim to quantify habitat use by bats in a national park in western Madagascar which has a complex system of caves and is therefore likely to be an area of high bat abundance (Kunz 1982). We assessed bat species composition, abundance and activity in three distinct forest habitats (interior, clearings and edges) during the dry season and early wet season in Parc National Tsingy de Bemaraha. We also surveyed a selection of caves in areas subject to either high or low tourism pressure. Very little sustained research has been attempted in the park but inventory surveys have revealed high levels of vertebrate endemism to the reserve (Rasoloarison and Paquier 2003). As elsewhere in Madagascar, pressures from local people on the resources within the protected area remain high and research is now required to assess the impact of human activity (e.g. tourism, forest clearance) in and around the park.

Study site

Parc National Tsingy de Bemaraha, Province de Mahajanga, lies approximately 200 km north of Morondava (18°12′–19°07′ S and 44°34′–44°56′ E) in western Madagascar. The climate has distinct wet and dry seasons which extend from November to April and May to October, respectively. Maximum monthly temperatures are recorded in October (max 41°C) and highest rainfall in February (456 mm; data from 1993 to 2000, Programme Bemaraha, Antsalova).

Situated within a limestone belt that stretches intermittently from the North West to the South West of the island, the area is famous for its water-eroded features, most notably its caves and ‘tsingy’ formations (sharp, needle-like limestone pinnacles). Since receiving its UNESCO World Heritage Site status in 1990 and being made a national park in 1997 the southern section of the Bemaraha plateau has become a major tourist attraction (Rasoloarison and Paquier 2003). Our study was conducted near Bekopaka village, on the northern bank of the Manambolo River. Land use outside of the park boundary is mainly rice cultivation and open pasture land and the improved access in recent years has led to increased immigration and growing levels of forest degradation on the fringes of the reserve.

Methods

The fieldwork was carried out over 9 weeks during 2003 from 12 July to 20 August in the austral winter and 7–30 October in the beginning of the austral summer. In most years the park is inaccessible by road from December until mid-May because of high water levels and our study periods were thus selected to represent a major seasonal contrast within the available time. Based on reconnaissance walks and conversations with park staff we selected three habitat types for the study (i) closed canopy deciduous forest, (ii) forest clearings of ‘tsingy’ rocks—a distinctive feature of the reserve and (iii) forest/agriculture edges containing a mixture of fruit trees, grazing pasture, rice fields, isolated large trees and patches of low shrubby vegetation.

Bat trapping

Bats were trapped in mist nets (6 or 9 m long) with the bottom pocket placed just above ground level. In intact forest, the nets were placed with the bottom shelf approximately 20 cm above the ground, across trails, small gaps and streams. We used five or six nets per night and the total length was either 36, 39 or 42 m, and varied in accordance with the size of the trap sites. Nets were open from 1800 h until 2200 h each night. During the first phase, 15 different netting sites, at least 500 m apart, were used each night. The nets were checked approximately every 5 min and bats were immediately extracted. Trapping was repeated at 14 of the same 15 sites during the second phase. Mist nets in the 'tsingy' clearings traversed the natural gaps, which were rarely larger than 200 × 100 m. Edge mist nets were placed either perpendicular to the forest edge or in vegetation gaps within the surrounding agricultural land and were always within 150 m of the forest edge.

After capture, bats were placed in cloth bags and retained for approximately 2 h for faeces collection (except bats caught within the first 15 min after sunset which were quickly identified and released). Species were identified using the keys and notes in Peterson et al. (1995) and Russ et al. (2003). For each individual bat the following information was recorded: time of capture, species, sex, age, reproductive condition, forearm length (mm) and weight (g). Voucher specimens were deposited at the Université d'Antananarivo.

Acoustic surveys

It is widely recognised that microchiropteran surveys should use both bat detectors and trapping devices to fully document the species composition or activity in a given area (Sedlock 2001). We assessed microchiropteran activity using 'Duet' bat detectors (Stag Electronics, UK) from point counts situated at approximately 300–400 m intervals. The closed-forest category was sub-divided into riparian (<50 m from water) and non-riparian (>50 m) habitats and bats were surveyed on trails and in the forest adjacent (c. 10 m) to trails. Listening with the 'Duet' detector in heterodyne mode, an observer swept through the frequency range until a bat was heard. Each 'bat pass' (defined as a sequence of at least two echolocation pulses of a passing bat) was counted and notes made on the maximum frequency and range of the pulse. We used the following system to classify bat passes heard in the field:

(i) <33 kHz—Molossidae and also *Taphozous mauritianus*
(ii) >33 kHz <60 kHz—Vespertilionidae
(iii) >60 kHz—Hipposideridae, but also includes *Myotis goudoti*

During each point count a second observer made simultaneous recordings with a Pettersson D980 bat detector in time-expansion mode (×10) which were recorded onto Sony mini-disks in the field for later analyses with Batsound Professional software (Pettersson Elektronik, Sweden). For trapped microchiropterans, we made reference recordings of flying individuals in a mesh-sided cage (3 × 3 × 3 m) and also recorded the bats as they were released. We identified species from recordings of their echolocation using notes and sonograms (Russ et al. 2003) and from comparisons with our own data. In Madagascar, the sonograms of Hipposideridae (*Triaenops rufus*, *T. furculus* and *Hipposideros commersoni*), *Myotis goudoti*, *Miniopterus manavi* and *Emballonura atrata* are straightforward to identify from field recordings (J. Russ *pers. comm.*; A. Kofoky unpubl. data). However, the echolocation calls of molossids, and the rarely caught *Scotophilus* spp., are less well described in Madagascar and cannot be used to identify free-flying bats from recorded echolocations. We calculated a frequency of occurrence for each species by the ratio of point counts with a species to total point counts for that habitat using determinations made from the time-expanded recordings.

Cave surveys

Diurnal surveys were made in 16 caves to assess bat species composition and abundance in caves visited by tourists and to look for species not trapped during the mist netting inventories. All but one of the caves were located near Bekopaka, with Anjohikinakina cave located approximately 20 km from the other caves in the area known as 'Grande Tsingy'. Using advice from park staff we selected seven caves that are frequented (perhaps daily between July and September) by tourists and a further nine that are less frequently visited by tourists. Direct counts were made of roosting bats using torches fitted with red filters. When bats were found in large groups we counted the number in a known area and extrapolated based on the total estimated area of the colony. In each cave chamber with bats we measured the relative humidity (Hygros H100 Hygrometer), temperature (Hanna Mini-therm HI8753 probe) and light levels (Testo 245 Lux meter) at breast height at number of points on a transect running along the chamber axis. The same measurements were taken for each group of roosting bats (defined as one or more individuals with clear spatial separation from other bats) to investigate the microclimate at each roost position.

Statistical analysis

Bat captures were standardised according to the length of mist net used each night (range 36–42 m) and values given are means and SE. Chi-square was used to test for differences in capture frequency between seasons. Counts of bat passes resulted in non-parametric data and we therefore used Mann–Whitney and Kruskal–Wallis to test for differences between habitat and season. Relative humidity (%) was arcsine transformed before analysis. ANOVA and Kruskal–Wallis were used to compare roost microclimate of the two bat species with transect points.

Results

Community composition

Mist netting resulted in the capture of 10 species over the two seasons (Table 1). The small fruit bat *Rousettus madagascariensis* was the most frequently netted species and made up 33% of all captures. Four vespertilionid species made up a further 38% of captures with *M. manavi* the most common. Three hipposiderid species contributed 28% (Table 1) with *Triaenops rufus* the most common. Two species, *Scotophilus tandrefana* and *Emballonura* sp. nov., were represented by singletons.

Our cave surveys revealed the presence of two large microchiropteran bats that were not trapped in mist nets, *Miniopterus gleni* and *Otomops madagascariensis*. Three other bat species, *Pteropus rufus*, *Eidolon dupreanum* and *T. mauritianus*, were neither trapped, detected acoustically nor observed in caves. *Pteropus rufus* and *E. dupreanum* were observed feeding on kapok trees *Ceiba pentandra* near Andadoany village. Two *E. dupreanum* roosts were found, located high up in rock faces. We also discovered a small roost (<20 animals visible in the day) of *T. mauritianus* in a rocky fissure in a gorge along the Manambolo River.

Table 1 Species composition of Chiroptera in Parc National Tsingy de Bemaraha caught with mist nets during two different seasons in 2003

Taxa	July	October
*R. madagascariensis***	114	37
Emballonura nov. sp.	1	0
*T. furculus***	19	5
*T. rufus**	52	16
Hipposideros commersoni	0	37
*M. manavi**	85	49
*Myotis goudoti**	30	7
S. robustus	0	3
S. tandrefana	1	0
Chaerephon leucogaster	4	0
Total	308	155
Species richness	8	7

Significance levels are based on chi-squared statistics (*<0.05, **<0.01) for between season comparisons

Our time-expanded recordings of echolocation calls revealed six microchiropteran species and we also identified a sonogram similar to that of *Miniopterus majori*, but since that species was not trapped we have excluded it from our results. A medium-sized *Miniopterus* sp., which may have been *M. majori* or *M. fraterculus* was also observed during the cave surveys but was not trapped.

Seasonality

For the species caught in large numbers, all were significantly more abundant in July than October, with the notable exception of *Hipposideros commersoni* (Table 1). *H. commersoni* was not trapped or detected during July but made up 24% of captures and was detected on 3% of point counts in October (Table 1, Fig. 1). It is noteworthy however that a spot-survey conducted 25 km north of the Bekopaka site trapped two adult *H. commersoni* in July. Species richness was similar in both seasons, but visits made only in July would have missed three species and a further two different species

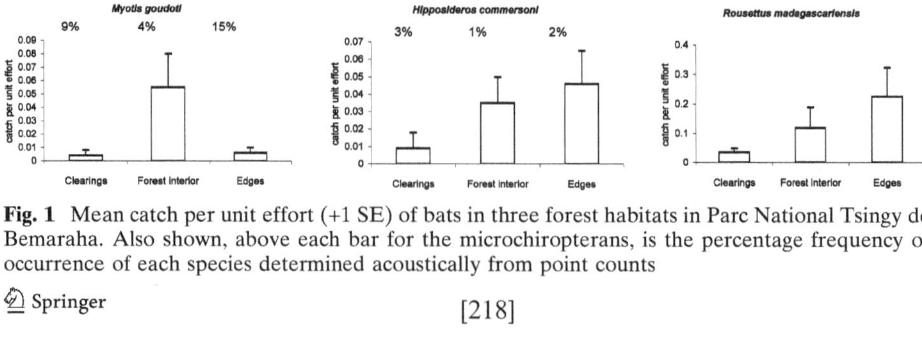

Fig. 1 Mean catch per unit effort (+1 SE) of bats in three forest habitats in Parc National Tsingy de Bemaraha. Also shown, above each bar for the microchiropterans, is the percentage frequency of occurrence of each species determined acoustically from point counts

were only recorded in October (Table 1). The mean number of bat passes was not significantly different between seasons (Mann–Whitney $U = 12{,}711$, ns).

Habitat use

The capture frequency of six species differed between habitats (Fig. 1) but the comparison was only statistically significant for *M. manavi* (Kruskal–Wallis $H = 9.07$, $P < 0.01$). Based on capture data alone, four species were caught most often in the forest—*Triaenops rufus* (71% of captures in the forest interior), *T. furculus* (92%), *M. manavi* (90%) and *Myotis goudoti* (92%). *Rousettus madagascariensis* and *H. commersoni* were caught less frequently in clearings but used edges and forest interior to a similar degree. All four individuals of *Chaerephon leucogaster* were caught in tsingy clearings. *S. robustus* was trapped in clearings and the forest edge, whilst the single *S. tandrefana* and *Emballonura* nov. sp. individuals were caught on the forest/agriculture interface.

A total of 320 point counts were made, 161 in July and 159 during October. Most point counts (184) were in the forest interior, with a further 104 in forest edge and 32 in clearings. Total bat activity was significantly different between habitats (Kruskal–Wallis H=29.6, $P < 0.01$). We detected highest activity along the forest/agriculture interface (mean = 15.7 ± 1.6 bat passes), lower activity in clearings (13.1 ± 3.7 bat passes) and the least activity inside the forest (7.5 ± 0.9 bat passes). Using bat passes classified into frequency categories there was also a significant difference between habitats (Fig. 2) for all groups (>60 kHz: $H = 19.0$, $P < 0.001$; 33–60 kHz: $H = 17.5$, $P < 0.01$; <33 kHz: $H = 7.5$, $P < 0.05$) and activity was consistently lower on forest-trails. Bat passes attributed to vespertilionids (33–60 kHz) were most commonly heard along the forest/agriculture interface, whilst the hipposiderids and *M. goudoti* (>60 kHz) used both edges and clearings.

Activity was significantly higher in riparian areas than non-riparian areas (Fig. 2) for >60 kHz passes (Mann–Whitney $U = 5{,}842$, $P < 0.001$) and 33–60 kHz passes (Mann–Whitney $U = 4{,}347$, $P < 0.001$) but not for the lower frequency <33 kHz

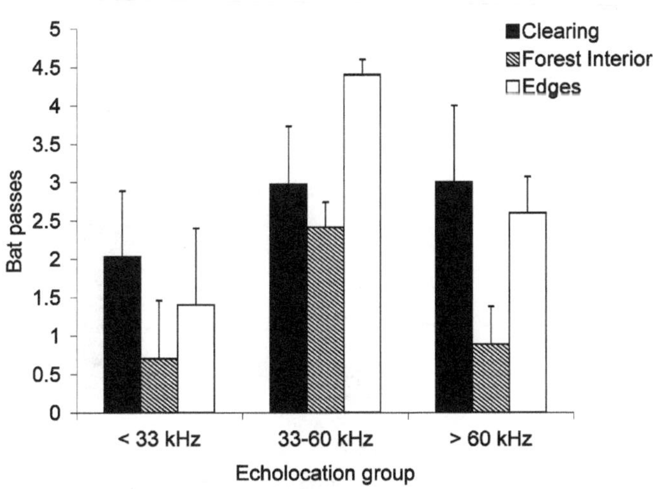

Fig. 2 Mean activity (+1 SE) of bats in three echolocation groupings in Parc National Tsingy de Bemaraha during July, August and October 2003 in three types of forest habitat

passes (Mann–Whitney $U = 6{,}987$, ns; Fig. 3a). A comparison of mist net captures in these habitats found significantly more microchiropterans in non-riparian habitats (Chi-squared = 65, $P < 0.01$) whilst megachiropterans were caught most often netted in the drier parts of the forest (Fig. 3b).

A comparison of bat activity between point counts on the forest-trails and in the forest-interior revealed significantly higher mean bat passes on the former (mean = 5.0 ± 0.6) than the latter (mean = 2.97 ± 0.8; Mann–Whitney $U = 2{,}915$, $P < 0.001$).

Our time-expanded recordings of free-flying bats show slightly different results to the capture data (Fig. 1). Most notable is that *Miniopterus manavi* was recorded in 41% of the point counts made in clearings but was not trapped in this habitat (Fig. 2). A high proportion of the recordings made during point counts from edges revealed the presence of *Triaenops rufus* and *M. manavi*, whereas trapping data suggested low use of edges by both species. *Emballonura* nov. sp., which was trapped only once, was most frequently recorded (11%) using the forest/edge interface.

We divided the capture and bat pass data into two categories (inside and outside the forest) for each 15-min period to investigate whether proximity to the cave roosts, which were located inside the forest, influenced temporal patterns of activity. Inside the forest, mist nets captures peaked between 1800 and 1814 h whilst captures outside the forest peaked between 1815 and 1829 h (Fig. 4a). Bat

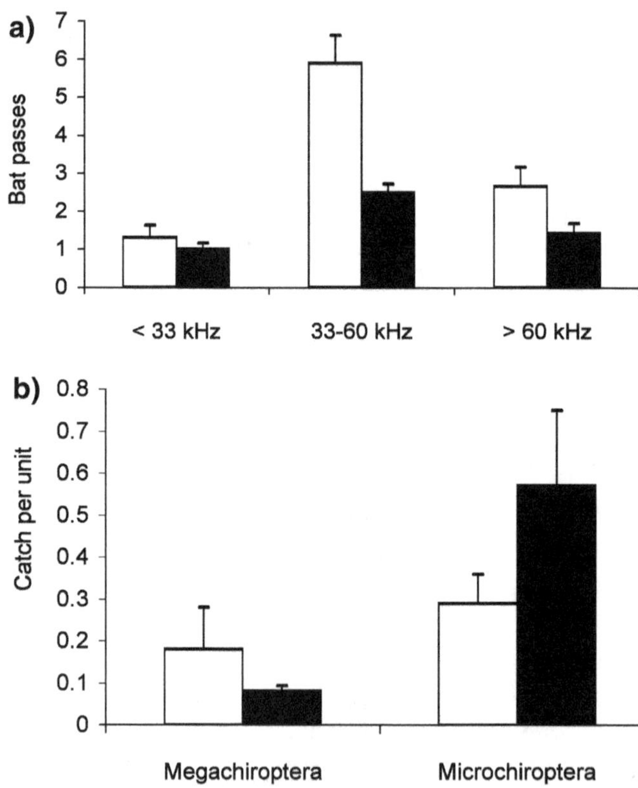

Fig. 3 (**a**) Mean activity of bats in three echolocation groupings and (**b**) catch per unit effort (+1 SE) for Megachiroptera and Microchiroptera in riparian (white) and non-riparian forest (black) in Parc National Tsingy de Bemaraha during July, August and October 2003

Table 2 Bat species composition and abundance in 16 caves, Parc National Tsingy de Bemaraha, frequented regularly or rarely by tourists

Cave name	Abundance		Species
	July	October	
Higher tourism			
Anjohifipetrahana	0	0	–
Anjohimanapaka	16	6	Em, Mm
Anjohimanitsikoa	1	4	Mm, My
Anjohimanitsy	38	9	Mg, Mm, My
Anjohimboro	0	0	–
Anjohitantely	31	19	Em
Gorge II	18	5	Mm
Lower or no tourism			
Anjohiatsimo I	24	0	Mm
Anjohiatsimo II	0	7	Em, Mm
Anjohiatsimo-nord	12	22	Mg, Mm
Anjohibemoka	50	–	Mm, My
Anjohikinakina	1,712	9,174	Mg, Mm, Ot, Ro, Tr
Anjohisiramamy I	9	6	Em, Mm
Anjohisiramamy II	12	–	Mm
Anjohitrombastimo	0	0	–
Gouffre cave	151	20	Mg, Mm, My

– not visited in October

Species abbreviations: *Miniopterus manavi* (Mm), *M. gleni* (Mg), *Myotis goudoti* (My), *Emballanura* nov. sp. (Em), *Rousettus madagascariensis* (Ro), *Otomops madagascariensis* (Ot), *Triaenops rufus* (Tr)

activity peaked between 1845 and 1859 h inside and outside the forest, although activity at the forest edge only exceeded that of forest-trails after 1815 h (Fig. 4b).

Cave surveys

Over 2,000 bats were observed in the 16 caves in July and over 9,000 in October (Table 2). *Triaenops rufus* was the most abundant, making up 54% (n = 1,127) of observations in July and 90% (n = 8,306) in October. *Rousettus madagascariensis* made up 19% (n = 400) and 6% (n = 620) of total observations in July and October respectively. *M. manavi* was the most widespread species and was found in 75% of the caves, but *O. madagascariensis*, *R. madagascariensis* and *T. rufus* were found only in Anjohikinakina. Extraction of *O. madagascariensis* from holes in the ceiling at this site accounted for 88 and 114 individuals in July and October, respectively. This cave contained 83% of all bats counted in July and 99% in October. Numbers of *R. madagascariensis* and *T. rufus* were higher in October than in July, but the opposite was observed for *M. gleni*.

Three caves contained no bats (including two that are frequently visited by tourists) in either season and another cave only contained *Emballonura* nov. sp. during October. In terms of species richness and number of individuals, Anjohikinakina Cave is the most important for bat conservation (Table 2).

Roost site selection

In many caves, the bats used small, round, vertical holes in the ceiling, and in Anjohikinakina *O. madagascariensis* was only observed in these features.

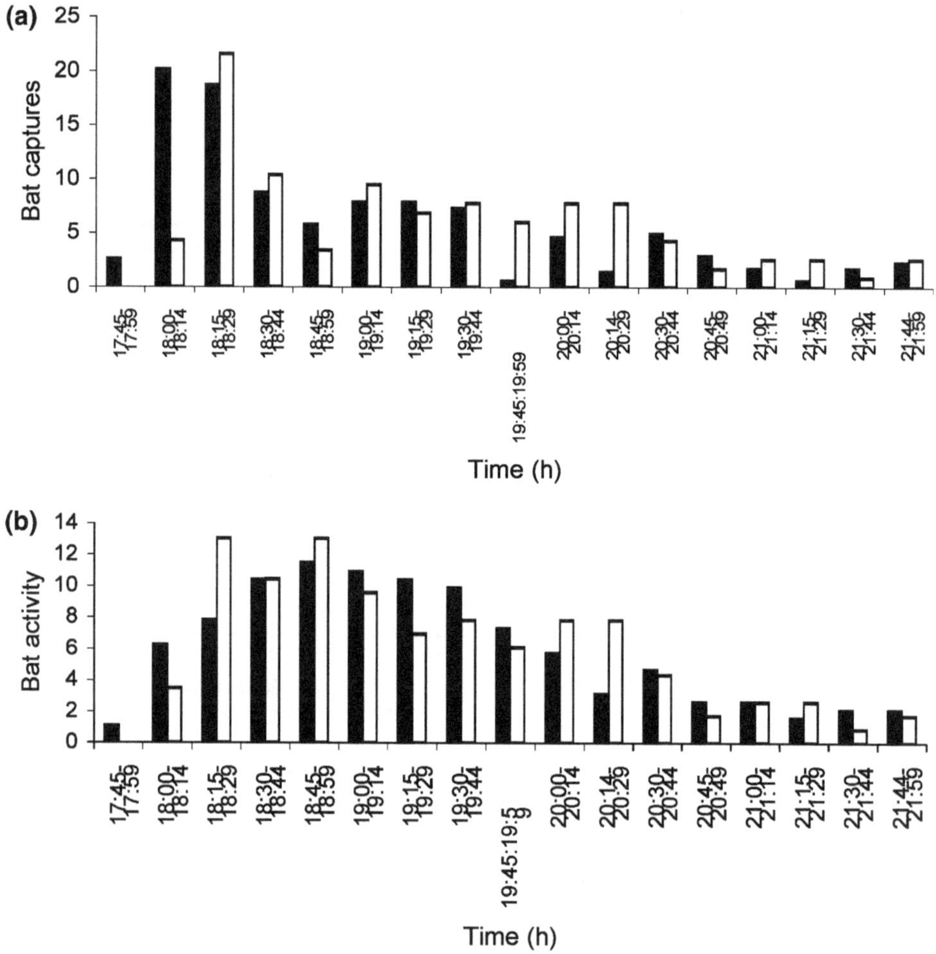

Fig. 4 (**a**) Bat captures and (**b**) activity in 15 min time periods from sunset to 22:00 in Bemaraha National Park during July, August and October 2003 from inside (black) and outside (white) the forest

Miniopterus manavi also sometimes used these holes, but were also observed suspended from over-hanging cave walls. *R. madagascariensis* and *Triaenops rufus* roosted in single, large colonies in Anjohikinakina. *Myotis goudoti* frequently roosted alone and was sometimes seen hanging from long roots dangling from the cave ceilings.

Microclimate measurements at the roost sites of *M. manavi* and *Emballonura* nov. sp. were significantly different from each other, and different to the transect (Table 3). No other species was common enough to allow statistical validation of roost site selection. *Emballonura* nov. sp. roost sites were significantly warmer than both the random points and *M. manavi* roosts. Both bat species roosted in areas of similar humidity levels, which were significantly lower than the levels recorded on the random points. High light values for the transect points reflect the starting points at cave entrances. However, *Emballonura* nov. sp. roosts were in localities that received significantly more sunlight at the time of the surveys.

Table 3 Roost site selection in caves by two Malagasy microchiropterans in Parc National Tsingy de Bemaraha

	Light (lux)	Temperature (°C)	Humidity (%RH)
Roosts			
Emballonura nov. sp.	1.8 ± 0.5	23.7 ± 0.3	80.9 ± 1.1
M. manavi	0.06 ± 0.1	20.5 ± 0.3	79.2 ± 1.2
Transect points	8.8 ± 2.33	21.8 ± 0.1	83.3 ± 0.6
Statistic	$H = 22.9$	$F = 36.2$	$H = 17.7$
P value	<0.001	<0.001	<0.05

The data are combined for two seasons and from 12 caves

Discussion

These results emphasize the important contribution that bats make to the island's endemic mammal fauna. Parc National Tsingy de Bemaraha has the highest chiropteran species richness of any known site in Madagascar (Goodman et al. 2005a, b) and must therefore be considered as a site of national importance for mammal conservation. Failure to incorporate bats into surveys of important conservation sites in Madagascar is therefore likely to significantly underestimate the true mammalian species richness.

Bat habitat use

Surveys to establish inventories of bat species for important wildlife areas traditionally involve either assessments of roosting populations in caves and buildings (e.g. Petit 1996) or intensive mist netting programs in conjunction with harp traps (e.g. Clarke and Downie 2001). Although some surveys incorporate trapping and the use of bat detectors (e.g. Bernard and Fenton 2002) many biologists rely solely on mist net captures to study habitat use by bats (e.g. Angelici et al. 2000). Harps traps are routinely employed in bat surveys in Madagascar (Goodman et al. 2005a, b) but would need to be deployed in prohibitively large quantity to provide information on habitat use. Foraging height, flight speed, type of echolocation and body size are amongst the many factors that can influence the capture rate of a given species in a mist net. The majority of studies from South and Central America use only mist nets because the chiropteran fauna of these regions are dominated by phyllostomid bats that have quiet echolocations which render them difficult to detect with bat detectors (Fenton et al. 1992). By contrast Madagascar's bat community is made up mostly of vespertilionid species and recent progress in the description of the echolocation of this family, as well as Hipposideridae, Myzopodidae and Emballonuridae (Russ et al. 2003) is good reason to begin to incorporate acoustic sampling into bat surveys in Madagascar.

Interpretation of either trapping or echolocation results alone would have resulted in markedly different conclusions from our study. For example, results from the bat detector surveys showed lowest microchiropteran activity in the forest interior, but trapping revealed the opposite result with four species caught most often inside the forest. Mist nets in the forest traversed trails and streams and were probably more efficient at intercepting passing bats than nets placed in the more open areas of the edge and clearings. Another important factor was that the rocks and crevices in the study area were mainly located in the forest, resulting in a large numbers of bats emerging around sunset in the forest and flying along the trails. This point is supported by our analysis of temporal patterns in bat activity that showed the first peak in the forest, followed 30 min later by a peak outside of the forest. Mist nets have been used in Madagascar to establish inventories of bats in the eastern rainforests (e.g. Pont and Armstrong 1990; Bayliss and Hayes 1999; Goodman 1999) but few attempts have been made to assess their habitat preferences. Our results demonstrate the important influence that roost location can have on bat surveys and further emphasizes the inherent biases in using ground level mist nets. We recommend that bat surveys report results for roost and foraging site captures/observations separately.

Assessments of activity using bat detectors are less susceptible than mist nets to variation from habitat features and we would not expect a significant bias between trails, edge or clearings using acoustic survey methods. We are therefore confident that the lower activity, measured as 'bat passes', detected inside the forest compared to edges and clearings is representative of habitat use by microchiropterans.

Eger and Mitchell (2003) suggested that both *T. rufus* and *T. furculus* have similar requirements for forest habitats and this is supported by our results. *Myotis goudoti* and *M. manavi* are both distributed widely across Madagascar and are found in forests near suitable cave roosting sites (Eger and Mitchell 2003). Both species were caught most frequently in the forest but detected most frequently in edges and clearings. Of the four individuals of *Scotophilus* captured, representing two different species, two were at the agriculture/forest interface and two were in the tsingy clearings. Given the success of our nets at catching other vespertilionid species inside the forest, this suggests that *Scotophilus* is not strongly associated with either forests or caves. Information on this genus of bats is particularly lacking in Madagascar, and the discovery of a new species during this survey (Goodman et al. 2005a) should provide an incentive for further comprehensive bat surveys in Tsingy de Bemaraha and other protected areas in western Madagascar.

The high capture rates of four small bat species in the forest appears to be explained by the presence of cave roost sites in the forest and the efficiency of nets across narrow trails at catching hungry, emerging bats at sunset. Many studies on habitat use of bats have demonstrated a close link between bat activity and linear or aquatic habitat features (Verboom and Spoelstra 1999; Law and Chidel 2002; Russ and Montgomery 2002). Forest trails in the park appear to be used mainly by bats as thoroughfares to access the edge habitats where foraging occurs; the forest maybe an important source of insect prey in these areas. Riparian habitats were used mainly by vespertilionids, but overall microchiropteran activity was higher away from water, possibly reflecting the importance of the edges to foraging bats.

Although many of Madagascar's small land mammals are forest dependent (Ganzhorn et al. 2003) and successful conservation is closely linked with the preservation of intact forest, bat conservation may demand a change in approach to recognize the value of non-pristine habitats adjacent to the forest. Estrada et al. (1993) found that agricultural habitats contained a high bat species richness and abundance and they suggested that areas of mixed plantation and isolated forest trees are important habitat to bats because they reduce the distance between blocks of remaining rainforests. Similarly, Bernard and Fenton (2002) recorded most species and highest capture rates in savanna habitats located in a mosaic of rainforest fragments in Brazil. Similar studies in Africa are rare, but Angelici et al. (2000)

found a higher species richness and abundance of bats in secondary forest than either bush or primary forest habitats. Although the change in microclimate associated with forest edges is often detrimental to forest species (Lethinen et al. 2003), we show in this study that forest edges are important for habitats foraging bats. Associations between bats and the forest are less clear, and although they appear to be mainly associated with the presence of suitable roosts, the forest vegetation is likely to be an important source of insect food and to provide the structural formations (e.g. tree-lined edges) necessary for bats with short, wide-wings to forage alongside. A preference for forest edges over the forest interior may make bats less susceptible to the detrimental effects from forest fragmentation, although this area of research clearly requires further work.

Cave conservation in Tsingy de Bemaraha

Although caves are key habitats for bats, our survey revealed that most of the potential cave roosts in Bemaraha contained small populations of the common and abundant *M. manavi* and *M. goudoti*. The capture of obligate caves species such as *T. furculus* with mist nets, but their absence from the 16 caves studied demonstrates that other bat roost caves remain undiscovered.

Tourism at its current levels is unlikely to be presenting a significant threat to bat populations in Bemaraha, especially as the park is closed to visitors from December to April when water levels are high, and which coincides with the period of chiropteran parturition and lactation. However, we propose some conservation recommendations for both the management of caves and the monitoring of species. Anjohikinakina is clearly a cave of national importance to bats in Madagascar because it contains relatively large roosting colonies of five species. Although *R. madagascariensis* has a wide distribution in Madagascar only a small number of roost sites are known from the protected area network (MacKinnon et al. 2003). *O. madagascariensis* is associated with Madagascar's limestone and sandstone deposits and has been recorded from sites across the dry portions of the island (Goodman et al. 2005b); however only a handful of roost sites have been identified.

Anjohikinakina Cave is off the main tourist circuit and although it receives some visits by intrepid groups from July until September, current visitation rates are unlikely to threaten the bats. Complacency in this respect should be avoided we recommend that the cave is excluded from any future expansion of the tourist circuits. Occasional monitoring by park staff is, however, recommended to deter potential hunters.

Goodman et al. (2005b) list *Emballonura* nov. sp. as one of only five bat species that are possibly dependent on intact forest. We frequently observed the species roosting near cave entrances and it readily took to flight when approached. Other studies have shown that tour group visits to caves can have a detrimental impact on roosting bats by provoking increased levels of activity and flight during the day (Lacki 2000), presumably leading to increased demands on the energy budget. The roost site preference and behaviour of this species therefore appear to make it vulnerable to disturbance by tour group visits to caves and we strongly recommend that the frequency of such visits are at least monitored, and preferably limited, to manage noise, light levels and human behaviour (Mann et al. 2002).

Acknowledgements We would like to thank Olga Ramilijaona and Chantal Andrianarivo for their assistance throughout the project. The work was conducted through collaboration between ANGAP, the University of Aberdeen and the Universities of Antananarivo and Toliara in Madagascar, and the Ministry of the Environment kindly granted us permission. We are very grateful to the staff of Project Bemaraha for their support, in particular Rindra Rakotoharifetra, Hery Lala Ravelomanantsoa, Bruno Rasoanaivo, Tsimahoary Tsiverila and Eugene Solofo Rakotonirina. Steve Goodman was generous with his time and helped us with taxonomy. We also thank Steve Goodman and Jon Russ for helpful their comments. Invaluable assistance in the field was provided by Tsibara Mbohoahy, Andrianajoro Rakotoarivelo and Felicien Randrianandrianina. Funding for the work was from the British Government's Darwin Initiative (#162/10/024) and additional support was kindly provided by Fauna and Flora International, National Geographic Society (#C23-02), Rufford Foundation, British Ecological Society (#2159) and a Bat Conservation International Scholarship to AK.

References

Angelici FM, Wariboko SM, Luiselli L, Politano E (2000) A long-term ecological survey of bats (Mammalia: Chiroptera) in the eastern Niger Delta (Nigeria). Ital J Zool 67:169–174

Bayliss J, Hayes B (1999) The status and distribution of bats, primates and butterflies from the Makira Plateau, Madagascar. Unpublished. University of Aberdeen

Bernard E, Fenton MB (2002) Bat mobility and roosts in a fragmented landscape in central Amazonia, Brazil. Biotropica 35:262–277

Bollen A, Van Elsacker L (2002) Feeding ecology of *Pteropus rufus* (Pteropodidae) in the littoral forest of Sainte Luce, SE Madagascar. Acta Chiropt 4:33–47

Clarke FM, Downie JR (2001) A bat (Chiroptera) survey of Mora rainforest in Trinidad's Victoria-Mayaro Forest Reserve. Biodivers Conserv 10:725–736

Eger JL, Mitchell L (2003) Chiroptera, Bats. In: Goodman SM, Benstead JP (eds) The natural history of Madagascar. University of Chicago Press, Chicago, pp 1287–1298

Estrada A, Coates-Estrada R, Meritt D (1993) Bat species richness and abundance in tropical rainforest fragments and in agricultural habitats at Los Tutlas, Mexico. Ecography 16:309–318

Fenton MB, Acharya L, Audet D, Hickey MBC, Merriman C, Obrist MK, Syme DM (1992) Phyllostomid bats (Chiroptera: Phyllostomidae) as indicators of habitat disruption in the Neotropics. Biotropica 24:440–446

Ganzhorn JU, Goodman SM, Dehgan A (2003) Effects of forest fragmentation on small mammals and lemurs. In: Goodman SM, Benstead J (eds) The natural history of Madagascar. University of Chicago Press, Chicago, pp 1228–1234

Goodman SM (1999) Notes on the bats of the Réserve Intégrale d'Andohahela and surrounding areas of southeastern Madagascar. Fieldiana: Zoology, new series 94:251–257

Goodman SM, Benstead JP (2003) The natural history of Madagascar. The University of Chicago Press, Chicago

Goodman SM, Cardiff SG (2004) A new species of *Chaerephon* (Molossidae) from Madagascar with notes on other members of the family. Acta Chiropt 6:227–248

Goodman SM, Ranivo J (2004) The taxonomic status of *Neoromicia somalicus malagasyensis*. Mamm Biol 69:434–438

Goodman SM, Jenkins RKB, Ratrimomanarivo FH (2005a) A review of the genus *Scotophilus* (Mammalia, Chiroptera, Vespertilionidae) on Madagascar, with the description of a new species. Zoosys 27:867–882

Goodman SM, Andriafidison D, Andrianaivoarivelo R, Cardiff SG, Ifticene E, Jenkins RKB, Kofoky A, Mbohoahy T, Rakotondravony D, Ranivo J, Ratrimomanarivo F, Razafimanahaka J, Razakarivony V, Racey PA (2005b) The distribution and conservation of bats in the dry regions of Madagascar. Anim Conserv 8:153–165

Hutcheon JM (2003) Frugivory by Malagasy bats. In: Goodman SP, Benstead JP (eds) The natural history of Madagascar. University of Chicago Press, Chicago, pp 1205–1207

Hutson AM, Mickleburgh SP, Racey PA (2001) Microchiropteran bats—Global status survey and conservation action plan. IUCN/SSC Chiroptera Specialist Group, Gland, Switzerland, 258 pp

Kappeler PM, Rasoloarison RM (2003) *Microcebus*, Mouse Lemurs, Tsidy. In: Goodman SM, Benstead J (eds) The natural history of Madagascar. University of Chicago Press, Chicago and London, pp 1310–1315
Kunz TH (1982) Roosting ecology of bats. In: Kunz TH (eds) Ecology of bats. Plenum Press, New York and London, pp 1–52
Lacki MJ (2000) Effect of trail users at a maternity roost of Refinesque's big-eared bats. J Cave Karst Stud 62:162–168
Law BS, Chidel M (2002) Tracks and riparian zones facilitate the use of Australian regrowth forest by insectivorous bats. J Appl Ecol 39:605–617
Lethinen RM, Ramanamanjato J-B, Raveloarison JB (2003) Edge effects and extinction proneness in a herpetofauna in Madagascar. Biodivers Conserv 12:1357–1370
MacKinnon JL, Hawkins CE, Racey PA (2003) Pteropodidae, Fruit Bats, *Fanihy*, *Angavo*. In: Goodman SM, Benstead JP (eds) The natural history of Madagascar. University of Chicago Press, Chicago, pp 1299–1302
Mann SL, Steidl RJ, Dalton VM (2002) Effects of cave tours on breeding *Myotis velifer*. J Wildlife Manage 66:618–623
Peterson RL, Eger JL, Mitchell L (1995) Chiroptères. Muséum National d'Histoire Naturelle. 84 Faune de Madagascar, Paris
Petit S (1996) The status of bats on Curacao. Biol Conserv 77:27–31
Pont SM, Armstrong JD (1990) A study of the bat fauna of the reserve Naturelle Integral de Marojejy in North East Madagascar. University of Aberdeen, Unpublished
Ramanamanjato J-B, Ganzhorn GU (2001) Effects of forest fragmentation, introduced *Rattus rattus* and the role of exotic tree plantations and secondary vegetation for the conservation of an endemic rodent and a small lemur in littoral forests of southeastern Madagascar. Anim Conserv 4:175–183
Rasoloarison V, Paquier F (2003) Tsingy de Bemaraha. In: Goodman SM, Benstead JP. (eds) The natural history of Madagascar. University of Chicago Press, Chicago and London, pp 1507–1512
Russ J, Bennett D, Ross K, Kofoky A (2003) The bats of Madagascar: a field guide with descriptions of Echolocation Calls. Viper Press, Glossop, UK, 96 pp
Russ JM, Montgomery WI (2002) Habitat associations of bats in Northern Ireland: implications for conservation. Biol Conserv 108:49–58
Sedlock JL (2001) Inventory of insectivorous bats on Mount Makiling, Philippines using echolocation call signatures and a new tunnel trap. Acta Chiropt 3:163–178
Stephenson PJ, Randriamahazo H, Rakotoarison N, Racey PA (1994) Conservation of mammalian species diversity in Ambohitantely Special Reserve, Madagascar. Biol Conserv 69:213–218
Verboom B, Spoelstra K (1999) Effects of food abundance and wind on the use of tree lines by an insectivorous bat, *Pipistrellus pipistrellus*. Can J Zool 77:1393–1401

Biodivers Conserv (2007) 16:1055–1061
DOI 10.1007/s10531-006-9062-5

ORIGINAL PAPER

Territoriality and survivorship of the Sierra Madre sparrow in La Cima, México

Adán Oliveras de Ita · Héctor Gómez de Silva

Received: 6 April 2005 / Accepted: 24 April 2006 / Published online: 12 July 2006

Abstract We studied the life-history of the Sierra Madre sparrow (*Xenospiza baileyi*) in a subalpine grassland-agriculture mosaic south of Mexico City. From March to early September 1999 we captured and color-banded 53 adults with mist-nets and mapped the breeding territories of 21 males. We again mapped territories in the same spot in April and May 2000. The number of breeding territories was found to be the same in the two consecutive years and interannual survivorship was found to be relatively high. Breeding territories were restricted to the bunchgrass-covered areas. We used the density of territories and the amount of remaining habitat to estimate a total population size of 5,380–6,150 adults for this species. Using this and other recent data, we recommend raising the Sierra Madre sparrow to the status of critically endangered using BirdLife International criteria.

Keywords Mexico · Restricted range · Critically endangered · Sierra Madre sparrow · Survivorship · Territoriality · *Xenospiza baileyi*

Introduction

The Mexican endemic Sierra Madre sparrow (*Xenospiza baileyi*) is considered endangered by BirdLife International (2004) and has one of the highest vulnerability

A. O. de Ita · H. G. de Silva
Instituto de Ecología, UNAM, Ciudad Universitaria, Apartado Postal 70-275, C.P. 04510 México, D.F. México

A. O. de Ita
Centro de Investigaciones en Ecosistemas (CIECO), UNAM, Antigua Carretera a Patzcuaro No. 8701, C.P. 58190 Morelia, Michoacán, México

A. O. de Ita (✉)
Instituto de Ecología, Priv.Tecolapa#35.Casa4.Tepepan,Xochimilco, C.P.16020 México, D.F.México
e-mail: oliveras@laneta.apc.org

scores of any Mexican bird species following the criteria of Partners in Flight (H. Berlanga et al. unpublished data). This species possesses many of the indicators of high susceptibility to extinction discussed by Collar (1999). It has narrow habitat requirements, being found only where there are subalpine bunchgrass meadows (Collar et al. 1992). It is known in modern times only from two disjunct regions separated by 600 km (Oliveras de Ita et al. 2003). This relict distribution is evidence of a prehistoric population decline. It has further been subject to alarming population declines in recent times due to human activities (reduction of potential breeding habitat by severe overgrazing and conversion to agricultural fields, Collar et al. 1992; BirdLife International 2004). Nesting success is reported to be low at least at one site, La Cima, whereas population density within suitable breeding habitat is high (Oliveras de Ita et al. 2001; note that a printing error gave an incorrect number of territories and territorial males in p. 45 of that work, the correct number appearing in p. 46). We studied the territoriality and annual survivorship of this species at that site in an attempt to facilitate the development of adequate conservation plans.

The Sierra Madre sparrow's restricted range, narrow habitat requirements and high population density is an unusual combination of features among North American land birds (Brown 1995).

Methods

The study area is a 2 km^2 meadow located 20 km from the southern end of Mexico City (19°05′N, 99°12′W, 3,050 m asl). Around 37% has been converted to agricultural fields. Our study plot was a rectangle that measured 12 ha, of which 7.6 ha were bunchgrass and the rest were cultivated fields (Fig. 1).

In 1999, we carried out 72 visits to the study plot from March to early September: color-banding was carried out on 15 visits (790 net-hours) and territory-mapping was carried out on 52 visits, with most information on territory location being gathered from June to early September, when territorial males had already been banded.

Individuals were fitted with a unique combination of 1–2 color bands (6 colors), the presence of a cloacal protuberance or brood patch was noted, and wing chord, tail length, beak width at base, beak height at base, beak length and weight were measured on almost all birds captured. A discriminant analysis using SPSS 10.0 for Windows was performed on the body measurements of birds sexed by cloacal protuberance/brood patch to determine a canonical function that discriminated between the sexes. This was done to recognize males additional to birds already identified by the presence of a cloacal protuberance. Territorial males were considered to be those with clusters of registrations.

Territories were mapped using a Kowa TSN-2 with a 20–60× zoom eyepiece. On each visit, the position and movements of all individuals were plotted on a 1:2000 map of the study plot obtained from an aerial photograph (Instituto de Geografía 1997). We surveyed the entire study plot on every visit, although we usually concentrated on certain sections on each visit. A territory map of the study plot was made by combining the 52 daily maps (Fig. 1, modified from Oliveras de Ita et al. 2001). Territories were mapped by drawing a line between the peripheral points of occurrence. Males with more than six observations from early June on were considered informative with respect to site-fidelity.

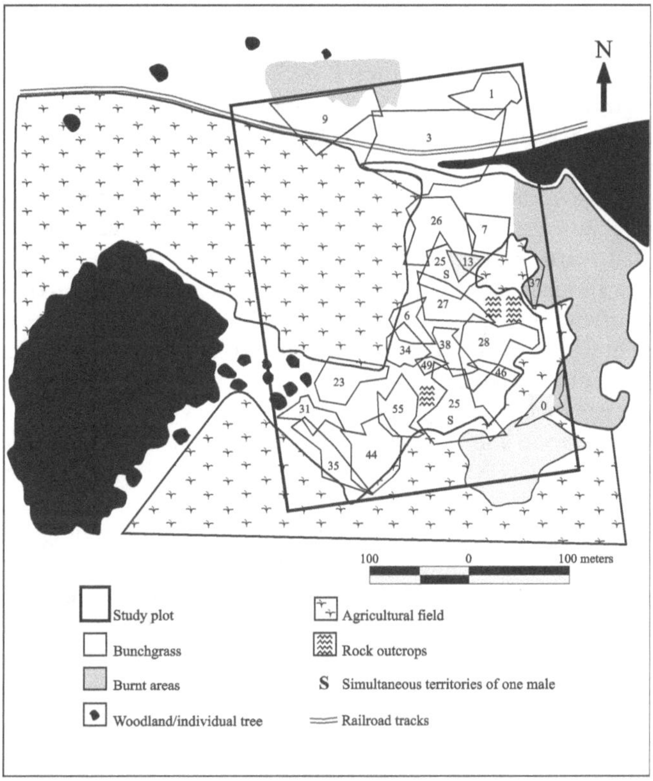

Fig. 1 Study plot showing the 22 Sierra Madre sparrow territories established in 1999. Numbers correspond to those assigned to each male

The software EstimateS (Colwell 1997) was used to estimate the actual number of adult Sierra Madre sparrows using the study plot, employing individuals in the place of species. The jacknife, Chao and Michaelis-Menten means estimators were used to produce a range of estimates. Michaelis-Menten Runs was not used because of its erratic behavior.

An additional 14 visits were made in April and May 2000 to determine the number and identity of territorial males in the study plot and infer their interannual survivorship.

Results

Seventy percent of mistnetted individuals (26 males and 15 females) were sexed by cloacal protuberance/ brood patch (the other 30% were mistnetted only once or twice, most of them early in the season). We obtained measurements for 37 individuals of known sex (26 males and 11 females). By performing discriminant analyses on different combinations of variables, we found that wing chord and tail length, combined, perform just as well as or better than any other combination of variables. Using wing chord and tail length, 84.6% of the males and 81.8% of the females were correctly classified by the discriminant function: $D = 4.753$ (wingchord) + 1.835 (tail length)–39.184; individuals with values of 0.3 and above could be identified with certainty as males and individuals with values of –1.2 or lower as females. This analysis allowed us to sex eight additional individuals, including two which showed clusters of registrations (one of which sang only a few times).

We mapped 22 breeding territories. Breeding territories were restricted to the 7.6 ha of bunchgrass-covered areas in the study plot. Territories frequently overlapped at their edges and at least 69% of the grassy area was occupied (Fig. 1). Territories were missing from four areas. Two of these were adjacent to pine woodland or had scattered trees.

No females were recorded singing, whereas most territorial males sang frequently. However, identifying territorial males independently of their singing behavior allowed us to detect a few that did not sing or did not sing frequently. Three birds that sang infrequently (recorded singing only on one or two out of the 7–16 days they were observed) were recognized as males by the presence of cloacal protuberance and another one was recognized as a male by its body measurements. One of these infrequent singers was unusual in that it was observed feeding nestlings at two nests that were active concurrently.

Not all territorial males were site-faithful throughout the breeding season. Twelve of the fourteen 'informative' males held their territories at least from the beginning of June, but the other two established their territories later. One had been caught and banded in April outside the study plot, 450 m from where he ended up establishing a territory in late June. The other was singing on territory in early June and from late July on, but during most of the rest of June was seen intermittently outside his territory.

Two-thirds of the grassy area in our study plot had been completely burned 2 years prior to our study (aerial photograph in Instituto de Geografía 1997; we have no data for 1998). Two small portions at the edge of our study plot were burned on 4 May 1999, yet one territory was established in each of these areas by July. One territory was partly in unburned grass and partly in the recently burned area; however, the other was almost entirely within the recently burned area.

Thirteen to seventeen adult male Sierra Madre sparrows (four individuals could not be sexed) did not have territories and were detected only once or twice. The coverage estimators indicated that the 53 adults that were banded were between 45% and 81% of the adults that use or pass through the study plot. These observations suggest that there was much movement through the study plot of either non-territorial individuals or ones whose territories were outside the study plot ('floaters'). Some of these individuals established territories in the study plot in the following year (see below). According to the number of adults estimated to use the study plot, there were roughly 1–2 'floaters' per breeding pair (assuming the usual one breeding pair per territory).

In the year 2000 we found the same number of territories in the study plot as in 1999, including 16 of that year's territorial males (76%). Fourteen males had their territories in roughly the same spots as in the previous breeding season (Fig. 2). The territories of two 1999 males that were not detected in 2000 were occupied by four males: two former floaters and two which had territories in other parts of the study plot in 1999. Additionally, the northern part of the study plot, which held three territories in 1999, including the largest territory, was subdivided in 2000 between the three former owners, one former floater and three "new" unbanded males (sexed by their singing behavior).

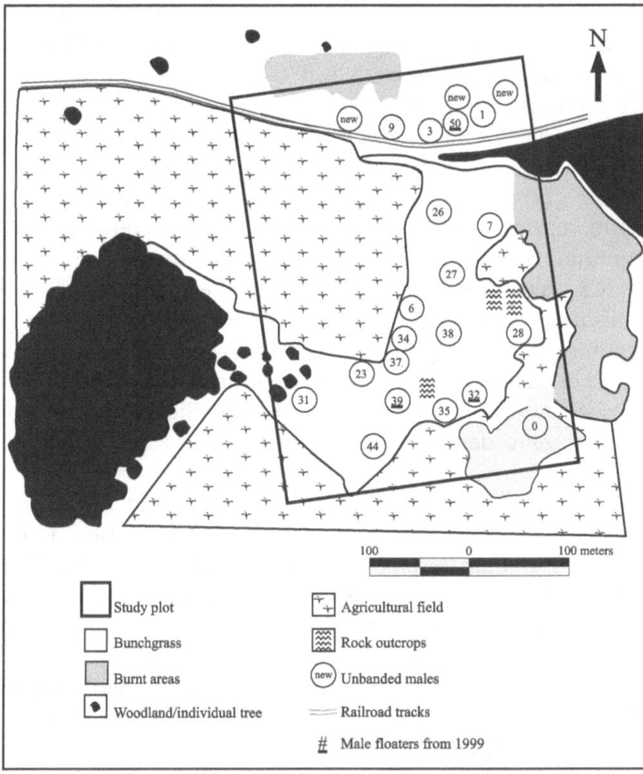

Fig. 2 Location of the Sierra Madre sparrow territories in 2000. Numbers correspond to those assigned to each male. Underlined numbers refer to floaters from 1999

Discussion

The territorial males which sang infrequently would probably not have been detected by quick auditory surveys, nor would all of them have been recognized as males without the morphometric analysis. A less intensive mapping effort or point counts would have yielded an inaccurate estimate of the number of breeding territories. This contrasts with the lower amount of mapping effort that Ryder (1986), Koskimies and Vaisanen (1990), and Bibby et al. (2000) considered sufficient for an accurate estimate.

The number of breeding territories being the same in two consecutive years suggests that where habitat persists population size may be stable without the large fluctuations observed from 1 year to the next in certain sparrow populations (e.g., Nice 1964). We do not know if some banded males undetected in 2000 established breeding territories outside the study area, but even assuming their absence in 2000 was due to death, the interannual survivorship was higher than reported for other grassland sparrows (Wheelwright and Rising 1993; Martin 1995; Vickery 1996; DeSante et al. 1998; Lockwood et al. 2001; Perkins and Vickery 2001; Sandercock and Jaramillo 2002). This may be related to the non-migratory nature of the Sierra Madre sparrow, whereas the other grassland sparrows that have been studied are migratory, because mortality risk appears to be associated with migration and dispersal (Dobson 1990; Daniel and Walters 2000; Sillet and Holmes 2002). The relatively high annual survivorship compared to other grassland sparrows, in contrast to the relatively high nest failure (no chick fledged in 64% of 14 nests monitored, Oliveras de Ita et al. 2001), suggests that mortality during the non-breeding season may not be a severe threat to the Sierra Madre sparrow.

Even though the Sierra Madre sparrow is known to forage in agricultural fields, a widespread and increasing habitat (Collar et al. 1992; Oliveras de Ita et al. 2001), our observation that breeding territories and nest sites were restricted to the bunchgrass-covered areas, particularly away from the forest edge, are more consistent with specialization on a shrinking habitat. The observation that there are 1–2 floaters per breeding pair, together with the estimate of 2.9 breeding pairs/ha in the study plot, and the estimate of 790 ha of suitable breeding habitat in 1997, allow us to estimate the population size as 5,380–6,150 adult Sierra Madre sparrows (2,300 breeding pairs), assuming density in the study plot is representative and that no habitat has been lost.

One of the criteria of BirdLife (2004) for considering a species as critically endangered is that its population be lower than 250 mature individuals. This threshold is much lower than our population estimate for Sierra Madre sparrow. However, the Sierra Madre sparrow falls within the critically endangered thresholds of BirdLife's criteria B2a and B2b. Its southern population has a severely fragmented area of occupancy <8 km^2 (González 2000), which area has decreased by more than 50% in the last 10 years (Cabrera 1999). The northern population is even more critical, having recently found to be absent from its three historical localities and apparently now restricted to a single locality (Oliveras de Ita and Rojas-Soto 2006). We recommend raising the Sierra Madre sparrow to the status of critically endangered.

Despite the Sierra Madre sparrow's shrinking habitat (subalpine grasslands with no trees), the observation that it can breed in recently burned areas suggests that the species can benefit from management actions intended to maintain or even expand the area occupied by treeless bunchgrass, including controlled burning of grasslands prior to the breeding season.

Acknowledgments Financial support for this project was awarded by the Neotropical Bird Club Small Grants Program. Dr. L. Bojórquez and his students assisted us in obtaining the aerial photographs of the study area; Adriana Garza helped with figures; Dr. R. Medellín, UNAM, and Fundación UNAM provided A.O.I. with a scholarship; B. Milá of the Point Reyes Bird Observatory provided mistnets, vernier, and one pair of binoculars; We give special thanks to E. Núñez and Don Chava, who live 1 km from out study plot and kindly hosted us when we had to spend the night at La Cima; and to V. Smith of Eagle Eye Tours who lent us a telescope for the territory mapping. I. Azcona, C. Bueno and Manuel Grosselet assisted in the fieldwork on several occasions.

References

Brown J (1995) Macroecology. University of Chicago Press, Chicago, Illinois

Bibby C, Burgess N, Hill D, Mustoe S (2000) Bird Census Techniques, second edition. Academic Press, London

BirdLife International (2004) Threatened Birds of the World 2004. CD-ROM. BirdLife International, Cambridge, UK

Cabrera L (1999) La avifauna del sur del Valle de México: aplicación de un enfoque sinecológico-paisajístico para su conservación. Dissertation, Universidad Nacional Autónoma de México

Collar N, Gonzaga L, Krabbe N, Madroño Nieto A, Naranjo L, Parker III T, Wege D (1992) Threatened birds of the Americas. The ICBP/IUCN Red Data Book. Smithsonian Institution Press /International Council for Bird Preservation, Cambridge, United Kingdom

Collar N (1999) Risk indicators and status assessment in birds. In: del Hoyo J, Elliott A, Sargatal J (eds) Handbook of the Birds of the World, Barn-owls to Hummingbirds, vol 5. Lynx Edicions, Barcelona, pp 3–28

Colwell R (1997) EstimateS: statistical estimation of species richness and shared species from samples. Version 5. http://www.viceroy.eeb.uconn.edu/EstimateS. Cited 4 April 2006

Daniel S, Walters J (2000) Between-year breeding dispersal in Red-cockaded woodpeckers: multiple causes and estimated cost. Ecology 81:2473–2484

DeSante D, Grady D, Burton K, et al (1998) The monitoring avian productivity and survivorship (MAPS) program sixth and seventh annual report (1995 and 1996). Bird Popul 4:69–122

Dobson A (1990) Survival rates and their relationship to life-history traits in some common British birds. Curr Ornithol 7:115–147

González C (2000) Identificación de las áreas críticas para la conservación del gorrión serrano (*Xenospiza baileyi*). Dissertation, Universidad Nacional Autónoma de México

Instituto de Geografía (1997) Videografía aérea de la Ciudad de México. Instituto de Geografía, UNAM. México, D.F

Koskimies P, Vaisanen R (1990) Monitoring bird populations; a manual of methods applied in Finland. Zoological Museum, Finnish Museum of Natural History. University of Helsinki

Lockwood J, Fenn K, Caudill J, et al (2001) The implications of Cape Sable Seaside sparrow demography for everglades restoration. Anim Conserv 4:275–281

Martin T (1995) Avian life history evolution in relation to nest sites, nest predation, and food. Ecol Monogr 65:101–127

Nice M (1964) Studies in the life history of the song sparrow. Dover Publications, Inc, New York

Oliveras de Ita A, Gómez De Silva H, Grosselet M (2001) Population dynamics and natural history of the Sierra Madre sparrow (*Xenospiza baileyi*) at La Cima, México. Cotinga 15:43–47

Oliveras de Ita A, et al (2003) El Gorrión Serrano (*Xenospiza baileyi*). In: Gómez de Silva H, Oliveras de Ita A (eds) Conservación de Aves en México. CIPAMEX, NFWF, CONABIO, México, D.F., pp 65–167

Oliveras de Ita A, Rojas-Soto, O (2006) A survey for the Sierra Madre sparrow (*Xenospiza baileyi*), with its rediscovery in the state of Durango, Mexico. Bird Conservation International. DOI 10.1017/S0959270905000687

Perkins D, Vickery P (2001) Annual survivorship of an endangered passerine: Florida grasshopper sparrow. Wilson Bull 113:211–216

Ryder, R (1986) Songbirds. In: Cooperrider A, Boyd R, Hanson S (eds) Inventory and Monitoring of Wildlife Habitat, pp. 291–312

Sandercock B, Jaramillo A (2002) Annual survival rates of wintering sparrows: assessing demographic consequences of migration. Auk 119:149–165

Sillet T, Holmes R (2002) Variation in survivorship of a migratory songbird throughout its annual cycle. J Anim Ecol 71:296–308

Vickery P (1996) Grasshopper sparrow (*Ammodramus savannarum*). In: Poole A, Gill F (eds) The Birds of North America Vol 239. Academy of Natural Sciences and American Ornithologists Union, Washington, DC

Wheelwright N, Rising J (1993) Savannah sparrow (*Passerculus sandwichensis*). In: Poole A, Gill F (eds) The Birds of North America Vol 45. Academy of Natural Sciences and American Ornithologists Union, Washington, DC

Biodivers Conserv (2007) 16:1063–1086
DOI 10.1007/s10531-006-9069-y

ORIGINAL PAPER

Analysis of the species description process for a little known invertebrate group: the limnoterrestrial tardigrades (Bilateria, Tardigrada)

Noemí Guil · Francisco J. Cabrero-Sañudo

Received: 18 May 2005 / Accepted: 19 May 2006 / Published online: 23 August 2006

Abstract This study was designed to examine the entire species description process of limnoterrestrial tardigrades up until the present, and to establish the effects of several variables on this process. Information from monographs, revisions and published papers related to limnoterrestrial tardigrades was assessed using generalized linear models (GLM) and variance partition. Our findings reveal the tardigrade species description process fails to show an asymptotic tendency. That is, at the current description rate, we will not get to know all the species in the short- or mid-term. The variables examined served to explain about half of the process variance. Species distribution areas and variables related to their presence in each region were the factors found to have the greatest explanatory capacity. Our analysis indicates that insufficient efforts have been invested in the tardigrades compared to other invertebrates (such as Insecta: Coleoptera), and that there is a need for more taxonomists and more extensive sampling areas. Studies such as this will be useful for planning future species sampling campaigns, and help us to improve our methodological approach to investigations aimed at conservation and the preservation of biodiversity.

Keywords Factors determining description process · Limnoterrestrial · Species description process · Tardigrada

N. Guil (✉)
Museo Nacional de Ciencias Naturales, Dpto Biodiversidad y Biología Evolutiva, José Gutiérrez Abascal 2, 28006 Madrid, Spain
e-mail: nguil@oeb.harvard.edu.

F. J. Cabrero-Sañudo
Instituto de Ecología, A.C. Dpto, Biodiversidad y Ecología Animal, Km 2,5, Ctra antigua a Coatepec 351, Congregación El Haya 91070 Xalapa, Veracruz, México

Present Address:
N. Guil
Department of Organismic and Evolutionary Biology, Harvard University, BioLabs 1113, Divinity Avenue, Cambridge, Massachusetts 02138, USA

Abbreviations

GLM	Generalized linear models
LR	Linear regression analysis
Pal	Palaearctic region
Afr	Afrotropical region
Ori	Oriental region
Aus	Australian region
Nea	Nearctic region
Neo	Neotropical region
Ant	Antarctic region
TAX	taxonomic variables
Class	categorical variables related to the species taxonomic class
DIST	distributional variables
NReg	continuous variables representing the number of biogeographic regions in which each species exists
LOC	a group of categorical variables related to location of each species
RegType	variable referred to the biogeographic region in which the species was described
ORIG	variable related to the biogeographic region of author's place of origin.

Introduction

The field of systematics has two interdependent and simultaneous methodological components (Dirzo and Gómez 1996): taxonomy—that is, the descriptive study of biological diversity—and analysis of the origins, evolution and upkeep of biological diversity. A systematic scientific process has a particular order that begins with a taxonomic study, which is essential for future conservation (Margules and Pressey 2000; García-Valdecasas and Camacho 2003) or other biological studies. Systematics, in any of its forms, is a science that requires worldwide attention, owing to the current extinction rates of several animals (see, for example, Raven 1987; Wilson 1988; Dobson and Gentry 1991; Ricciardi and Rasmussen 1999; Purvis et al. 2000; Myers and Knoll 2001; Novacek and Cleland 2001; Woodruff 2001; Pin Koh et al. 2004; Sekercioglu et al. 2004).

Indeed, we are profusely ignorant of the total number of species on Earth (Diamond 1985; May 1986, 1988, 1989, 1990; Hammond 1992, 1995; Groombridge and Jenkins 2002), or even its order of magnitude (see Erwin 1991; Gaston 1991a). This problem is aggravated by the large proportion of species not yet described (Groombridge and Jenkins 2002; Mackay 2002), above all invertebrate species. As a consequence, the majority of conservation studies only consider taxonomically well-known groups, that is, vertebrates and/or plants, and exclude invertebrates (Rosas et al. 1992; Galante and Verdú 2000).

Moreover, the economic resources and time dedicated to systematic descriptions are insufficient (Cotterill 1995) and taxonomy is an undervalued discipline (García-Valdecasas et al. 2000; Krell 2000). The result of all this will be that part of the present biodiversity will disappear before its identification (Mackay 2002). There is therefore a critical need to plan sampling efforts using the available resources to describe biodiversity and to observe the distribution patterns of several taxa directly in nature.

In the present study, we evaluate the description process of an unknown group of microscopic invertebrates, the limnoterrestrial tardigrades (Bilateria, Tardigrada), so we can monitor future description work focusing on this biological group. The first documented observation of a tardigrade dates back to 1773, when Goeze described an animal as a waterbear (*kleiner wasser bär*). In the following decades, a few researchers studied tardigrades, mainly describing new taxa (orders, families, genera and species) and discussing their position in the Animal Kingdom (Ramazzotti and Maucci 1982). However, in the 20th century their appeal seemed to increase and several researchers exclusively dedicated their work to the study of tardigrades and understanding the biology of these animals. This renewed interest gave rise to the first International Symposium on Tardigrada in 1974. Since then, meetings have been held every three years with more diverse studies presented each time by an increasing number of participants.

The *phylum* Tardigrada, representing a new design of life in the Animal Kingdom, probably emerged during the Cambrian explosion (Maas and Waloszek 2001), and has been relatively neglected by invertebrate zoologists, especially in terms of its taxonomy. At present, the phylogenetic relationships between Tardigrada and other invertebrate *phyla* are unknown, although recent papers ascribed it to the *superphylum* Ecdysozoa that included all molting invertebrate taxa (Garey and Schmidt-Rhaesa 1998; Giribet and Ribera 1998). However, tardigrades align with: (a) Arthropoda, Onycophora or the Achelminthes complex, depending on the morphological characters studied (Nielsen 1995); (b) or, appear as a basal sister group of Arthropoda as well as celomate protostomids (Moon and Kim 1996), between Priapulida and Arthropoda (Winnepenninckx et al. 1995); (c) or, as a sister group of Arthropoda (Garey et al. 1996; Giribet et al. 1996; Garey and Schmidt-Rhaesa 1998), depending on the molecular sequence analyzed. Hence, this group could be evolutionary important and possibly be the key to resolving some of the phylogenetic relationships between pararthropoda taxa (Garey et al. 1996; Giribet et al. 1996; Garey and Schmidt-Rhaesa 1998; Zrzavy et al. 1998; Peterson and Eernisse 2001). Moreover, tardigrades may even be of economic interest due to their ability to undergo cryptobiosis, an environmentally resistant state, when conditions are unfavourable; the substances involved in cryptobiosis have potential applications in biomedicine and biotechnology (Crowe and Crowe 2000; Eroglu et al. 2000; Guo et al. 2000).

Limnoterrestrial tardigrades were initially investigated with respect to marine tardigrades (Ramazzotti and Maucci 1982) and behave as a group although they are not a phylogenetic self-entity (Ramazzotti and Maucci 1983). In fact, specialists on limnoterrestrial tardigrades do not use to devote their efforts to marine tardigrades. Moreover, the latest taxonomic monographs and revisions have centred on limnoterrestrial tardigrades, such that there is more ecological and biogeographic information available on this group. The life cycle of limnoterrestrial tardigrades requires the presence of water so they can be found in: freshwater and associated interstitial habitats and terrestrial habitats, such as mosses, lichens, leaf litter, soil, etc. (Ramazzotti and Maucci, 1983). To date, about 800 limnoterrestrial Tardigrada species have been described, of which 2/3 have been ascribed to the class Eutardigrada ('naked' tardigrades) and the rest to the class Heterotardigrada, or 'armoured' tardigrades (Guidetti and Bertolani, 2005). The description of Tardigrada species has been an opportunist process. Most new species descriptions have been made in the Palaearctic region, albeit the majority of limnoterrestrial tardigrade species are widely distributed (McInnes 1994).

The present paper analyses several topics related to species description processes for the limnoterrestrial tardigrades. Our first aim was to try to establish the state of knowledge of the inventory of limnoterrestrial tardigrade species. This number of species can be estimated by analysing the increase in the cumulative number of species with time as a function of the year of description (Steyskal 1965; Arnett 1967; White 1975; Frank and Curtis 1979; O'Brien and Wibmer 1979; White 1979; Medellín and Soberón 1999; Cabrero-Sañudo and Lobo 2003). We also explore the effects of taxonomy, location, geographic distribution area of the species and the place of origin of the authors describing each species on the description process. Analysis of the effects of several variables on species descriptions has been undertaken in other animal groups (Gaston 1991b, 1993, 1994, 1996; Hammond 1992; Gaston and Mound 1993; Gaston and Blackburn 1994; Patterson 1994; Blackburn and Gaston 1995; Gaston et al. 1995a, b; Allsopp 1997; Patterson 2000; Cabrero-Sañudo and Lobo 2003). The main factors that seem to affect the species description process are those linked to the body size and geographic location of the species and their geographic distribution range. Thus, the main questions arising are: How could we describe the limnoterrestrial tardigrade species description process? Why have so few species been described? Is there really such a small number of species or are there many yet to describe? What might future tardigrade studies hold? The answers to these questions could help us guide future taxonomic efforts invested in the Tardigrada.

The main objectives of this study were thus: (a) to describe and analyze the species description process since the first description up to the year 2004 based on a compilation of biogeographic data, (b) to establish a possible relationship between the tardigrade species description process and a set of variables related to the taxonomy of species and their geographic distribution, as well as the describing author, and (c) to identify causal or accidental factors affecting the description process.

Material and methods

Data collection

A database gathering all taxonomic and biogeographic distribution information on limnoterrestrial Tardigrada in the literature was created. The list of taxa was based on that of McInnes (1994) and completed using names provided in subsequent published articles and arising from an exhaustive bibliographic revision of the Zoological Record database (Appendix 1). The information obtained covered the period since the first description of a limnoterrestrial tardigrade in 1833 to 2004. This data corresponds to a total of 792 limnoterrestrial Tardigrada species including only currently recognized species and ignoring synonyms. For each species, information on its taxonomy, describing author, year of description, geographical distribution, and type locality was compiled. The corresponding biogeographic region of each species was identified from information on the type locality and denoted: Palaearctic (Pal), Afrotropical (Afr), Oriental (Ori), Australian (Aus), Nearctic (Nea),

Neotropical (Neo), or Antarctic (Ant). In addition, using the available information on species distribution, a presence/absence matrix of species in the different biogeographic regions was constructed. Figure 1 shows the limits of these regions.

We also obtained information on the biogeographic region of the authors describing the species. When several authors described a species, we only considered the place of origin of the first author.

Analysis of the species description process

To fit the species description process to a curve, we used the logarithm of the cumulative number of species described in the successive years 1833 to 2004. Cumulative curves may be used to estimate the number of species not yet discovered within a group (Steyskal 1965). Its use is based on the following assumptions (Cabrero-Sañudo and Lobo 2003): (a) a finite number of species exists on Earth (Steyskal 1965; O'Brien and Wibmer 1979); (b) no generation of new species has occurred since the beginning of the species description process (Frank and Curtis 1979); (c) it is possible to progressively approach a state in which all species are known (Steyskal 1965; Arnett 1967; White 1975; Frank and Curtis, 1979; Diamond 1985); and, (d) the rate of the species description process gradually diminishes as we approach full knowledge of extant species (White 1979; Medellín and Soberón 1999).

We therefore plotted logarithms of cumulative species numbers against the year of species description, trying to fit the different types of species-accumulation functions found in the literature (Soberón and Llorente 1993; Flather 1996), as well as functions up to third degree (linear, quadratic, cubic). The value of the asymptote (whenever it is achieved) reflects the total number of species in a group and may thus give an idea of the number of species not yet described. The curvilinear function was fitted by the *Simplex and Quasi-Newton* method using Statistica software (StatSoft

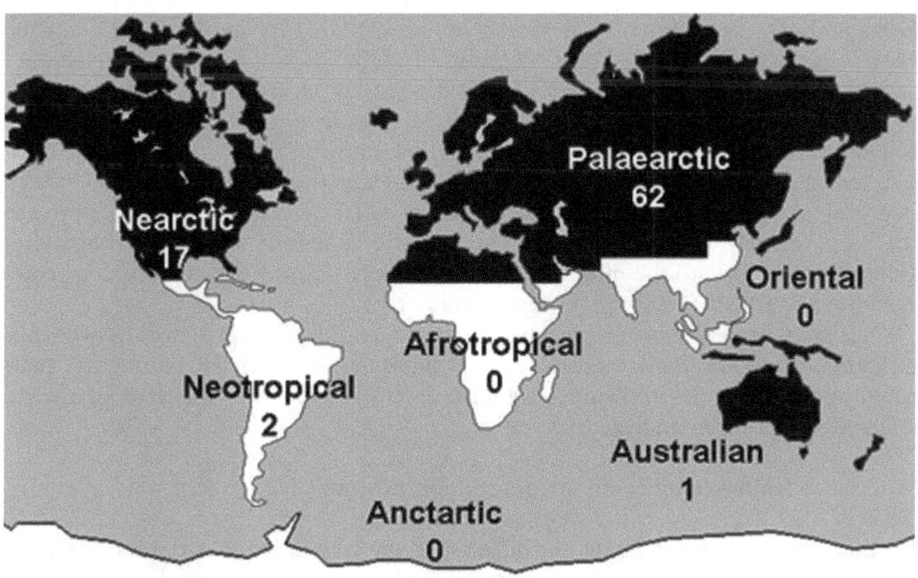

Fig. 1 Map with the different regions considered and the number of specialist of each one

Inc. 2001). Best fit was obtained through the use of non-asymptotic functions, including cubic ($P < 0.05$; $R^2 = 95.27$), quadratic ($P < 0.05$; $R^2 = 95.26$) or rational functions ($P < 0.05$; $R^2 = 93.57$; Ratkowski, 1990).

To explore the different tendencies throughout the species description process for the limnoterrestrial Tardigrada, a piecewise linear regression of the cumulative number of species against the year of description was also conducted. The cutoff point was automatically estimated by the programme and corresponding linear functions were searched for by the *Simplex and Quasi-Newton* method.

Analysis of explanatory variables

To explore relationships between year of species description and the possible explanatory variables considered, generalized linear models were used (GLM; McCullagh and Nelder 1989; Crawley 1993; Dobson, 1999). Four different variable types were considered: taxonomic (TAX) categorical variables related to the species taxonomic class (*Class*); distributional (DIST) continuous variables representing the number of biogeographic regions (*NReg*) in which each species exists; a group of categorical variables related to location (LOC), seven of which were the variables (*Pal*, *Afr*, *Ori*, *Nea*, *Neo*, *Aus*, and *Ant*) describing the presence/absence of each species in each biogeographic region, and a further one (*RegType*) referring to the biogeographic region in which the species was described; and, lastly, categorical variables (*RegAuth*) related to the biogeographic region of author's place of origin (ORIG).

Every possible relationship between the explanatory variables and the year of species description was explored (described in detail in Cabrero-Sañudo and Lobo 2003). We assumed all variables were related via an identity link function and were distributed normally and used a GLM rather than the classic linear regression (LR) models, since this type of procedure is less restrictive and include LR analysis. Each possible relationship was explored, measuring the percentage of explained deviance (Dobson 1999). Subsequently, a search for an explanatory model was conducted (MacNally 2000, 2002) for the LOC group of variables and a general explanatory model for all the variables together (TAX, DIST, LOC, and ORIG). Thus, we used a stepwise regression procedure (Nicholls 1989; Austin et al. 1996; Heikkinen and Neuvonen 1997) in which the variable function showing the most significant deviance change was included, and the other variable functions were added step by step according to the explained percentage of deviance (forward selection). This process was repeated until no more significant variable functions were available. After adding the variables to the model, a backward removal procedure was performed to exclude any terms becoming non-significant.

Since interactions between variables usually have better explanatory power than the explanatory variables themselves (Margules et al. 1987), all interaction pairs including the variables showing a significant relationship with the year of species description were also analyzed.

Analysis of factors influencing the description process

We then went on to evaluate the effects of the explanatory variables upon the species description process, to discriminate between causal variables and simple accidental relationships (Legendre and Legendre 1998; Guisan and Zimmermann

2000; MacNally 2000, 2002). Two different approaches are usually used to distinguish the effects of every variable or variable groups: variance partition (Legendre and Legendre 1998) and hierarchical variance decomposition (MacNally 2000, 2002). We selected the variance partition method based on the procedure described by Cabrero-Sañudo and Lobo (2003), modifying the equation system developed for three groups of variables (Borcard et al. 1992; Qinghong and Bråkenhielm 1995; Anderson and Gribble 1998) for use with four groups of variables (TAX, DIST, LOC, and ORIG). The equation system can be summarised as follows:

$$e+f+g+h+i+j+k=\mathrm{TAX}-a$$
$$e+h+i+k+l+m+n=\mathrm{DIST}-b$$
$$f+h+j+k+l+n+o=\mathrm{LOC}-c$$
$$g+i+j+k+m+n+o=\mathrm{ORIG}-d$$

and,

$$e=(\mathrm{TAX}+\mathrm{DIST})-(f+g+h+i+j+k+l+m+n)-(a+b)$$
$$f=(\mathrm{TAX}+\mathrm{LOC})-(e+g+h+i+j+k+l+n+o)-(a+c)$$
$$g=(\mathrm{TAX}+\mathrm{ORIG})-(e+f+h+i+j+k+m+n+o)-(a+d)$$
$$h=(\mathrm{TAX}+\mathrm{DIST}+\mathrm{LOC})-(e+f+g+i+j+k+l+m+n+o)-(a+b+c)$$
$$i=(\mathrm{TAX}+\mathrm{DIST}+\mathrm{ORIG})-(e+f+g+h+j+k+l+m+n+o)-(b+d)$$
$$j=(\mathrm{TAX}+\mathrm{LOC}+\mathrm{ORIG})-(e+f+g+h+i+k+l+m+n+o)-(a+c+d)$$
$$k=(\mathrm{TAX}+\mathrm{DIST}+\mathrm{LOC}+\mathrm{ORIG})-(e+f+g+h+i+j+l+m+n+o)-(a+b+c+d)$$
$$l=(\mathrm{DIST}+\mathrm{LOC})-(e+f+h+i+j+k+m+n+o)-(b+c)$$
$$m=(\mathrm{DIST}+\mathrm{ORIG})-(e+g+h+i+j+k+l+n+o)-(b+d)$$
$$n=(\mathrm{DIST}+\mathrm{LOC}+\mathrm{ORIG})-(e+f+g+h+i+j+k+l+m+o)-(b+c+d)$$
$$o=(\mathrm{LOC}+\mathrm{ORIG})-(f+g+h+i+j+k+l+m+n)-(c+d)$$

Results

Analysis of the species description process

Figure 2 is a plot of the species description process. Since the curvilinear fit of the description process did not adjust to an asymptotic function nor a function with biological sense (it fitted a cubic function at 95%), the limnoterrestrial Tardigrada species description process appears as a constant and progressive task (Fig. 2). This indicates that, at the present rate of species descriptions, complete knowledge of the species will not be acquired in the short- or mid-term. The species description process has been on the whole very irregular. Just after their discovery, taxonomic studies on tardigrades were sporadic or even lacking. Then around 1900, the number of species descriptions increased to slightly diminish around 1915 (Fig. 2). A little earlier than 1930, attention increased once again and this rekindled interest has

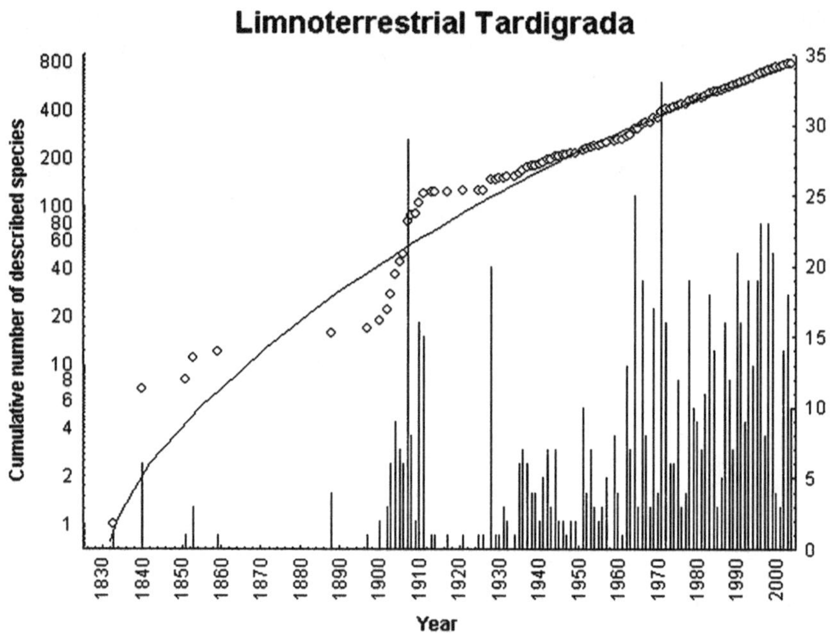

Fig. 2 Number of species described (bars) and temporal variation in the logarithm of the cumulative number of limnoterrestrial tardigrade species described from 1833 to 2004. The cumulative curve (continuous line) was fitted to a third degree equation using the Simplex and Quasi-Newton method

continued to the present (Fig. 2). If we compare the description process for the different biogeographic regions, it can be observed that descriptions in the Palaearctic region appeared half a century before those for the other regions (Fig. 3). Tardigrades were subsequently described in the Neotropical region, and more than a decade later descriptions in other regions emerged coinciding with the time—1900–1915—of the increased effort noted above (Table 1). In every region, the number of described species rises from 1910 onwards, shortly after an inactive period (Fig. 3). It was not until the late 1920s, that descriptions steadily increased, particularly in Palaearctic and Neotropical regions (Fig. 3). From the 1940s to the present, the number of described species steadily rises in all areas, except the Oriental region, where it is stagnant (Fig. 3). It may also be noted that the species description process in the Neotropical region occurs as abrupt steps, rather than the gradual steady increase shown in the remaining zones (Fig. 3).

The piecewise linear regression fit reveals two different trends, before and after the year 1958 ($P < 0.05$; $R^2 = 98.66$). The worldwide species description rate before and after this year show a 1:6 proportional relationship (1.97–11.83 species per year; Table 1; significant differences between rates; $t_{170} = 11.08$; $P < 0.01$). Before 1958, the highest species description rate corresponds to the Palaearctic region (1.33; a significantly higher rate compared to those of the other regions; all t_s significant; all $P_s < 0.01$), and lowest rates correspond to the Antarctic, Australian, and Oriental regions (0.11, 0.18, and 0.19, respectively; Table 1). Subsequent to 1958, the Palaearctic region continues to yield the highest species description rate (6.37; significantly higher than the other regions; all t_s significant; all $P_s < 0.01$), while the lowest rates correspond to the Oriental, Antarctic, and Afrotropical regions (0.30, 0.67, and 0.78,

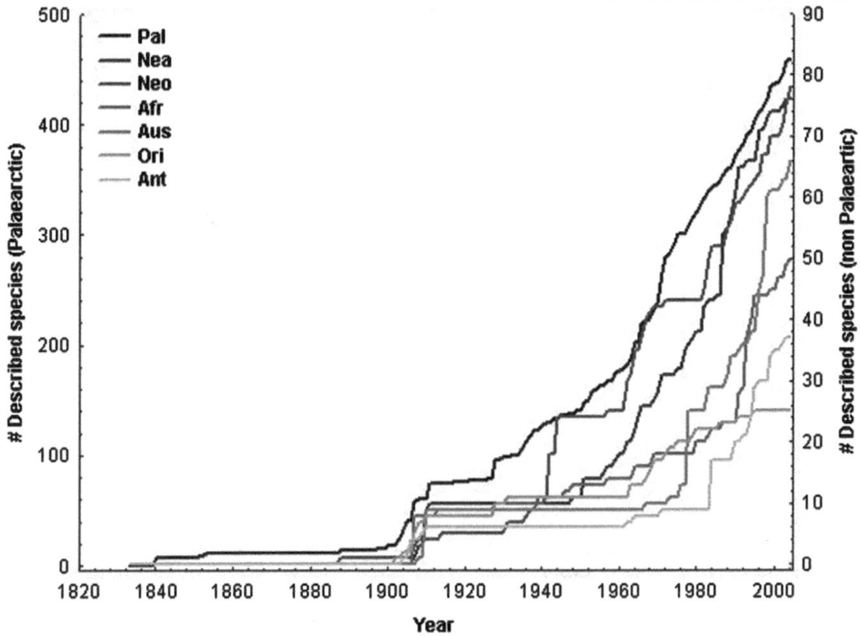

Fig. 3 Cumulative number of limnoterrestrial tardigrade species described in each biogeographic region from 1833 to 2004. Regions: Pal (Palaearctic), Nea (Nearctic), Neo (Neotropical), Afr (Afrotropical), Aus (Australian), Ori (Oriental) and Ant (Antarctic)

respectively; Table 1; only the rate for the Oriental zone differs significantly from rates emerging from the Palaearctic, Nearctic, Neotropical, and Australian regions; all t_s significant; all $P_s < 0.03$). The highest increases in the species description rates correspond to those of the Australian region (0.18–1.24; relationship 1:6.89; Table 1; $t_{95} = 2.50$; $P = 0.01$, difference between the two periods significant) and the Antarctic (0.11–0.67; relationship 1:6.09; Table 1; $t_{99} = 2.55$; $P = 0.01$, difference between the two periods significant). The rest of the regions undergo discrete rates of growth: i.e., Palaearctic (1.33–6.37; relationship 1:4.79; Table 1; $t_{170} = 8.14$; $P < 0.01$, difference between the two periods significant); Nearctic (0.31–1.30; relationship 1:4.19; Table 1; $t_{96} = 3.40$; $P < 0.01$, difference between the two periods significant); and Neotropical (0.35–1.15; relationship 1:3.29; Table 1; $t_{115} = 3.30$; $P = 0.001$; difference between the two periods significant). Lowest growth rates were recorded for the Afrotropical region (0.27–0.78; relationship 2.89; Table 1; $t_{96} = 1.86$; $P = 0.07$, difference between the two periods not significant) and Oriental regions (0.19–0.30; relationship 1:1.58; Table 1; $t_{101} = 0.99$; $P = 0.33$, difference between the two periods not significant).

In over half the years of the species description process, at least one species has been described worldwide (Table 1). The situation for the Palaearctic region is practically the same, whereas in the other regions only 10–20% of the years yielded at least one species description (Table 1). On the worldwide scale, more than one species is described at the same time (48.26% of years with more than one species description and 30.23% of years with more than five descriptions; Table 1). In the Palaearctic, generally more than one species is described per year (in 53.49% of years; Table 1), while for the Oriental and Antarctic regions, descriptions are more

Table 1 Characteristics of the species description process for the limnoterrestrial tardigrades on the regional and global scales

	Biogeographical region of the type locality							
	Pal	Afr	Ori	Nea	Neo	Aus	Ant	Total
# Species described	460	50	25	76	78	66	37	792
% Total species	58.01	6.31	3.15	9.58	9.84	8.32	4.67	100
Year of first description	1833	1907	1902	1907	1888	1908	1904	1833
Year of last description	2004	2004	1995	2003	2004	2004	2003	2004
Time range of descriptions	172	98	94	97	117	97	100	172
Description rate from first description to 2004	2.67 ± 4.23	0.51 ± 1.38	0.24 ± 0.57	0.78 ± 1.52	0.67 ± 1.33	0.68 ± 2.14	0.37 ± 1.14	4.60 ± 6.76
Description rate from first description to 1958	1.33 ± 2.84	0.27 ± 1.14	0.19 ± 0.58	0.31 ± 0.98	0.35 ± 1.16	0.18 ± 1.13	0.11 ± 0.46	1.97 ± 4.12
Description rate from 1959 to 2004	6.37 ± 5.16	0.78 ± 1.58	0.30 ± 0.55	1.30 ± 1.84	1.15 ± 1.44	1.24 ± 2.79	0.67 ± 1.56	11.83 ± 7.33
% Years with descriptions	53.49	12.79	11.05	20.93	20.35	10.47	9.88	56.98
% Years with more than one description	40.70	5.81	2.91	9.30	11.05	6.98	3.49	48.26
% Years with more than five descriptions	18.60	1.16	0.00	1.16	1.16	2.33	0.58	30.23
# Specialist per region	61	0	0	15	2	1	0	79
% Total specialists	77.22	0.00	0.00	18.99	2.53	1.27	0.00	100
% Descriptions made by non-native specialists	0.43	100.00	100.00	56.58	93.59	77.27	100.00	35.48
% Descriptions made by Palaearctic specialists	99.57	100.00	100.00	56.58	82.05	54.55	94.59	89.77

Regions: Pal (Palaearctic), Nea (Nearctic), Neo (Neotropical), Afr (Afrotropical), Aus (Australian), Ori (Oriental) and Ant (Antarctic). Description rate and corresponding standard deviation sporadic (more than one description in 11.05% and 9.88% of years, respectively; Table 1).

The number of specialists describing a limnoterrestrial tardigrade species is highest for the Palaearctic region (61 specialists), followed by the Nearctic (15; Table 1). Conversely, in Afrotropical, Oriental, and Antarctic regions no specialists have described a limnoterrestrial tardigrade species (Table 1). The percentage of species in each region described by a non-native specialist is almost negligible in the Palaearctic region, while for Afrotropical, Oriental, and Antarctic areas all descriptions correspond to non-native experts (Table 1). On the other hand, in regions such as the Neotropical, the Australian, and above all, the Nearctic, a number of species descriptions by native authors appear (Table 1). Nevertheless, Palaearctic authors account for a large number of descriptions made not only in the Palaearctic region but also in Afrotropical, Oriental, Antarctic, and Neotropical zones; with less descriptions in Nearctic and Australian regions, since these areas count on the work of several autochthonous authors or specialists from other non-Palaearctic regions (Table 1).

A positive correlation was observed between the number of specialists in each region and the percentage of descriptions made by autochthonous authors ($R = 0.96$; $t_{10} = 7.83$; $P < 0.001$). Hence, the more specialists there are in a region, the higher number of species descriptions. We were also able to correlate the percentage of descriptions made by native authors and the yearly species description rate ($R = 0.96$; $t_{10} = 8.06$; $P < 0.001$), such that an increased number of descriptions by autochthonous specialists is related to a larger number of described species for that region. In general it may be stated that the more specialists there are in a region the higher rate of species descriptions emerging from that particular region.

Analysis of the explanatory variables

Table 2 summarises the relationships between the species description process and the explanatory variables considered.

The taxonomic variable *Class* served to explain a small proportion of the descriptive process variation (around 5%), such that in the last years, a few more species of the class Eutardigrada have been described than of Heterotardigrada (Table 2).

The distributional variable *NReg* has considerable explanatory capacity (around 40% of the total variation in the descriptive process; Table 2); in the early years of the descriptive process, species with larger distribution ranges were described, whereas species showing a smaller distribution seem to have been described more recently (Table 2).

Any variable in the location related group can explain part of the variation of the species description process (Table 2). All the presence/absence variables (*Pal, Afr, Ori, Nea, Neo, Aus*, and *Ant*) are negatively related to the species description process, explaining 4–20% of its variation. These variables indicate that, early on in the description process, relatively more species were described in each biogeographical region compared to later on in the process, perhaps as a consequence of the larger distribution ranges. The *RegType* variable emerges as significant for explaining a small part of the variation in the species description process, indicating an increased advance of the description process in the Australian region compared to the Palaearctic over the last years (Table 2). The explanatory model obtained using

Table 2 Relationships between the species description process for the limnoterrestrial tardigrades and the explanatory variables considered in this study

Group	Variables	Abbreviation	Selected terms	Deviance	d.f	Change in deviance	F	Sign	% Explained deviance
	Null model			908,713.72	791				
TAX	Taxonomy								
	Taxonomic class	Class	Class	867,480.31	790	41,233.0	37.55***	Eutardigrada (+) Heterotardigrada (–)	4.54
DIST	Distribution								
	# Biogeographical regions	NReg	NReg + $NReg^2$	547,643.65	789	361,070.07	520.20***	– +	**39.73**
LOC	Location								
	Palaearctic region	Pal	Pal	837,408.20	790	71,305.52	67.27***	–	7.85
	Afrotropical region	Afr	Afr	850,441.86	790	58,271.86	54.13***	–	6.41
	Oriental region	Ori	Ori	720,657.87	790	188,055.84	206.15***	–	20.69
	Nearctic region	Nea	Nea	761,120.30	790	147,593.42	153.19***	–	16.24
	Neotropical region	Neo	Neo	806,038.47	790	102,675.25	100.63***	–	11.30
	Australian region	Aus	Aus	870,300.85	790	38,412.86	34.87***	–	4.23
	Antarctic region	Ant	Ant	872,356.63	790	36,357.09	32.92***	–	4.00
	Type region	RegType	RegType	862,651.82	785	46,061.90	41.92***	Pal (–) Aus (+)	5.07
	Location explanatory model		Pal + Ori + Nea + Neo + Aus + Ant + RegType	516,004.05	779	392,709.66	592.87***		43.22
ORIG	Authors'place of origin								
	Author's region	RegAuth	RegAuth	867,513.35	788	41,200.37	37.42***	Pal (–)	4,53
	General explanatory model		Class + NReg + Pal + Ori + Nea + RegAuth	466,831.32	783	441,882.40	741.15***		**48.63**

Regions: Pal (Palaearctic), Nea (Nearctic), Neo (Neotropical), Afr (Afrotropical), Aus (Australian), Ori (Oriental) and Ant (Antarctic). Taxonomic variables (TAX): *Class* (taxonomic class of each species). Geographic distribution variables (DIST): *NReg* (number of biogeographic regions in which each species has been found). Geographic location variables (LOC): *Pal*, *Afr*, *Ori*, *Nea*, *Neo*, *Aus* and *Ant* (presence/absence of each species in each biogeographic region), and *RegType* (type region in which each species has been described). Origins of specialist variables (ORIG): *RegAuth* (place of origin of the first author for each species description)

*** = variable significant at $P < 0.001$

these location variables accounts for over 40% of the variation in the species description process (Table 2).

Finally, the variable related to the author's place of origin, *RegAuth* is significant at explaining around 5% of the variation in the descriptive process. Thus, it seems that in recent years the number of non-Palaearctic specialists has risen (Table 2).

A general explanatory model derived from all the observed significant variables serves to explain around 50% of the species description process (Table 2).

Analysis of factors influencing the description process

The variance partition indicates that it is the joint effect of distribution and location variables which captures most of the variation of the species description process (explaining up to 40% of all variation; Table 3). This effect is the consequence of the wide distribution range of the first described species, which appeared in the Palaearctic (i.e., having a small distribution and not inhabiting the Palaearctic are associated characteristics). The pure effect of species taxonomy carries some weight (a little more than 5%; Table 3) and reveals that recently described species belong to the Eutardigrada class

Discussion and conclusions

Among the main applications of studies designed to explain and describe species description processes, is to create a tool for use in systematic studies. Using the tool established, species sampling campaigns can be monitored, maximizing the sampling effort, as argued by Soberón and Llorente (1993).

Table 3 Results of the decomposition and hierarchical partition of variance analysis for the year of each species description

Pure effects of each group of variables	
TAX	5.41
IST	–0.68
LOC	1.50
ORIG	0.25
Joint effects of two groups of variables	
TAX + DIST	0.25
TAX + LOC	0.25
TAX + ORIG	0.25
DIST + LOC	37.92
DIST + ORIG	0.25
LOC + ORIG	1.99
Joint effects of three groups of variables	
TAX + DIST + LOC	0.25
TAX + DIST + ORIG	0.25
TAX + LOC + ORIG	0.25
DIST + LOC + ORIG	0.25
Joint effects of four groups of variables	
TAX + DIST + LOC + ORIG	2.41

Taxonomy (TAX), Geographic distribution (DIST), Geographic location (LOC) and Origin of author (ORIG)

The first paper to deal with cumulative species curves was written by Steyskal in 1965. This author represented cumulative species numbers in a general sample of several thousands of species from the Animal Kingdom and a sample of several groups of insects by decades. Similar studies on insects were to follow Steyskal's work (Arnett, 1967; White 1975), but were merely representations. Later on, cumulative species curves were fitted to mathematic models so that we could predict and extrapolate results (Clench 1979). The species description rate is specific to each organism (see, for example, Soberón and Llorente 1993; Gaston et al. 1995b; Allsopp, 1997; Dirzo and Gómez, 2002; Cabrero-Sañudo and Lobo 2003), such that these studies are useful for: (a) establishing the state of knowledge of the group and identify its taxonomic and biogeographic deficiencies, (b) undertaking more rigorous and quantitative comparisons between lists of fauna and flora, (c) providing a scheduling tool for future collection expeditions, and (d) providing a tool for conservation and biodiversity studies (Steyskal 1965; White 1979; O'Brien and Wibmer 1979; Frank and Curtis 1979; Soberón and Llorente 1993).

After fitting the cumulative species curve obtained for the limnoterrestrial tardigrades to several mathematic models, it emerges that there are many species to be described, since the cumulative species curve is not yet stabilised (Fig. 2) and neither does it seems that it will reach an asymptote in the short- or mid-term at the current description rate. This could be the consequence of two complementary factors: (a) the total species number of limnoterrestrial tardigrades is high and/or (b) the taxonomic sampling effort (number of specialist, sampled areas, etc.) is low relative to the real number of species. It is also true, that the description process of tardigrades in general suffered a delay of half a century with respect to other animal groups (see Table 1).

The species description pattern for the limnoterrestrial tardigrades can be related to certain historical events. Thus, the appearance in 1900 of the specialists, Richters and Murray, increased the number of descriptions and extended the regions examined to, for example, Africa or the Orient (Ramazzotti and Maucci 1982). In the first half of the 20th century, the number of authors working on tardigrades rose from about 10 to over 60, although most of their works on these organisms were sporadic. This surge contributed to the increased number of species described. The most significant improvement in the species description rate was the result of the spectacular increase in the number of specialists (by more than 100) occurring between 1940 and 1960 (Table 1). This is the reason why our piecewise linear regression analysis revealed a change in the slope of the species description process from 1958 onwards. In fact, most of today's tardigradologists are scientifically productive; some of whom have described a large number of species and/or expanded the sampled regions. The reason for this surge in the number of specialists could be the result of the interest generated by the second Tardigrada monograph (Ramazzotti 1962) written by Marcus in 1929. This highlights the significant impact of monographs and revisions of the different animal groups, although this impact is not recognized by a large proportion of the scientific community (Godfray 2002; Mallet and Willmott 2003; García-Valdecasas and Camacho 2003).

The specific size of an animal is the most common factor considered to affect the species description process (Gaston 1991b; Gaston and Blackburn 1994; Patterson 1994; Allsopp 1997; Cabrero-Sañudo and Lobo 2003). In general, larger species are discovered before smaller ones in other taxonomic groups, although the influence of size on the different species description processes is not the same. We did not take into account the size of the species herein because of the intraspecific variation between tardigrades. Before analysing the importance of size in the Tardigrada, a likely mean length for each species will need to be determined. However, we think that in the case of the Tardigrada, the size factor loses importance, since a microscope is needed for any new discoveries rather than simple eyesight.

The taxonomic group of each limnoterrestrial tardigrade species does nevertheless have a certain influence on the species description process. Thus, relatively more Eutardigrada than Heterotardigrada species were described during the initial years of the process. In effect, three quarters of all described limnoterrestrial species belong to the Eutardigrada on a global and regional scale. This could mean there are more eutardigrades in limnoterrestrial habitats (unlike marine habitats which are mostly inhabited by heterotardigrades) or that habitats preferred by eutardigrades were more intensely explored in the earliest years of the description process.

A further factor considered in our study was where each specialist came from. Thereby, our results show that the number of descriptions published by Palaearctic specialists is higher than descriptions published by authors from any other region throughout the species description process. However, the number of Palaearctic specialists has recently diminished, while authors from other regions have increased in number. This has led to an increased species description rate in regions outside the Palaearctic. Moreover, the augmented specialist numbers in a particular region will invariably increase the number of species descriptions in that region. Thus, specialist training in the less examined regions (Africotropical, Oriental areas, etc.) should be promoted or specialists from well-explored regions (such as the Palaearctic) should focus on the lesser studied regions. In other words, if the number of specialists in a particular region increases, then the species description rate in that region will be expedited. A greater sampling effort over time invested in the tardigrades will improve our knowledge of the *phylum* diversity and biology of the Tardigrada.

Factors related to distribution range and geographic site were found to be the most influential in the limnoterrestrial tardigrade species description process. In effect, species with a broad geographic range were described before those with a more restricted range. This has happened in other biological groups (Patterson, 1994, 2000; Blackburn and Gaston 1995; Gaston et al. 1995a; Allsopp 1997; Cabrero-Sañudo and Lobo 2003). The geographic location of a species has also favoured the description process, as in many other groups of organisms (Gaston, 1993; Patterson 1994; Gaston et al. 1995b; Allsopp 1997; Cabrero-Sañudo and Lobo 2003). Commonly, species inhabiting areas, or regions close to areas of some taxonomic tradition are described earlier (Gaston and May 1992; Gaston 1994). Most of the variance in the species description process of the limnoterrestrial tardigrades is the combined effect of distribution range and geographic location. Hence, limnoterrestrial tardigrade species with a broad distribution range in the Palaearctic region were first described.

In conclusion, our findings point to a need for: (a) a greater sampling effort by training specialists worldwide and (b) surveying small enclaves in regions other than the Palaearctic and sampling other possible habitats.

In the present analyses we have considered the described Tardigrada species as valid extant species. However, we must not forget that some of the widespread problems in taxonomy are present too in the Tardigrada, and our analyses could have not been able to manage them. On the one hand, some genera contain a (probably) overestimated number of species because similar morphotypes are described as different species and grouped into clusters. This happen for example in the *hufelandi*, the *tenuis*, and the *richtersi* groups of the *Macrobiotus* genus (Macrobiotidae family) and in the *pingue* group of the *Diphascon* genus (Hypsibiidae family) (Bertolani and Rebecchi 1993; Pilato and Binda 1999). Nevertheless, up to date it is impossible to handle this problem until alternative molecular techniques or developmental analyses are applied to Tardigrada studies. On the other hand, many synonymies exist. These can come from the similar morphotypes mentioned above, or from poor descriptions, mainly the oldest ones. These could be revised in the light of that new technologies applied to taxonomy.

So, at present the total number of limnoterrestrial Tardigrada species is unknown, as is the time needed to describe all these species or an acceptable proportion of these. Nevertheless, as for any other biological group, if we want to expand our knowledge of these invertebrates and explore the possible biotechnological and biomedical applications of their biology, more economic and human resources need to be devoted to this group. The taxonomical crisis is patent both in handling extant information (Godfray 2002) and in generating taxonomic information, molecular (Mallet and Willmott, 2003) and morphological (García-Valdecasas and Camacho 2003). The trained taxonomists or parataxonomists who generate and update taxonomic information (García-Valdecasas and Camacho 2003) are in decline; this fact is lessening biological advances that may surprise us in the future.

Acknowledgements This work was made possible by a postdoctoral fellowship from the *INECOL* (*Instituto de Ecología, A. C.*, Xalapa, Veracruz) awarded to FJCS, a pre-doctoral grant from the *Comunidad de Madrid*, *Consejo Superior de Investigaciones Científicas* (CSIC) and *Museo Nacional de Ciencias Naturales de Madrid* awarded to NG, and a *Comunidad de Madrid* project (number: 07M/0125/2000). We also thank Dr. Jorge M. Lobo, whose comments served to improve our paper, and Dr. Annie Machordom Barbé for further financial support.

Appendix

Appendix Reference list of papers from which geographic information related to the limnoterrestrial tardigrades was obtained for the period 1994 (last compilation) to 2004

Reference list from which data were compiled		
Abe and Ito, 1994	Dastych and Harris, 1994	Pilato and Binda, 1998b
Abe et al., 1998	Dastych and Harris, 1995	Pilato and Binda, 1999
Abe et al., 2000	Dastych and Kristensen, 1995	Pilato and Binda, 2003
Abe and Takeda, 2000a	Dastych and McInnes, 1995	Pilato et al., 1998
Abe and Takeda, 2000b	Dastych and McInnes, 1996	Pilato et al., 1999
Beasley, 1998	Degma, 2003	Pilato et al., 2001
Bertolani et al., 1994a	Degma and Pecalkova, 2003	Pilato et al., 2002a
Bertolani et al., 1994b	Dudichev and Biserov, 2000	Pilato et al., 2002b
Bertolani et al., 1996	Dudichev et al., 1998	Pilato et al., 2003a
Bertolani and Rebecchi, 1996	Guidetti, 1998	Pilato et al., 2003b
Binda et al., 2001	Guidi and Grabowski, 1996	Pilato et al., 2003c
Binda and Pilato, 1995a	Iharos and Vargha, 1995	Pilato et al., 2004
Binda and Pilato, 1995b	Ito, 1995	Pilato and Lisi, 2003
Binda and Pilato, 1995c	Ito, 1997	Pilato and Lisi, 2004
Binda and Pilato, 1995d	Kaczmarek et al., 2002	Pilato and Patanè, 1997

Appendix continued

Reference list from which data were compiled		
Binda and Pilato, 1995e	Kaczmarek and Michalczyk, 2002	Rho et al., 1997
Binda and Pilato, 1995f	Kaczmarek and Michalczyk, 2003	Schuster, 2000
Binda and Pilato, 1999a	Kaczmarek and Michalczyk, 2004a	Semeria, 1994
Binda and Pilato, 1999b	Kaczmarek and Michalczyk, 2004b	Semeria, 2003
Binda and Pilato, 2000	Kendall-Fite and Nelson, 1996	Strayer et al., 1994
Biserov, 1994	Maucci, 1996	Tumanov, 1997a
Biserov, 1996a	McInnes, 1994	Tumanov, 1997b
Biserov, 1996b	McInnes, 1995	Tumanov, 2003a
Biserov, 1998	Michalczyk and Kaczmarek, 2003a	Tumanov, 2003b
Biserov, 1999a	Michalczyk and Kaczmarek, 2003b	Tumanov, 2004
Biserov, 1999b	Michalczyk and Kaczmarek, 2003c	Van-Rompu et al. 1995
Biserov et al., 2001	Miller et al., 1995	Van-Rompu and De-Smet, 1996
Biserov and Gerlach, 1998	Moon et al., 1994	Van-Rompu and De-Smet, 1998
Chang and Rho, 1996	Moon and Won, 1994	Vargha, 1995
Claps and Rossi, 1997	Nelson et al., 1999	Vargha, 1998
Claxton 1998	Nelson and McGlothlin, 1996	Vargha, 2000
Claxton 1999	Pilato, 1996	Vasquez and Vargas, 2003
Claxton, 2001	Pilato and Binda, 1994	Yang, 1999
Dastych, 1997	Pilato and Binda, 1996a	Yang, 2002a
Dastych, 1999	Pilato and Binda, 1996b	Yang, 2002b
Dastych, 2000	Pilato and Binda, 1997	Yang, 2003
Dastych, 2002	Pilato and Binda, 1998a	

References

Abe W, Ito M (1994) *Itaquascon globuliferum*, a new species of Tardigrada (Eutardigrada: Hypsibiidae) from Japan. Proc Jpn Soc Syst Zool 51:8–11

Abe W, Takeda M (2000a) A new record of *Cornechiniscus madagascariensis* Maucci, 1993 (Tardigrada: Echiniscidae) from India. Proc Biol Soc Washington 113:480–485

Abe W, Takeda M (2000b) A new *Calcarobiotus* (Tardigrada: Macrobiotidae) from the Imperial Palace of Japan. Zool Sci 17:259–263

Abe W, Utsugi K, Takeda M (1998) *Pseudechiniscus asper*, a new Tardigrada (Heterotardigrada: Echiniscidae) from Hokkaido, northern Japan. Proc Biol Soc Washington 111:843–848

Abe W, Ito M, Takeda M (2000) First record of *Echiniscus hoonsooi* (Tardigrada: Echiniscidae) from Japan. Species Divers 5:103–110

Allsopp PG (1997) Probability of describing an Australian scarab beetle: influence of body size and distribution. J Biogeogr 24:717–724

Anderson MJ, Gribble NA (1998) Partitioning the variation among spatial, temporal and environmental components in a multivariate data set. Aust J Ecol 23:158–167

Arnett RH (1967) Present and future systematics of the Coleoptera in North America. Ann Entomol Soc Am 60:162–170

Austin MP, Pausas JG, Nicholls AO (1996) Patterns of tree species richness in relation to environment in south-eastern New South Wales, Australia. Aust J Ecol 21:154–164

Beasley CW (1998) A new species of *Echiniscus* (Tardigrada, Echiniscidae) from northern Yunnan Province, China. Zoologischer Anzeiger 238:135–138

Bertolani R, Rebecchi L (1993) A revision of Macrobiotus hufelandi group (Tardigrada, Macrobiotidae), with some observations on the taxonomic characters of eutardigrades. Zool Scr 22(2):127–152

Bertolani R, Rebecchi L (1996) The Tardigrades of Emilia (Italy). II. Monte Rondinaio. A multihabitat study on high altitude valley of the northern Appennines. Zool J Linn Soc 116:3–12

Bertolani R, Guidetti R, Rebecchi L (1994a) Tardigradi dell'Appennino umbro-marchigiano. Biogeographia 17:113–124

Bertolani R, Guidetti R, Rebecchi L (1994b) Ulteriore contributo alla conoscenza dei Tardigradi delle marche e dell'umbria. Atti della Società Toscana di Scienze Naturali Residente in Pisa Memorie Serie B 101:21–34

Bertolani R, Guidi A, Rebecchi L (1996) Tardigradi della Sardegna e di alcune-piccole isole circumsarde. Biogeographia 18:229–247

Binda MG, Pilato G (1995a) *Echiniscus peruvianus*, nuova specie di eterotardigrado del Sudamerica. Animalia 21:93–96

Binda MG, Pilato G (1995b) Notizie sui tardigradi delle Isole Hawaii con descrizione de due pecie nuove. Animalia 21:57–62

Binda MG, Pilato G (1995c) *Macrobiotus mottai*, nuova specie di eutardigrado dell'Antartide. Animalia 21:53–56

Binda MG, Pilato G (1995d) Ridescrizione di *Echiniscus reticulatus* Murray, 1905 e descrizione di *Echiniscus cirinoi*, nuova specie di tardigrado della Tanzania. Animalia 20:55–58

Binda MG, Pilato G (1995e) Some notes on African tardigrades with a description of two new species. Trop Zool 8:367–372

Binda MG, Pilato G (1995f) Remarks on tardigrades from the Seychelles, with a description of two new species. Trop Zool 8:1–6

Binda MG, Pilato G (1999a) *Dactylobiotus lombardoi* sp. n. (Eutardigrada: Macrobiotidae) from Tierra del Fuego, with a key to the *Dactylobiotus*-species. Zoologischer Anzeiger 238:147–155

Binda MG, Pilato G (1999b) *Macrobiotus erminiae*, new species of eutardigrade from southern Patagonia and Tierra del Fuego. Entomologische Mitteilungen aus dem Zoologischen Museum Hamburg 13:151–158

Binda MG, Pilato G (2000) *Diphascon* (*Adropion*) *tricuspidatum*, a new species of eutardigrade from Antarctica. Polar Biol 23:75–76

Binda MG, Pilato G, Moncada E, Napolitano A (2001) Some tardigrades from Central Africa with the description of two new species: *Macrobiotus ragonesei* and *M. priviterae* (Eutardigrada Macrobiotidae). Trop Zool 14:233–242

Biserov VI (1994) Some tardigrades from the Seychelles with description of three new species. Trop Zool 7:181–189

Biserov VI, (1996a) Tardigrades of the Taimyr peninsula with descriptions of two new species. Zool J Linn Soc 116:215–237

Biserov VI (1996b) *Macrobiotus lorenae* sp. n., a new species of Tardigrada (Eutardigrada Macrobiotidae) from the Russian Far East. Arthropoda Selecta 5:145–149

Biserov VI (1998) The Tardigrada of the Komandorskiye Islands, with a description of *Dactylobiotus dervizi* sp. n. (Eutardigrada, Macrobiotidae). Entomologische Mitteilungen aus dem Zoologischen Museum Hamburg 12:327–336

Biserov VI (1999a) A review of the Tardigrada from Novaya Zemlya, with descriptions of three new species, and an evaluation of the environment in this region. Zoologischer Anzeiger 238:169–182

Biserov VI (1999b) Tardigrada of Turkmenistan, with description of three new species. Zoologischer Anzeiger 238:157–167

Biserov VI, Gerlach J (1998) New records of Seychelles tardigrades. Phelsuma 6:85–87

Biserov VI, Dudichev AL, Biserova NM (2001) Preliminary data on tardigrades of Lake Biwa (Japan). Arthropoda Selecta 10:307–310

Blackburn TM, Gaston KJ (1995) What determines the probability of discovering a species?: a study of South American oscine passerine birds. J Biogeogr 22:7–14

Borcard D, Legendre P, Drapeau P (1992) Partialling out the spatial component of ecological variation. Ecology 73:1045–1055

Cabrero-Sañudo FJ, Lobo JM (2003) Estimating the number of species not yet described and their characteristics: the case of Western Palaearctic dung beetle species (Coleoptera, Scarabaeoidea). Biodivers Conserv 12:147–166

Chang CY, Rho HS (1996) *Isohypsibius granditintinus*, a new psammobenthic tardigrade from Estuary of South Korea. Korean J Syst Zool 12:245–252

Claps MC, Rossi GC (1997) Tardígrados de Uruguay, con descripción de dos nuevas especies (Echiniscidae, Macrobiotidae). Iheringia Serie Zoologia 83:17–22

Claxton SK (1998) A revision of the genus *Minibiotus* (Tardigrada: Macrobiotidae) with descriptions of eleven new species from Australia. Rec Aust Mus 50:125–160

Claxton SK (1999) *Milnesioides exsertum* gen. n. sp. n., a new tardigrade from Australia (Tardigrada: Milnesiidae). Zoologischer Anzeiger 238:183–190

Claxton SK (2001) *Antechiniscus* in Australia: description of *Antechiniscus moscali* sp. n. and redescription of *Antechiniscus parvisentus* (Horning & Schuster, 1983) (Heterotardigrada: Echiniscidae). Zoologischer Anzeiger 240:281–289
Clench H (1979) How to make regional lists of butterflies: some thoughts. J Lepidopteirsts' Soc 33:216–231
Cotterill FPD (1995) Systematics, biological knowledge and environmental conservation. Biodivers Conserv 4:183–205
Crawley MJ (1993) GLIM for ecologists. Blackwell Scientific Publications, Oxford
Crowe JH, Crowe LM (2000) Preservation of mammalian cells learning nature's tricks. Nat Biotechnol 18:145–146
Dastych H (1997) A new species of the genus *Echiniscus* (Tardigrada) from New Zealand. Entomologische Mitteilungen aus dem Zoologischen Museum Hamburg 12:209–215
Dastych H (1999) *Mopsechiniscus frenoti* sp. n., a new water-bear (Tardigrada) from Iles Crozet, the Sub-Antarctic. Entomologische Mitteilungen aus dem Zoologischen Museum Hamburg 13:49–57
Dastych H (2000) A new species of the genus *Mopsechiniscus* DuBois-Reymond Marcus, 1944 (Tardigrada) from the Venezuelan Andes. Acta Biol Benrodis 10:91–101
Dastych H (2002) A new species of the genus *Macrobiotus* Schultze, 1834 from Iles Kerguelen, the sub-Antarctic (Tardigrada). Mitteilungen aus dem Hamburgischen Zoologischen Museum und Institut 99:11–27
Dastych H, Harris JM (1994) A new species of the genus *Hebesuncus* from the Antarctic nunataks, Dronning Maud Land (Tardigrada). Entomologische Mitteilungen aus dem Zoologischen Museum Hamburg 11:139–145
Dastych H, Harris JM (1995) A new species of the genus *Macrobiotus* from inland nunataks in western Dronning Maud Land, continental Antarctica (Tardigrada). Entomologische Mitteilungen aus dem Zoologischen Museum Hamburg 11:175–182
Dastych H, Kristensen RM (1995) *Echiniscus ehrenbergi* sp. n., a new water bear from the Himalayas (Tardigrada). Entomologische Mitteilungen aus dem Zoologischen Museum Hamburg 11:221–230
Dastych H, McInnes S (1995) *Hexapodibius boothi* sp. n., a new species of semi-terrestrial tardigrade from the Maritime Antarctic. Entomologische Mitteilungen aus dem Zoologischen Museum Hamburg 11:111–117
Dastych H, McInnes SJ (1996) A new species of the genus *Diphascon* (Tardigrada) from the Maritime Antarctic. Entomologische Mitteilungen aus dem Zoologischen Museum Hamburg 12:35–41
Degma P (2003) First records of six Hypsibiidae species (Tardigrada, Eutardigrada) in Slovakia. Biologia 58:1003–1005
Degma P, Pecalkova M (2003) First records of Tardigrada in Slovakia from Stuzica National Nature Reserve (Bukovske vrchy Mts, NE Slovakia). Biologia 58:274, 286
Diamond JM (1985) How many unknown species are yet to be discovered? Nature 315:538–539
Dirzo R, Gómez G (1996) Ritmos temporales de la investigación taxonómica de plantas vasculares en México y una estimación del número de especies conocidas. Ann Mo Bot Gard 83:396–403
Dobson A (1999) An introduction to generalized linear models. Chapman and Hall/CRC, London
Dobson CH, Gentry AH (1991) Biological extinction in western Ecuador. Ann Mo Bot Gard 78:273–295
Duditchev AL, Golubev AI, Biserov VI (1998) A new data about fauna of Tardigrada from west Tatarstan. In: Utkin NA (ed) Invertebrates of the Southern Transrurals and adjacent territories: materials of the All-Russian. Izd-vo Kurganskogo Universiteta, Kurgan, pp 129–133
Dudichev AL, Biserov VI (2000) Tardigrada from Iturup and Paramushir Islands (the Kuril Islands). Zoologicheskii Zhurnal 79:771–778
Eroglu A, Russo MJ, Bieganski R, Fowler A, Cheley S, Bayley H, Torner M (2000) Intracellular trehalose improves the survival of cryopreserved mammalian cells. Nat Biotechnol 18:163–167
Erwin TL (1991) How many species are there?: revisited. Conserv Biol 5:330–333
Flather CH (1996) Fitting species-accumulation functions and assessing regional land use impacts on avian diversity. J Biogeogr 23:155–168
Frank JH, Curtis GA (1979) Trend lines and the number of species of Staphylinidae. Coleopt Bull 33:133–168
Galante E, Verdú JR (2000) Los artrópodos de la "Directiva Hábitat" en España. Ministerio de Medio Ambiente, Dirección General de Conservación de la Naturaleza, Madrid

García-Valdecasas A, Camacho AI (2003) Conservation to the rescue of taxonomy. Biodivers Conserv 12:1113–1117

García-Valdecasas A, Castroviejo S, Marcus LF (2000) Reliance on the citation index undermines the study on biodiversity. Nature 403:698

Garey JR, Krotec M, Nelson DR, Brooks J (1996) Molecular analysis supports a tardigrade-arthropod association. Invertebr Biol 115:79–88

Garey JR, Schmidt-Rhaesa A (1998) The essential role of 'minor' phyla in molecular studies of animal evolution. Am Zool 38:907–917

Gaston KJ (1991a) The magnitude of global insect species richness. Conserv Biol 5:283–296

Gaston KJ (1991b) Body size and probability of description: the beetle fauna of Britain. Ecol Entomol 16:505–508

Gaston KJ (1993) Spatial patterns in the description and richness of the Hymenoptera. In: Lasalle J, Gould ID (eds) Hymenoptera biodiversity. C. A. B. International, Wallingford

Gaston KJ (1994) Spatial Patterns of species description: how is our knowledge of the global insect fauna growing? Biol Conserv 67:37–40

Gaston KJ, (1996) Species richness: measure and measurement. In: Gaston KJ (ed) Biodiversity. A Biology of numbers and difference. Blackwell Science, Oxford

Gaston KJ, Blackburn TM (1994) Are newly described bird species small-bodied? Biodivers Lett 2:16–20

Gaston KJ, Blackburn TM, Loder N (1995a) Which species are described first?: the case of North American butterflies. Biodivers Conserv 4:119–127

Gaston KJ, May RM (1992) Taxonomy of taxonomists. Nature 356:281–282

Gaston KJ, Mound LA, (1993) Taxonomy, hypothesis testing and the biodiversity crisis. Proc Roy Soc Lond B 251:139–142

Gaston KJ, Scoble MJ, Crock A, (1995b) Patterns in species description: a case study using the Geometridae (Lepidoptera). Biol J Linn Soc 55:225–237

Giribet G, Carranza S, Baguna J, Riutort M, Ribera C (1996) First molecular evidence for the existente of a Tardigrada + Arthropoda clade. Mol Biol Evol 13:76–84

Giribet G, Ribera C (1998) The position of arthropods in the animal kingdom: a search for a reliable outgroup for internal arthropod phylogeny. Mol Phylogenet Evol 9:481–488

Godfray HCJ (2002) Challenges for taxonomy. Nature 417:17–19

Goeze JAE (1773) Uber den kleinen wasserbär. In: Bonnet, H. K. (ed), Abhanlungn aus der Insectologie. Ubers. Usw, 2. Beobachtg. Pps. 67

Groombridge B, Jenkins MD (2002) World atlas of biodiversity. Earth's living resources in the 21st century. UNEP World Conservation Monitoring Centre. University of California Press, Berkeley

Guidetti R (1998) Two new species of Macrobiotidae (Tardigrada: Eutardigrada) from the United States of America, and some taxonomic considerations of the genus Murrayon. Proc Biol Soc Washington 111:663–673

Guidetti R, Bertolani R (2005) Tardigrade taxonomy: an update check list of the taxa and a list of characters for their identification. Zootaxa 845:1–46

Guidi A, Grabowski B (1996) A new species of *Isohypsibius* (Tardigrada, Eutardigrada, Hypsibiidae) from Germany. Acta Biol Benrodis 8:145–148

Guisan A, Zimmermann NE (2000) Predictive habitat distributions models in ecology. Ecol Modell 135:147–186

Guo N, Puhlev I, Brown DR, Mansbridge J, Levine F, (2000) Trehalose expression confers desiccation tolerance on human cells. Nat Biotechnol 18:168–171

Hammond P (1992) Species inventory. In: Groombridge B (ed) Global biodiversity. World Conservation Monitoring Centre. Chapman & Hall, London

Hammond P (1995) The current magnitude of biodiversity. In: Heywood VH, Watson RT (eds) Global biodiversity assessment. United Nations Environment Program. Cambridge University Press, Cambridge, pp 113–138

Heikkinen RK, Neuvonen S (1997) Species richness of vascular plants in the subarctic landscape of northern Finland: modelling relationships to the environment. Biodivers Conserv 6:1181–1201

Iharos G, Vargha B (1995) Tardigrades from the Zselic Downs, south Hungary. Janus Pannonius Muzeum Evkonyve 39:33–38

Ito M (1995) Taxonomic study on the Eutardigrada from the northern slope of Mt. Fuji, central Japan, 2. Family Hypsibiidae. Proc Jpn Soc Syst Zool 53:18–39

Ito M (1997) Taxonomic study on the Eutardigrada from northern slope of Mt. Fuji, central Japan. III. Families Macrobiotidae and Milnesiidae. Species Divers 2:167–178

Kaczmarek L, Michalczyk L (2002) *Echiniscus barbarae*, a new species of tardigrade from Cuba Island (Tardigrada: Heterotardigrada, Echiniscidae, 'arctomys group'). Zootaxa 53:1–4

Kaczmarek L, Michalczyk L (2003) *Platicrista horribilis*, a new species of Tardigrada (Eutardigrada: Hypsibiidae) from Mongolia. Zootaxa 132:1–8

Kaczmarek L, Michalczyk L (2004a) Notes on some tardigrades from South Africa, with the description of *Diphascon* (*Diphascon*) *zaniewi* sp. nov. (Eutardigrada: Hypsibiidae). Zootaxa 576:1–6

Kaczmarek L, Michalczyk L (2004b) New records of Tardigrada from Cyprus with a description of the new species *Macrobiotus marlenae* (*hufelandi* group) (Eutardigrada: Macrobiotidae). Genus 15:141–152

Kaczmarek L, Michalczyk L, Weglarska B (2002) *Itaquascon mongolicus*, a new species of Eutardigrada from Mongolia (Eutardigrada: Hypsibiidae). Genus 13:1–3

Kendall-Fite K, Nelson DR (1996) Two new species of tardigrades from Short Mountain, Tennessee, U.S.A. Zool J Linn Soc 116:205–214

Krell FT (2000) Impact factors aren't relevant to taxonomy. Nature 405:507–508

Legendre P, Legendre L (1998) Numerical ecology. 2nd edn. Elsevier, Amsterdam

Maas A, Waloszek D, (2001) Cambrian derivates of the early Arthropod stem lineage, Pentastomids, Tardigrades and Lobopodians- An 'Orsten' perspective. Zoologische Anzeiger 240:451–459

Mackay R (2002) The Penguin atlas of endangered species. A worldwide guide to plants and animals. Penguin Books Ltd. Harmondsworth, Middlesex

Macnally R (2000) Regression and model-building in conservation biology, biogeography and ecology: The distinction between – and reconciliation of – 'predictive' and 'explanatory' models. Biodivers Conserv 9:655–671

Macnally R (2002) Multiple regression and inference in ecology and conservation biology: further comments on identifying important predictor variables. Biodivers Conserv 11:1397–1401

Mallet J, Willmott K (2003) Taxonomy: renaissance or Tower of Babel. Trends Ecol Evol 18:57–59

Marcus, E., 1929. Tardigrada: In: Bronn's Klassen und Ordnungen des Tierrichs. Leipzig, 5. 608 pp

Margules CR, Nicholls AO, Austin MP (1987) Diversity of Eucalyptus species predicted by a multivariable environment gradient. Oecologia 71:229–232

Margules CR, Pressey RL (2000) Systematic conservation planning. Nature 405:243–253

Maucci W (1996) Tardigrada of the Arctic tundra with descriptions of two new species. Zool J Linn Soc 116:185–204

May RM (1986) How many species are there? Nature 324:514–515

May RM (1988) How many species are there on Earth? Science 241:1441–1449

May RM (1989) How many species are there? In: Friday L, Laskey R (eds) The fragile environments. Cambridge University Press, Cambridge, EEUU, pp 61–81

May RM (1990) Taxonomy as destiny. Nature 347:129–130

McCullagh P, Nelder JA (1989) Generalized Linear Models. 2nd edn. Chapman and Hall, London

McInnes SJ (1994) Zoogeographic distribution of terrestrial/freshwater tardigrades from current literature. J Nat Hist 28:257–352

McInnes SJ (1995) Tardigrades from Signy Island, South Orkney Islands, with particular reference to freshwater species. J Nat Hist 29:1219–1445

Medellín RA, Soberón J (1999) Predictions of mammal diversity on four land masses. Conserv Biol 13:143–149

Michalczyk L, Kaczmarek L (2003a) A description of the new tardigrade *Macrobiotus reinhardti* (Eutardigrada: Macrobiotidae, harmsworthi group) with some remarks on the oral cavity armature within the genus Macrobiotus Schultze. Zootaxa 331:1–24

Michalczyk L, Kaczmarek L (2003b) Water bears (Tardigrada) of the Bieszczady Mountains (Poland, west Carpathians). Zootaxa 242:1–24

Michalczyk L, Kaczmarek L (2003c) *Minibiotus constellatus*, a new species of Tardigrada from Peru (Eutardigrada: Macrobiotidae). Genus 14:295–305

Miller WR, Horning DS, Dastych H (1995) Tardigrades of the Australian Antarctic: description of two new species from Macquarie Island, Subantarctica. Entomologische Mitteilungen aus dem Zoologischen Museum Hamburg 11:231–239

Moon SY, Won K (1994) New species of *Echiniscus* (Heterotardigrada: Echiniscoidea: Echiniscidae) from Korea. Proc Biol Soc Washington 107:511–513

Moon SY, Won K, Bertolani R (1994) *Doryphoribius koreanus*, a new species of Tardigrada from Korea. Proc Biol Soc Washington 107:514–516

Moon SY, Kim W (1996) Phylogenetic position of yhe Tardigrada based on the 18S ribosomal RNA gene sequence. Zool J Linn Soc 116:61–62

Myers N, Knoll AH (2001) The Biotic Crisis and the Future of Evolution. Proc Nat Acad Sci USA 98:5389–5392
Nelson DR, McGlothlin KL (1996) A new species of *Calohypsibius* (*phylum* Tardigrada, Eutardigrada) from Roan Mountain, Tennessee-North Carolina, U.S.A. Zool J Linn Soc 116:167–174
Nelson DR, Marley NJ, Bertolani R (1999) Re-description of the genus *Pseudobiotus* (Eutardigrada, Hypsibiidae) and of the new type species *Pseudobiotus kathmanae* sp. n. Zoologischer Anzeiger 238:311–317
Nicholls AO (1989) How to make biological surveys go further with generalised linear models. Biol Conserv 50:51–75
Nielsen C (1995) Animal evolution. Interrelationship of the living phyla. Oxford University Press, Oxford
Novacek MJ, Cleland EE (2001) The current biodiversity extinction event: scenarios for mitigation and recovery. Proc Nat Acad Sci USA 98:5466–5470
O'Brien CW, Wibmer GJ (1979) The use of trend curves of rates of species descriptions: examples from the Curculionidae (Coleoptera). Coleopt Bull 33:151–166
Patterson BD (1994) Accumulating knowledge on the dimensions of biodiversity: systematic perspectives on Neotropical mammals. Biodivers Lett 2:79–86
Patterson BD (2000) Patterns and trends in the discovery of new Neotropical mammals. Divers Distrib 6:145–151
Peterson KJ, Eernisse DJ (2001) Animal phylogeny and the ancestry of bilaterians: inferences from morphology and 18S rDNA gene sequences. Evol Develop 3:170–205
Pilato G (1996) Redescription of *Isohypsibius wilsoni* and description of two new species of *Isohypsibius* (Eutardigrada) from New Zealand. NZ J Zool 23:67–71
Pilato G, Binda MG (1994) *Dactylobiotus caldarellai*, nuova specie di eutardigradi della Terra del Fuoco. Animalia 21:87–91
Pilato G, Binda MG (1996a) Two new species and new records of *Macrobiotus* (Eutardigarda) from New Zealand. NZ J Zool 23:375–379
Pilato G, Binda MG (1996b) *Mixibius fueginus*, nuova specie di eutardigrado della Terra del Fuoco. Bollettino dell'Accademia Gioenia di Scienze Naturali 29:27–32
Pilato G, Binda MG (1997) Remarks on some tardigrades of New Zealand with the description of *Hypsibius novaezeelandiae* n. sp. (Eutardigrada). NZ J Zool 24:65–67
Pilato G, Binda MG (1998a) A comparison of *Diphascon (D.) alpinum* Murray, 1906, *D. (D.) chilenense* Plate, 1889 and *D. (D.) pingue* Marcus, 1936 (Tardigrada), and description of a new species. Zoologischer Anzeiger 236:181–185
Pilato G, Binda MG (1998b) Two new species of *Diphascon*(Eutardigrada) from New South Wales, Australia. NZ J Zool 25:171–174
Pilato G, Binda MG (1999) Three new species of *Diphascon* of the *pingue* group (Eutardigrada, Hypsibiidae) from Antarctica. Polar Biol 21:335–342
Pilato G, Binda MG (2003) *Hexapodibius christenberryae*, a new species of tardigrade from North America (Eutardigrada, Calohypsibiidae). Zootaxa 140:1–6
Pilato G, Lisi O (2003) *Echiniscus walteri*, new species of tardigrade from Madagascar. Bollettino del Museo Civico di Storia Naturale di Verona Botanica Zoología 27:65–70
Pilato G, Lisi O (2004) *Doryphoribius neglectus* sp. n. and *Parascon nichollsae* sp. n., new species of eutardigrades from Australia. Zootaxa 545:1–7
Pilato G, Patanè M (1997) *Macrobiotus ovostriatus*, a new species of eutardigrade from Tierra del Fuego. Bollettino dell'Accademia Gioenia di Scienze Naturali 30:263–268
Pilato G, Binda MG, Qualtieri F (1998) *Diphascon* (*Diphascon*) *mitrense*, new species of eutardigrade from Tierra del Fuego. Bollettino dell'Accademia Gioenia di Scienze Naturali 31:101–105
Pilato G, Binda MG, Moncada E (1999) *Itaquascon biserovi*, new species of eutardigrade from central Africa. Bollettino del Museo Civico di Storia Naturale di Verona 32:171–176
Pilato G, Binda MG, Napolitano A, Moncada E, (2001) Notes on South American tardigrades with the description of two new species: Pseudechiniscus spinerectus and Macrobiotus danielae. Trop Zool 14:223–231
Pilato G, Binda MG, Napolitano A, Moncada E (2002a) Tardigrades from Ecuador, with the description of two new species: *Mixibius ornatus* n. sp. and *Diphascon* (*Adropion*) *onorei* n. sp. (Eutardigrada, Hypsibiidae). Stud Neotrop Environ 37:175–179
Pilato G, Binda MG, Claxton S (2002b) *Itaquascon unguiculum* and *Itaquascon cambewarrense*: two new species of eutardigrades from Australia. NZ J Zool 29:87–93
Pilato G, Kaczmarek L, Michalczyk L, Lisi O (2003a) *Macrobiotus polonicus*, a new species of Tardigrada from Poland (Eutardigrada: Macrobiotidae, '*hufelandi* group'). Zootaxa 258:1–8

Pilato G, Binda G, Lisi O (2003b) Remarks on some species of tardigrades from South America with the description of *Minibiotus sidereus* n. sp. Zootaxa 195:1–8
Pilato G, Binda MG, Lisi O (2003c) Notes on some tardigrades from Central Africa, with the description of a new species of Hypsibiidae. Zootaxa 241:1–7
Pilato G, Binda MG, Napolitano A, Moncada E (2004) Remarks on some species of tardigrades from South America with the description of two new species. J Nat Hist 38:1081–1086
Pin Koh L, Dunn RR, Sodhi NS, Colwell RK, Proctor HC, Smith VS (2004) Species coextinctions and the biodiversity crisis. Science 305:1632–1634
Purvis A, Agapow PM, Gittleman JL, Mace GM (2000) Nonrandom extinction and the loss of evolutionary history. Science 288:328–330
Qinghong L, Bråkenhielm S (1995) A statistical approach to decompose ecological variation. Water Air Soil Pollut 85:1587–1592
Ramazzotti G (1962) Il phylum Tardigrada. Memorie dell'Istituto Italiano di Idrobiologia. 16:1–595
Ramazzotti G, Maucci W (1982) A history of tardigrade taxonomy. Proceedings of the III international symposium on the Tardigrada. East Tennessee State University Press, Tennessee, pp 11–30
Ramazzotti G, Maucci W (1983) Il phylum Tardigrada. III edizione riveduta e aggiornata. Memorie dell'Istituto Italiano di Idrobiologia Dott. Marco de Marchi 41:1–1012
Ratkowski DA (1990). Handbook of nonlinear regression models. Marcel Dekker Press, New York
Raven PH (1987) We're killing our world: the global ecosystem in crisis. McArthur Found, Occas. Pap
Rho HS, Moon SY, Chang CY, Kim W (1997) Four freshwater eutardigrades from Korea with description of *Isohypsibius brevitubulatus* n. sp. Korean. J Biol Sci 1:271–279
Ricciardi A, Rasmussen JB (1999) Extinction rates of North American freshwater fauna. Conserv Biol 13:1220–1222
Rosas G, Ramos MA, García-Valdecasas A (1992) Invertebrados españoles protegidos por Convenios Internacionales. Editorial Madrid, ICONA, Madrid
Schuster R (2000) A new species of the genus *Diphascon* (Tardigrada) from Germany. Acta Biol Benrodis 10:21–25
Sekercioglu CH, Daily GC, Ehrlich PR (2004) Ecosystem consequences of bird declines. Proc Nat Acad Sci USA 101:18042–18047
Semeria Y (1994) Une espece nouvelle de tardigrade de Taiwan: *Echiniscus pseudelegans*, n. sp. (Heterotardigrada Echiniscidae). Bulletin Mensuel de la Societe Linneenne de Lyon 63:28–30
Semeria Y (2003) Une espece nouvelle de tardigrade pour l'Ile de La Reunion: *Cornechiniscus madagascariensis* Maucci. Bulletin Mensuel de la Societe Linneenne de Lyon 72:233–234
Soberón J, Llorente J (1993) The use of species accumulation functions for the prediction of species richness. Conserv Biol 7:480–488
StatSoft, Inc., 2001. Statistica (data analysis software system), version 6.0. http://www.statsoft.com
Steyskal GC (1965) Trend curves of the rate of species description in zoology. Science 149:880–882
Strayer D, Nelson DR, O'Donnell EB (1994) Tardigrades from shallow groundwaters in southeastern New York, with the first record of *Thulinia* from North America. Trans Am Microsc Soc 113:325–332
Tumanov DV (1997a) *Hypsibius iskandarovi* sp. n., a new species of Tardigrada from fresh waters of north-west Russia (Tardigrada: Hypsibiidae). Zoosystematica Rossica 5:219–220
Tumanov DV (1997b) Tardigrades in freshwater benthic communities in northwestern Russia. Trudy Zoologicheskogo Instituta 272:330–333
Tumanov DV (2003a) Four new *Isohypsibius* species from Russian fresh waters (Tardigrada, Hypsibiidae). Bulletin de l'Institut Royal des Sciences Naturelles de Belgique Biologie 73:183–189
Tumanov DV (2003b) *Isohypsibius borkini*, a new species of Tardigrada from Tien Shan (Kirghizia) (Eutardigrada: Hypsibiidae). Genus 14:439–441
Tumanov DV (2004) *Macrobiotus kovalevi*, a new species of Tardigrada from New Zealand (Eutadigrada, Macrobiotidae). Zootaxa 406:1–8
Van-Rompu EA, De-Smet WH (1996) Freshwater tardigrades from Hopen, Svalbard 76[degree]31[minute]N). Fauna Norvegica Ser A 17:1–9
Van-Rompu EA, De-Smet WH (1998) Records of Tardigrada new for the Belgian fauna. Natuurwetenschappelijk Tijdschrift 77:12–19
Van-Rompu EA, De-Smet WH, Coomans A (1995) Some terrestrial tardigrades from Zimbabwe. Biologisch Jaarboek 62:48–55

Vargha B (1995) Three new tardigrade species from Hungary. Acta Zoologica Academiae Scientiarum Hungaricae 41:301–313

Vargha B (1998) Adatok a Duna-Drava Nemzeti Park medveallatka (Tardigrada) faunajahoz. Dunantuli Dolgozatok (A) Termeszettudomanyi Sorozat 9:73–80

Vargha B (2000) Adatok a Villanyi-hegyseg medveallatka (Tardigrada) faunajahoz. Data to the water bear (Tardigrada) fauna of the Villany Hills, Hungary. Dunantuli Dolgozatok (A) Termeszettudomanyi Sorozat 10:121–125

Vasquez JH, Vargas MV (2003) Tardigrades (Tardigrada) from the western part of the central valley, Costa Rica with some ecological annotations. Brenesia 59/60:69–76

White RE (1975) Trend curves of the rate of species description for certain North American Coleoptera. Coleopt Bull 29:281–295

White RE (1979) Response to the use of trend curves by Erwin, Frank and Curtis, and O'Brien and Wibmer. Coleopt Bull 33:167–168

Wilson EO (1988) The current state of biological diversity. In: Wilson EO (ed) Biodiversity, National Academy Press, Washington, D. C

Winnepenninckx B, Backeljau T, Mackley LY, Brook JM, de Wachter R, Kumar S, Garey JR (1995) 18 rRNA data indicate that Aschelminthes are polyphyletic in origin and consist of at least three distinct clades. Mol Biol Evol 12:1132–1137

Woodruff DS (2001) Declines of biomes and biotas and the future of evolution. Proc Nat Acad Sci USA 98:5471–5476

Yang T (1999) Three new species and six new records of the class Eutardigrada (Tardigrada) from China (Parachela: Macrobiotidae: Hypsibiidae). Acta Zootaxonomica Sinica 24:444–453

Yang T (2002a) Three new species and one new record of the Tardigrada from China. Acta Hydrobiologica Sinica 26:505–508

Yang T (2002b) The tardigrades from some mosses of Lijiang County in Yunnan Province (Heterotardigrada: Echiniscidae; Eutardigrada: Parachela: Macrobiotidae, Hypsibiidae). Acta Zootaxonomica Sinica 27:53–64

Yang T (2003) Two new species and three new records of the Tardigrada (Heterotardigrada, Echiniscidae; Eutardigrada, Milnesiidae, Macrobiotidae, Hypsibiidae). Acta Zootaxonomica Sinica 28:235–240

Zrzavy J, Hypso V, Vlaskova M (1998) Arthropod phylogeny: taxonomic congruence, total evidence and conditional combination approaches to morphological and molecular data sets. In: Fortey RA, Thomas RH (eds) Arthopod relationships. Chapman and Hall, London, pp 97–107

Biodivers Conserv (2007) 16:1087–1102
DOI 10.1007/s10531-006-9070-5

ORIGINAL PAPER

Environmental correlates for species richness among amphibians and reptiles in a climate transition area

Claudia Soares · José Carlos Brito

Received: 16 June 2005 / Accepted: 19 May 2006 / Published online: 12 July 2006

Abstract Correlations between environmental factors and the distribution of amphibian and reptile species richness were investigated in a climate transition area, Peneda-Gerês National Park (PNPG), in North-Western Portugal. Using presence-data at a local-scale (1 × 1 km), Ecological-Niche Factor Analysis (ENFA) identified a mixture of climatic (precipitation and number of days with fog), topographical (altitude and relief) and habitat factors (number of water-courses and water surfaces, the type of the largest water surface and tree diversity cover), as accurate predictors of species occurrence. Three factors were common for both taxonomic groups, and consistently presented a positive relation with species occurrence: precipitation, number of water surfaces, and tree diversity cover; suggesting a strong coincidence in the environmental correlates that influence amphibian and reptile species richness. Distribution patterns of observed and predicted species richness were compared using a Geographical Information System. Overall, three high species richness areas were predicted in common for both taxonomic groups and two additional areas for amphibians only. These areas matched with the observed species richness but suggested larger areas of high species richness. The location of the PNPG in a biogeographic crossroad, between Euro-Siberian and Mediterranean provinces, emphasised species richness of amphibians and reptiles and suggests a high priority conservation status for this protected area. Most of Central-Northern Portugal is located in a climatic transition area; therefore, increased species richness should be expected for other areas. Local scale studies for other protected areas should be

C. Soares (✉) · J. C. Brito
CIBIO, Centro de Investigação em Biodiversidade e Recursos Genéticos da Universidade do Porto, Campus Agrário de Vairão, Rua Padre Armando Quintas, 4485-661 Vairão, Portugal
e-mail: csoares@mail.icav.up.pt

C. Soares
Departamento de Zoologia e Antropologia, Faculdade de Ciências da Universidade do Porto, Praça Gomes Teixeira, 4050 Porto, Portugal

planned as a framework for the development of multi-scale conservation planning by Portuguese authorities.

Keywords Amphibians · Climate transition area · Ecological-Niche Factor Analysis (ENFA) · Environmental factors · Geographical Information Systems · Habitat suitability models · Peneda-Gerês National Park · Portugal · Reptiles · Species richness

Abbreviations

a.s.l.	above sea level
EGV	Ecogeographical variables
ENFA	Ecological-Niche Factor Analysis
PNPG	Peneda-Gerês National Park
UTM	Universal Transverse of Mercator

Introduction

One of the main goals of conservation nowadays is the maintenance of biological diversity (Myers et al. 2000). The socio-economic hardship and the increase of human population contribute for eroding biological diversity very rapidly in relation to historical extinction rates (Wilson 1992). For instance, one of the most important biodiversity hotspots in the world, the Mediterranean Basin, exhibits a high percentage of endemic species of several taxonomic groups, but is threatened by habitat loss and by the collapse of the traditional agro-sylvo-pastoral system (Mittermeier et al. 1998; Médail and Quézel 1999). This scenario is worrying from a conservation viewpoint. Therefore, the identification and conservation of specific important areas is urgently needed to reduce the loss of biological diversity at different levels.

Large biogeographic intersections, or biogeographic crossroads, constitute potentially attractive areas for long-term conservation of biodiversity, as they allow conserving evolutionary processes such as speciation and coevolution (Spector 2002). Most importantly, they constitute regions of rapid turnover (or high beta diversity) of species and habitats, leading to exceptionally high levels of species richness and creating the opportunity to meet goals of representativeness and complementarity in protected-areas systems (Araújo et al. 2002). In the north-west of the Iberian Peninsula there is a marked transition between the Euro-Siberian and Mediterranean biogeographic provinces (Rivas-Martínez 1987). Correspondingly, there is a clear transition between the Atlantic Mediterranean and the Continental Mediterranean climates in the series of mountains in the northern border area between Portugal and Spain (Goday 1953). The consequences of this climate transition on the geographical distribution of plants, invertebrates and vertebrates are remarkable, as they allow the co-existence of typical species from southern and northern Iberian Peninsula in a relatively small area, resulting in increased alpha, beta and gamma species richness (Serra and Carvalho 1989; Pimenta and Santarém 1996; Maravalhas 2003; Soares et al. 2005). For instance, the amphibian and reptile communities of North-Western Portugal are noticeably rich, leading to the international recognition of the area as important for the conservation of the herpetofauna (Malkmus 2004; Mateo 2006).

Biogeographical studies on species richness can be considered at various scales. However, they are usually focused on broad scales, using grid sizes of 10 × 10 km or larger (Araújo 1999; Kier and Barthlott 2001; Rensburg et al. 2002). These scales are usually used for species distribution atlas, to relate species distribution areas with environmental factors, or to analyse biological interactions at regional (Inger and Voris 2001; Luoto et al. 2002), continental (Cumming 2000; Franklin et al. 2000; Kier and Barthlott 2001; Rahbek and Graves 2001) and intercontinental levels (Schall and Pianka 1978). However, relatively little attention has been given to species richness patterns and the underlying environmental factors at smaller scales, namely with grid sizes ranging from 500 × 500 m to 2 × 2 km. These scales are more suitable for studying species richness in some taxonomic groups, such as beetles, passerines, or small mammals (Okland et al. 1996; Rensburg et al. 2002; Gallego et al. 2004), and plants (Heikkinen and Neuvonen 1997; Bullock et al. 2000). Additionally, local scales are far more useful for land-use planning and management decisions (Grand et al. 2004).

For conservation purposes, the study of the distribution patterns of species richness in diverse habitats or regions and their ecological determinants are vital steps for understanding the processes that affect the spatial distribution of biological diversity as well as to predict the response of ecosystems to global changes (Stoms and Estes 1993; Peterson et al. 2002). The recent advances in ecological modelling techniques and their combination with Geographical Information Systems have allowed the development of more robust and reliable models, relating biological diversity and environmental factors (Jones et al. 1997). Thus, nowadays, they are fundamental tools for the establishment of conservation strategies and evaluation of management options.

This paper aims to evaluate the effects of climate transition on the distribution of species richness at a local-scale. The Peneda-Gerês National Park, in the North-Western fringe of Portugal, is a privileged area for studying this subject since it is located in an area of transition between two biogeographic provinces. This transition influences particularly the distribution of amphibians and reptiles, since these taxonomic groups are strongly dependent on environmental parameters, and usually exhibit strong associations with climatic factors (Brito et al. 1999; Teixeira et al. 2001). Additionally, the small dispersal capability and relatively small home range size of amphibians and reptiles, in relation with birds and large mammals, emphasises their usefulness for correlating environmental factors with species richness distribution at a local scale. The main objectives of this study were to identify correlations between environmental factors and the distribution of amphibian and reptile species richness, and predict species richness within the study area for these taxonomic groups.

Methods

Study area

The Peneda-Gerês National Park (PNPG) is included in the "Natura 2000" network of European priority conservation areas. It covers an area of about 72,000 ha (latitudes 41° 36′ to 42° 07′ N and longitudes 7° 44′ to 8° 27′ W) along the border with Spain (Fig. 1). The PNPG consists of a series of complex mountains, Peneda, Soajo,

Amarela and Gerês, delimited by deep river valleys, such as Lima, Homem, Gerês, Cávado, and Beredo. The altitude ranges from 50 to 1,500 m a.s.l.. The predominantly Atlantic Mediterranean climate is characterised by high levels of precipitation and drainage, with an average annual rainfall and soil drainage above 2,800 mm/year and 2,000 mm/year, respectively (C.N.A. 1983). In the Continental Mediterranean enclaves, located mostly on the south facing slopes of the mountains and in the low altitude river valleys, average annual rainfall and soil drainage drops to 1,600 mm/year and 1,100 mm/year, respectively (C.N.A. 1983). Additionally, the diversified orientation of the relief and altitude variations provide a wide variety of microclimates. Correspondingly, deciduous oak-forests (*Quercus robur* and *Q. pyrenaica*) and mixed deciduous and coniferous forests occupy most of the Atlantic Mediterranean areas whereas ever-green oak forests (*Q. suber*) occupy the Continental Mediterranean climatic-influenced areas. Pastures, with small bushes and low arboreous cover occur in the high altitude plateaus (>1000 m), located in the extreme North-Western (Castro Laboreiro) and North-Eastern (Mourela) regions of the study area (Fig. 1).

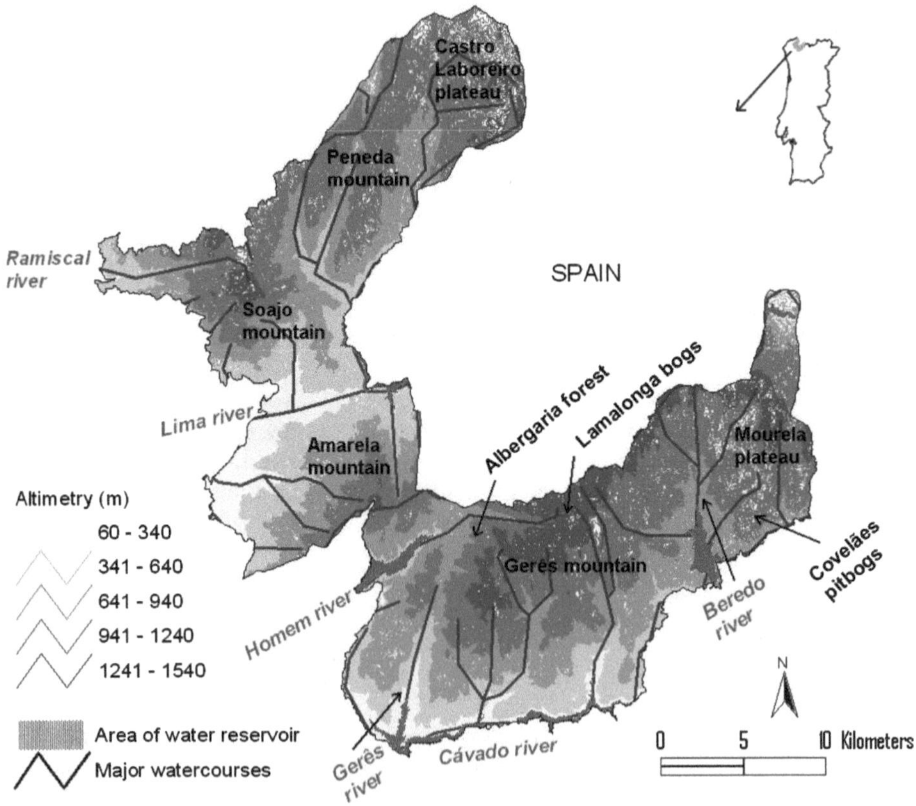

Fig. 1 Study area, Peneda-Gerês National Park, and its location in Portugal

Fieldwork and data collection

To determine the distribution of species, fieldwork was carried out from 1998 to 2002 for reptiles and from 2001 to 2003 for amphibians. The collection of data involved 360 man/days fieldwork for amphibians and 420 man/days fieldwork for reptiles, approximately. Amphibian breeding sites were surveyed using dip-netting and egg searching on water vegetation and margins, complemented with night searching for calls. Reptiles were searched using visual encounter surveys. Both taxonomic groups were also searched for in potential shelters, such as rocks and fallen logs. Published data (Malkmus 1986a, 1986b, 2004) and ad-hoc observations (road-kills and live specimens) collected by the authors and National Park staff were also recorded.

The geographic location of the amphibians and reptiles was inscribed in a georeferenced database, using the UTM 1 × 1 km squares as reference grid (n = 814 squares) (Fig. 2). A total of 490 UTM squares were sampled, representing 60.2% of the study area. The sampling was designed in order to cover the diversity of habitats, climates, and topographical conditions available in the study area. A total of 13 amphibian and 20 reptile species were identified in the study area (Soares et al. 2005) (Table 1). In this study, species density, i.e. the number of species per 1 × 1 km UTM square, was used as a measure of species richness.

Information on 18 ecogeographical variables (hereafter EGV) describing each 1 × 1 km UTM square of the study area were gathered for the analyses (Table 2). Four of the EGVs (ROCK, AGRI, WPAS, SHRU) were collected during the fieldwork, and the remaining 14 EGVs were derived from governmental data bases and military cartography.

Data analyses

Geographical Information Systems and ecological modelling techniques are increasingly being used to model wildlife distributions, to identify suitable habitats and to predict species potential distribution (e.g. Austin et al. 1996; Corsi et al. 1999; Teixeira et al. 2001; Nally and Fleishman 2003). The majority of these methods are based on presence-absence species' data sets, and they make the intuitive assumption that the presence of a species is an indicator of suitable habitat and its absence an indicator of unsuitable habitat (Hirzel et al. 2004a). In this study, reliable absence data was not available for certain species. For instance, some species were scarce (e.g. *Hyla arborea*), others had secretive habits (e.g. the fossorial *Blanus cinereus*), and others had relatively short activity periods (e.g. *Vipera seoanei*) which affected their detectability. Therefore, Ecological-Niche Factor Analysis (hereafter ENFA), a modelling technique that only employs presence data, (Hirzel et al. 2004b), was used to identify the correlations between EGVs and distribution patterns of species richness and determine habitat suitability areas. ENFA has been used to determine habitat suitability areas for several taxonomic groups, such as birds (Brotons et al. 2004), mammals (Dettki et al. 2003), insects (Gallego et al. 2004) and plants (Zaniewski et al. 2002).

The ENFA principle is to compare the distributions of the EGVs between the presence data set and the whole study area. The ENFA summarises several EGVs into a few uncorrelated factors retaining most of the information. Therefore, this type of analysis quantifies the niche occupied by a species comparing its distribution in the ecological space ("the species distribution") with the distribution of all squares ("the global distribution") (Hirzel et al. 2002).

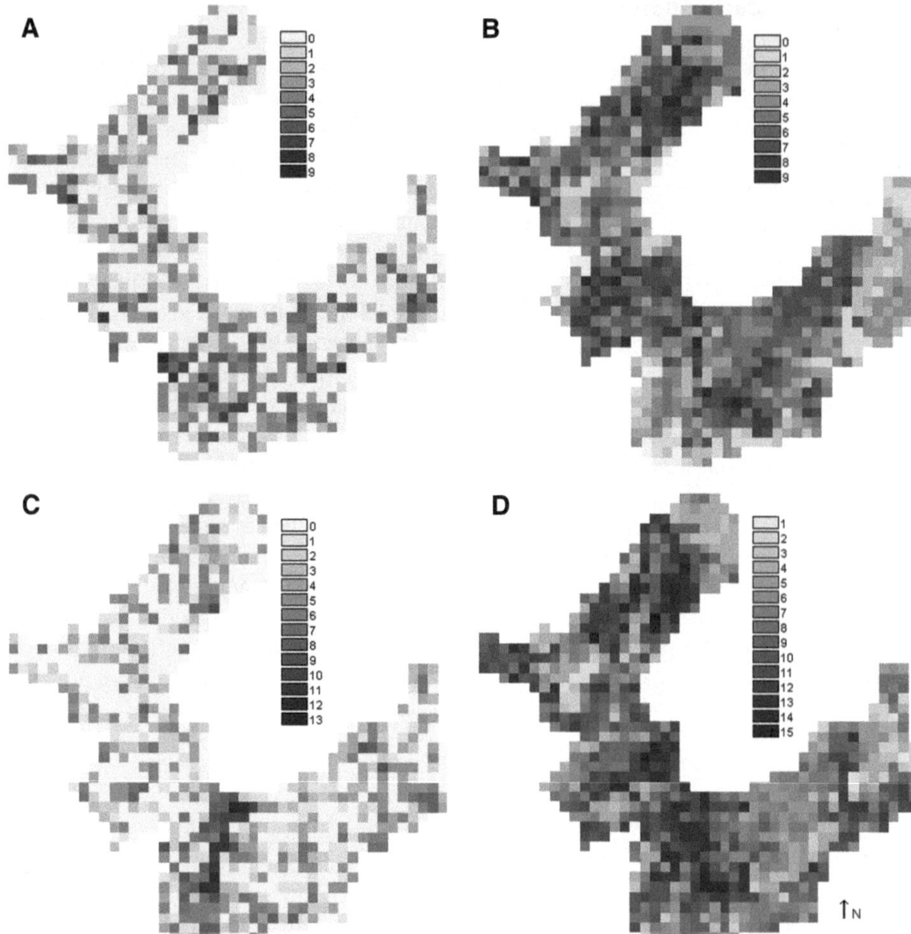

Fig. 2 Patterns of species richness distribution (number of species per 1 × 1 km UTM square represented by grey scale) in the Peneda-Gerês National Park: observed richness for amphibians (A) and reptiles (C) and predicted richness for amphibians (B) and reptiles (D). The value 0 corresponds to squares not surveyed. Predicted species richness was computed by Ecological-Niche Factor Analysis (ENFA). The rare and introduced amphibians (n = 4) and reptiles (n = 5) which were not analysed by ENFA are absent from the observed species richness maps to allow a direct comparison between maps (see Methods section for details)

The outputs of the ENFA include factor scores and eigenvalues. The first factor, called Marginality, is defined as the standardised difference between the species mean and the global mean on all descriptors, i.e., it describes the direction in which the species niche differs at most from the available conditions in the study area (Hirzel et al. 2002). The coefficients of the scores matrix related to the marginality factor indicate the correlation between each EGV and the factor. The marginality coefficients range from –1.0 to +1.0 and positive values mean that the species "prefers" the high values of this EGV, while negative values mean that species "prefers" the low values (Hirzel et al. 2002).

Table 1 Taxonomic list of amphibians and reptiles occurring in the Peneda-Gerês National Park. NUTM - Number of 1 × 1 km UTM squares with observed presence

Amphibian species	NUTM	Reptile species	NUTM
*Chioglossa lusitanica**	106	*Emys orbicularis***	1
Salamandra salamandra	217	*Mauremys leprosa***	3
*Triturus boscai**	212	*Blanus cinereus**	1
Triturus helveticus	10	*Anguis fragilis*	70
Triturus marmoratus	136	*Chalcides bedriagai**	2
Alytes obstetricans	173	*Chalcides striatus*	75
*Discoglossus galganoi**	6	*Tarentola mauritanica***	3
Pelobates cultripes	4	*Lacerta lepida*	289
Hyla arborea	6	*Lacerta schreiberi**	354
Bufo bufo	183	*Podarcis bocagei**	307
Bufo calamita	71	*Podarcis hispanica*	162
*Rana iberica**	272	*Psammodromus algirus*	150
Rana perezi	124	*Elaphe scalaris*	29
		Coronella austriaca	54
		Coronella girondica	65
		Malpolon monspessulanus	86
		Natrix maura	121
		Natrix natrix	147
		Vipera latastei	85
		*Vipera seoanei**	35

*Endemic species to the Iberian Peninsula. **Introduced species in the study area and excluded from the analyses (Soares et al. 2005).

Table 2 Ecogeographical variables (EGVs) used to characterize the 814 1 × 1 km UTM squares of the Peneda-Gerês National Park and to derive habitat suitability models

Category	Variable	Description (units)	Code
Topographical	Altitude	Average altitude a.s.l.(m)	ALTI
	Relief	Number of 50 m altitude isolines that intersect the square	RELI
	Orientation	0; 1–S, SW, SE; 2–E; 3–W; 4–N, NE, NW	ORIE
Climatic	Precipitation	Average annual total precipitation (mm/year)	PREC
	Insolation	Average annual insolation (h/year)	INSO
	Fog	Average annual number fog days (days/year)	FOG
Aquatic habitat	Lotic systems	Number of watercourses	NWAT
	Largest watercourse	Width: 1-≤1.5 m; 2- [1.5–4 m]; 3- [4–8 m]; 4-≥8 m	LWAT
	Lentic systems	Number of water surfaces	NWSU
	Largest water surface	1- quarry; 2- spring or tank; 3- wells; 4-pond; 5- water dam	LWSU
Terrestrial habitat	Soil type*	1- sand or clay; 2-xist; 3- granite	SOIL
	Stone wall*	Available amount of stone walls (m)	STON
	Rock outcrops	Area occupied by rock outcrops (%)	ROCK
	Agriculture areas*	Area occupied by agriculture fields (%)	AGRI
	Wet pasture*	Area occupied by wet herbaceous cropland (%)	WPAS
	Shrubs	Area occupied by shrubs (%)	SHRU
	Shrubs diversity	Inverse of Simpson's Index for diversity of shrubs	DSHR
	Tree diversity	Inverse of Simpson's Index for diversity of trees	DTRE

*EGVs excluded for the Ecological-Niche Factor Analysis because they were nearly-Boolean (see Data analyses section for details)

ENFA was developed using Biomapper 3.0 (Hirzel et al. 2004b), and followed the procedures outlined by Hirzel et al. (2002). EGVs were previously tested for linear association using the Pearson's correlation coefficient (two-tailed) with the SPSS software (LEAD Technologies 2003). The database with the observations of the species and the EGVs were converted into Idrisi-formatted maps. EGVs were checked for variability, and four variables (SOIL, STON, AGRI, WPAS) were removed as they were nearly-boolean, i.e. they presented almost no variability (Table 2). Then, EGVs were normalised using the Box-Cox algorithm, and factor scores were computed using a factor analysis.

Ecological models were not developed for three reptiles (*Emys orbicularis*, *Mauremys leprosa*, and *Tarentola mauritanica*) as they were most likely introduced in the study area (Soares et al. 2005), thus not suitable for modelling species richness.

In four amphibians (*Triturus helveticus*, *Discoglossus galganoi*, *Pelobates cultripes*, and *Hyla arborea*) and two reptiles (*Chalcides bedriagai* and *Blanus cinereus*) the models produced presented extremely large eigenvalues, precluding the use of ENFA as a modelling technique. These species were scarce in the study area and were detected in less than 10 UTM squares, each (Table 1). In these cases, small sample size can affect the model computation, since there should never be more EGVs than species records (Hirzel et al. 2004b). Despite several efforts made to produce the models for these scarce species, namely by using fewer variables, no habitat suitability models could be produced. Therefore, habitat suitability models were derived for 70 and 88% of the amphibian and reptile species, respectively.

Predicted species richness

Habitat suitability maps for each species were derived using ENFA with the distance geometric mean algorithm, following Brotons et al. (2004) and Hirzel and Arlettaz (2003). This algorithm makes no assumption on the shape of the species distribution, and takes into account the density of observation points in environmental space by computing the geometric mean to all observation points. The habitat suitability maps were presented in the form of a grid in which each square has a value ranging from 0 to 100, corresponding to no suitability and high habitat suitability, respectively. For the reasons outlined in the previous section, habitat suitability maps were not derived for the scarcer species.

The individual habitat suitability maps were then overlaid, by taxonomic group, in Idrisi for Windows (Clark Labs 2003), and the result was a habitat suitability map for species richness of amphibians and reptiles (Cumming 2000; Gioia and Piggott 2000) Subsequently, a likelihood grid was defined for each species by adopting a cut-off point at the likelihood value corresponding to the fifty percentile. This value was defined arbitrarily based on visual assessment of the models, taking into account the expectations regarding the field experience. The outcome is a map with the predicted distribution of species richness for each taxonomic group.

To evaluate the accuracy of the predicted species richness models, a cross-tabulation between observed and predicted values for amphibian and reptile species richness was performed with a confusion matrix, using Idrisi for Windows. Each column of the matrix represented the predicted species richness and each row represented the observed species richness. The number of squares within each class of observed species richness was compared against the number of squares of each class of predicted species richness. In this study, 10 and 16 classes of predicted and of observed species richness for amphibians and reptiles, respectively, were used.

Results

Environmental correlates for species richness

The ecogeographical variables (EGVs) that are mostly correlated with the occurrence of amphibian and reptile species according to the first factor (Marginality) included topographic, climatic and habitat characteristics (Tables 3 and 4).

For amphibians, precipitation, number of watercourse and surfaces, and tree diversity cover, were consistently positively correlated with species occurrence, whereas number of days with fog was negatively related (Table 3). For reptiles, precipitation, largest watercourse, number and largest water surfaces, and tree diversity cover, were consistently positively correlated with species occurrence, whereas insolation and area occupied by shrub were negatively related (Table 4).

Three EGVs related with species occurrence were common for both taxonomic groups, and consistently presented a positive relation with species occurrence: precipitation, number of water surfaces, and tree diversity cover; suggesting a strong coincidence in the environmental correlates that influence amphibian and reptile species richness (Tables 3 and 4).

Predicted species richness

The distribution of the observed amphibian species richness showed one large area, Gerês-Homem river valleys and Albergaria forest, and four smaller sites of high richness (1) transition area between Peneda mountain and Castro Laboreiro plateau (2) Ramiscal river valley (3) Lamalonga bogs, and (4) Covelães pit bogs and Beredo river valley (Fig. 2A, see Fig. 1 for toponomy). The distribution of the predicted amphibian species richness identified five large areas of high richness, which generally correspond to the areas where higher species richness was observed (Fig. 2B). However, Ecological-Niche Factor Analysis (ENFA) suggested an area of high species richness, Amarela mountain, that was not observed in the fieldwork and failed to identify an area, Covelães pit bogs, in which it was observed high species richness.

The distribution of the observed reptile species richness identified one large and continuous area, located in the Homem-Gerês river valleys and Albergaria forest (Fig. 2C). Two smaller areas of high species richness were also identified in the transition area between Peneda mountain and Castro Laboreiro plateau, and Covelães pit bogs and Beredo river valley. The distribution of the predicted reptile species richness identified five large areas of high richness, which correspond to the areas where generally higher number of species was observed (Fig. 2D). Two areas of high species richness, Ramiscal river valley and Amarela mountain, were not detected in the fieldwork but were also suggested by ENFA.

The high species richness areas predicted by ENFA were spatially common to amphibian and reptile species (Figs. 2B and 2D). The cross-tabulation of predicted against observed amphibian and reptile species richness showed a total of 12 and 6%

Table 3 Ecogeographical variables (EGVs) and scores of the first factor (Marginality) obtained by the Ecological-Niche Factor Analysis for individual amphibian species

	ALTI	RELI	ORIE	PREC	INSO	FOG	NWAT	LWAT	NWSU	LWSU	ROCK	SHRU	DSHR	DTRE
A. obstetricans	0.078	–0.190	–0.074	0.534*	–0.161	–0.333*	0.638*	–0.089	–0.012	0.070	0.258*	0.149	–0.073	0.129
B. bufo	–0.437*	0.020	0.005	–0.092	–0.028	–0.343	0.012	0.186	0.385*	0.385*	–0.181	–0.328	–0.076	0.452*
B. calamita	0.438*	–0.374*	0.095	–0.036	–0.207	0.039	–0.301	–0.372*	–0.317	–0.369*	–0.149	0.328	–0.093	0.092
C. lusitanica	–0.375*	0.446*	0.001	0.229	–0.135	–0.418*	0.295	0.239	0.236	0.080	–0.005	–0.231	–0.233	0.317*
R. iberica	0.308*	–0.051	–0.106	0.711*	–0.346*	–0.258	0.347*	0.032	0.145	–0.092	0.195	–0.013	0.030	0.100
R. perezi	–0.023	–0.568*	–0.186	–0.172	0.280	0.192	–0.096	0.214	0.284*	0.447*	–0.366*	–0.131	–0.078	0.063
S. salamandra	0.045	0.140	0.022	0.378	–0.161	–0.407*	0.432*	0.103	0.395*	0.028	0.066	–0.265	0.09	0.455*
T. boscai	0.107	–0.350	–0.126	0.430*	–0.174	–0.421*	0.451*	0.055	0.422*	0.060	0.131	0.037	0.021	0.224
T. marmoratus	0.168	–0.469*	–0.141	0.295*	–0.062	–0.247	0.279	0.068	0.565*	0.333*	–0.078	–0.114	0.044	0.215
%	44.4	44.4	0	44.4	11.1	44.4	44.4	11.1	55.5	44.4	22.2	0	0	33.3

*First four most explaining EGVs for each species model. % Percentage of occurrence of each EGV as important explaining factors of the distribution of individual species

Table 4 Ecogeographical variables (EGVs) and scores of the first factor (Marginality) obtained by the Ecological-Niche Factor Analysis for individual reptile species

	ALTI	RELI	ORIE	PREC	INSO	FOG	NWAT	LWAT	NWSU	LWSU	ROCK	SHRU	DSHR	DTRE
A. fragilis	–0.101	0.190	–0.077	0.242	–0.117	–0.115	0.079	0.591*	0.383*	0.204	–0.093	–0.416*	–0.206	0.307*
C. austriaca	0.500*	–0.078	–0.136	0.577*	–0.509*	0.262*	–0.127	–0.001	–0.110	–0.005	0.123	0.051	0.020	–0.134
C. girondica	0.002	0.174	–0.053	0.487*	0.013	–0.029	0.013	0.443*	0.461*	0.317	0.209	–0.091	–0.363*	0.190
C. striatus	0.531*	–0.534*	–0.125	0.210	–0.217	–0.038	–0.292*	0.071	0.135	0.105	–0.203	–0.145	–0.371*	–0.073
E. scalaris	–0.468*	0.202	0.142	–0.172	0.107	–0.257	0.259	0.082	0.414*	0.249	–0.208	–0.310*	0.019	0.407*
L. schreiberi	0.106	–0.343*	–0.317	0.284	–0.018	–0.036	0.378*	0.222	0.420*	0.389*	–0.074	–0.250	–0.091	0.303
L. lepida	–0.313*	0.076	–0.050	0.014	0.039	–0.213	0.230	0.292	0.444*	0.340	–0.097	–0.403*	0.013	0.472*
M. monspessulanus	–0.392*	0.177	0.008	0.016	0.013	–0.460*	0.116	0.179	0.409	0.350*	–0.080	–0.219	0.056	0.459*
N. maura	–0.331*	0.067	–0.048	0.082	–0.075	–0.304	0.009	0.390*	0.453*	0.467*	–0.098	–0.312	–0.036	0.308
N. natrix	0.097	0.072	–0.238	0.432*	–0.430*	–0.198	0.028	0.237	0.282*	0.259	–0.199	–0.260	–0.187	0.414*
P. algirus	–0.486*	0.196	–0.026	–0.029	0.092	–0.392*	0.341*	0.216	0.315	0.228	–0.065	–0.244	0.009	0.430*
P. bocagei	0.469*	–0.358*	–0.426*	0.426*	–0.288	0.260	0.075	0.147	0.236	0.165	–0.026	–0.034	–0.016	0.164
P. hispanica	–0.243	0.150	–0.103	0.205	0.185	–0.302*	0.574*	0.136	0.138	0.344*	0.293	–0.187	–0.028	0.366*
V. latastei	–0.028	0.375*	–0.100	0.580*	–0.402*	–0.437*	0.064	0.067	0.061	0.068	0.240	–0.097	–0.251	0.111
V. seoanei	0.282	–0.448*	–0.057	–0.394*	0.287	0.316*	–0.142	0.033	0.303	0.131	–0.425*	–0.203	–0.120	–0.113
%	53.3	33.3	6.6	40.0	20.0	40.0	26.6	20.0	53.3	20.0	6.6	20.0	13.3	46.6

* First four most explaining EGVs for each species model. % Percentage of occurrence of each EGV as important explaining factors of the distribution of individual species

correct classification rate, respectively, indicating a low correspondence between squares with equal observed and predicted richness, and suggesting that observed species richness was largely underestimated.

Discussion

This study showed that a mixture of environmental factors is the best predictor of amphibian and reptile species occurrence at a local scale, in the Peneda-Gerês National Park. These factors included (1) habitat variables, such as the number of watercourses and water surfaces, the type of the largest water surface and tree diversity cover (2) topographical variables, such as altitude and relief; and (3) climatic variables, such as precipitation and number of days with fog.

Environmental factors, such as altitude, relief or precipitation, are widely known to influence ecological processes in organisms (e.g. Brown and Lomolino 1998). Several other studies, modelling the distribution of amphibians and reptiles, and herpetofauna species richness, have identified these factors as significant for explaining the observed distribution patterns (Brito et al. 1996; Sequeira et al. 2001; Teixeira et al. 2001; Morales et al. 2002; Guisan and Hofer 2003).

The environmental factors related with water availability exhibited the highest percentages of occurrence among the ecogeographical variables (EGVs) that explained the distribution of individual species. Climatic factors, such as precipitation, presented positive coefficients for all amphibians, whereas reptiles presented positive or negative coefficients, according to the biogeographic traits of species. Reptiles typical from the Euro-Siberian biogeographic province had positive coefficients, e.g. *Coronella austriaca*, whereas characteristic species from the Mediterranean biogeographic province had negative coefficients, e.g. *Psammodromus algirus*. Habitat factors, such as the number of watercourses and water surfaces, also presented positive coefficients for most species. Interestingly, the number of water surfaces was the most frequent explanatory EGV for the distribution of both taxonomic groups.

Topographical variables, such as altitude and relief, were also identified as very important factors explaining the distribution of the species, particularly for the reptiles. For instance, altitude presented positive or negative coefficients according to the biogeographic traits of species. Typical Euro-Siberian species presented positive coefficients, e.g. *Vipera seoanei*, whereas typical Mediterranean species had negative coefficients, e.g.*Elaphe scalaris* and *Malpolon monspessulanus*. Altitude is usually correlated with climatic variables since it is known to influence, for instance precipitation or evapotranspiration (C.N.A. 1983). Relief exhibited negative scores for most of the species belonging to both taxonomic groups, which could be due to a decrease in water availability in the areas of accentuated slope and higher availability of thermoregulation spots offered by the flat areas. However, *Chioglossa lusitanica* and *Vipera latastei* stand as exceptions presenting positive scores in this EGV, the former because it inhabits clear and oxygenated running streams (Sequeira et al. 2001), and the latter because it inhabits rock outcrops with dense bush cover (Brito and Crespo 2002).

Three ecogeographical variables – precipitation, number of water surfaces, and tree diversity cover – were common for both taxonomic groups and consistently presented a positive relation with species occurrence. Therefore, this association suggested a strong coincidence in the environmental correlates that influence amphibian and reptile species occurrence and ultimately species richness. In fact, high species richness areas for amphibians and reptiles were located in squares with at least one to six water surfaces, at least one to four species of trees, and with precipitation levels between 1,800 and 2,700 mm/year. The high number of water surfaces should be related with water availability which favours the existence of numerous amphibian breeding sites. High levels of tree cover correspond to diversified lowland habitats that account for most of the autochthonous forests and probably offer multiple microhabitats with numerous shelters (e.g. fallen logs) and ample prey availability. Precipitation is related with both tree diversity cover and water availability, thus it could be considered as the ultimate factor for high levels of amphibian and reptile species richness in the study area.

The distribution patterns of observed species richness were quite similar in amphibians and reptiles, and comprised one large area, Gerês-Homem river valleys and Albergaria forest, and two smaller areas: Peneda mountain and Castro Laboreiro plateau, and Covelães pit bogs/Beredo river valley. Two additional smaller areas were identified for the amphibians: the Ramiscal river valley and the Lamalonga bogs. The distribution patterns of predicted species richness were mostly similar for both taxonomic groups and included five large and continuous areas. These areas are mostly located along steep river valleys associated with marked altitudinal changes where a climatic transition occurs. The lower river valleys present a Mediterranean climate, whereas the high river valleys usually present an Atlantic climate. Thus, river valleys act as corridors for Mediterranean and Euro-Siberian species to attain high and low altitude areas, respectively, and allow the sympatric coexistence of biogeographically distinct species. For instance, the Euro-Siberian *C. austriaca* occurs in sympatry with the Mediterranean *E. scalaris* in the Gerês river valley (Soares et al. 2005).

Local scales, such as the 1 × 1 km squares used in this study, proved to be suitable for studying species richness patterns of amphibians and reptiles. This scale allowed the discrimination of particular river valleys, such as the Gerês and Beredo, with high species richness. Therefore, special attention should be given to riparian habitats and ponds in mountain habitats in future conservation measures and management in the National Park. The 1 × 1 km scale gives insights for important areas for amphibians and reptiles that would be interesting to assess regarding other taxonomic groups for future conservation planning at local-scale. However, at least three constrains can arise when working at local scales. First, small sample size of species presence can affect model production. Indeed, the scarcer species were excluded from the analysis because the low number of observations precluded model development. This may constitute a serious problem when the target species are exactly the scarcer ones. Models for species with exceptionally low number of observations have been developed using Ecological-Niche Factor Analysis, e.g. for the Iberian-rare carnivore *Martes martes* (Álvares and Brito 2006). However, this species restricts habitat selection to an extremely narrow range of habitat conditions, enabling the use of ecological modelling techniques that use presence-only data. However, for rare and/or secretive species but with wide-range and occurring in different ecological conditions, models are extremely difficult to develop (Santos et al. 2006). Second, when dealing with species richness, sub-sampling can strongly affect classification rates. In this study, a high percentage of misclassification arose mainly because the squares with low number of observed species were systematically classified by the model as having high number of species. Thus, ecological modelling suggested that observed species richness was largely underestimated and that more fieldwork would be needed to detected actual species richness in the study area. Lastly, large-scale EGVs may affect the predictive capacity of models. For instance, the amphibians are strongly dependent on the aquatic habitat type apart from the general climatic, topographic and habitat variables usually considered in these types of studies. The inadequate selection of variables can lead to failure in finding consistent correlations between environmental factors and species diversity (Busack and Jaksic 1982). Additionally, increasing the accuracy and resolution of environmental variables allows their use as predictors for studies of several ranges of geographical extents (Guisan and Hofer 2003). Thus, fine-scale EGVs, such as land-cover data from remote sensing, are advisable for species or taxonomic groups with particular biological characteristics.

Nowadays, it is broadly recognised that the identification and conservation of specific important areas is a primordial measure to reduce the loss of biological diversity at different levels. Biogeographic crossroads with increased beta diversity are among the priority areas for conservation worldwide (Spector 2002). The geographic location of Peneda-Gerês National Park in the Iberian Peninsula, and the high number of amphibians and reptiles species present in the area justify its value as an important area for conservation. But, most of Central-Northern Portugal is located in the transition area between the Atlantic Mediterranean and the Continental Mediterranean climates (Rivas-Martínez 1987). Therefore, increased species richness should be expected for other mountain areas, such as Alvão or Estrela mountains. Local scale studies for these protected areas, including also other taxonomic groups, should be planned as a framework for the development of multi-scale conservation planning by Portuguese authorities.

Acknowledgments We thank N. Sillero for useful advices with the use of Geographical Information Systems, and M. Carretero, J. Harris, F. Álvares, F. Sequeira and A. Loureiro for helpful comments on the manuscript. Field work was funded by Instituto de Conservação da Natureza (PNPG/ICN) and Plano Operacional do Ambiente. J.C. Brito was supported by a Post-doctoral Grant (SFRH/BPD/11542/2002) from Fundação para a Ciência e Tecnologia (FCT).

References

Álvares F, Brito JC (2006) Habitat requirements and potential areas of occurrence for the Pine Marten in North-western portugal: conservation implications. In: Santos-Reis M, Birks JD, O'Doherty EC, Proulx G (eds), *Martes* in carnivore communities. Alpha Wildlife Publications, Alberta, Canada, *in press*

Araújo MB (1999) Distribution patterns of biodiversity and the design of a representative reserve network in Portugal. Diversity and Distribution 5:151–163

Araújo MB, Williams PH, Fuller RJ (2002) Dynamics of extinction and the selection of nature reserves. Proc Roy Soc, Biolog Sci 269:1971–1980

Austin GE, Thomas CJ, Houston DC, Thompson DB (1996) Predicting the spatial distribution of buzzard *Buteo buteo* nesting areas using geographical information systems and remote sensing. J Appl Ecol 33:1541–1550

Brito JC, Brito e Abreu F, Paulo OS, Rosa HD, Crespo EG (1996) Distribution of Schreiber's green lizard (*Lacerta schreiberi*) in Portugal: a predictive model. Herpetol J 6:43–47

Brito JC, Crespo EG, Paulo OS (1999) Modelling wildlife distributions: logistic multiple regression vs overlap analysis. Ecography 22:251–260
Brito JC, Crespo EG (2002) Distributional analysis of two vipers (*Vipera latastei* and*V. seoanei*) in a potential area of sympatry in the north-western Iberian Peninsula. In: Schuett GW, Höggren M, Douglas ME, Greene HW (eds), Biology of the Vipers. Eagle Mountain Publishing, pp 129–138
Brotons L, Thuiller W, Araújo MB, Hirzel AH (2004) Presence–absence versus presence-only modelling methods for predicting bird habitat suitability. Ecography 27:437–448
Brown JH, Lomolino MV (1998) Biogeography. Sinauer Associates, MA
Bullock JM, Edwards RJ, Carey PD, Rose RJ (2000) Geographical separation of two *Ulex* species at three spatial scales: does competition limit species' range?. Ecography 23:257–271
Busack SD, Jaksic FM (1982) Ecological correlates of Iberian herpetofaunal diversity: an analysis at regional and local levels. J Biogeogr 9:289–302
Clark Labs (2003) Idrisi Kilimanjaro vs 14.01. Clark University, Worcester MA
C.N.A. (1983) Atlas do Ambiente. Comissão Geral do Ambiente, Lisboa
Corsi F, Duprè E, Boitani L (1999) A large-scale model of wolf distribution in Italy for conservation planning. Conserv Biol 7:150–159
Cumming GS (2000) Using habitat models to map diversity: pan-African species richness of ticks (Acari: Ixodida). J Biogeogr 27:425–440
Dettki H, Lofstrand R, Edenius L (2003) Modelling habitat suitability for Moose in coastal Northern Sweden: empirical *vs* process-orientated approaches. Ambio 32:549–556
Franklin DC, Woinarski JC, Noske RA (2000) Geographical patterning of species richness among granivorous birds in Australia. J Biogeogr 27:829–842
Gallego D, Cánovas F, Esteve MA, Galián J (2004) Descriptive biogeography of *Tomicus* (Coleoptera: Scolytidae) species in Spain. J Biogeogr 31:2011–2024
Gioia P, Pigott JP (2000) Biodiversity assessment: a case study in predicting richness from the potential distributions of plant species in the forests of south-western Australia. J Biogeogr 27:1065–1078
Goday SR (1953) Apreciación sintética de los grados de vegetación de la Sierra de Gerês. In: A Flora e a Vegetação da Serra do Gerês. Simpósio da I Reunião de Botânica Peninsular. Instituto de Alta Cultura, Lisboa, pp 449–480
Grand J, Buonaccorsi J, Cushman SA, Griffin LR, Neel MC (2004) A multiscale landscape approach to predicting bird and moth rarity hotspots in a threatened pitch pine-scrub oak community. Conserv Biol 18:1063–1077
Guisan A, Hofer U (2003) Predicting reptile distribution at the mesoscale: relation to climate and topography. J Biogeogr 30:1233–1243
Heikkinen RK, Neuvonen S (1997) Species richness of vascular plants in the subartic landscape of northern Finland: modelling relationships to the environment. Biodivers Conserv 6:1181–1201
Hirzel AH, Hausser J, Chessel D, Perrin N (2002) Ecological-Niche Factor Analysis: how to compute habitat-suitability maps without absence data?. Ecology 83:2027–2036
Hirzel AH, Arlettaz R (2003) Modelling habitat suitability for complex species distributions by environmental-distance geometric mean Environ Manage 32:614–623
Hirzel AH, Posse B, Oggier PA, Crettenands Y, Glenz C, Arlettaz R (2004a) Ecological requirements of reintroduced species and the implications for release policy: the case of bearded vulture. J Appl Ecol 41:1103–1116
Hirzel AH, Hausser J, Perrin N (2004b) Biomapper 3.0. Division of Conservation Biology, University of Bern. URL: http://www.unil.ch/biomapper
Inger RF, Voris HK (2001) The biogeographical relations of the frogs and snakes of Sundaland. J Biogeogr 28:863–891
Jones PG, Beebee SE, Tohme J, Galwey NW (1997) The use of geographical information systems in biodiversity exploration and conservation. Biodivers Conserv 6:947–958
Kier G, Barthlott W (2001) Measuring and mapping endemism and species richness: a new methodological approach and its application on the flora of Africa. Biodivers Conserv 10:1513–1529
LEAD Technologies 2003. SPSS for Windows vs 12.0. LEAD Technologies
Luoto M, Kuussaari M, Toivonen T (2002) Modelling butterfly distribution based on remote sensing data. J Biogeogr 29:1027–1037
Malkmus R (1986a) Herpetologische beobachtungen im Nationalpark Peneda-Gerês/Portugal. Herpetofauna 8:18–25
Malkmus R (1986b) Zur verbreitung der amphibien und reptilien im Parque Nacional da Peneda-Gerês. Nachrichten des Naturwissenschaftlichen Museums 94:37–87

Malkmus R (2004) Amphibians and Reptiles of Portugal, Madeira and the Azores-Archipelago. Gantner Verlag K.G., Germany

Maravalhas E (eds) (2003) As Borboletas de Portugal. Vento Norte, Porto

Mateo JA (2006) Áreas importantes para la herpetofauna española In: Pleguezuelos JM, Márquez R, Lizana M (eds) Atlas e Libro Rojo de los Anfibios y Reptiles de España. Dirección General de Conservación de la Naturaleza – A.H.E., pp 485–500

Médail F, Quézel P (1999) Biodiversity hotspots in the Mediterranean Basin: setting global conservation priorities. Conserv Biol 13:1510–1513

Mittermeier RA, Myers N, Thomsen JB, Da Fonseca GAB, Oliveieri S (1998) Biodiversity hotspots and major tropical wilderness areas: approaches to setting conservation priorities. Conserv Biol 12:516–520

Myers N, Mittermeier RA, Mittermeier CG, Da Fonseca GAB, Kent J (2000) Biodiversity hotspots for conservation priorities. Nature 403:853–858

Morales JJ, Lizana M, Arco C (2002) Análises de la distribuición altitudinal de la herpetofauna en las sierra de Segovia Revista Española de Herpetología 16:29–42

Nally RM, Fleishman E (2003) A successful predictive model of species richness based on indicator species. Conserv Biol 18:646–654

Okland B, Bakke A, Hagvar S, Kvamme T (1996) What factors influence the diversity of saproxylic beetles? A multiscaled study from a spruce forest in southern Norway. Biodivers Conserv 5:75–100

Peterson AT, Ortega-Huerta MA, Bartley J, Sánchez-Cordero V, Soberón J, Buddemeier RH, Stockwell DR (2002) Future projections for Mexican faunas under global change scenarios. Nature 416:626–629

Pimenta M, Santarém L (1996) Atlas das Aves Nidificantes no Parque Nacional da Peneda-Gerês. P.N. Peneda-Gerês/ICN, Lisboa

Rahbek C, Graves GR (2001) Multiscale assessment of patterns of avian species richness. Proc Natl Acad Sci USA 98:4534–4539

Rensburg BJ, Chown SL, Gaston KJ (2002) Species richness, environmental correlates, and spatial scale: a test using South African Birds. The American Naturalist 159:566–577

Rivas-Martínez S (1987) Nociones sobre fitosociologia, biogeografia y bioclimatologia. In: Peinda M, Rivas-Martínez S (eds) La Vegetación de España. Universidad de Alcalán de Henares, Madrid, pp 19–45

Santos X, Brito JC, Sillero N, Pleguezuelos JM, Llorente GA, Fahd S, Parellada X (2006) Inferring habitat-suitability areas with ecological modelling techniques and GIS: a contribution to assess the conservation status of *Vipera latastei*. Biological Conservation, *in press*

Schall JJ, Pianka ER (1978) Geographical trends in numbers of species Science 201:679–686

Serra MG, Carvalho ML (1989) A Flora e a Vegetação do Parque Nacional da Peneda-Gerês. Contribuição para o Plano de Ordenamento desta Área Protegida. S.N.P.R.C.N., Lisboa

Sequeira F, Gonçalves H, Faria MM, Meneses V, Arntzen JW (2001) Habitat-structural and meteorological parameters influencing the activity and local distribution of the golden-striped salamander, *Chioglossa lusitanica*. Herpetological J 11:85–90

Soares C, Álvares F, Loureiro A, Sillero N, Arntzen JW, Brito JC (2005) Atlas of the amphibians and reptiles of Peneda-Gerês National Park, Portugal. Herpetozoa 18 (3/4)

Spector S (2002) Biogeographic crossroads as priority areas for biodiversity conservation. Conserv Biol 16:1480–1487

Stoms DM, Estes JE (1993) A remote sensing research agenda for mapping and monitoring biodiversity. Int J Remote Sens 14:1839–1860

Teixeira J, Ferrand N, Arntzen JW (2001) Biogeography of the golden-striped salamander *Chioglossa lusitanica*: a field survey and a spatial modelling approach. Ecography 24:618–624

Wilson EO (1992) The Diversity of Life. W.W. Norton and Co., New York

Zaniewski AE, Lehmann A, Overton JM (2002) Predicting species spatial distributions using presence-only data: a case study of native New Zealand ferns. Ecological Modell 157:261–280

Biodivers Conserv (2007) 16:1103–1118
DOI 10.1007/s10531-006-9074-1

ORIGINAL PAPER

Amphibian diversity in East African biodiversity hotspots: altitudinal and latitudinal patterns

J. C. Poynton · S. P. Loader · E. Sherratt · B. T. Clarke

Received: 3 October 2005 / Accepted: 19 May 2006 / Published online: 12 July 2006

Abstract The Eastern Arc mountain chain and adjoining coastal forests of Tanzania and Kenya have been listed as world biodiversity hotspots. We report on an ongoing attempt to estimate amphibian diversity on the three best studied mountains of the Eastern Arc, the East Usambara, Uluguru and Udzungwa mountains of Tanzania, complemented by an estimate of diversity on the adjoining coastal lowland. This proves to be a complex task, which introduces a note of caution into evaluating global biodiversity estimates. Most amphibian species in eastern Tanzania occur on the coastal lowlands and are widely distributed, extending at least north or south of Tanzania and, to a variable extent, westwards to the elevated interior. Diversity patterns along the length of the lowlands are complex, with the presence of a Sahelian element in the extreme north. On the three Eastern Arc mountains studied, species turnover associated with rising altitude is greater than turnover associated with latitudinal distance between the mountain blocks, leading to greater altitudinal than latitudinal diversity in this equatorial region. A long-standing divergence is indicated between montane and lowland endemics. Although forest-associated species are not the largest contributor to the eastern Tanzanian total species diversity (some 48%), the uniqueness of these species both in lowland and montane forests, combined with their evident vulnerability to disturbance, makes them a subject for particular conservation concern, and justifies hotspot status for both montane and lowland forests.

Keywords Amphibia · East Africa · Eastern Arc mountains · Tanzania · Diversity · Altitudinal and latitudinal species turnover · Hotspot

J. C. Poynton (✉) · S. P. Loader · E. Sherratt · B. T. Clarke
The Natural History Museum, Cromwell Road, London SW7 5BD, UK
e-mail: poynton@ukonline.co.uk

Introduction

The Eastern Arc mountain chain and the coastal forests of Tanzania and Kenya were listed by Myers et al. (2000) as one of the world's "hottest" biodiversity hotspots. Conservation International (Mittermeier et al. 2004) more recently recognised the Eastern Arc and the coastal lowlands as belonging to two different world biodiversity hotspot regions. The Eastern Arc was included in an Eastern Afromontane hotspot, while the coastal lowlands were included in a Coastal Forests of Eastern Africa hotspot. The criteria for evaluating biodiversity hotspots were based on several factors: numbers of endemic species, endemic species/area ratios, and habitat loss. We present an attempt to investigate more closely the diversity of amphibians on the Tanzanian Eastern Arc and the adjoining coastal lowlands (Fig. 1), but we place emphasis on the word "attempt" because of the difficulties encountered, which introduce a note of caution into evaluating global estimates such as those given in Myers et al. (2000) and Mittermeier et al. (2004).

An evaluation of diversity in East Africa has to deal with an extremely complex ecological, topographical and historical situation. A number of disjunct mountain blocks constitute the Eastern Arc (Fig. 1), which influence rainfall and temperature, and consequently the development of a great variety of natural vegetation types (Mittermeier et al. 2004). Major formations are arid shrubland, savanna, dry forest, rain forest and highland grassland. Lovett (1996) has shown a relationship between elevation and species composition in the Eastern Arc vegetation, which can be represented as two gradients, one from high elevation and the other from lower elevation. This replacement of species with change in elevation contributes substantially to the diversity of the region. The same pattern is evident in amphibian distribution, where there is grouping into highland and lowland sets, with opposing species gradients (Poynton 2003).

Evaluation of diversity also has to take into account the fact that much of the natural vegetation, especially forest, has been replaced by various kinds of agricultural crops (e.g. Newmark 1998; Burgess et al. 2000; Mittermeier et al. 2004). The full effects of this disturbance on amphibian diversity are not yet clear, but available evidence suggests that the response of amphibian species to habitat disturbance is not uniform (Poynton 1999).

The first attempt at a general survey of amphibian diversity in East Africa was published by Loveridge (1937). On the basis of the herpetofauna, he divided East Africa into a number of what he termed "ecological life zones", and made the important observation that 83% of the anurans listed as occurring below an altitude of 1,000 ft (c. 300 m) were widely distributed, whereas only 48% in the highest zone between 5,000 ft to 12,000 ft (c. 1500–3650 m) were widespread. This may not seem surprising since the lowland terrain is continuous whereas the highland terrain is markedly interrupted, but it does draw attention to one underlying cause of high diversity, namely fragmentation associated with altitude, which results in geographically restricted high-altitude endemics. Another cause of diversity that Loveridge noted is the zonation of climate from the torrid lowlands to the cool highlands. Associated with this is an altitudinal turnover in species composition, which Poynton (1962) suggested conformed to a tropical—transitional—temperate zonation that is evident over the whole of southern and eastern Africa. The zonation is altitudinal in the tropics but it becomes more latitudinal south of the Tropic of

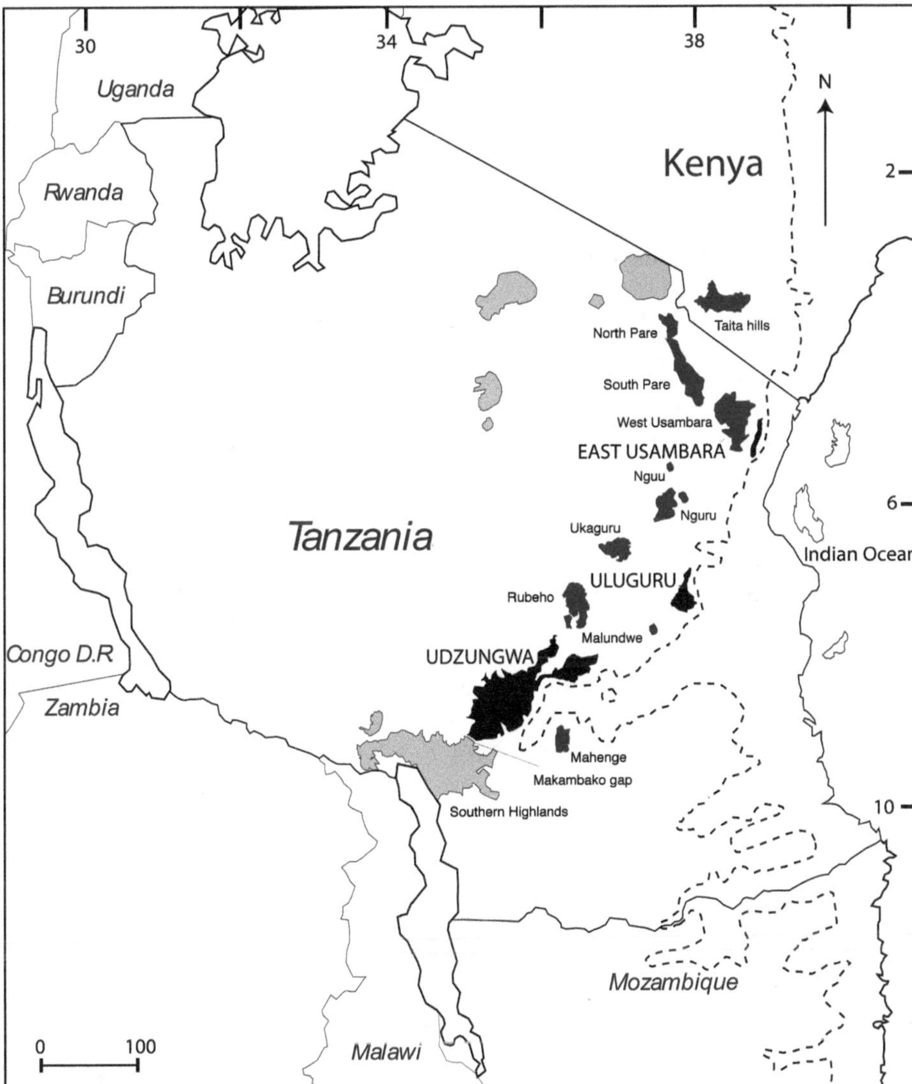

Fig. 1 Eastern Tanzania. The units of study are the coastal lowlands, demarcated by the dashed 400 m contour line, and the East Usambara, Uluguru and Udzungwa mountains shown in black. The remainder of the Eastern Arc mountain chain is shown in dark grey, and the northern volcanic mountains and the Southern Highlands in light grey

Capricorn. While the temperate element is relatively widespread in southern latitudes, it is highly fragmented in East Africa owing to its confinement to the discontinuous highlands.

Despite topographical discontinuity, Loveridge (1933) noted a marked similarity in species composition between the two then best studied mountains of the Eastern Arc, the Usambaras and Ulugurus in Tanzania, some 215 km apart (Fig. 1). An extensive mountain range about 100 km southwest of the Ulugurus, the Udzungwas, was assessed by him to show greater faunal differences, but it had been only poorly sampled, and subsequent extensive collecting has shown that he underestimated the similarities. Sampling on the Eastern Arc is currently active, but the degree of sampling along the mountain chain is still very uneven. We have therefore restricted analysis to the three ranges that are the best studied and which have been a subject of comparative study for some eighty years (Barbour and Loveridge 1928; Loveridge 1933, 1937), namely the East Usambara complex *c.* 5° S, the Uluguru mountains *c.* 7° S, and the Udzungwa range *c.* 8° S (Fig. 1). Species accumulation curves (Fig. 2) indicate that the sampling history of each mountain range is different, but the sampling on all three sites is considered to be adequate for the purpose of diversity evaluations. In this paper the amphibian assemblages recorded on the East Usambaras, Ulugurus and Udzungwas are compared with each other, and contrasted with assemblages occurring on the adjoining coastal lowland. The Tanzanian lowlands are also regarded as being adequately sampled (Fig. 2). The coastal assemblages are heterogeneous in distributional makeup: they include species restricted to the lowlands, species mainly limited there, and more widespread species with ranges of varying size but reaching the lowlands at least in the latitude of Tanzania. Table 2 indicates different distributional categories included in the list of species occurring in the lowland. The demarcation of 'lowland' and 'highland' or montane zones will be discussed below in connection with the various units of study.

According to an estimate made by a working group in 1997 (Anon. 1998), the East Usambaras had at that time a forest area of 413 km^2 (235 km^2 closed forest), the Ulugurus 248 km^2 (180 km^2 closed), and the Udzungwas 1118 km^2 (416 km^2 closed). These values for total forest area need to be treated with caution, since species/area ratios should take account of the fact that several species are very localised within the total area, such as *Nectophrynoides asperginis*, limited to parts of

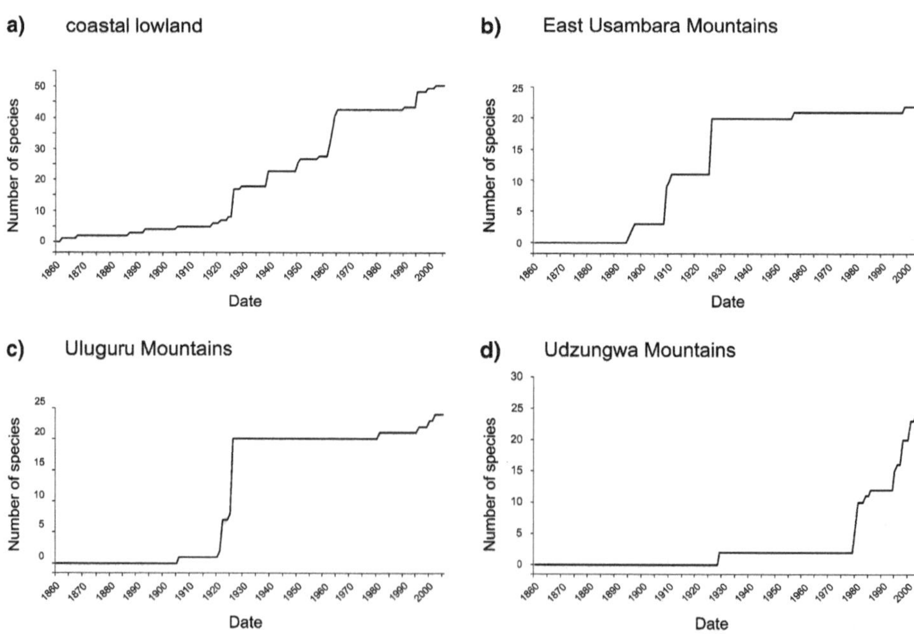

Fig. 2 Species accumulation curves of amphibians collected in eastern Tanzania, (**a**) coastal lowland, (**b**) East Usambara Mountains, (**c**) Uluguru Mountains, (**d**) Udzungwa Mountains

a single gorge in the extensive Udzungwa range (Poynton et al. 1998). Exact species ranges are still inadequately gauged, consequently species/area calculations are not attempted in this paper.

The taxonomy of East African amphibians is still unsettled, but it is reasonable to admit a total of 105 described species present in the four units of study, representing the orders Anura and Gymnophiona. These are assignable to 12 families and 33 genera. The species lists given below are based on specimens accessioned in the Natural History Museum, or else on specimens examined by us or in a few cases by other authorities. Available material suggests there are in addition several undescribed or poorly understood species; these are not included in this study apart from one *Scolecomorphus* species and a problematic upland reedfrog conventionally assigned to *Hyperolius puncticulatus* from Zanzibar, here referred to as *Hyperolius* sp. The number of species recorded in the four units of study is listed in Table 1.

Methods

In estimating biodiversity in eastern Tanzania, it is helpful to follow a distinction that Loveridge (1937) made between what he termed "ubiquitous, widespread species that occur in several habitats", and species limited to restricted areas, species that serve in some respect to characterise those areas. In Loveridge's ecological life zone paper (1937), all species, whether restricted or widespread in distribution, were listed in each zone of diverse habitats, providing a measure of gamma diversity. Yet in his comparison of the Usambara, Uluguru and Udzungwa forest faunas (1933), he explicitly excluded species judged by him not to be "almost entirely dependent on the rain forest," thereby giving a measure of alpha or within-habitat diversity. This distinction clearly has a bearing on diversity assessment. In general, widespread species breed in sites that may be described as open, but a few species with limited ranges are also open-site breeders, and it should be borne in mind that distinctions made between forest and open-site breeders are not the same as distinctions between mountain-dwelling and plains-dwelling breeders: even if forest may be associated mainly with montane conditions, open situations occur on highlands that carry characteristic open-site species, and the lowland plain includes forest patches with a few forest-associated species (Poynton 2000). On mountains there is a greater diversity of localised species, including both open-site and forest breeders in the larger mountain masses, added to which is the presence of wide-ranging species that exploit open sites where present.

Table 1 Number of species recorded in the four units of study

Study unit	No. currently recognised and recorded species		
	Total	Concentrated above 600 m	%
Coastal lowlands	51	0	0
East Usambara mountains	40	23	58
Uluguru mountains	38	25	66
Udzungwa mountains	64	37	58

In her *Measuring Biological Diversity*, Magurran (2004) adopted a definition of biological diversity that may be taken as standard: "the variety and abundance of species in a defined unit of study." Key terms in this definition are "species" and "unit of study". The first begs the question of an adequate definition of the term 'species'. As de Queiroz (2005) has pointed out, species concepts differ according to the primary interests of different biologists, and are often in conflict. The choice of species concept has a direct bearing on biodiversity estimates, as Agapow et al. (2004) have shown; phylogenetic species concepts tend to result in higher biodiversity values for a given area than do other concepts, especially when molecular phylogenies are used. The study of Tanzanian amphibians has followed a general pragmatic approach of recognising and characterising species as perceived unique sets of morphological, and occasionally behavioural, characters. This in effect sets species up as hypotheses about the presence of reproductively inclusive and mainly exclusive assemblages of individuals (Poynton and Broadley 1985), thereby conforming to a biological species concept. It accepts the reality that species may not be groups with discrete boundaries, but are "fuzzy sets" with blurred borders (Agapow et al. 2004). Yet in the case of Tanzanian brevicipitids, microhylids and caecilians, cladistic analysis based on DNA sampling is starting to apply a phylogenetic species concept to the Tanzanian fauna (de Sá et al. 2004, Loader et al. 2004a; Wilkinson et al. 2003). So far this has not led to a significant increase in species numbers, but it is likely to do so in the future.

Magurran's second key term, "unit of study", involves recognising that diversity does not allow an absolute measure, since it is affected by many different factors to do with units of study, notably the geographical extent of an area being investigated, habitat diversity, the kind and degree of sampling in an area, and by the adequacy of the taxonomy, which is especially compromised by clinal and other kinds of variation. All this complicates the comparison of diversity in assemblages recorded in different areas.

Many methods have been devised to arrive at an estimation of diversity especially when, as is usually the case, a study area is considered not to be fully sampled. Where collecting has been unsystematic or subject to a variety of sampling efforts, as in most of Africa, estimates have to be treated with circumspection. This situation was pointed out most recently by O'Hara (2005). Different estimators, as O'Hara concluded, give different estimates of species richness, with no standard value against which to evaluate different estimates. Consequently, methods used in obtaining diversity values should be explicitly stated, and weaknesses in the methods clarified. In this study, a diversity value is gained directly from the number of recognised species recorded in a unit of study, either a single habitat type (alpha diversity) or a whole landscape (gamma diversity). The reliability of the values obtained may be assessed by examining the species accumulation curve for each unit of study, which provides some idea of the completeness or incompleteness of sampling (Fig. 2). Beta diversity, comparing values from different areas, is estimated in this paper using a standard similarity index which is not adjusted for variation in geographical extent, or degree of sampling. The index used is Sørensen's Quotient of Similarity (QS): 2 × number of taxa common to both areas/sum of totals of taxa from both areas, expressed as a percentage (Sørensen 1948). We believe that Sørensen's quotient provides a convenient preliminary measure of beta diversity; its reliability will be evaluated for each pair of areas.

Results

The coastal lowlands

The "coastal lowlands of East Africa" was taken to be delimited by the 300 m contour in a study by Poynton (1990) on the composition and subtraction patterns of the amphibian fauna of this zone. Loveridge (1937) used the same altitude to demarcate his coastal plain zone. On botanical grounds, Clarke (2000) recognised 400 m to delimit the coastal forests of the Usambara and Uluguru Mountains. A list of amphibians on the East Usambara mountain complex shows a change in characteristic local bufonid genera from *Stephopaedes* and *Mertensophryne* to *Nectophrynoides* at about 400 m. The former two genera, associated with lowland forest, deposit eggs in tree holes or snail shells and have distinctive tadpoles; *Nectophrynoides* is associated with montane forest and is ovoviviparous (Grandison and Ashe 1983; Poynton 2000). The altitudinal separation of these highly adapted genera around 400 m confirms this altitude as a useful limit to the lowland fauna in northeastern Tanzania. In a study of amphibians on an outlier of the Udzungwa range, Mahenge Mountain (Loader et al. 2004b), the Kilombero Valley below 400 m was taken to constitute a lowland area. However, an analysis of anuran distribution carried out by means of a transect through southeastern Tanzania (Poynton 2003) showed the rate of species turnover to become more pronounced at higher altitudes. Menegon and Salvidio (2004) found evidence of marked species turnover at around 800 m on the Udzungwa scarp, which is in agreement with Clarke's (2000) delimitation of coastal forest on the Udzungwa Mountains at 750 m, not 400 m as on the Usambara and Uluguru Mountains. Difficulties associated with choice of a contour to demarcate the lowlands in the southeast are noted in the Discussion, but for the sake of uniformity, 400 m is taken as the lowland limit on all three mountains.

Regarding sampling, the accumulation rate of new or newly recorded species on the lowlands appears to be levelling off (Fig. 2a), even though recent meticulous sampling especially by Frontier-Tanzania has taken place. The lowlands may therefore be regarded as having been adequately sampled for diversity study purposes, although future taxonomic revisions may alter diversity values.

The 51 lowland species currently recognised are listed in Table 2. This list excludes some forest-associated species whose ranges are largely confined to uplands but which have been collected below 400 m at one or two forested localities, notably *Bufo brauni, Callulina kreffti, Hoplophryne rogersi, Afrana angolensis, Arthroleptides martiensseni, Phrynobatrachus ukingensis, Arthroleptis affinis, Leptopelis uluguruensis* and *Leptopelis vermiculatus*.

The five asterisked species are considered to be Tanzanian lowland endemics. *Hyperolius puncticulatus* is provisionally treated as a lowland species with uncertain status and distribution, not conspecific with a montane frog conventionally given that name. The total number of eleven lowland-limited species comprise only 22% of the total number of 51 species recorded on the Tanzanian lowlands; the 78% majority agrees with Loveridge's estimate that 83% of amphibians occurring on the lowlands are widely distributed. Latitudinally, 82% extend at least to the north or the south of Tanzania, most (57%) in both directions. Altitudinally, 78% occur at least to some extent above 400 m, and 45% extend onto the plateau west of the Eastern Arc.

Distribution and diversity along the length of the lowlands shows substantial complexity, as indicated by the need to have the categories T, Tn, Ts, and Us, Ul, Ud in Table 2. The presence of *Bufo xeros* and *Kassina somalica* in the extreme northeast (Cherry et al. 1998) indicates a Sahelian element in that area, consequently the lowland fauna cannot be regarded as simply 'East African'. It seems that only a few species in the lowlands are closely associated with true forest, notably *Mertensophryne micranotis, Stephopaedes usambarae, S. howelli, Arthroleptis xenodactyloides, Afrixalus sylvaticus*, possibly *Spelaeophryne methneri*, and less closely associated, *Stephopaedes loveridgei* and *Leptopelis flavomaculatus*. The remaining anurans breed in bodies of water in open areas as far as is known, even if in close proximity to forest. Species ranges are therefore not continuous over the length of the lowlands, but are fragmented according to the availability of breeding sites. As far as is known, the ranges of forest-associated species are particularly fragmented.

Table 2 Species considered to contribute to a lowland assemblage. Names of lowland species endemic to Tanzania are asterisked

Species	Altitudinal			Latitudinal				Eastern Arc		
	L	L+	H	T	Tn	Ts	T+	Us	Ul	Ud
Xenopus muelleri			x				x			
Bufo gutturalis			x				x	x	x	x
Bufo lindneri		x				x				x
Bufo maculatus			x				x		x	x
**Bufo reesi*	x			x						
Bufo xeros			x		x					
Schismaderma carens			x			x				x
Mertensophryne micranotis	x				x					
**Stephopaedes howelli*	x			x						
Stephopaedes loveridgei		x		x						x
**Stephopaedes usambarae*	x			x						
Phrynomantis bifasciatus			x				x			x
Spelaeophryne methneri		x		x						x
Breviceps mossambicus			x			x				x
Hemisus marmoratus			x				x			x
Pyxicephalus edulis			x				x			x
Hildebrandtia ornata			x				x			
Amnirana galamensis			x				x			
Ptychadena anchietae			x				x	x		x
Ptychadena mascareniensis			x				x	x		
Ptychadena mossambica			x				x			x
Ptychadena oxyrhynchus			x				x	x		
Ptychadena schillukorum			x				x			
Ptychadena taenioscelis		x					x			x
Phrynobatrachus acridoides		x					x	x	x	x
Phrynobatrachus mababiensis			x			x				
Phrynobatrachus natalensis		x					x		x	x
Arthroleptis stenodactylus			x				x	x	x	x
Arthroleptis xenodactyloides		x				x		x	x	x
Chiromantis xerampelina		x					x	x	x	x
**Leptopelis argenteus*	x			x						
Leptopelis concolor	x				x					
Leptopelis flavomaculatus		x					x	x	x	x
Kassina maculata		x					x	x		x
Kassina senegalensis			x			x				x

Table 2 continued

Species	Altitudinal			Latitudinal				Eastern Arc		
	L	L+	H	T	Tn	Ts	T+	Us	Ul	Ud
Kassina somalica	x				x					
Afrixalus delicatus		x					x		x	x
Afrixalus fornasini			x				x	x	x	x
Afrixalus stuhlmani		x		x				x	x	
Afrixalus sylvaticus	x				x					
Hyperolius argus		x					x	x		
Hyperolius mariae		x		x				x		
Hyperolius mitchelli			x			x		x	x	x
Hyperolius nasutus			x				x			x
Hyperolius parkeri		x					x	x		x
Hyperolius puncticulatus	x						x			
Hyperolius pusillus		x					x			
**Hyperolius reesi*	x			x						
Hyperolius tuberilinguis		x					x	x		x
Hyperolius viridiflavus			x				x	x	x	x
Schistometopum gregorii	x				x					

Altitudinal- L = limited to the lowland. L+ = lowland and at least a portion of Eastern Arc. H = recorded also on the highland plateau west of the Eastern Arc

Latitudinal- T = limited to a portion of Tanzanian lowlands only. Tn = all or portion of Tanzanian lowlands continuing northwards. Ts = all or portion of Tanzanian lowlands continuing southwards. T+ = full extent of Tanzanian lowlands and also north and south

Eastern Arc occurrence- Us = common to Usambaras and surrounding lowland. Ul = common to Ulugurus and surrounding lowland. Ud = common to Udzungwas and surrounding lowland

The East Usambara mountain complex

Collecting has been carried out for over a century in the Amani region of the East Usambara Mountains. Recent major collections made on this mountain and its foothills by Frontier-Tanzania and others have added only five more species to Loveridge's (1933) list of 17 amphibians "associated with mountain rain forest" in the East Usambaras, which suggests comprehensive sampling (Fig. 2b). Taxonomic refinements in the future may nevertheless alter the diversity value presented here.

The West Usambara massif, separated from the East Usambaras by the narrow but deep Lwengera Valley, appears to have some vicariant species but it has not been as well collected as the East Usambaras. Consequently it is not covered in this paper.

Table 3 lists the 22 forest-associated East Usambaran species whose ranges appear to be concentrated above an altitude of 600 m. These are all Tanzanian endemics apart from *Afrana angolensis, Hyperolius spinigularis* and *H.* sp. The open-site *Xenopus laevis* has also been collected on the East Usambaras; it is a widespread upland vicariant of *X. muelleri* within the tropics (Poynton and Broadley 1991). Open habitat is in fact very limited on the East Usambaras, particularly highland grassland, in contrast to the West Usambaras and the Udzungwas. The composition of the East Usambaran upland fauna differs markedly from the forest-associated fauna of the lowlands, and is much more taxonomically diverse than the small group of forest-associated species limited to the lowlands. The change-over from the lowland bufonids *Stephopaedes* and *Mertensophryne* to the upland *Nectophrynoides* has already been noted; the presence of the microhylid and brevicipitid genera *Callulina, Hoplophryne, Parhoplophryne* and *Probreviceps* in the uplands is also noteworthy. These distinctive genera have the

Table 3 Forest-associated species found centred above 600 m on the East Usambara, Uluguru and Udzungwa mountains, Tanzania. 1 = endemic to one unit, 2 = endemic to two units, 3 = endemic to three units, 4 = all or part of Eastern Arc plus highland further south

Species	East Usambara	Uluguru	Udzungwa
Bufo brauni	3	3	3
Nectophrynoides asperginis	–	–	1
Nectophrynoides cryptus	–	1	–
Nectophrynoides frontierei	1	–	–
Nectophrynoides laevis	–	1	–
Nectophrynoides minutus	–	1	–
Nectophrynoides poyntoni	–	–	1
Nectophrynoides pseudotornieri	–	1	–
Nectophrynoides tornieri	3	3	3
Nectophrynoides viviparus	–	4	4
Nectophrynoides wendyae	–	–	1
Callulina kreffti	3	3	3
Hoplophryne rogersi	1	–	–
Hoplophryne uluguruensis	–	2	2
Parhoplophryne usambarica	1	–	–
Probreviceps macrodactylus	3	3	3
Probreviceps rungwensis	–	–	4
Probreviceps uluguruensis	–	1	–
Afrana angolensis	4	4	4
Arthroleptides martiensseni	1	–	–
Arthroleptides yakusini	–	2	2
Phrynobatrachus krefftii	1	–	–
Phrynobatrachus parvulus	–	–	4
Phrynobatrachus rungwensis	–	–	4
Phrynobatrachus uzungwensis	–	4	4
Arthroleptis affinis	3	3	3
Arthroleptis reichei	–	–	4
Arthroleptis xenodactylus	1	–	–
Leptopelis barbouri	2	–	2
Leptopelis parkeri	3	3	3
Leptopelis uluguruensis	3	3	3
Leptopelis vermiculatus	2	–	2
Afrixalus uluguruensis	3	3	3
Hyperolius kihangensis	–	–	1
Hyperolius sp	4	4	4
Hyperolius spinigularis	4	4	4
Boulengerula boulengeri	1	–	–
Boulengerula uluguruensis	–	1	–
Scolecomorphus kirkii	–	4	4
Scolecomorphus uluguruensis	–	1	–
Scolecomorphus vittatus	2	2	–
Scolecomorphus sp	1	–	–
Total 42	22	24	26

Table 4 Quotients of similarity obtained from comparing total upland East Usambara, Uluguru and Udzungwa assemblages with adjacent lowland assemblages, gamma diversity in each case

	East Usambara	Uluguru	Udzungwa
Lowland	43%	32%	52%

appearance of being ancient elements in the Tanzanian fauna, therefore a long-standing difference between lowland and upland faunas can be assumed.

The list of lowland species given in Table 2 indicates that at least 18 species present in the lowlands extend up the Usambara slopes (Us). The similarity between the list of all species found in the northeastern Tanzanian lowlands, and the list of all species collected above 600 m on the East Usambaras, gives a similarity quotient (QS) of 43% (Table 4). This is below the value of 54% obtained from comparing the amphibians on the eastern Zimbabwean highlands with those on the Mozambique coast (Poynton and Broadley 1991), but almost the same as the 42% obtained from comparing amphibians on the highland and lowland of Swaziland (Poynton and Boycott 1996). A substantial turnover of species from lowland to upland in north-eastern Tanzania is indicated, resulting in a high diversity for the whole region.

The Uluguru Mountains

Loveridge (1933) listed 20 species he had collected from the Uluguru Mountains. He was followed by several collectors mostly associated with the Tanzania Forest Conservation Group (Doggart et al. 2004) and the University of Dar es Salaam (Fig. 2c). The 24 forest-associated species found on the Ulugurus concentrated above 600 m are listed in Table 3. The highland grassland *Strongylopus fuelleborni* has been collected, and 13 widespread but forest-tolerant species are recorded from the Ulugurus as well as being included in the eastern lowland assemblage, as indicated in the Table 2 lowlands list (Ul).

The concentration of *Nectophrynoides* species and of caecilians is extraordinary, and currently defies explanation. The appearance of species distributed on the Southern Highlands further south, notably *Nectophrynoides viviparus, Phrynobatrachus uzungwensis, Scolecomorphus kirkii* and *Strongylopus fuelleborni* gives the Uluguru list a mix of Eastern Arc and Southern Highlands elements.

Applying a similarity index measure to the lists of forest-associated species from the Usambaras and Ulugurus gives a QS of 52% (Table 5). This value might be lowered by further taxonomic adjustments, for example a re-evaluation of *Callulina* (de Sá et al. 2004). The current value is substantially more than the QS of 32% obtained from comparing the list of species known in the adjacent Tanzanian lowlands with the list of all species collected above 600 m on the Ulugurus (Table 4). This again emphasises the greater altitudinal versus latitudinal turnover in eastern Tanzania. The particularly high lowland—upland turnover (low similarity) in the Uluguru area deserves further investigation.

The Udzungwa range

While not intensively collected by Loveridge, more recent collections, notably by members of the University of Dar es Salaam and Frontier-Tanzania, and by Michael

Table 5 Quotients of similarity obtained from comparing the forest-associated assemblages of the East Usambara, Uluguru and Udzungwa mountains, alpha diversity in each case

	Uluguru	Udzungwa
Usambara	52%	54%
Uluguru	–	64%

Klemens, whose collections have been deposited in the American Museum of Natural History, have allowed a fairly detailed transect to be run from the coast to the crest of the Udzungwa Range (Poynton 2003). Amphibians on the Mahenge range, situated across the Kilombero valley from the Udzungwa range, have more recently been studied by Loader et al. (2004b); the distribution of species on the Mahenge range conforms with distribution on the Udzungwa range. Altitudinal species turnover detected in these studies has been confirmed by Menegon and Salvidio (2004). Nevertheless, substantial areas of the Udzungwas have not yet been thoroughly explored, making an increase in the current species list more likely than in the lists for the Usambaras and Ulugurus (Fig. 2d).

Species of the Udzungwas found concentrated above 600 m are listed in Table 3. Eleven open-site breeders may be added; of this total, 18 species are distributed also on the Southern Highlands, which gives the Udzungwas an even richer mix of Eastern Arc and Southern Highlands elements than the Ulugurus. A QS of 52% results from comparing all species recorded on the Udzungwas above 600 m with those recorded on the adjacent lowland (gamma diversity, Table 4). This value is greater than the values found comparing the Uluguru and Usambara lists with their adjacent lowland lists (43% and 32% respectively, Table 4).This disparity is considered in the Discussion.

Comparison of the upland faunal lists from Udzungwas with that of the East Usambara is complicated by 11 Udzungwan species being more or less open-site breeders, compared with one on the Usambaras. If lists of essentially forest-associated species (alpha diversity) of the Udzungwas and Usambaras are compared, the QS is 54%, which is close to the Uluguru–Usambara comparison value (QS of 52%, Table 5). Comparison of the forest-associated lists for the Ulugurus and Udzungwas gives a QS of 64% (Table 5), which may reflect the closer proximity of these two mountains.

Discussion

Comparison of the QS values of the assemblages on the East Usambara, Uluguru and Udzungwa mountains with their adjoining lowlands (Tables 4 and 5) introduces the complication that alpha diversity estimates are more appropriate for highland comparisons with similar habitats, while gamma diversity estimates suit highland—lowland comparisons with a diversity of habitats. This is an inherent complication that was noted in connection with Loveridge's estimates of diversity and distribution, a complication that is partly a result of the relatively small number of forest-associated species in the lowlands. Table 6 shows the comparison of upland assemblages based on gamma diversity (all species in all habitats above 600 m). The similarity values are lower than alpha diversity values in the two southern blocks, largely because different widespread species range up the slopes of

Table 6 Quotients of similarity obtained from comparing the total upland assemblages of East Usambara, Uluguru and Udzungwa mountains, gamma diversity in each case.

	Uluguru	Udzungwa
Usambara	56%	47%
Uluguru	–	56%

the different mountains. This is particularly evident on the Udzungwas, where the lowland—highland QS is 52% (Table 4). The reason for the higher lowland—highland similarity values in the Udzungwa complex cannot lie in a lack of distinctiveness or diversity of the rich upland fauna, rather it appears to lie in the greater abundance of widespread species, which occur also on the lowland. It is not clear why this should be so, but it seems paralleled by Clarke's (2000) observation noted earlier of a higher reach of 'lowland' plant species in this area. Loader et al. (2004b) reported a QS of 62% when comparing the assemblages of the Kilombero Valley and Mahenge Mountain; the higher QS value in this comparison largely results from the restricted areas sampled.

As taxonomic analysis becomes more refined, greater taxonomic differences may emerge between assemblages on the different montane areas, resulting in lower QS values. Also, as the mapping of species ranges becomes more precise, the application of general dividing lines such as the 400 and 600 m contours may prove to be unsound because each mountain has different characteristics that will affect the range of species up and down its slopes. Future estimates of diversity may therefore encounter even greater complications than were found in the present study. For the moment, however, the conclusion based on a limited sampling of the Eastern Arc chain seems clear: that there is greater similarity between assemblages on isolated montane elements of the Eastern Arc than between lowland and upland assemblages on a single mountain slope, in other words, in this equatorial region there is greater altitudinal diversity than latitudinal diversity; latitudinal turnover becomes marked only south of the tropics. The overall picture is indicative of distinct amphibian assemblages that can be characterised as temperate and tropical in a climatic, not cartographic sense (Poynton 1999). For this reason it could seem more useful to treat the Eastern Arc and coastal lowlands as two separate biodiversity hotspot regions, as in Mittermeier et al. (2004), rather than as one, as in Myers et al. (2000).

The differences indicated between montane and lowland assemblages in eastern Tanzania seem to be long-standing. As several authors have concluded (e.g. Burgess et al. 1998), the region has experienced exceptionally stable climatic regimes, but there has been sufficient climatic fluctuation in the past to create ecological links between the montane forest blocks, which have allowed migration from one block to the other. Fjeldså and Lovett (1997) have drawn a distinction between forest blocks that are speciation centres and blocks where species accumulate because they are maintained by landscape dynamics. Mainly the latter condition appears to apply to East African forest-associated amphibians, which are old species in their terminology. Molecular analysis may however show this assessment to be an over-simplification. The distinct upland and lowland species assemblages appear to be the result of settled distribution rather than local elimination of formerly more widespread species, therefore destruction of coastal forest, for example, will probably result in extinction of the species peculiar to it as there are no means of replacing them.

In whatever way the diversity of amphibians in eastern Tanzania is measured, the richness and distinctiveness of the forest-associated fauna of this area still emerges as a major feature, constituting some 48% of the total amphibian population. Yet destruction of forest already troubled Loveridge in his 1933 paper, a concern that continues to the present (e.g. Newmark 1998; Lawson and Klemens 2001; Myers et al. 2000; Mittermeier et al. 2004). Regrettably little is known about the effects of habitat disturbance on species diversity, whether the disturbance is caused by human settlement or whether it occurs naturally, with the result that at present little can be said that is definite. In general, wide-ranging, open-site breeding species seem less vulnerable to human settlement than forest-associated species, but there is small consolation in the fact that open-site breeders account for slightly over half the total species diversity in eastern Tanzania. The uniqueness of the forest-associated fauna combined with its evident vulnerability to disturbance makes it a subject for conservation concern.

Of the 42 species listed in Table 3 as being forest-associated and centred above 600 m, 19 species (45%) are endemic to one unit only. A further 5 (12%) are endemic to only two units. No species is endemic to a Usambara—Uluguru pairing, which could suggest that the Udzungwa and Usambara blocks have been connected more directly along a Rubeho—Ukaguru—Nguru chain than via the isolated Uluguru block. Given a total of 85 species recorded on the three mountains, 26% are endemic to one mountain only. The restriction of so many endemics to a single mountain with steadily diminishing habitat justifies including the Eastern Arc among the world's major biodiversity hotspots.

The situation in the Tanzanian lowlands differs in that there are relatively fewer endemics, only five among the 51 recorded species (10%). Two of the endemics appear to be closely associated with forest, *Stephopaedes howelli* and *S. usambarae*, to which the Tanzanian—southeastern Kenyan endemics *Mertensophryne micranotis* and *Hyperolius sylvaticus* can be added. Agricultural clearing and logging continue to reduce the extent of the fragmented coastal forest (Burgess et al. 2000), resulting in ever-diminishing habitat. Two endemic open-site breeders, *Bufo reesi* and *Hyperolius reesi*, appear also to have very restricted ranges but they have been found in less threatened swampy habitats in the Kilombero valley. The remaining Tanzanian endemic, *Leptopelis argenteus*, is basically an open-site breeder that is relatively widely spread in southern Tanzanian lowlands. The low proportion of endemics in the coastal lowlands compared to the highlands could lead to the conclusion that the coastal lowlands do not qualify for hotspot status. Yet against this view is the situation that the forest-associated assemblage, while small, is at least as vulnerable as the montane forest-associated assemblage. Moreover, the species and indeed genera constituting it are unique to this area and are of particular biological interest. This can be taken to justify the inclusion of the coastal forests of Tanzania and Kenya among the world's hotspots.

Acknowledgements Material and data used in this study came in large part from Professor Kim Howell and his associates in the Department of Zoology, University of Dar es Salaam, and from many associates of The Society for Environmental Exploration (Frontier Tanzania) and the Tanzania Forest Conservation Group. We thank the Tanzania Commission for Science and Technology (COSTECH research permit RCA 2110-272) and the Wildlife Division for granting permission to conduct research in Tanzania to SPL. Field work was funded by a grant from the Systematics Association and a NERC studentship (NER/S/A/2000/3366) to SPL. Part of this research was supported by the Natural History Research Fund and the Department of Zoology Research Fund of the Natural History Museum. An early version of this paper received a valuable airing at the 5th World Congress of Herpetology at the University of Stellenbosch, South Africa, 2005. Dr Arne Schiøtz is thanked for comments on species distribution.

References

Agapow P-M, Bininda-Edmonds ORP, Grandall KA, Gittleman JL, Mace GM, Marshall JC, Purvis A (2004) The impact of species concept on biodiversity studies. Quart Rev Biol 79:161–179

Anon (1998) Status of remaining forest resources. In: Burgess ND, Nummelin M, Fjeldså J, Howell KM, Lukumbyzya K, Mhando L, Phillipson P, Vandenberghe E (eds) Biodiversity and conservation of the eastern arc mountains of Tanzania and Kenya. J East Afr Nat Hist 87: ix

Barbour T, Loveridge A (1928) A comparative study of the herpetological fauna of the Uluguru and Usambara Mountains, Tanganyika Territory. Mem Museum Comp Zool, Harvard 50:87–265

Burgess ND, Fjeldså J, Botterweg R (1998) Faunal importance of the eastern arc mountains of Kenya and Tanzania. Journal of East African Nat Hist 87:37–58

Burgess ND, Clarke GP, Madgwick J, Robertson SA, Dickinson A (2000) Distribution and status. In: Burgess ND, Clarke GP (eds) The coastal forests of eastern Africa. IUCN, Gland, pp 71–81

Cherry M, Stander M, Poynton JC (1998) Amphibians of Mkomazi. In: Coe M, McWilliam N, Stone G, Packer M (eds) Mkomazi: the ecology, biodiversity and conservation of a Tanzanian savanna. Royal Geographical Society, London, pp 405–410

Clarke GP (2000) Defining the eastern African coastal forests. In: Burgess ND, Clarke GP (eds) The coastal forests of eastern Africa. IUCN, Gland, pp 9–26

De Queiroz K (2005) A unified concept of species and its consequences for the future of taxonomy. Proc Calif Acad Sci 56(Supplement 1(18)):196–215

De Sá RO, Loader SP, Channing A (2004) A new species of *Callulina* (Anura: Microhylidae) from the West Usambara Mountains, Tanzania. J Herpetol 38:219–224

Doggart NJ, Lovett J, Mhoro B, Kiure J, Burgess ND (2004) Biodiversity surveys in the forest reserves of the Uluguru Mountains: an overview of the biodiversity of the Uluguru Mountains. Tanzania Forest Conservation Group, Dar es Salaam

Fjeldså J, Lovett JC (1997) Geographical patterns of old and young species in African forest biota: the significance of specific montane areas as evolutionary centres. Biodivers Conserv 6:325–346

Grandison ACG, Ashe S (1983) The distribution, behavioural ecology and breeding strategy of the Pygmy Toad, *Mertensophryne micranotis* (Lov.). Bull Brit Museum Nat Hist (Zoology) 45:85–93

Lawson DP, Klemens MW (2001) Herpetofauna of the African rain forest: overview and recommendations for conservation. In: Weber W, White JT, Vedder A, Naughton-Treves L (eds) African rain forest ecology and conservation: an interdisciplinary perspective. Yale Universitry Press, New Haven & London, pp 291–307

Loader SP, Gower DJ, Howell KM, Doggart N, Rödel M-O, Clarke BT, de Sá RO, Cohen BL, Wilkinson M (2004a) Phylogenetic relationships of African Microhylid frogs inferred from DNA sequences of mitochondrial 12S and 16S ribosomal rRNA genes. Org, Divers Evol 4:227–235

Loader SP, Poynton JC, Mariaux J (2004b) Herpetofauna of Mahenge Mountain, Tanzania: a window on African biogeography. Afr Zool 39:71–76

Loveridge A (1933) Reports on the scientific results of an expedition to the south-western highlands of Tanganyika Territory. I. Introduction and zoogeography. Bull Museum Comp Zool, Harvard 75:1–43

Loveridge A (1937) Scientific results of an expedition to rain forest regions of eastern Africa. IX. Zoogeography and itinerary. Bull Museum Comp Zool, Harvard 79:481–541

Lovett JC (1996) Elevational and latitudinal changes in tree associations and diversity in the Eastern Arc mountains of Tanzania. J Trop Ecol 12:629–650

Magurran AE (2004) Measuring biological diversity. Blackwell Science Ltd., Oxford

Menegon M, Salvidio S (2004) Amphibians and reptiles diversity in the southern Udzungwa Scarp Forest Reserve, south-eastern Tanzania. In: Huber BA, Sinclair BJ, Lampe K-H (eds) African biodiversity: molecules, organisms, ecosystems. Springer, New York, pp 205–212

Mittermeier RA, Gil PR, Hoffman M, Pilgrim J, Brooks T, Mittermeier CG, Lamoreux J, da Fonesca AB (2004) Hotspots revisited: Earth's biologically richest and most threatened terrestrial ecoregions. Connex, Mexico City

Myers N, Mittermeier RA, Mittermeier CG, da Fonesca GAB, Kent J (2000) Biodiversity hotspots for conservation priorities. Nature 403:853–858

Newmark WD (1998) Forest area, fragmentation, and loss in the Eastern Arc Mountains: implications for the conservation of biological diversity. J East Afr Nat Hist 87:29–36

O'Hara RB (2005) Species richness estimators: how many species can dance on the head of a pin? J Anim Ecol 74:375–386

Poynton JC (1962) Zoogeography of eastern Africa: an outline based on anuran distribution. Nature 194:1217–1219

Poynton JC (1990) Composition and subtraction patterns of the East African lowland amphibian fauna. In: Peters G, Hutterer R (ed) Vertebrates in the tropics. Museum Alexander Koenig, Bonn, pp 285–296

Poynton JC (1999) Distribution of amphibians in sub-Saharan Africa, Madagascar, and Seychelles. In: Duellman WE (ed) Patterns of distribution of amphibians. Johns Hopkins University Press, Baltimore, pp 483–540

Poynton JC (2000) Amphibians. In: Burgess ND, Clarke GP (eds) The coastal forests of eastern Africa. IUCN, Gland, pp 201–210

Poynton JC (2003) Altitudinal species turnover in southern Tanzania shown by anurans: some zoogeographical considerations. Syst Biodivers 1:117–126

Poynton JC, Boycott RC (1996) Species turnover between Afromontane and eastern African lowland faunas: patterns shown by amphibians. J Biogeogr 23:669–680

Poynton JC, Broadley DG (1985) Amphibia Zambesiaca 1 Scolecomorphidae, Pipidae, Microhylidae, Hemisidae, Arthroleptinae. Ann Nat Museum 26:503–554

Poynton JC, Broadley DG (1991) Amphibia Zambesiaca 5. Zoogeography. Ann Natal Museum 32:221–277

Poynton JC, Howell KM, Clarke BT, Lovett JC (1998) A critically endangered new species of *Nectophrynoides* (Anura: Bufonidae) from the Kihansi Gorge, Udzungwa Mountains, Tanzania. Afr J Herpetol 47:59–67

Sørensen T (1948) A method of establishing groups of equal amplitude in plant sociology based on similarity of species content and its application to analyses of the vegetation on Danish commons. Biologiske Skrifter 5:1–34

Wilkinson M, Loader SP, Gower DG, Sheps JA, Cohen BL (2003) Phylogenetic relationships of African caecilians (Amphibia: Gymnophiona): insights from mitochondrial rRNA gene sequences. Afr J Herpetol 52:83–92

Biodivers Conserv (2007) 16:1119–1130
DOI 10.1007/s10531-006-9080-3

ORIGINAL PAPER

China subregional avian endemism and biodiversity conservation

Fu-Min Lei · Guo-An Wei · Hong-Feng Zhao · Zuo-Hua Yin · Jian-Li Lu

Received: 11 April 2006 / Accepted: 30 May 2006 / Published online: 12 July 2006

Abstract Richness of endemic species is considered important for biodiversity conservation and avifaunal regionalization, but no detailed studies concerning the subregional endemism have yet been done in China. Here we investigate the assemblage of China avian endemics of each avifaunal subregion, and analyze the faunal subregional species diversity, subspecific differentiation, and their relationships. The endemic species richness is found to be the highest in the Southwest Mountainous subregion where 60 species account for 57.1% of China's 105 endemics. Seventy seven species accounting for 73.3% are monotypic. Twenty eight species (26.7%) occurred in only one subregion; the mean was 3.14±1.929 subregions, while 3 species were found widespread over 8 subregions. The number of species distributed only at one specific subregion (EOSR) is the highest in Taiwan subregion. The Southwest Mountainous subregion has the second highest EOSR and the highest richness of monotypic species, but the ratio of numbers of monotypic species or EOSR to the numbers of its subregional overall endemic species is not high in this subregion (68.3%; 10.0%). On the contrary, Taiwan subregion does not have higher richness of overall endemic species and monotypic species, but the EOSR is the highest, while the ratio of numbers of monotypic species or EOSR to the numbers of its subregional overall endemic species is also the highest among all subregions (76.2%; 71.4%). From the evidence of comparing subregional distribution of overall endemic species, EOSR, monotypic species and subspecific diversification, we may conclude that geographical isolation might be the main effect factor contributing to both subspecific diversification and avian subregional endemism in China. The higher endemic and monotypic richness in the Southwest Mountainous subregion may also indicate that the subregion's avifaunal evolutionary and ecological isolation results from the highly diversified habitats and geographical environments as well as the historical effects from the primitive avifauna.

F. M. Lei (✉) · G. A. Wei · H. F. Zhao · Z. H. Yin · J. L. Lu
Institute of Zoology, Chinese Academy of Sciences, Beijing 100080, China
e-mail: leifm@ioz.ac.cn

Keywords Avifaunal subregion · Avian endemic species · Richness · Subregional endemism

Introduction

Endemic species distribution and regional richness are very useful for designing biodiversity conservation projects (Jablonski 1986; Sepkoski 1988; Balmford et al. 1996a, b, 2000; Kerr 1997; Lei et al. 2003a, b; Sfenthourakis 2001; Orme et al. 2005). Both Myers' "biodiversity hotspots" and BirdLife International's "Global Endemic Bird Areas" (EBA) are used for mapping the global and regional priority biodiversity conservation areas or spots on the basis of endemism (Stattersfield et al. 1998; Myers et al. 2000). As one of the 'megadiversity' countries of the world, and containing one of 25 global biodiversity hotspots (Myers et al. 2000; Xu et al. 2000), China plays a very important role in global biodiversity conservation for having high species richness and high endemicity (Lei et al. 2003a; SEPA et al. 1998). Avian endemism in China has been used as an indicator for setting biodiversity conservation priorities (Lei et al. 2003a, b). Chen and Peterson (2002) selected 90 threatened bird species of China to prioritize areas for conservation on evaluating the 20 areas based on the existing biosphere reserve system. Even though richness of avian endemic species has been mentioned by Lei et al. (2002a), Chen and Peterson (2002), and endemics' distribution has also been considered very important for China avifaunal regionalization (Cheng et al. 1997), no detailed studies concerning the subregional endemism have been done in sofar.

In this paper, we first compare the assemblage of each avifaunal subregion's endemicity, and finally analyze the faunal subregional species diversification. This research may be useful in conservation action plan and management from a zoogeographical perspective.

Methods

In the paper, we analyzed 105 endemic bird species. Category was followed Lei et al. (2002a) except for Chinese Merganser (*Mergus squamatus*), Western Tragopan (*Tragopan melanocephalus*), Grey-bellied Tragopan (*T. blythii*), Chinese Bulbul (*Pycnonotus sinensis*), and Red-winged Laughingthrush (*Garrulax formosus*), but 5 more species are added including the Sichuan Wood Owl (*Strix davidi*), Chestnut Bulbul (*Hemixos castanonotus*), Taiwan Bush Warbler (*Bradypterus alishannensis*) and Sillem's Mountain Finch (*Leucosticte sillemi*).

The geographic distribution database of these endemics was from Lei et al. (2002b) with references to MacKinnon et al. (2000) and Zheng (2005). The species taxonomic status followed Cheng (2000). The similarity of subregional endemic species distribution was analyzed by hierarchical cluster analysis using average linkage (within groups). Dendrogram was made from Jaccord similarity on matrix of species presence (1) and absence (0) through SPSS 13.0. The avifaunal regionalization followed Cheng et al. (1997) as in Fig. 1 and Table 1, and the features on vegetation and climate followed Zhang (1999) as in Table 1. There are currently 19 subregions belonging to 7 regions, 4 subrealms and 2 realms. Cheng's avifaunal regionalization was based on the results of avifaunal surveys throughout the country,

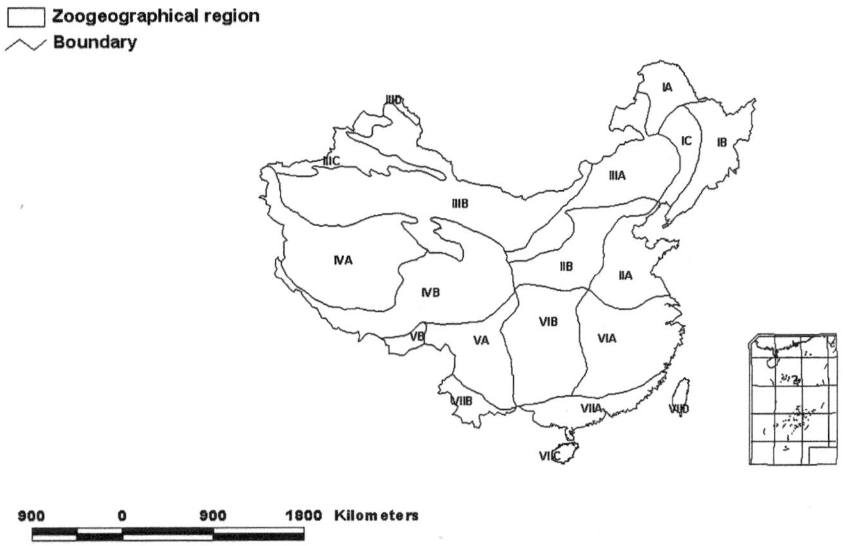

Fig. 1 The sketch map of avifaunal subregion in China

the analysis of endemic species, dominant species as well as the main economic species. Such division was made with respected not only to avifaunal diversity but also to the boundary line of soils, vegetation and climate (Cheng et al. 1997; Zhang 1999). The histograms showing numbers of subspecies and subregions were also produced through SPSS 13.0. The relationship between the subregional occupation and the subspecific diversification was tested using Spearman nonparametric correlation and Pearson correlations.

Results

Endemic species assemblage of avifaunal subregions

The distribution similarity of 105 endemic species among all these 18 subregions is made by clustering analyzing as shown in Fig. 2. VIIE (South-Sea Island) was not involved for clustering because of data shortage.

As shown in Fig. 2, all subregions can be clustered into 3 main branches. IB, IC, IIA, IIIA, IIIB, IIIC and IA are clustered in one main branch (branch 1). In this branch, IB and IC are clustered in a very near terminal branch, and then clustered with IIA, IIIB, IIIC and IA. IVB, VA, IIB, VB and IVA are clustered in one main branch (branch 2). In this branch, IVB and VA are in a terminal branch, while IVB and VA are in a terminal branch, and then with others. VIA, VIIA, VIB, VIIB, VIIC and VIID are clustered in one main branch (branch 3). In this branch, VIA and VIIA are clustered in a terminal branch, and then with others.

The dendrogram based on avian endemics assemblage reveals that most subregional distribution similarity are accordant with Cheng's avifaunal regionalization in the level of regions (Cheng et al. 1997). However, IIB is clustered with IVB and VA rather than with IIA. VIA is not first clustered with VIB but with VIIA. Evidently,

Table 1 Avifaunal regionalization of China (Ref. Cheng et al. 1997) and climate and vegetation features of subregion (Ref. Zhang et al. 1999)

Realms	Subrealms	Region	Subregion	Main temperature zone	Humidity	Main vegetation
Palaearctic Realm	East-Asian Subrealm	Northeast Region	IA: Da Xingganling Mt.	Northern temperate zone	Humid	Taiga
			IB: Changbai Mt.	Mid-temperate zone	Humid	Temperate coniferous and deciduous broad-leaved mixed forests
			IC: Song-Liao Plain	Mid-temperate zone	Sub-humid	Forests-steppe and meadow-steppe
		North-China Region	IIA: Huang-Huai Plain	Southern temperate zone	Sub-humid	Cropland, deciduous broad-leaved forests and forests-steppe
			IIB: Loess Plateau	Southern temperate zone	Sub-arid	Deciduous broad-leaved forests, forests-steppe
	Eremian Subrealm	Mongo-Xinjiang Region	IIIA: East Meadow	Mid-temperate zone	Sub-arid	Arid steppe
			IIIB: West Desert	Mid-temperate zone and southern temperate zone	Arid	Semi-desert and desert
			IIIC: Tianshan Hilly	Mid-temperate zone	Sub-arid and sub-humid	Mountain forest-steppe
	Central-Asian Subrealm	Qinghai-Xizang Region	IVA: Qiangtang Plateau	Plateau climate region (frigid and sub-frigid)	Arid and sub-arid	Steppe-meadow, meadow and highland desert
			IVB: Qinghai-Zangnan	Plateau climate region (super-frigid)	Sub-humid and sub-arid	Forest-meadow and meadow-steppe
Riental Realm	Sino-Indian Subrealm	Southwest Region	VA: Southwest Mountainous	Mid-subtropical zone (high frigid)	Humid	Mountain meadow and mountain forest (sub-highland coniferous forest, scrub-meadow, sub-tropic mixed forest, temperate deciduous forest and tropical evergreen forest)
			VB: Southeast Himalayan Slope	Mid-subtropical zone	Humid	Mountain meadow and mountain forest (mountain scrub-meadow, mountain coniferous forest, coniferous broad-leaved mixed forests and ever-green broad-leaved forest)
		Mid-China Region	VIA: Eastern Hillock-Plain	Northern tropical zone	Humid	Deciduous-evergreen broad leaf forest, evergreen broad-leaved forest and farmland
			VIB: Western Mountainous Plateau	Mid-subtropical zone	Humid	Deciduous-evergreen broad-leaved forest, evergreen broad-leaved forest and farmland

Table 1 continued

Realms	Subrealms	Region	Subregion	Main temperature zone	Humidity	Main vegetation
		South-China Region	VIIA: Min-Guang Coastal	Southern subtropical zone	Humid	Evergreen broad leaf forest, tropic monsoon rain forest and farmland
			VIIB: Southern Yunnan Hilly	Southern subtropical zone and northern tropical zone	Humid	Tropic monsoon rain forest
			VIIC: Hainan	Northern tropical zone and Mid-tropical zone	Humid	Tropic monsoon rain forest
			VIID: Taiwan	Southern subtropical zone and Northern tropical zone	Humid	Tropic monsoon rain forest and subtropical ever-green broad leaf forest
			VIIE: South-Sea Island	Mid-tropical zone	Humid	Coral island forest-scrub

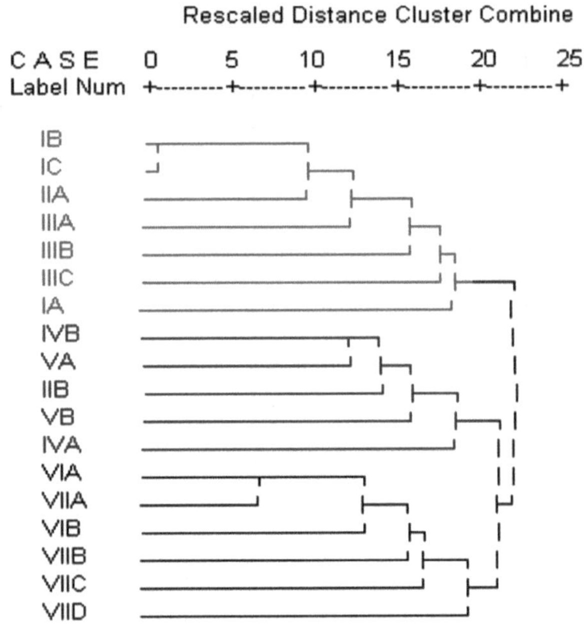

Fig. 2 Dendrogram of avifaunal subregional similarity based on species distribution

Changbai Mt. and Song-Liao Plain subregions are more similar. Eastern Hillock-Plain and Min-Guang Coastal subregions are similar. Qinghai-Zangnan and Southwest Mountainous subregions are similar. These similarities well reflect the subregional endemism and its relationship even though different subregions are different in area size.

From the temperature zone and vegetation type, subregions in branch 1 are located in the temperate zone expect for IIB, having a variety of habitats as follows: coniferous and deciduous broad-leaved mixed forests, deciduous broad-leaved forests, forests-steppe, meadow, arid steppe, cropland, semi-desert and desert. Subregions in branch 2 are located from the sub-tropic zone to the frigid Zone around the Tibet Plateau having the typical habitats of deciduous-evergreen broad-leaved forest, evergreen broad-leaved forest, tropic monsoon rain forest and farmland. And subregions in branch 3 are located in the tropic and sub-tropic zones, mostly possessing the typical sub-tropic habitats, such as evergreen broad-leaved forest, tropic monsoon rain forest and farmland (Table 1).

Subregional species diversification

Figure 3 reveals that among all avifaunal subregions, 28 species distributed in only one subregion account for 26.7% of the total 105 endemic species. 18 species are distributed in 2 subregions, 19 species are distributed in 3 subregions, 16 species are distributed in 4 subregions, 7 species are distributed in 5 subregions, 12 species in 6 subregions, 2 species in 7 subregions, and 3 species in 8 subregions, respectively. And species are distributed in averaging 3.14 ± 1.929 subregions. From the current data, *Garrulax canorus* and *Rhopophilus pekinensis* are the most widespread residential endemic species and *Grus nigricollis* is the most widespread migrant bird with the

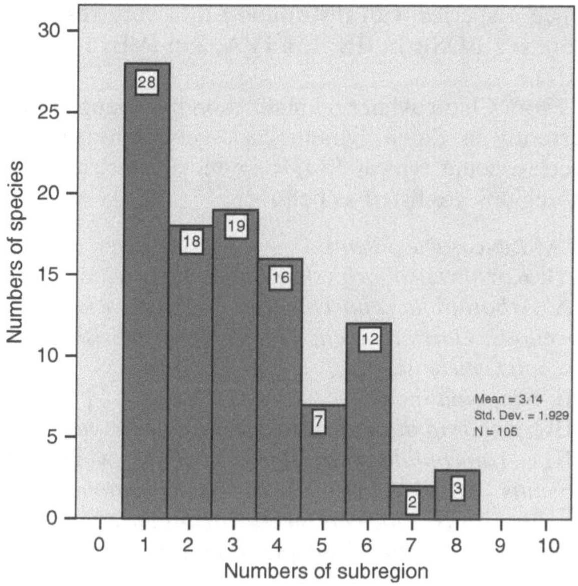

Fig. 3 Histogram of subregional occupation by endemic species

widest range of subregional distribution, these 3 species are distributed up to 8 subregions.

Figure 4 shows that endemic species are mainly distributed in VA (60), IVB (42) and VIB (35), especially in VA for housing over half of the total endemics (57.1%). The richness is also very high in IIB (33), VB (24) and VIID (21).

Species only distributed at one specific subregion reflects that it has more restricted subregional scale (area) than other widespread species which are distributed at more than one subregions. We prefer considering this species with only one subregion distribution as "EOSR" (means the endemic species encountered at one subregion) so as to simplify the expressing and analyzing in this paper. EOSR is

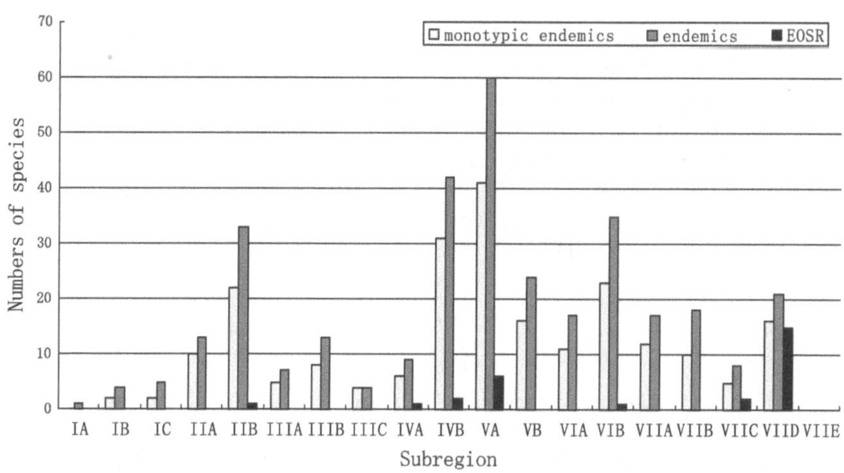

Fig. 4 Species richness (total endemic, EOSR and monotypic) at different subregions

hence a species with distribution in a very restricted area. If so, Fig. 4 show that there is 1 EOSR in IIB, 1 in IVA, 2 in IVB, 6 in VA, 1 in ViB, 2 in VIIC and 16 in VIID.

From China avian regionalization by Cheng et al. (1997), there are 19 subregions currently in China avifaunal zoo-geography (Table 1). For restricted distribution species, some typical EOSR or mostly restricted subregional species at different subregions are listed as bellow:

IVA: *Leucosticte sillemi*
IVB: *Kozlowia roborowskii*, *Babax koslowi* and *Emberiza koslowi*
VA: *Arborophila rufipectus*, *Lophophorus sclateri*, *Crossoptilon harmani*, *Moupinia poecilotis*, *Garrulax bieti*, *Liocichla omeiensis* and *Paradoxornis zappeyi* and *Phylloscopus emeiensis*
VB: *Nipponia nippon*,
VIIC: *Arborophila arden*, and *Phylloscopus hainanus*
VIID: *Arborophila crudigularis*, *Lophura swinhoii*, *Syrmaticus mikado*, *Pycnonotus taivanus*, *Urocissa caerulea*, *Tarsiger johnstoniae*, *Myiophoneus insularis*, *Garrulax morrisonianus*, *Liocichla steerii*, *Actinodura morrisoniana*, *Heterophasia auricularis*, *Yuhina brunneiceps*, *Bradypterus alishannensis*, *Regulus goodfellowi* and *Parus holsti*

Subspecific diversification

Figure 4 also shows a higher richness of monotypic species in VA. We can induce that VA has higher species richness and EOSR besides monotypic species, but the ratio of numbers of monotypic species or EOSR to the numbers of subregional overall species are lower in this subregion (68.3%; 10.0%). On the contrary, VIID has not the higher richness of overall endemic species and monotypic species, but the EOSR richness is the highest among all subregions, and the ratio of numbers of monotypic species or EOSR to the numbers of overall endemic species are also the highest among all subregions (76.2%; 71.4%).

Subspecific differentiation and subregional diversification

Among all these endemics, 77 species accounting for 73.3% are monotypic, while 19 species with 2 subspecies, 7 species with 3 subspecies (Fig. 5). The relationship between the subregional occupation and the subspecies differentiation is correlated. Pearson Correlation is 0.286, Sig. (2-tailed) = 0.003; Nonparametric Correlation—Spearman's correlation coefficient is 0.238, Sig. (2-tailed) = 0.015. These tests reveal that the correlation is significant at the 0.05 (2-tailed). This means that the larger the distribution scale (numbers of subregion), the larger degree of subregional distribution to some degree (Fig. 5).

Discussion

The clustering dendrogram reflect the similarity or assemblage of endemic species distribution among different avifaunal subregions. It is accordant with Cheng's avifaunal regionalization in the level of regions (Cheng et al. 1997). This dendrogram with three main clustering branches based on avian endemics assemblage

Fig. 5 Histogram of numbers of subspecies in a species (1 means a species with no subspecific differentiation, 2 means a species with 2 subspecies...)

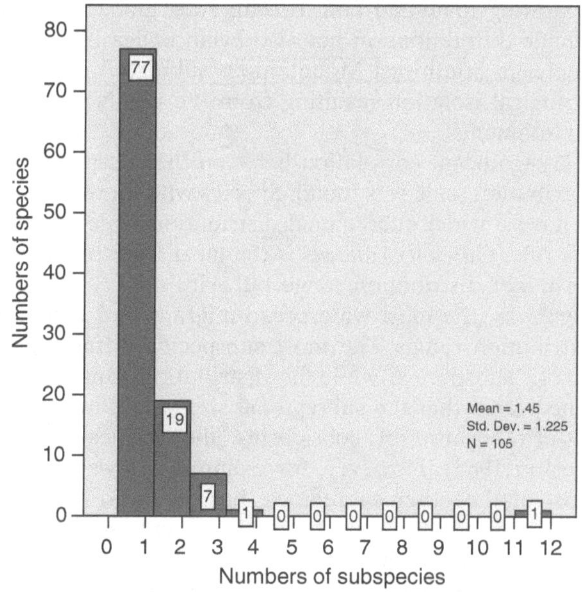

suggest that South-Sea Island subregion has no rich endemic species distributed similar to other subregions of South-China Region, while Da Xingganling Mt. subregion has also no rich endemic species distributed similar to other subregions of Northeast Region. The lacking of endemic species or data in these two subregions as well as Qiangtang Plateau Subregion is the main factor to influence the topology of clustering result.

Lei et al. (2002b) reported the taxa diversity of different subregions, the hierarchical order from highest to lowest is "VIIA-IVB-VA-VIA-VIB-IIA..."for order, "VA-IVB-VIB-IIIB-VIA-VIIA-IIIA..." for family, "VA-VIB-IVB-IIB-IIIB-VIID..." for genus and "VA-VIB-IVB-IIB-VIID..." for species. Moreover, the current paper reveals the order of EOSR as "VIID-VA-IVB & VIIC-IIB & IVA...". These suggest that the Taiwan subregion has more EOSR than any other subregions but has fewer overall endemics than the Southwest Mountainous, Qinghai-Zangnan, Western Mountainous Plateau Loess Plateau and Southeast Himalayan Slope subregions. As for distribution of avian endemics, Bird Life International has developed the Global Endemic Bird Areas (EBA) system for identifying areas of high avian biodiversity (Stattersfield et al. 1998). The EBA system provides a good platform to identify the area where the distribution of two or more of these restricted-range (smaller than 50,000 km^2) species overlaps. In this paper, we alternatively suggest the EOSR diversity distribution pattern which considers not only species geographical distribution but also ecology and local avifaunal regionalization (Cheng et al. 1997).

Monotypic taxa were valued highly for their taxonomic or evolutionary isolation (IUCN 1980). The paper shows that 77 species accounting for 73.3% of all China avian endemics are monotypic. Furthermore, this high richness of monotypic species in China reveals a highly evolutionary isolation pattern. From comparing the subregional distribution of overall endemic species, EOSR and subspecific diversification, we can draw a preliminary conclusion that the geographical isolation (such as that separating mainland and Taiwan subregion) is the main factor for not only subspecific differentiation but also avian endemism. A higher incidence of monotypic species in Southwest Mountainous subregion may also reveal the evolutionary and ecological isolation resulting from the highly diversified habitats and geographical environments.

A significant correlation between the subspecific differentiation and subregional distribution scale was found. Species with more subspecific differentiation was found to have a wider subregional distribution scale (Fig. 6); but there are exceptions to this rule. *Garrualx canorus* is the most widespread residential endemic species with the larger distribution range but with only 3 subspecies, while exceptionally *Grus nigricollis*, the most widespread migrant bird is monotypic species with the largest distribution range. The most subspecific differentiation species, *Ithaginis cruentus* has 12 subspecies, while its distribution range has 7 subregions. Generally it is conceivable that the subregional size would influence the richness and distribution pattern because of considering the species-area relationship (Arrhenius 1921; Preston 1962). However, for example, Taiwan subregion has a smaller area than most of the other subregions, but the ratio of numbers of monotypic species or EOSR to the numbers of overall endemic species is very high. Therefore, the species–area relationship and isolation effect should be considered integratively for the further research on subregional endemism.

The formation of endemism is complicated, and closely related to geology, climate, and the process of bio-evolution (Lei et al. 2003a). By comparing and analyzing Taiwan island and Hainan island, high species richness in Taiwan is considered most possibly due to Taiwan's isolation from the mainland earlier than Hainan as well as to the diversified habitats generated by Taiwan's steep and complex topography (Zhang 1999). The richness of EOSR in Taiwan is much higher than in Hainan, which confirm a longer history of independent evolution. For this reason, Lei et al. (2003a) argued that the endemism could reflect the historical geographical and geological events and the process of their speciation. From this viewpoint, the higher richness in Southwest Mountainous subregion may possibly reveal the subregion's isolation as being ecological isolation or "ecological island" even if its size is larger than Taiwan and there seems to be no geographical isolation.

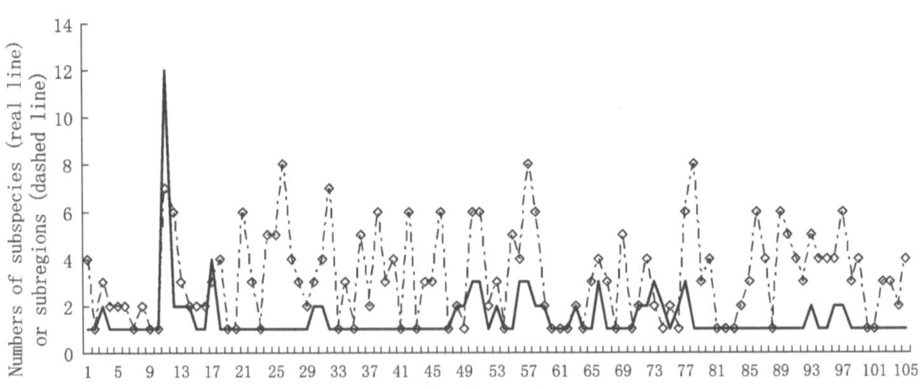

Fig. 6 The correlation between numbers of subspecies and numbers of subregion

From a zoogeographical and avifaunal viewpoint, a faunal region, which has more endemic genera, should be biogeographically older than those with fewer or no endemic genera (Cheng et al. 1981). Also, endemic taxa may indicate past Teritary or Quaternary refugia (Brown 1991). For example, in African (Crowe and Crowe 1982) and South American (Haffer 1969; Cracraft and Prum 1988) the occurrence of endemic birds with restricted distribution, are considered indicative of past refugia as evidence. The past studies of China avian endemism revealed that distribution center of endemic genera in Hengduanshan Mts. were possibly reservoirs of genetic diversity, and endemism, and avian evolutionary history (Lei et al. 2003b). Furthermore, the higher richness of endemic species (Lei et al. 2003a), the higher richness of endemic genera (Lei et al. 2003b), the higher richness of monotypic species and a higher richness of EOSR may most possibly reflect a longer evolutionary history which is often accompanied by geological effect and complementary evolutionary and ecological isolation to the subregion in producing such a new refugia of the current avifauna.

The southeastern part of the Tibet Plateau has the highest richness of EOSR and monotypic species, which is only next to the Taiwan subregion. The Plateau is in the vicinity of Southwest Mountainous, Southeast Himalayan Slope, Loess Plateau, Qiangtang Plateau and Qinghai-Zangnan subregions. This high richness area is located at the mountainous area from the Himalayan Mts to Hengduan Mts. The uplifting of the Tibet Plateau was earlier considered making the Himalayan Mts a boundary between the Palaearctic and Oriental realms fauna, while the transitional zone from east Himalayan Mts to Hengduan Mts had ever been refugia for the animals of Quaternary (Cheng et al. 1981). Our current studies reveal that Southwest Mountainous subregion and its vicinities as the transitional zone may be distributed not only the primitive relics but also the newcomers. At the same time, from a conservation perspective, conservation efforts should "focus on both hotspots of biodiversity and on associated transitional zones" (Smith et al. 2001), the Southwest Mountainous subregion and its vicinities should have a weighed importance in the biodiversity conservation, which is support of Lei's biodiversity "hotspot" (Lei et al. (2003a, b). The highly endemism subregion was thus discovered, however, endemic—rich areas sometimes corresponding with centers that may be genetically rich in species, or may be with high percentage of endemism but low species richness (UNEP 1995). Hot spots of species richness do not always show the same geographical distribution with endemism (Orme et al. 2005). More studies concerning this subregional overall biodiversity are needed.

Acknowledgements Many thanks to Carol Yeung for revising the paper and to two anonymous reviewers for their valuable comments. This paper was supported by NSFC (30170126, 30270182) and CAS Innovation Program (KSCX3-IOZ-01) to LEI F.M., and by NSFC for Fostering Talents (NSFC-J0030092) to Wei G.A., ZHAO H.F. and LU J.L.

References

Arrhenius O (1921) Species and area. J Ecol 9:95–99

Balmford A, Green MJB, Murray MG (1996a) Using highertaxon richness as a surrogate for species richness: I. Regional tests. Proc Royal Soc, London B 263:1267–1274

Balmford A, Jayasuriya AHM, Green MJB (1996b) Using higher-taxon richness as a surrogate for species richness: II. Local applications. Proc Roy Soc, Lond B 263:1571–1575

Balmford A, Lyon AJE, Lang RM (2000) Testing the highertaxon approach to conservation planning in a megadiverse group: the macrofungi. Biol Conserv 93:209–217
Brown KS (1991) Conservation of neotropical environments: insects as indicators. In: Collins NM, Thomas JA (eds) The conservation of insects and their habitats. Academic Press, London, pp 350–404
Chen GJ, Peterson AT (2002) Prioritization of areas in China for the conservation of endangered birds using modeled geographical distributions. Bird Conserv Int 12:197–209
Cheng TH, Feng ZJ, Zhang RZ, Hu SQ (1981) On the land-vertebrate fauna of Qinghai-xizang plateau with considerations concerning its history of transformation. Mem Beijing Nat Hist Museum 9:1–17 (In Chinese)
Cheng TH, Zheng GM, Zhang FY (1997) Fauna Sinica. Aves Vol l. Science Press, Beijing, China, pp 1–199 (In Chinese)
Cheng TH (2000) A complete checklist of species and subspecies of the Chinese birds. Science Press, Beijing, China, pp 1–322 (In Chinese)
Cracraft J, Prum RO (1988) Patterns and processes of diversification: specification and historical congruence in some neotropical birds. Evolution 42:603–620
Crowe TM, Crowe AA (1982) Patterns of distribution, diversity and endemism in Afrotropical birds. J Zool, London 198:417–442
Haffer J (1969) Speciation in Amazonia birds. Science 165:131–137
IUCN (1980) World conservation strategy: living resources conservation for sustainable development. IUCN, UNEP and WWF, Gland, Switzerland
Jablonski D (1986) Background and mass extinction: the alternation of macro-evolutionary regimes. Sciences (Wash) 231:129–133
Kerr JT (1997) Species richness, endemism, and the choice of areas for conservation. Conserv Biol 11(5):1094–1100
Lei FM, Lu JL, Liu Y, Qu YH, Yin ZH (2002b) Endemic birds species to China and their distribution. Acta Zoologica Sinica 5:599–608 (In Chinese)
Lei FM, Qu YH, Lu JL, Yin ZH (2003a) Conservation on diversity and distribution patterns of endemic birds in China. Biod Conserv 12(2):239–254
Lei FM, Qu YH, Lu JL, Yin ZH, Lu TC (2002a) A revision of China's endemic bird list. Acta Zootaxonomic Sinica 27(4):857–864 (In Chinese)
Lei FM, Qu YH, Tang QQ, An SC (2003b) Priorities for the conservation of avian biodiversity in China based on the distribution patterns of endemic bird genera. Biod Conserv 12(12):2487–2501
MacKinnon J, Phillipps K, He FQ (2000) A field guide to the birds of China. Henan Education Press, Changsha, China, pp 1–571
Myers N, Mittermeier RA, Mittermeier CG, Fonseca GAB, Kent J (2000) Biodiversity hotspots for conservation priorities. Nature 403:853–858
Orme CDL, Davies RG, Burgess M et al (2005) Global hotspots of species richness are not congruent with endemism or threat. Nature 436:1016–1019
Preston FW (1962) The canonical distribution of commonness and rarity: part I. Ecology 43(2):185–215
Sepkoski JJ Jr (1988) Alpha, beta, or gamma: Where does all the diversity go? Paleobiology 14:221–234
Sfenthourakis P (2001) Hotspots of endemic terrestrial invertebrates in southern Greece. Biod Conserv 10(8):1387–1417
Smith TB, Kark S, Schneider CJ, Wayne RK, Moritz C (2001) Biodiversity hotspots and beyond: the need for preserving environmental transitions. TREE 16:431
State Environmental Protection Administration (SEPA) et al (1998) China's biodiversity: a country study. China Environmental Science Press, Beijing, China, pp 1–479
Stattersfield AJ, Crowby MJ, Long AJ, Wege DC (1998) Endemic bird areas of the world: priorities for biodiversity conservation. Birdlife International, Cambridge (Conservation series 7)
UNEP (1995) Global biodiversity assessment. Cambridge University Press, Cambridge, pp 1–1140
Xu HG, Wang DH, Sun XF (2000) Biodiversity clearing-house mechanism in China: present status and future needs. Biod Conserv 9(3):361–378
Zhang RZ (1999) Zoogeography of China. Science Press, Beijing, China, pp 1–502 (In Chinese)
Zheng GM (2005) A checklist on the classification and distribution of the birds of China. Science Press, Beijing, China, pp 1–426 (In Chinese)

Biodivers Conserv (2007) 16:1131–1159
DOI 10.1007/s10531-006-9095-9

ORIGINAL PAPER

How resilient are Andean montane forest bird communities to habitat degradation?

Niall O'Dea · Robert J. Whittaker

Received: 5 August 2005 / Accepted: 3 July 2006 / Published online: 27 October 2006

Abstract The threatened forest habitats of the tropical Andes are reportedly being modified and destroyed 30% faster than their lowland tropical counterparts, but impacts on the hyper-diverse resident avifauna have received little systematic study. We present a baseline analysis of the effects of habitat modification on birds in a lower montane forest landscape in Ecuador, comparing avian community composition in landscape elements subjected to different levels of human modification: primary forest, secondary forest, edge habitat and agricultural land. We use data from a point count survey of 300 counts at 150 sites to test whether community composition and density of birds with different reported habitat preferences and foraging strategies change among landscape elements. Species richness and diversity were lowest in agricultural land, but on some measures, equally low in primary forest. Richness and diversity peaked in secondary forest and edge habitat, but ordination and density analysis revealed clear differences in their species composition. While secondary forest contained mostly forest-preferring species, edge habitat harboured a mix of forest and open-land birds. There was a clearly structured gradient in species composition across landscape elements, with densities of habitat specialists, foraging guilds and families varying considerably from primary forest to agricultural land. Agricultural land was characterised by an assemblage of widespread, abundant species very different from that in core forest habitats. As such, while the majority of montane forest birds appear resilient to a certain level of habitat modification, they cannot persist, and are displaced, where forest has been cleared outright. We argue that, for Andean montane forests, preservation of mature secondary forest offers flexibility in supplementing preserved primary forest areas to provide sufficient habitat for the persistence of this incredibly diverse but severely threatened bird community.

Keywords Ecuador · Human habitat modification · Montane forest birds · Population density · Species richness

N. O'Dea (✉) · R. J. Whittaker
Biodiversity Research Group, Centre for the Environment, University of Oxford, South Parks Road, Oxford OX1 3QY, United Kingdom
e-mail: niallodea@gmail.com

Introduction

The remaining forests of northwest Ecuador, the Choco–Andean region, are among the most severely threatened on earth, in terms of the risk of biological extinction as a result of human habitat modification. Ecuador has the highest rate of deforestation in South America and rates within the Choco–Andes exceed this national average by close to 30% (FAO 2001). For the diversity and level of endemism of its birds, plants and other organisms, the Choco–Andes is recognised as a high priority area for biodiversity conservation by all major international conservation organisations, including Conservation International, BirdLife International, the World Conservation Union (IUCN), and the World Wildlife Fund (WWF-US) (Dinerstein et al. 1995; Stattersfield et al. 1998; Myers et al. 2000; Brooks et al. 2001). This paper examines the effects of forest clearance and agricultural development on the hyper-diverse bird communities in lower montane forests of this region by observing differences in species richness and diversity, community composition and bird population densities across a gradient of human habitat modification from relatively pristine primary forest to active agricultural land.

Despite their biological importance and rapid disappearance, Andean montane forests have received very little systematic study (Kattan and Alvarez-Lopez 1996; Renjifo 2001). Hence, our baseline research derives its hypotheses primarily from studies conducted in the Neotropical lowlands, particularly Amazonia. In common with the Neotropical lowland avifauna, the threatened Andean forest avifauna is extremely diverse and is characterised by species that are resident and exhibit high levels of endemism and low individual species densities (Gentry 1986; Fjeldså et al. 1999; Bush 2002; Kattan and Franco 2004). However, there are good reasons to suspect that montane forest bird assemblages might differ from lowland counterparts in their habitat associations and sensitivity to habitat modification. First, they are richer in species: at the 1-degree latitude–longitude scale, Andean bird species richness exceeds that of the central Amazon basin by between 30% and 250% (Rahbek and Graves 2001). Second, the montane avifauna differs from that of the lowlands in both trophic structure and taxonomic composition (Renjifo 1999). Third, the present-day topography and climate of montane and lowland landscapes differ markedly, as do their climatic histories (Gentry 1986; Fjeldså et al. 1999; Bush 2002; Kattan and Franco 2004). Finally, although the Amazon basin is a dynamic landscape (Andrade and Rubio-Torgler 1994) and has long been occupied by humans (Denevan 1992; Willis et al. 2004), the frequency, persistence and pervasiveness of both natural (tectonics, volcanism, landslides, channel erosion) and anthropogenic disturbances in the Andes has arguably been far greater (Denevan 1992; Fjeldså et al. 1999).

There is an urgent need to assess whether montane forest bird communities exhibit responses to habitat modification divergent from their lowland neighbours. This assessment is of great practical relevance to conservation. It is estimated that 60% of the world's remaining forests are secondary or otherwise degraded (Chazdon 2003); assessing the potential value of partly degraded habitats in regions where pristine habitats are either unavailable, or insufficient to the task of created properly integrated reserve networks is a critical task.

Although often not the case for temperate regions, modification of tropical habitats has most often been associated with declines in species richness (e.g., Thiollay 1992, 1995, 1997; Cresswell et al. 1999; Gascon et al. 1999; Daily et al. 2001). More specifically, bird species preferring the forest understorey are frequently cited casualties (e.g., Thiollay 1992, 1995, 1997; Bierregaard and Dale 1996; Barlow et al. 2002; Sekercioglu et al. 2002) and among these understorey species, insectivores often have been identified as the foraging guild most sensitive to habitat modification (e.g., Johns 1991; Kattan et al. 1994; Thiollay 1995, 1997; Laurance et al. 2002b). Large frugivores also have been found by various authors (e.g., Thiollay 1995; Beier et al. 2002) to be sensitive to habitat modification; moreover, this has been shown to apply in studies conducted in the montane Neotropics (e.g., Kattan and Alvarez-Lopez 1996; Renjifo 1999). Other studies, however, have indicated that many bird species are capable of using—and may prosper in—secondary growth (e.g., Blake and Loiselle 2001; Laurance et al. 2002b), including hummingbirds (Stouffer and Bierregaard 1995b) and even certain endemic species (Cresswell et al. 1999; Welford 2000).

Grounded in the existing literature on the effects of habitat modification in the tropics, we developed the following research aims for our Andean montane forest system:

(1) to test whether bird species richness, diversity and density decline on a gradient from primary forest to open agricultural land;
(2) to test whether declines in species richness and density are particularly notable for species known preferentially to occupy the forest interior, while edge-tolerant and openland species exhibit greater resilience;
(3) to test whether bird foraging guilds have differential sensitivities to habitat modification.

Our overarching aim is to provide baseline empirical evidence to inform conservation strategies for Andean forest birds and other montane avifaunas.

Materials and methods

Study area

Our data derive from two areas located on the western slope of the Andes in the Pichincha province of northern Ecuador. The first study area was located in the Maquipucuna and Santa Lucia Reserves and adjacent lands (0°7′ N, 78°36′ W). The reserves cover an area of approximately 6,000 ha and an altitudinal range between 1,000 m and 2,900 m. The second study area, 25 km away, is located in the Mindo Valley (0°4′ S, 78°46′ W). This valley covers 4,058 ha and an altitudinal range between 1,200 m and 2,100 m. Annual precipitation is 3,200 mm and daily minimum and maximum air temperature average 17–26°C at 1,200 m, with little seasonal change (Rhoades and Coleman 1999). Following Holridge et al. (1971), the natural vegetation would be classified as lower montane wet forest. In both areas the landscape matrix is primarily forested, with agricultural land occurring in patches or strips.

Bird species data

Quantitative bird species data were collected in a point count survey. Point counts are a powerful method of measuring relative abundances efficiently (Whitman et al. 1997) and, combined with distance estimation, provide a means of estimating bird population densities (Buckland et al. 2001). They are also the preferred avian survey method where inferences are to be drawn about habitat selection because data collected can be directly related to the habitat measured (Remsen and Good 1996; Bibby et al. 2000; O'Dea et al. 2004), and the technique has been widely used in related studies on tropical forests elsewhere, including Central America (Gillespie and Walter 2001), Brazilian Atlantic forests (Marsden et al. 2000), Western Australia (Luck 2002), India (Raman and Sukumar 2002) and Madagascar (Watson et al. 2004).

Three hundred point counts were performed at 150 sites, 75 sites in each of the two study areas, in May and June 2003, the beginning of the dry season. Point count survey sites ranged from 1,100 m to 2,000 m in altitude and were selected without prior knowledge of bird distributions. Due to the extremely rugged terrain and access limitations, sites were located on existing trails (van Rensburg et al. 2000; Hiby and Krishna 2001; Naidoo 2004). The landscape is a mosaic of primary forest, secondary forest, and agricultural land. We stratified sites among these landscape elements, with 50 sites in each. Recognising the influence of altitude on species distribution in the Andes (Kattan and Beltran 2002), we also stratified sampling of different landscape elements across the altitudinal gradient. Due to prevailing patterns of clearance in the region, agricultural land often occurred in narrow strips between forested areas and initial surveying indicated that such sites had substantially different bird species composition from sites in more open land. Consequently, we decided to further stratify the 50 agricultural sites into 25 sites with centres < 50 m from the forest edge (hereafter 'edge sites') and 25 sites with centres ≥ 50 m from the forest edge (hereafter 'agricultural sites'). Ideally, we would have matched numbers of edge and agricultural sites to forest sites, but we were constrained by the accessibility of such sites within the survey area.

Forest site centres were located a minimum of 100 m from the forest edge. Primary forest sites were located in forested areas showing no evidence of cutting and undisturbed by humans for a minimum of 50 years; they have likely remain undisturbed far longer, but with no historical records, 50 years is a conservative estimate based on the living memory of local residents who assisted in site designation. Secondary forest sites were 15- to 20-year-old stands regenerating from abandoned agricultural lands and were characterised by remnant species from former cultivation and various successional species. They differed from primary forest sites in tree density but had similar canopy height and overall vegetation structure (N. O'Dea, unpublished data). Sites in land under active cultivation or in use as pasture were classified as agricultural. Edge sites were similar to agricultural land sites in vegetation structure, but typically had more remaining trees. Apart from these pasture trees, the edge was abrupt rather than shrubby.

Before commencing fieldwork and during a five-week pilot study in 2002, the first author practised distance estimation and became familiar with visual and auditory identification of resident birds, assisted by sound recordings (Moore et al. 1999). All distance estimation and identification was performed by the first author, while one field assistant recorded the data, and another helped locate additional bird contacts. All point counts were conducted by the first author between 06:15 and 10:00, the peak period of bird activity, For the second set of counts, the order in which sites were visited was reversed to control for any remaining time-of-day bias in detectability. Each count was 10 min in duration, balancing a maximisation of the probability of detecting all species present, while minimising the probability of multiple detection (van Rensburg et al. 2000). During this period, all bird species seen and heard within a 50 m radius were registered and their abundance recorded. This radius is considered appropriate for the detection of vocalisations (Schieck 1997). The distance to each individual bird contact was estimated to one of six radial bands (0–5 m, 5–10 m, 10–15 m, 15–20 m, 20–30 m, 30–50 m). Sites were located at least 200 m apart to minimise the risk of counting the same individual twice, although the distance sampling methodology (outlined below) is very robust to violations of this particular assumption. Differences in visual detectability exist between open and forested habitats (Bibby and Buckland 1987), making auditory identification important. Sixty-four percent of encounters were exclusively auditory. Although visual detections increased from 35% to 58% from forest to agricultural landscape elements, detection functions did not differ significantly (N. O'Dea, unpublished data).

Statistical methodology and analysis

For each bird species, data on habitat preference, foraging guild, range size and conservation status (Appendix Table 4) were extracted from Ridgely and Greenfield (2001), the most comprehensive and current source of information on Ecuadorian birds. Species were each allocated to one of three habitat preference classes: (i) understorey species were those reported to occupy ground, lower and middle strata of the forest interior; (ii) edge-tolerant species were those reported to occupy any forest stratum, including the canopy but also the forest border, as well as ground and canopy dwelling species that occasionally venture into semi-open habitat; (iii) openland species were those reported to occupy open or semi-open habitat and those with no specific habitat preference. Each species was allocated to one of five main foraging guilds based on principal food source: granivores, frugivores, omnivores, nectarivores and insectivores. Insectivores were further divided by habitat preference as above. Geographic range and conservation status, as reported in Ridgely and Greenfield (2001), follow Stattersfield et al. (1998).

Birds in flight recorded were excluded from all analyses. Species accumulation curves were used to determine the degree to which survey effort captured the bird species present in each landscape element. Four hundred and fifty-seven species are reported for the reserved area as a whole and, with such a high diversity of species, surveying was unlikely to produce plateaus in species accumulation (Blake and Loiselle 2001). Consequently, for each landscape element we calculated species richness estimates using EstimateS version 7.5 (Colwell 2005) and alpha-diversity indices using Species Diversity & Richness version 3.02 (Pisces Conservation 2002). These procedures allowed direct comparison of diversity and richness between landscape elements for a sampling intensity of 25 point count sites.

The accuracy of species richness estimators is known to be inconsistent and dependent on sample coverage, which was very difficult to determine *a priori* (Brose et al. 2003). Following the recommendation of recent studies, a suite of estimators are used herein (Keating et al. 1998; Walther and Martin 2001; Brose et al. 2003; Chiarucci et al. 2003; Melo et al. 2003). Sampling with replacement over 1,000 randomisation trials was used to generate meaningful standard deviation values for each estimator, with the exception of the Michaelis–Menten estimators. To correct for differences in sample size among landscape elements, we used rarefacted richness estimates for primary and secondary forest at the 25-sample level (Appendix Table 5 reports unrarefacted results). We tested significance of pairwise differences at $\alpha = 0.05$ using *t*-tests for two independent samples. For diversity indices, the significance of differences between landscape elements was tested by Bootstrap analysis with 10,000 random partitions. Full details and evaluations of each estimator and index are provided in Colwell and Coddington (1994), Magurran (1988) and Colwell (2005).

We calculated Chao's (Chao et al. 2005) Jaccard similarity index and it's standard deviation using EstimateS 7.5 (Colwell 2005) to examine changes in species composition among landscape elements. This estimator is designed to correct the negative bias affecting traditional measures of similarity, particularly for inexhaustive samples of rich communities (Chao et al. 2005). A detrended correspondence analysis (DECORANA, Hill 1979), conducted using CANOCO 4.5 (ter Braak and Smilauer 2002), was used to display the relationship of the sites in terms of their bird species composition. We used the untransformed quantitative bird species data, detrending by segments to correct the arch effect.

To obtain improved density estimates all point count data were analysed in the Distance 4.1 computer program (Buckland et al. 2001). The program models the decline in detectability with distance from the observer of a species or group of species to generate an estimate of population density. We stratified the analysis by landscape element to allow for any differences in visual and aural detectability between habitats. Density analyses are presented for the bird community as a whole, for species grouped by habitat preference, for foraging guilds, and for the insectivore guild subdivided by habitat preference. At these levels of aggregation, the numbers of individual observations were sufficient to generate robust density estimates for landscape elements, except where noted. The Chao 2 species richness estimator (Chao 1984; Colwell and Coddington 1994), argued to be among the most robust (Walther and Martin 2001; Herzog et al. 2002), was also applied to these subgroups for the purpose of comparison with density estimates. Significance of pairwise differences at ($\alpha = 0.05$ was tested by *z*-tests (density) and *t*-tests (richness) for two independent samples.

Results

Patterns in the bird community as a whole

We encountered 200 species belonging to 34 families during the point count survey (Appendix Table 4). One hundred and twenty-five species were recorded in primary forest, 141 within secondary forest, 132 in edge habitat and 99 in agricultural land (150 for the 50 edge and agricultural land sites combined; Table 1). As anticipated, species accumulation curves do not plateau for individual landscape elements, indicating that new species would be expected if sample size were increased (Fig. 1). The curve for the dataset as a whole, levels off considerably towards the end of the series, but did not plateau completely either.

For the two rarefaction and nine total species richness estimators, species richness for the rarefacted samples was in most cases not significantly different for edge habitat and secondary forest, but these were significantly richer than primary forest, which, in turn, was significantly richer than agricultural land (Table 1, Fig. 2a). For the eight diversity indices, by contrast, secondary forest was significantly less diverse than edge habitat in most cases, while primary forest was not significantly more diverse than agricultural land in most cases (Table 1, Fig. 2b). As such, on the whole, diversity was highest for edge habitat, followed by secondary forest, and finally primary forest and agricultural land.

Chao's Jaccard similarity index indicated a strong gradient of species composition across the gradient of habitat modification (Table 2). While primary and secondary forest were highly similar in species composition, similarity declined substantially to edge and further to agricultural land. The DCA ordination reflected these differences in species composition at the site level (Fig. 3). The first DCA axis, the strongest gradient of species composition among sites, was over five standard deviation units in length, indicating a complete turnover in species composition across this axis (Jongman et al. 1995). Primary forest sites occupy the left-most extreme of the compositional axis, while agricultural sites, with the

Table 1 Comparison of richness estimates and diversity indices for birds in primary forest (P), secondary forest (S), edge habitat (E) and agricultural land (A) in the Ecuadorian montane forest zone

Sobs	*P* 125	*S* 141	*E* 132	*A* 99
Rarefacted species richness estimates[1]				
Sample-based rarefaction	80 ± 5^{c}	107 ± 4^{b}	117 ± 6^{a}	82 ± 6^{c}
Individual-based rarefaction	82 ± 3^{c}	110 ± 3^{b}	116 ± 1^{a}	81 ± 1^{c}
Michaelis–Menten runs	111	132	140	99
Michaelis–Menten mean	101	134	141	98
Abundance-based coverage estimator	107 ± 13^{b}	125 ± 10^{a}	124 ± 11^{a}	94 ± 10^{c}
Incidence-based coverage estimator	116 ± 17^{b}	135 ± 13^{a}	132 ± 14^{a}	99 ± 12^{c}
Chao 1	106 ± 11^{b}	127 ± 11^{a}	126 ± 8^{a}	95 ± 8^{c}
Chao 2	118 ± 15^{b}	133 ± 12^{a}	132 ± 10^{a}	103 ± 11^{c}
First-order Jackknife	114 ± 8^{b}	137 ± 7^{a}	137 ± 9^{a}	102 ± 8^{c}
Second-order Jackknife	127 ± 19^{b}	148 ± 16^{a}	143 ± 20^{a}	108 ± 18^{c}
Bootstrap	99 ± 8^{b}	121 ± 7^{a}	124 ± 9^{a}	93 ± 8^{c}
Diversity indices[2]				
Shannon–Weiner *H*	3.93^{c}	4.14^{b}	4.24^{a}	3.82^{c}
Simpson's *D*	30.65^{c}	37.87^{b}	46.04^{a}	27.59^{c}
Margalef *D*	17.45^{b}	18.83^{b}	19.25^{a}	14.45^{c}
1/Berger Parker *D*	11.40^{b}	14.58^{a}	14.33^{a}	8.30^{c}
McIntosh *D*	0.84^{c}	0.86^{b}	0.88^{a}	0.83^{d}
Brillouin *D*	3.77^{c}	3.99^{b}	4.01^{a}	3.64^{c}
Fisher's Alpha	34.86^{c}	36.61^{b}	42.56^{a}	28.62^{d}
Q Statistic	33.91^{a}	37.11^{a}	39.31^{a}	26.06^{b}

Richness and diversity of landscape elements are ranked ($^{a\text{-}d}$); where values did not differ significantly at $\alpha = 0.05$, their rank is the same

[1] Means and standard deviations are reported for 1,000 randomised trails with rarefaction to equalise effort among landscape elements ($N = 25$); significance of difference in richness tested by *t*-test for all except Michaelis–Menten estimators, which do not report variance

[2] Calculated for all site samples ($N = 50$ for primary and secondary forest; $N = 25$ for edge and agricultural land); significance of difference tested by a bootstrap analysis with 10,000 randomisations

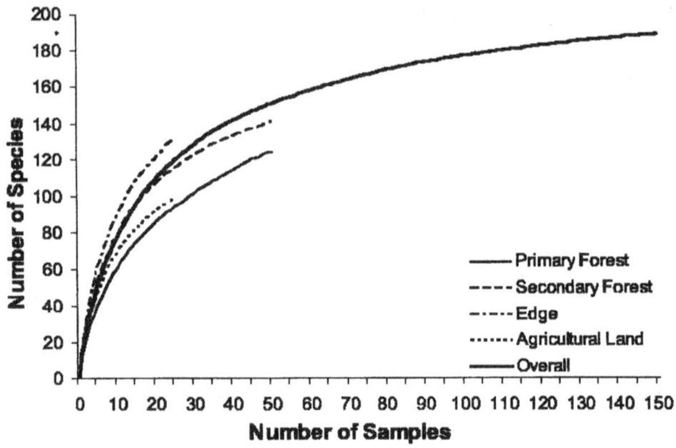

Fig. 1 Species accumulation curves, generated with EstimateS version 7 (Colwell 2004), derived from bird census data in each of four habitat elements and the dataset as a whole. Sample order is randomised to the maximal number of randomisations possible for the number of samples and the curves represent average values of over all randomisations

Fig. 2 Illustration of landscape element ranking by: (**a**) species richness estimators; and (**b**) diversity indices standardised by dividing the raw values by the maximum score for each diversity index. Raw values of rarefacted richness estimators, calculated with EstimateS for $N = 25$ counts version 7 (Colwell 2004), and of diversity indices, calculated with Species Diversity & Richness version 3.02 (Pisces Conservation 2002), are presented in Table 1. P = primary forest; S= secondary forest; E = edge; A = agricultural land

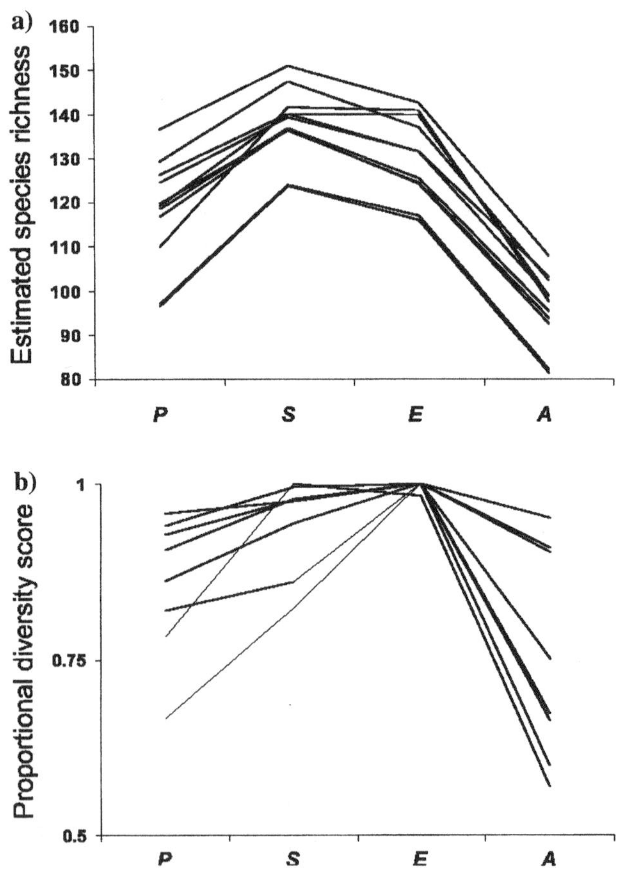

Table 2 Similarity of Ecuadorian montane forest birds among pairwise combinations of landscape elements: raw numbers of shared species are given above the diagonal and Chao's Jaccard similarity index values ± 1SD below

	P	S	E	A
P	–	103	81	52
S	94 ± 3	–	99	63
E	69 ± 9	82 ± 7	–	81
A	41 ± 11	65 ± 10	88 ± 8	–

P = primary forest (N = 50); S = secondary forest (N = 50); E = edge (N = 25); A = agricultural land (N = 25)

exception of a single site, cluster tightly at the right-most extreme. Secondary forest sites occupy an intermediary position between these extremes along the primary axis of the DCA, showing a high degree of overlap with primary forest sites but none with agricultural sites. Edge sites, by contrast, were scattered evenly across the first axis of the DCA, virtually from one extreme to the other. The second strongest axis of the DCA represented a relatively short compositional gradient for the majority of sites, at about two standard deviation units. Of the two outlying secondary forest sites, the lower was characterised by a lek (male display aggregation) of Stripe-throated Hermits (*Phaethornis striigularis*), occurring at otherwise very low densities in the area. At the higher site, a flock of 14 Band-tailed Pigeons (*Columba fasciata*) were encountered; only a single individual of this species was otherwise detected in the survey.

Comparing forest-interior, edge-tolerant and open-land birds

Overall bird density was significantly higher in secondary forest than in primary forest or agricultural land, while edge habitat harboured significantly higher bird density than agricultural land (Table 3, Fig. 4a). In contrast, forest-interior birds

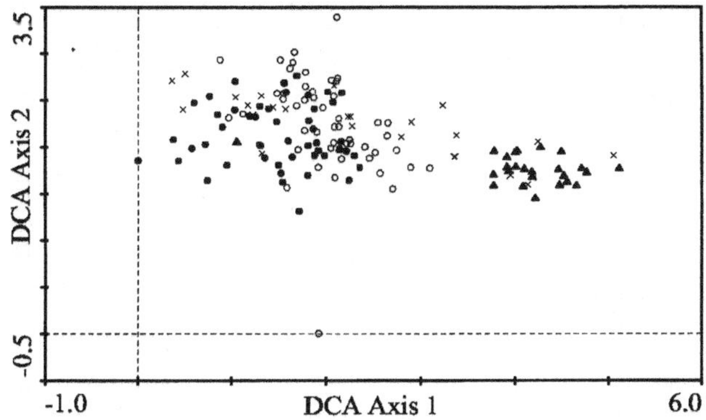

Fig. 3 Biplot of detrended correspondence analysis (DCA) of quantitative species data showing distribution of site scores along the first and second ordination axes. Plot generated using CANOCO 4.5 (ter Braak and Smilauer 2002). *Filled circles*: primary forest sites; *empty circles*: secondary forest sites; 'x's: edge sites; *filled triangles*: open agricultural land sites

Table 3 Density estimates in individuals per hectare and Chao 2 species richness estimates of birds in four landscape elements in the Ecuadorian montane forest zone

	Density[1]				Richness[2]			
	P	S	E	A	P	S	E	A
All species	90.6 ± 7.4^{bc}	128.4 ± 6.1^{a}	111.5 ± 12^{ab}	83.2 ± 7.2^{c}	118 ± 15^{b}	133 ± 12^{a}	132 ± 10^{a}	103 ± 11^{c}
Habitat preference								
Forest-interior	30.5 ± 4^{a}	26.9 ± 2.7^{a}	12.1 ± 2.4^{b}	2.6 ± 0.9^{c}	34 ± 5^{b}	39 ± 6^{a}	20 ± 4^{c}	9 ± 2^{d}
Edge-tolerant	57.1 ± 5.5^{b}	90.5 ± 5^{a}	55.7 ± 5.8^{b}	21.1 ± 4.8^{c}	68 ± 9^{b}	73 ± 7^{a}	73 ± 6^{a}	42 ± 6^{c}
Open-land	2.4 ± 0.6^{d}	12.4 ± 19^{c}	32.1 ± 5.6^{b}	71.8 ± 9.4^{a}	9 ± 3^{d}	18 ± 3^{c}	38 ± 4^{b}	47 ± 4^{a}
Foraging guild								
Frugivores	5.7 ± 1^{b}	12.3 ± 2.4^{a}	9.2 ± 2.6^{ab}	2.2 ± 1^{c}	15 ± 4^{b}	19 ± 4^{a}	14 ± 2^{b}	8 ± 2^{c}
Omnivores	19.8 ± 3.6^{c}	31.1 ± 3.7^{b}	43 ± 7.4^{a}	15.9 ± 4.2^{c}	27 ± 5^{b}	32 ± 5^{a}	34 ± 4^{a}	24 ± 4^{c}
Nectarivores	69.2 ± 16.1^{a}	30.9 ± 5.7^{b}	21.5 ± 5.1^{b}	2.4 ± 0.9^{c}	13 ± 3^{a}	13 ± 3^{a}	12 ± 2^{a}	7 ± 2^{b}
Insectivores	32.7 ± 2.8^{b}	60.6 ± 6.3^{a}	34 ± 6.3^{b}	51.5 ± 5.2^{a}	56 ± 8^{b}	63 ± 7^{a}	63 ± 5^{a}	50 ± 4^{c}
Insectivores by habitat preference								
Forest-interior	19.4 ± 3.2^{a}	13.1 ± 1.8^{b}	2.6 ± 0.9^{c}	1 ± 0.5^{c}				
Edge-tolerant	17.2 ± 1.7^{b}	38.9 ± 4.8^{a}	19.3 ± 4.4^{b}	13.4 ± 3.6^{b}				
Open-land	1.4 ± 0.4^{d}	9.3 ± 1.8^{c}	17.1 ± 3.4^{b}	49.6 ± 7.7^{a}				

Figures are reported for all species together, and as categorised by reported habitat preference and foraging guild. Density and richness in landscape elements are ranked ($^{a-d}$); where values did not differ significantly at $\alpha = 0.05$, their rank is the same. P = primary forest ($N = 50$); S = secondary forest ($N = 50$); E = edge ($N = 25$); A = agricultural land ($N = 25$)

[1] Individuals/ha ± 1SD; significance of differences among landscape elements tested by z-test

[2] Estimated species number ± 1SD (generated from 1,000 randomised trails); significance of differences among landscape elements tested by t-test

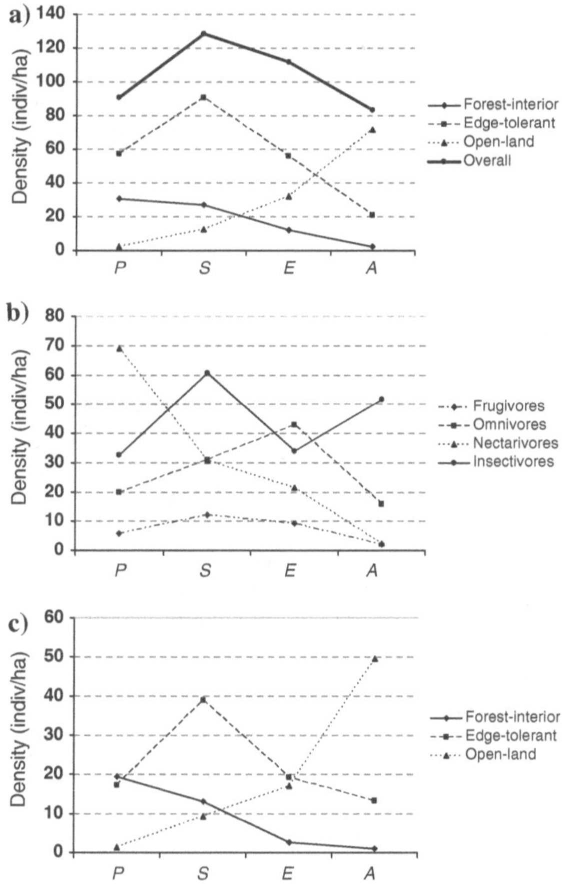

Fig. 4 Comparative density analyses of individual birds in four landscape elements: (**a**) for the dataset as a whole, and for species categorised by reported habitat preference; (**b**) for the four main bird foraging guilds; and (**c**) for insectivorous birds. Density analyses performed in Distance version 4.1 (Buckland et al. 2001). P = primary forest (N = 50 point counts); S = secondary forest (N = 50); E = edge (N=25); A = agricultural land (N = 25)

achieved highest density in primary and secondary forest, but did not differ significantly in density between them. Their density was significantly lower in edge sites, and significantly lower again in open agricultural land. In contrast, edge-tolerant species occurred at highest density in secondary forest. They occurred at similar and significantly lower densities in primary forest and edge habitat, and, like forest-interior species, occurred at lowest density in agricultural land. Finally, open-land species occurred at highest density in agricultural land. Their density decreased steeply and significantly from edge habitat to secondary forest to very low density in primary forest.

Species richness, as estimated by Chao 2, followed broadly similar patterns to density (Table 2). Forest-interior species, similar among the forest landscape elements but richest in secondary forest, declined significantly in edge habitat. Secondary forest and edge habitat had comparable numbers of edge-tolerant species, significantly more than in primary forest. By contrast, agricultural land harboured very few forest-interior species and relatively few edge-tolerant species. The species richness of open-land birds declined in a similar pattern to their density across landscape elements. Notably, however, their richness in edge habitat dropped proportionately less from that of open agricultural land compared with their density.

Comparing bird foraging guilds

Among foraging guilds overall, insectivores occurred at highest density (48.2 ± 2.2 individuals/ha), followed by nectarivores (38.5 ± 5.3 individuals/ha) and omnivores (26.7 ± 2.4 individuals/ha), while obligate frugivores were considerably less common (8.7 ± 1.3 individuals/ha). The granivores, largely comprised of seedeater species, did not occur in sufficient abundance to allow accurate estimation of their density. However, it is worth noting that granivores were absent from primary forest, and only five individuals were encountered in secondary forest (all members of the bamboo-specialising Blue Seedeater, *Amaurospiza concolor*), while in edge habitat eight individuals, belonging to a variety of species, were encountered. With 24 individuals at 17 of 25 sites, granivores clearly achieved highest abundance in agricultural land, but varied considerably in abundance among sites.

Foraging guilds exhibited different patterns of density across the gradient of habitat modification. Frugivores occurred at similar, relatively high density in secondary forest and edge habitat, at lower density in primary forest and lowest in agricultural land (Fig. 4b). Omnivores, by contrast, clearly peaked in density in edge habitat alone. Nectarivores were the only guild to exhibit a monotonic decline over the gradient of habitat modification, while the most abundant of the foraging guilds, insectivores, occurred at relatively high densities in all landscape elements, with highest densities in secondary forest and agricultural land.

Patterns of species richness were broadly similar to patterns of density for foraging guilds (Table 3). However, despite substantial differences in densities, nectarivores occurred at similar species richness for all but agricultural land, where they declined significantly. Further, insectivores were least species rich in agricultural land, despite their high density in this landscape element.

Insectivores exhibit a wide variety of feeding strategies and habitat preferences, which are not fully reflected in the above summary. Examining insectivores according to their reported habitat preferences, one can clearly differentiate the responses of separate assemblages, reflecting even more clearly the pattern seen for all species together. Forest-interior insectivores achieved peak density in primary forest, declining significantly to secondary forest, and further to edge and agricultural land (Fig. 4c). Edge-tolerant insectivores, in contrast, peaked in secondary forest. Meanwhile, insectivores reported to prefer open and semi-open habitats achieved by far their highest density in agricultural land, dropping steeply across the gradient of habitat modification to very low density in primary forest.

Discussion

Bird richness, diversity and density vary unimodally with habitat modification

Contrary to expectations based (mostly) on lowland tropical avifaunas, in the present study bird species richness did not decline monotonically on a gradient of increasing human habitat modification. Although estimated species richness for primary forest (the least-disturbed landscape element) was either indistinguishable or significantly higher than that for agricultural land (the most disturbed), the overall pattern was hump-shaped, peaking in secondary forest and edge sites (Table 1). Alpha-diversity indices reflected a similar hump-shaped pattern: a consistent peak in edge habitat, often closely matched by secondary forest, but declining for primary forest and more sharply for agricultural land. As such, secondary forest and edge habitat assemblages were more species rich with more even distributions of species abundance than either primary forest or agricultural land.

The expectation that reduced species richness will follow human habitat modification in tropical regions is based on both fragmentation and gradient-oriented studies. Fragmentation studies have traditionally treated the landscape as binary, i.e., as composed of forest and non-forest. If we had similarly treated our landscape, contrasting forest habitats together against open agricultural land, we would certainly have concluded that richness declines with habitat modification. However, it is increasingly recognised that this binary conception is insufficient to understanding processes determining species persistence and community composition (Gascon et al. 1999; Watson et al. 2004, 2005). Nevertheless, researchers examining graded effects of habitat modification have also reported linear declines in species richness. Thiollay (1992, 1995, 1997) found that selectively logged habitats in north-eastern Amazonia and agroforests in Sumatra harboured considerably lower richness and diversity of species than undisturbed forest. Creswell et al. (1999) report similar declines in species richness between pristine habitats, secondary scrub, and farmland in highland Ecuador.

The age of our secondary forest, at 15–20 years regrowth, may make it more similar to pristine habitat than the modified habitats examined in these other studies. In Uganda, Sekercioglu (2002) also found high degrees of similarity in forest bird composition between unlogged forest and forest that had been lightly logged during the 1960s, and in a recent review Dunn (2004) determined 20 years to be the mean recovery time for bird species richness in tropical secondary forests. This is, however, insufficient explanation of the hump-shaped response observed: in our study the richness and diversity of secondary forest and edge habitat exceeded that of primary forest, Blake and Loiselle (2001) note a similar pattern for montane forest in Costa Rica, as did Naidoo (2004) for rain forest in Uganda. In our study areas, overall bird densities also followed a similar hump-shaped pattern. Is this reflective of an assemblage of bird species adapted to the more intense disturbance regime and gap dynamics of montane forests? Examination of species composition dynamics provided deeper insight into this question.

Forest-interior species are less resilient

Similarity analysis and ordination of species composition, along with examination of the response of functional groups, indicated that overall patterns masked important underlying mechanisms, Similarity analysis indicated clear compositional differences among landscape elements with a decline in similarity over the gradient of habitat modification. Ordination illustrated the precise patterns of species turnover among sites. Notably, there was a clear disjunction between forest sites and open agricultural sites. The single agricultural site occurring within the forest cluster (Fig. 3) was an isolated site in an area otherwise characterised by small clearances in primary forest, indicating a sensitivity to wider landscape context (Watson et al. 2005) that we cannot explore further with present data. Primary forest and agricultural sites occupied either extreme of the first ordination axis, indicating that, despite sharing relatively low species richness, they do not share similar species compositions. Only the edge sites were distributed from one end of the ordination axis to the other, reflecting their ecotonal character. Thus, the richness and diversity of secondary forest and edge habitat were fundamentally different in origin; while the richness and diversity of secondary forest was comprised primarily of forest species, that of edge habitat resulted from a blending of forest and non-forest species components.

Examining species richness and density of birds with different reported habitat preferences allowed us to elucidate the structure of these changes in species composition. We found that forest-interior species occurred at highest richness and density in primary and secondary forest, but at much lower levels in edge sites and minimal levels in agricultural land. Birds of the forest interior—50 species including antbirds, trogons and manakins, among others—appeared unable to occupy cleared land and had only a limited capacity to utilise edge habitats (see Appendix Table 4). With the data available we can make no inferences about the willingness of these species to cross open land between forest blocks (Sekercioglu et al. 2002), but to the extent that our results reflect instantaneous movements, they indicate that open land may not just be a sub-optimal habitat for these birds, but also a barrier to their dispersal. In this regard, the result indicates that forest-interior birds respond to habitat modification in these Andean montane forests as they do elsewhere in the tropics (Bierregaard and Dale 1996; Renjifo 1999; Greenberg et al. 2000; Barlow et al. 2002; Sekercioglu et al. 2002). However, the richness and abundance of these species in secondary forest is unusual. While it may result, in part, from the proximity of primary and secondary forest in the landscape mosaic, it does, at the very least, suggest a capacity for these sensitive species to utilise secondary habitat where it has had sufficient time (approximately two decades) to regenerate.

Many of the most characteristic montane-forest birds are reported to be edge-tolerant. Among the 98 such species we encountered were quetzals, motmots, barbets, toucans and antpittas, as well as many tanagers, flycatchers and hummingbirds (Appendix Table 4). Edge-tolerators reached peak density in secondary forest, but unlike forest-interior species exhibited similar species richness in all of primary forest, secondary forest and edge habitat, indicating a greater resilience to habitat modification. However, despite being the most abundant and species-rich group, they exhibited both very low density and very low richness in open agricultural land. As such, they may tolerate a broad spectrum of forest quality, but they quite clearly are dependent on forest habitat.

The decreased richness and density of both forest interior and edge-tolerant species in agricultural land, importantly, reflects processes that may affect range-restricted endemic species: a subset of birds which did not occur in sufficient abundance to be modelled separately. Twenty-one of 23 endemic species encountered during the survey are reported as forest-interior or edge-tolerant species. In raw numbers, 169 individual endemic birds were encountered in primary forest, 129 in secondary forest, 70 in edge habitat and just 30 in agricultural land. Thus, habitat modification clearly has a negative impact on this high-priority group of species. Moreover, neither the at-risk Crested Guan (*Penelope purpurascens*), nor any of the five at-risk endemic species encountered—Dark-backed Wood-Quail (*Odontophorous melanotus*), Empress Brilliant (*Heliodoxa imperatrix*), Toucan Barbet (*Semornis ramphastinus*), Plate-billed Mountain Toucan (*Adigena laminirostris*) or Pacific Tuftedcheek (*Pseudocolaptes johnsoni*)—were encountered in agricultural land (Appendix Table 4).

Viewed proportionally, forest birds, either of the forest-interior or edge-tolerant subsets, represent the greatest fraction of birds in forest and edge habitats, but only a small proportion in agricultural land, where open-land birds predominate. Open-land birds—55 species including grassquits, seedeaters, flycatchers and several tanagers, among others—showed a pattern of density and richness precisely inverse to that of forest-interior species. They achieved highest density and richness in agricultural land, dropped markedly in secondary forest, and were virtually excluded from primary forest. Importantly, the enhanced species richness of secondary forest relative to primary forest resulted mostly from this contribution of open-land species. However, apart from agricultural land, only edge habitat retains a substantial proportion of open-land birds, emphasising its ecotonal character. Given that edge sites were in open land bordering forest, the proportional predominance of forest birds over open-land birds is surprising. It may indicate that what the birds perceive as an edge lies further from the forest border than expected.

Bird foraging guilds each respond differently to habitat modification

Certain foraging guilds have been repeatedly identified as particularly sensitive to habitat modification. Our third aim was to examine changes in density and richness of birds in different guilds over the habitat modification gradient. Granivores represent a foraging guild largely atypical of forest habitats and their patterns of occurrence in the point count survey reinforced this. None were encountered in primary forest and only the bamboo specialist Blue Seedeater was encountered in secondary forest. Unable to inhabit forest, the majority of these species may invade only once clearance has occurred, expanding their already broad geographic ranges.

Obligate frugivores such as guans, toucans, and parrots are often identified as highly forest dependent, particularly in the Andes (Kattan et al. 1994; Kattan and Alvarez-Lopez 1996; Renjifo 1999; Beier et al. 2002). In our survey, they were quite sparse in primary forest relative to the edge and particularly to secondary forest. Reasons why their density should be low in primary forest are unclear, but may be related to the availability of fruit during the survey period. Renjifo (1999) specifically identified large frugivores as extinction prone in fragmented forest. Although we can infer nothing directly about fragmentation effects, the persistence of frugivores in our unfragmented primary and secondary forest indicates that forest quality or successional status may be less important than patch size in limiting the distribution of these species. In turn, the persistence of some large frugivores, including toucans, in edge sites may result from their proximity to continuous forest or the presence of remnant trees. However, we must interpret findings related to frugivores with caution. These species pursue food resources at greater spatial scales and with higher temporal variability than most other species considered and may, thus, not be well captured by short-term, intensive survey work.

Omnivores were the only foraging guild for which the edge appeared to be the optimal landscape element, their density in edge sites being 50% greater than that in secondary forest and more than double that in either primary forest or agricultural land. Omnivores have also been identified as sensitive to habitat modification (Renjifo 1999; Greenberg et al. 2000; Raman and Sukumar 2002), but others have found them to be unaffected by or even to benefit from disturbance (Johns 1991; Thiollay 1992; Canaday 1996; Dale et al. 2000), Recalling the seminal work of Leopold (1933), the edge may represent an optimal habitat for omnivores because it affords access to spatially separated resources such as insects and fruit, or perhaps offers an optimal combined density of these two resources. Omnivores occurred at lowest density in agricultural land, but even this belies the degree to which the agricultural land community differs from that of the forest. The three most abundant species in agricultural land—the Shiny Cowbird (*Molothrus bonariensis*), Tropical Kingbird (*Tyrannus melancholicus*) and Blue-Grey Tanager (*Thraupis episcopus*)—occurred in very low numbers or not at all in forest landscape elements (Appendix Table 4), and the overall diversity of agricultural land omnivores was significantly lower (Table 3). As such, although omnivores may be thought to live more generalist lifestyles, the same omnivore community does not continue beyond the forest edge.

In other studies, nectarivores—almost exclusively comprised of hummingbirds—have been frequently identified as resilient to habitat modification and as often benefiting from edge creation (Stouffer and Bierregaard 1995a; Thiollay 1997, 1999; Renjifo 1999; Dale et al. 2000; Laurance et al. 2002b). However, in our study, nectarivores achieved their highest density in primary forest, dropping considerably to secondary forest and edge habitats and sparsely inhabiting agricultural land (Table 3, Appendix Table 4). Moreover, while secondary forest and edge habitat had similar nectarivore species richness to primary forest, agricultural land was relatively species poor. Of the 21 nectarivore species encountered, 6 were range-restricted endemics; the Violet-tailed Sylph (*Aglaiocercus coelestis*), White-whiskered Hermit (*Phaethornis yaruqui*), Brown Inca (*Coelogena wilsoni*), Purple-bibbed Whitetip (*Urosticte benjamini*), Empress Brilliant and Western Emerald (*Chlorostilbon melanorhychus*). Only two of these—the Violet-tailed Sylph and White-whiskered Hermit—were encountered in agricultural land. Although further study would be required to capture the full temporal variability of hummingbird distributions and their small-scale responses to resources (Duran and Kattan 2005), it is important to note that they appear to suffer when forest is modified and particularly when it is cleared.

The insectivore guild achieves dual peaks in density in secondary forest and agricultural land. However, looking only at the group as a whole, the abundant subset of widespread, generalist open-land insectivores obscures the negative response of understorey insectivores to habitat modification. Forest-interior insectivores, such as antbirds and woodcreepers, are perhaps the most commonly cited victims of habitat modification (Johns 1991; Kattan et al. 1994; Stouffer and Bierregaard 1995b; Thiollay 1995, 1997; Laurance et al. 2002a). In our study, this group of species peaked in density in primary forest but virtually disappeared in agricultural land. The dual peaks for insectivores overall were a result of edge-tolerant insectivores, such as warblers, in secondary forest, while in agricultural land, open-land insectivores, such as spinetails, comprised the majority. Indeed, spinetails alone accounted for almost 80% of insectivore density in our agricultural land point counts. As such, despite high insectivore densities, species richness was lowest in agricultural land, indicating an abundant but depauperate community.

Conclusion

General conclusions

Having tested in this primarily forest landscape whether bird species richness, diversity and density declined monotonically on a gradient from primary forest to open agricultural land, we found that it did not. Instead, although these values were generally higher in primary forest than agricultural land, they reached their peak in intermediary landscape elements—secondary forest and edge habitat. Looking at functional groups more closely, however, it is clear that forest-interior birds were negatively affected by habitat modification, both in terms of species richness and density. In contrast, edge-tolerant species were more resilient to a degree of habitat modification, but also declined sharply in open agricultural habitats. Open-land species unsurprisingly fared best in agricultural land, but penetrated to a very limited degree into forest habitats. As such, they benefit from clearance but are excluded when forest recovers. Bird foraging guilds exhibited idiosyncratic responses to habitat modification. Most notably, while forest-interior insectivores declined as anticipated, nectarivores also declined markedly, indicating a higher degree of sensitivity to habitat modification than has been observed in other areas (Stouffer and Bierregaard 1995a; Thiollay 1997; Renjifo 1999).

Conservation implications

It is clear that, in the tropical Andes, areas entirely deforested and left that way for the long term, through use as agricultural land, are of benefit to only a limited suite of species. On the whole, forest clearance leads to an impoverished avian community characterised by a few highly abundant, widespread, generalist species, and few, if any at-risk or endemic species. On the other hand, our data also indicate that a large fraction of this Andean montane forest bird community is resilient to the degree of habitat modification characterising the study system. Possible explanations for this resilience are contentious, rooted as they are in long-standing debates about the richness of the Andean avifauna as a whole (Fjeldså 1994; Fjeldså et al. 1999; Bush 2002). Nevertheless, comparative analysis of relevant studies could help to disentangle the contribution of elements such as anthropogenic impacts, climatic history and natural disturbance regimes to observed differences in the resilience of montane and lowland avifaunas. For these species, the threshold of tolerance appears to be slightly higher than for lowland tropical birds. At the edge, where forest and open agricultural land meet, a significant fraction of the avian community is retained. However, further from that edge, the negative effects of clearance on forest birds become apparent.

In a region where primary forest is rapidly disappearing, it may be consoling to managers attempting to conserve the Andean avifauna that regenerating secondary forest supported an equal, if not richer, assemblage of forest-dependent and edge-tolerant bird species to that of primary forest on a fairly short (15- to 20-year) timescale of recovery. Interestingly, examining the densities of birds in pristine and degraded habitats in the high Andes, Cresswell et al. (1999) found that most species, including a high proportion of restricted-range endemics, were similarly resilient to habitat degradation. In the Colombian Andes, Renjifo (2001) found that, relative to pastures, plantations enhanced the connectivity of forest patches. However, further research is needed to examine how birds in secondary forest fare when they are isolated from continuous primary forest. There is no doubt that primary forest should be conserved, where possible. Nevertheless, secondary forests may valuably serve other functions in the reserve network, for instance, as corridors connecting reserve areas.

The nature of community-scale analysis is that it necessarily under-represents processes affecting rare species. Only very focussed, long-term autecological studies can ascertain the needs of species occurring at very low densities. In the absence of such studies, and they are certainly absent for the tropical Andes, our conservation strategies for these species must be guided by looking either at related species or species that share similar behaviours or habitat requirements. Our hope is that the current study will contribute to that decision-making process.

How other Andean montane forest taxa are affected by habitat modification remains a key question and we would not suggest that the conclusions of this study could legitimately be extended to other taxa. While a recent study has demonstrated an equally sharp difference between forest and open agricultural habitats for montane-forest amphibians (Toral et al. 2002), other characteristic taxa, such as orchids or butterflies may exhibit different resiliencies. Nevertheless, in an area where mammals and reptiles occur in relatively low abundance and diversity, birds are the characteristic vertebrate of this system. Moreover, they serve crucial ecological roles as both pollinators and seed dispersers (Loiselle and Blake 1991). As such, their persistence is important not only in their own right, but for the survival of the diverse flora and fauna of the region as a whole.

Acknowledgements We are grateful to Kate Ballem, Charles Marsh, William Perez, Andy Taylor and Anna Whitfield for assistance with fieldwork in Ecuador. We thank Fundación Maquipucuna, Cooperativa Santa Lucia and the Mindo Biological Station for granting us access to their land and for their hospitality during our stay and Scott Henderson for help in initiating the project. We thank James Watson, Gustavo Kattan, and an anonymous referee for providing useful criticisms of earlier drafts of the manuscript. This research was funded in part by Fundación Maquipucuna, the Rhodes Trust, the Natural Science and Engineering Research Council of Canada (NSERC), the Vaughan Cornish Bequest, and Exeter College, Oxford.

Appendix

Table 4 Bird species encountered in Maquipucuna and Santa Lucia Reserves (0°7′ N, 78°36′ W) and in the Mindo Valley (0°4′ S, 78°46′ W) in the tropical Andes of northwest Ecuador

Species	Common name	Habitat preference[a]	Guild[b]	Endemic[c]	Risk status[d]	Primary forest	Secondary forest	Edge	Agricultural land
Crypturellus soui	Little Tinamou	E	F				1		1
Elanoides forficatus	Swallow-tailed Kite	E	I				2	10	
Buteo magnirostris	Roadside Hawk	O	R				5	3	1
Penelope purpurascens	Crested Guan	I	F		V		4		
Chamaepetes goudotii	Sickle-winged Guan	E	F			3	4	1	
Odontophorus melanonotus	Dark-backed Wood-Quail	I	F	Y	NT	3	4		
Laterallus albigularis	White-throated Crake	O	A				5	1	5
Columba fasciata	Band-tailed Pigeon	E	F			1	14		
Columba subvinacea	Ruddy Pigeon	E	F				3	8	
Columba plumbea	Plumbeous Pigeon	E	F			6	7	16	1
Leptotila verreauxi	White-tipped Dove	E	F				2	5	2
Geotrygon frenata	White-throated Quail-Dove	I	F			2	2		
Pionus sordidus	Red-billed Parrot	E	F			29	8	5	
Piaya cayana	Squirrel Cuckoo	E	I			1	10	3	7
Piaya minuta	Little Cuckoo	O	I						9
Crotophaga ani	Smooth-billed Ani	O	I						15
Tapera naevia	Striped Cuckoo	O	I						2
Glaucidium jardinii	Andean Pygmy-Owl	E	R			2			
Chordeiles minor	Common Nighthawk	E	I			1			
Phaethornis yaruqui	White-whiskered Hermit	E	N	Y		6	24	1	6
Phaethornis syrmatophorus	Tawny-bellied Hermit	E	N			12	2		
Phaethornis striigularis	Stripe-throated Hermit	E	N			5	31	1	
Doryfera ludovicae	Green-fronted Lancebill	I	N			1	2		
Florisuga mellivora	White-necked Jacobin	E	N				2	1	1
Colibri delphinae	Brown Violet-ear	O	N			1			
Chlorostilbon melanorhynchus	Western Emerald	O	N	Y				1	
Thalurania fannyi	Green-crowned Woodnymph	E	N			2	2	1	
Amazilia tzacatl	Rufous-tailed Hummingbird	O	N			1	10	6	12
Amazilia franciae	Andean Emerald	E	N				7	2	

Table 4 continued

Species	Common name	Habitat preference[a]	Guild[b]	Endemic[c]	Risk status[d]	Primary forest	Secondary forest	Edge	Agricultural land
Urosticte benjamini	Purple-bibbed Whitetip	E	N	Y		2	2		
Heliodoxa imperatrix	Empress Brilliant	E	N	Y	NT			1	
Heliodoxa jacula	Green-crowned Brilliant	E	N			2	1	4	1
Urochroa bougueri	White-tailed Hillstar	E	N			1			
Coeligena wilsoni	Brown Inca	I	N	Y		11	1		
Boissonneaua flavescens	Buff-tailed Coronet	E	N			2		14	4
Ocreatus underwoodii	Booted Racket-tail	E	N			65	74	24	
Aglaiocercus coelestis	Violet-tailed Sylph	I	N	Y		27	10	22	1
Schistes geoffroyi	Wedge-billed Hummingbird	E	N			1			2
Heliothryx barroti	Purple-crowned Fairy	I	N					1	
Calliphlox mitchellii	Purple-throated Woodstar	E	N				1		
Pharomachrus antisianus	Crested Quetzal	E	O			4			
Pharomachrus auriceps	Golden-headed Quetzal	E	O			21	16	5	
Trogon collaris	Collared Trogon	I	O			5	2	2	
Trogon personatus	Masked Trogon	I	O				5	2	
Electron platyrhynchum	Broad-billed Motmot	E	I			2	1		
Baryphthengus martii	Rufous Motmot	E	I			1	4	1	
Eubucco bourcierii	Red-headed Barbet	E	O			8	7	1	1
Semnornis ramphastinus	Toucan Barbet	E	O	Y	NT	14	11	2	
Aulachorhynchus haematopygus	Crimson-rumped Toucanet	E	F			15	26	5	1
Pteroglossus erythropygius	Pale-mandibled Aracari	E	F	Y		7	7	7	6
Andigena laminirostris	Plate-billed mountain Toucan	E	F	Y	NT	9	4		
Ramphastos brevis	Chocó Toucan	E	F	Y		2	9	4	
Ramphastos swainsonii	Chestnut-mandibled Toucan	E	F			1	1	1	
Picumnus olivaceus	Olivaceous Piculet	O	I					1	1
Piculus rivolii	Crimson-mantled Woodpecker	O	I			4			
Piculus rubiginosus	Golden-olive Woodpecker	O	I			7	4	9	3
Veniliornis fumigatus	Smoky-brown Woodpecker	E	I			7	8	2	4
Campephilus guayaquilensis	Guayaquil Woodpecker	I	I				1	1	
Campephilus pollens	Powerful Woodpecker	E	I			2	7	2	
Furnarius cinnamomeus	Pacific Hornero	O	I	Y					12
Synallaxis azarae	Azara's Spinetail	O	I				2	7	

Table 4 continued

Species	Common name	Habitat preference[a]	Guild[b]	Endemic[c]	Risk status[d]	Primary forest	Secondary forest	Edge	Agricultural land
Synallaxis brachyura	Slaty Spinetail	O	I				19	25	53
Cranioleuca erythrops	Red-faced Spinetail	E	I			5	16	15	24
Pseudocolaptes boissonneautii	Streaked Tuftedcheek	E	I					1	
Pseudocolaptes johnsoni	Pacific Tuftedcheek	E	I	Y	V	1		1	
Premnoplex brunnescens	Spotted Barbtail	I	I			20	3		
Syndactyla subalaris	Lineated Foliage-gleaner	I	I			15	19	6	3
Anabacerthia variegaticeps	Scaly-throated Foliage-gleaner	E	I			6	24		3
Philydor rufus	Buff-fronted Foliage-gleaner	E	I			8	28	4	1
Thripadectes virgaticeps	Streak-capped Treehunter	I	I			1			
Thripadectes ignobilis	Uniform Treehunter	I	I	Y			1		
Xenopus minutus	Plain Xenops	E	I				1		
Sclerurus mexicanus	Tawny-throated Leaftosser	I	I			6	1		
Dendrocincla tyrannina	Tyrannine Woodcreeper	E	I				1		
Dendrocincla fuliginosa	Plain-brown Woodcreeper	I	I			2	11	4	1
Glyphorynchus spirurus	Wedge-billed Woodcreeper	I	I			1	1		
Xiphocolaptes promeropirhynchus	Strong-billed Woodcreeper	I	I			2	1		
Xyphorhynchus erythropygius	Spotted Woodcreeper	E	I			9	9	2	2
Lepidocolaptes lachrymiger	Montane Woodcreeper	E	I			1	3	2	
Campylorhamphus pusillus	Brown-billed Scythebill	I	I				1		
Thamnophilus unicolor	Uniform Antshrike	I	I			8	8		
Thamnophilus atrinucha	Western Slaty Antshrike	I	I				2		
Dysithamnus mentalis	Plain Antvireo	I	I			8	10		
Mymotherula schisticolor	Slaty Antwren	I	I			12	11		
Drymophila caudata	Long-tailed Antbird	E	I			1		5	
Pyriglena leuconota	White-backed Fire-eye	E	I			1	3		
Myrmeciza immaculata	Immaculate Antbird	I	I			14	23	2	
Myrmeciza nigricauda	Esmeraldas Antbird	I	I	Y		12	5		
Myrmeciza exsul	Chestnut-backed Antbird	I	I				1	6	
Formicarius rufipectus	Rufous-breasted Antthrush	I	I			36	23	5	
Grallaria guatimalensis	Scaled Antpitta	I	I				2	1	
Grallaria ruficapilla	Chestnut-crowned Antpitta	E	I			2			
Grallaria flavotincta	Yellow-breasted Antpitta	E	I	Y		14		1	

Table 4 continued

Species	Common name	Habitat preference[a]	Guild[b]	Endemic[c]	Risk status[d]	Primary forest	Secondary forest	Edge	Agricultural land
Grallaricula flavirostris	Ochre-breasted Antpitta	E	I			1	2		
Scytalopus vicinior	Nariño Tapaculo	I	I	Y		18	2	5	
Scytalopus parkeri	Spillmann's Tapaculo	E	I			1			
Phyllomyias griseiceps	Sooty-headed Tyrannulet	O	I					2	1
Phillomyias cinereiceps	Ashy-headed Tyrannulet	E	I			2	4	1	
Zimmerius chrysops	Golden-faced Tyrannulet	O	O			1	24	15	8
Camptostoma obsoletum	Southern-beardless Tyrannulet	O	I						10
Elaenia pallatangae	Sierran Elaenia	O	O						1
Mecocerculus poecilocercus	White-tailed Tyrannulet	E	I			1	1	1	1
Serpophaga cinerea	Torrent Tyrannulet	O	I			1			1
Mionectes striaticollis	Streak-necked Flycatcher	E	O			1			1
Mionectes olivaceus	Olive-striped Flycatcher	E	O			1	4		
Leptopogon superciliaris	Slaty-capped Flycatcher	E	I			7	12	2	2
Pogonotriccus ophthalmicus	Marble-faced Bristle-Tyrant	E	I			1	4	1	1
Capsiempis flaveola	Yellow Tyrannulet	E	I				2	9	15
Pseudotriccus pelzelni	Bronze-olive Pygmy-Tyrant	I	I			6	3	1	
Lophotriccus pileatus	Scale-crested Pygmy-Tyrant	E	I			25	84	39	17
Todirostrum cinereum	Common Tody-Flycatcher	O	I					2	6
Myiotriccus ornatus	Ornate Flycatcher	E	I			16	35	8	1
Myiophobus flavicans	Flavescent Flycatcher	I	I			1			
Myiophobus fasciatus	Bran-coloured Flycatcher	O	I						2
Pyrrhomyias cinnamomea	Cinnamon Flycatcher	E	I			1		1	
Contopus fumigatus	Smoke-coloured Pewee	O	I				3	7	1
Sayornis nigricans	Black Phoebe	O	I					4	7
Myiarchus tuberculifer	Dusky-capped Flycatcher	E	I			1	9	6	11
Myiozetetes cayanensis	Rusty-margined Flycatcher	O	O				1		1
Myiozetetes similis	Social Flycatcher	O	O					1	8
Myiodynastes chrysocephalus	Golden-crowned Flycatcher	O	O			27	34	11	2
Tyrannus melancholicus	Tropical Kingbird	O	O					16	49
Pachyramphus versicolor	Barred Becard	E	I					1	
Pachyramphus cinnamomeus	Cinnamon Becard	O	I					1	
Pachyramphus polychopterus	White-winged Becard	O	I			2	16	5	5

Table 4 continued

Species	Common name	Habitat preference[a]	Guild[b]	Endemic[c]	Risk status[d]	Primary forest	Secondary forest	Edge	Agricultural land
Pachyramphus albogriseus	Black-and-white Becard	E	I			2	2	3	
Platypsaris homochrous	One-coloured Becard	E	I				2		
Tityra semifasciata	Masked Tityra	E	O			2	1	7	2
Tityra inquisitor	Black-crowned Tityra	E	O				1		
Pipreola rieferii	Green-and-black Fruiteater	E	F					3	
Lipaugus cryptolophus	Olivaceous Piha	I	O			1	1		
Rupicola peruviana sanguinolenta	Andean Cock-of-the-Rock	I	F			34	6		
Masius chysopterus	Golden-winged Manakin	I	F			1	4	1	
Machaeropterus deliciosus	Club-winged Manakin	I	F	Y			2		
Cyclarhis nigrirostris	Black-billed Peppershrike	E	I				1	1	2
Vireolanius leucotis	Slaty-capped Shrike-Vireo	E	I				2		
Vireo olivaceus	Red-eyed Vireo	O	I				7	3	6
Vireo leucophrys	Brown-capped Vireo	E	I			17	29	12	2
Myadestes ralloides	Andean Solitaire	E	O			3	2	3	1
Catharus dryas	Spotted Nightingale-Thrush	I	O			6	16		
Turdus obsoletus	Pale-vented Thrush	I	O			3	17		
Turdus maculirostris	Ecuadorian Thrush	O	O				2	5	5
Cinclus leucocephalus	White-capped Dipper	O	I						1
Notiochelidon cyanoleuca	Blue-and-white Swallow	O	I			1	2	2	25
Stelgidopterix ruficollis	Southern Rough-winged Swallow	O	I					4	10
Thryothorus nigricapillus	Bay Wren	E	I			7	10	9	35
Thryothorus mystacalis	Whiskered Wren	E	I				16	3	8
Troglodytes aedon	House Wren	O	I					12	43
Troglodytes solsititialis	Mountain Wren	I	I			2			
Henicorhina leucophrys	Grey-breasted Wood-Wren	E	I			101	116	35	8
Microcerculus marginatus	Southern Nightingale-Wren	I	I			6	8	3	4
Parula pitiayumi	Tropical Parula	E	I			56	104	31	42
Geothlypis semiflava	Olive-crowned Yellowthroat	O	I				1	1	16
Myioborus miniatus	Slate-throated Whitestart	E	I			59	63	46	13
Basileuterus tristriatus	Three-striped Warbler	I	I			107	101	8	2
Basileuterus coronatus	Russet-crowned Warbler	E	I			1	1	4	
Basileuterus fulvicauda	Buff-rumped Warbler	I	I			1			3

Table 4 continued

Species	Common name	Habitat preference[a]	Guild[b]	Endemic[c]	Risk status[d]	Primary forest	Secondary forest	Edge	Agricultural land
Coereba flaveola	Bananaquit	O	O			2	1	1	14
Diglossopis cyanea	Masked Flowerpiercer	O	O					7	
Diglossa albilatera	White-sided Flowerpiercer	E	O			2	1	8	1
Pipraeidea melanonota	Fawn-breasted Tanager	O	I					2	1
Euphonia laniirostris	Thick-billed Euphonia	O	F			1	4	5	5
Euphonia cyanocephala	Golden-rumped Euphonia	O	F						1
Euphonia xanthogaster	Orange-bellied Euphonia	I	F			63	79	36	11
Euphonia saturata	Orange-crowned Euphonia	O	F						3
Chlorochrysa phoenicotis	Glistening-green Tanager	I	I	Y		2	3		
Tangara rufigula	Rufous-throated Tanager	I	O	Y		1	7		
Tangara arthus	Golden Tanager	E	O			17	44	19	17
Tangara icterocephala	Silver-throated Tanager	E	O			1	2	1	
Tangara parzudakii	Flame-faced Tanager	E	O			4	13	13	
Tangara rufivertex	Golden-naped Tanager	E	O			3	10	7	
Tangara labradorides	Metallic-green Tanager	E	O			5	13	2	
Tangara nigroviridis	Beryl-spangled Tanager	E	O			13	7	3	
Tangara cyanicollis	Blue-necked Tanager	O	O					7	1
Tangara gyrola	Bay-headed Tanager	E	O			1	1	4	13
Anisognathus somptuosus	Blue-winged Mountain-Tanager	E	I			1	7	11	
Anisognathus notabilis	Black-chinned Mountain-Tanager	E	I	Y		1		1	2
Thraupis episcopus	Blue-grey Tanager	O	O				1	19	40
Thraupis palmarum	Palm Tanager	O	O					2	5
Ramphocelus icteronotus	Lemon-rumped Tanager	O	I				21	30	106
Piranga leucoptera	White-winged Tanager	E	O			7	5	1	
Chlorothraupis stolzmanni	Ochre-breasted Tanager	E	I	Y		3	22	2	3
Tachyphonus rufus	White-lined Tanager	O	I					5	2
Chlorospingus semifuscus	Dusky Bush-Tanager	E	O	Y		35	15	22	
Chlorospingus flavigularis	Yellow-throated Bush-Tanager	E	O			40	51	1	4
Saltator maximus	Buff-throated Saltator	E	O			4	4	10	20
Saltator atripennis	Black-winged Saltator	E	O				7	3	9
Pitylus grossus	Slate-coloured Grosbeak	I	O			1	3		
Volatinia iacarina	Blue-black Grassquit	O	G					1	6

Table 4 continued

Species	Common name	Habitat preference[a]	Guild[b]	Endemic[c]	Risk status[d]	Primary forest	Secondary forest	Edge	Agricultural land
Tiaris obscura	Dull-coloured Grassquit	O	G						4
Tiaris ofivacea	Yellow-faced Grassquit	O	G					1	4
Sporophila corvina	Variable Seedeater	O	G					4	2
Sporophila nigricollis	Yellow-bellied Seedeater	O	G					2	12
Amaurospiza concolor	Blue Seedeater	I	G				5	1	2
Atlapetes tricolor	Tricoloured Brush-Finch	I	O			3		1	1
Atlapetes brunneinucha	Chestnut-capped Brush-Finch	I	O			3	8	1	
Arremon aurantiirostris	Orange-billed Sparrow	I	O			2	2		
Zonotrichia capensis	Rufous-collared Sparrow	O	O					7	1
Psarocolius angustifrons	Russet-backed Oropendola	E	O				16	2	
Molothrus bonariensis	Shiny Cowbird	O	O					63	26

Nomenclature, habitat preference, feeding guild, endemism, and extinction risk status derived from Ridgely and Greenfield (2001). Also reported is the abundance of each species within each landscape element in the present study. Distribution of point counts: Primary forest, $N = 50$; Secondary forest, $N = 50$; Edge, $N = 25$; Agricultural land, $N = 25$

[a] Habitat preference abbreviations: I, forest-interior species; E, edge-tolerant species; O, open-land species

[b] Guild abbreviations: A, aquatic; F, frugivore; G, granivore; I, insectivore; N, nectarivore; O, omnivore; R, raptor

[c] 'Y' indicates endemic species, defined as those with geographic ranges less than 50,000 km^2

[d] IUCN risk status as reported in Stattersfield et al. (1998): NT, near-threatened; V, vulnerable

Table 5 Comparison of richness estimates for birds in four landscape elements in the Ecuadorian montane forest zone, including comparisons of Chao 2 values for habitat preference and foraging guilds

Sobs	Primary forest 125	Secondary forest 141	Edge 132	Agricultural land 99
Species richness estimators				
Sample-based rarefaction	97 ± 6[c]	124 ± 5[a]	117 ± 6[b]	82 ± 6[d]
Individual-based rarefaction	97 ± 1[c]	124 ± 1[a]	116 ± 1[b]	81 ± 1[d]
Michaelis–Menten runs	119	140	140	99
Michaelis–Menten mean	110	142	141	98
Abundance-based coverage estimator	120 ± 10[c]	136 ± 7[a]	124 ± 11[b]	94 ± 10[d]
Incidence-based coverage estimator	125 ± 12[c]	139 ± 8[a]	132 ± 14[b]	99 ± 12[d]
Chao 1	119 ± 8[c]	137 ± 7[a]	126 ± 8[b]	95 ± 8[d]
Chao 2	126 ± 11[c]	140 ± 8[a]	132 ± 10[b]	103 ± 11[d]
First-order Jackknife	129 ± 8[c]	147 ± 6[a]	137 ± 9[b]	102 ± 8[d]
Second-order Jackknife	137 ± 18[b]	151 ± 13[a]	143 ± 20[b]	108 ± 18[d]
Bootstrap	117 ± 7[c]	137 ± 6[a]	124 ± 9[b]	93 ± 8[d]
Chao 2 by habit preference				
Forest-interior	37 ± 4[b]	42 ± 4[a]	20 ± 4[c]	9 ± 2[d]
Edge-tolerant	75 ± 7[b]	77 ± 5[a]	73 ± 6[b]	42 ± 6[c]
Open-land	11 ± 3[d]	18 ± 2[c]	38 ± 4[b]	47 ± 4[a]
Chao 2 by foraging guild				
Frugivores	16 ± 2[b]	20 ± 3[a]	14 ± 2[c]	8 ± 2[d]
Omnivores	29 ± 41[b]	34 ± 3[a]	34 ± 4[a]	24 ± 4[c]
Nectarivores	14 ± 3[a]	13 ± 2[b]	12 ± 2[c]	7 ± 2[d]
Insectivores	62 ± 6[b]	67 ± 5[a]	63 ± 5[b]	50 ± 4[c]

Means and standard deviations are reported for 1,000 randomised trails without rarefaction ($N = 50$ for primary and secondary forest; $N = 25$ for edge and agricultural land); significance of difference in richness was tested by *t*-test for all Michaelis–Menten estimators which do not report variance. Richness of landscape elements is ranked ([a-d]); where values did not differ significantly at $\alpha = 0.05$, their rank is the same

References

Andrade GI, Rubio-Torgler H (1994) Sustainable use of the tropical rain forest: evidence from the avifauna in a shifting cultivation habitat mosaic in the Colombian Amazon. Conserv Biol 8:545–554

Barlow J, Haugaasen T, Peres CA (2002) Effects of ground fires on understorey bird assemblages in Amazonian forests. Biol Conserv 105:157–169

Beier P, Van Drielen M, Kankam BO (2002) Avifaunal collapse in West African forest fragments. Conserv Biol 16:1097–1111

Bibby CJ, Buckland ST (1987) Bias of bird census results due to detectability varying with habitat. Acta Oecol 8:103–112

Bibby CJ, Burgess ND, Hill DA (2000) Bird census techniques. Academic Press, London

Bierregaard RO, Dale VH (1996) Islands in an ever-changing sea: the ecological and socioeconomic dynamics of Amazonian rainforest fragments. In: Schelhas J, Greenberg R (eds) Forest patches in tropical landscapes. Island Press, Washington, DC, pp 187–204

Blake JG, Loiselle BA (2001) Bird assemblages in second-growth and old-growth forests, Costa Rica: perspectives from mist nets and point counts. Auk 118:304–326

Brooks T, Balmford A, Burgess N, Hansen LA, Moore J, Rahbek C, Williams P, Bennun LA, Byaruhanga A, Kasoma P, Njoroge P, Pomeroy D, Wondafrash M (2001) Conservation priorities for birds and biodiversity: do East African important bird areas represent species diversity in other terrestrial vertebrate groups? Ostrich (suppl 15):3–12

Brose U, Martinez ND, Williams RJ (2003) Estimating species richness: sensitivity to sample coverage and insensitivity to spatial patterns. Ecology 84:2364–2377
Buckland ST, Anderson DR, Burnham KP, Laake JL, Borchers DL, Thomas L (2001) Introduction to distance sampling: estimating abundance of biological populations. Oxford University Press, Oxford
Bush MB (2002) Distribution change and conservation on the Andean flank: a palaeoecological perspective. Global Ecol Biogeogr 11:463–473
Canaday C (1996) Loss of insectivorous birds along a gradient of human impact in Amazonia. Biol Conserv 77:63–77
Chao A (1984) Non-parametric estimation of the number of classes in a population. Scand J Stat 11:265–270
Chao A, Chazdon RL, Colwell RK, Shen T-J (2005) A new statistical approach for assessing compositional similarity based on incidence and abundance data. Ecol Lett 8:148–159
Chazdon RL (2003) Tropical forest recovery: legacies of human impact and natural disturbances. Perspect Plant Ecol Evol Syst 6:51–71
Chiarucci A, Enright NJ, Perry GLW, Miller BP, Lamont BB (2003) Performance of nonparametric species richness estimators in a high diversity plant community. Divers Distrib 9:283–295
Colwell RK (2005) EstimateS: statistical estimation of species richness and shared species from samples. Version 7.5. User's guide and application published at: http://www.purl.oclc.org/estimates
Colwell RK, Coddington JA (1994) Estimating terrestrial biodiversity through extrapolation. Philos Trans Roy Soc Lond Ser B 345:101–118
Cresswell W, Hughes M, Mellanby R, Bright S, Catry P, Chaves J, Freile J, Gabela A, Martineau H, Macleod R, McPhie F, Anderson N, Holt S, Barabas S, Chapel C, Sanchez T (1999) Densities and habitat preferences of Andean cloud-forest birds in pristine and degraded habitats in north-eastern Ecuador. Bird Conserv Int 9:129–145
Daily GC, Ehrlich PR, Sanchez-Azofeifa GA (2001) Countryside biogeography: use of human-dominated habitats by the avifauna of southern Costa Rica. Ecol Appl 11:1–13
Dale S, Mork K, Solvang R, Plumptre AJ (2000) Edge effects on the understory bird community in a logged forest in Uganda. Conserv Biol 14:265–276
Denevan WM (1992) The pristine myth – the landscape of the Americas in 1492. Ann Assoc Am Geogr 82:369–385
Dinerstein E, Olson DM, Graham DJ, Webster AL, Primm SA, Bookbinder MP, Ledec G (1995) A conservation assessment of the terrestrial ecoregions of Latin America and the Caribbean. World Bank, Washington, DC
Dunn RR (2004) Recovery of faunal communities during tropical forest regeneration. Conserv Biol 18:302–309
Duran SM, Kattan GH (2005) A test of the utility of exotic tree plantations for understorey birds and food resources in the Colombian Andes. Biotropica 37:129–135
FAO (2001) State of the world's forests 2001. Food and Agriculture Organisation of the United Nations, Rome
Fjeldså J (1994) Geographical patterns for relict and young species of birds in Africa and South America and implications for conservation priorities. Biodivers Conserv 3:207–226
Fjeldså J, Lambin E, Mertens B (1999) Correlation between endemism and local ecoclimatic stability documented by comparing Andean bird distributions and remotely sensed land surface data. Ecography 22:63–78
Gascon C, Lovejoy TE, Bierregaard RO, Malcolm JR, Stouffer PC, Vasconcelos HL, Laurance WF, Zimmerman B, Tocher M, Borges S (1999) Matrix habitat and species richness in tropical forest remnants. Biol Conserv 91:223–229
Gentry AH (1986) Species richness and floristic composition of the Choco region plant communities. Caldasia 15:71–79
Gillespie TW, Walter H (2001) Distribution of bird species richness at a regional scale in tropical dry forest of Central America. J Biogeogr 28:651–662
Greenberg R, Bichier P, Angon AC (2000) The conservation value for birds of cacao plantations with diverse planted shade in Tabasco, Mexico. Anim Conserv 3:105–112
Herzog SK, Kessler M, Cahill TM (2002) Estimating species richness of tropical bird communities from rapid assessment data. Auk 119:749–769
Hiby L, Krishna MB (2001) Line transect sampling from a curving path. Biometrics 57:727–731
Hill MO (1979) DECORANA: a FORTRAN program for detrended correspondence analysis and reciprocal averaging. Cornell University Press, Ithaca, NY

Holdridge LR, Grenke WC, Hatheway WH, Liang T, Tosi JA Jr (1971) Forest environments in tropical life zones: a pilot study. Pergamon Press, Oxford
Johns AD (1991) Responses of Amazonian rain-forest birds to habitat modification. J Trop Ecol 7:417–437
Jongman RHG, ter Braak CJF, van Tongeren OFR (1995) Data analysis in community and landscape ecology. Cambridge University Press, Cambridge
Kattan GH, Alvarez-Lopez H (1996) Preservation and management of biodiversity in fragmented landscapes in the Colombian Andes. In: Schelhas J, Greenberg R (eds) Forest patches in tropical landscapes. Island Press, Washington, DC, pp 3–18
Kattan GH, Beltran JW (2002) Rarity in antpittas: territory size and population density of five *Grallaria* spp. in a regenerating habitat mosaic in the Andes of Colombia. Bird Conserv Int 12:231–240
Kattan GH, Franco P (2004) Bird diversity along elevational gradients in the Andes of Colombia: area and mass effects. Global Ecol Biogeogr 13:451–458
Kattan GH, Alvarez-Lopez H, Giraldo M (1994) Forest fragmentation and bird extinctions – San-Antonio 80 years later. Conserv Biol 8:138–146
Keating KA, Quinn JF, Ivie MA, Ivie LL (1998) Estimating the effectiveness of further sampling in species inventories. Ecol Appl 8:1239–1249
Laurance WF, Albernaz AKM, Schroth G, Fearnside PM, Bergen S, Venticinque EM, Da Costa C (2002a) Predictors of deforestation in the Brazilian Amazon. J Biogeogr 29:737–748
Laurance WF, Lovejoy TE, Vasconcelos HL, Bruna EM, Didham RK, Stouffer PC, Gascon C, Bierregaard RO, Laurance SG, Sampaio E (2002b) Ecosystem decay of Amazonian forest fragments: a 22-year investigation. Conserv Biol 16:605–618
Leopold A (1933) Game management. C. Scribner's Sons, London
Loiselle BA, Blake JG (1991) Resource abundance and temporal variation in fruit-eating birds along a wet forest elevational gradient in Costa Rica. Ecology 72:180–193
Luck GW (2002) The habitat requirements of the rufous treecreeper (*Climacteris rufa*). 1. Preferential habitat use demonstrated at multiple spatial scales. Biol Conserv 105:383–394
Magurran AE (1988) Ecological diversity and its measurement. Princeton University Press, Princeton, NJ
Marsden SJ, Whiffin M, Sadgrove L, Guimaraes P (2000) Parrot populations and habitat use in and around two lowland Atlantic forest reserves, Brazil. Biol Conserv 96:209–217
Melo AS, Rodrigo AS, Santos AJ, Shepherd GJ, Machado G, Medeiros HF, Saways RJ (2003) Comparing species richness among assemblages using sample units: why not use extrapolation methods to standardize different sample sizes? Oikos 101:398–410
Moore JV, Coopmans P, Ridgely RS, Lysinger M (1999) The birds of northwest Ecuador. Volume I: the upper foothills and subtropics. John V. Moore Nature Recordings, San Jose, CA
Myers N, Mittermeier RA, Mittermeier CG, da Fonseca GAB, Kent J (2000) Biodiversity hotspots for conservation priorities. Nature 403:853–858
Naidoo R (2004) Species richness and community composition of songbirds in a tropical forest-agricultural landscape. Anim Conserv 7:93–105
O'Dea N, Watson JEM, Whittaker RJ (2004) Rapid assessment in conservation research: a critique of avifaunal assessment techniques illustrated by Ecuadorian and Madagascan case study data. Divers Distrib 10:55–63
Pisces Conservation (2002) Species diversity and richness. Pisces Conservation Limited, Lymington
Rahbek C, Graves GR (2001) Multiscale assessment of patterns of avian species richness. Proc Natl Acad Sci USA 98:4534–4539
Raman TRS, Sukumar R (2002) Responses of tropical rainforest birds to abandoned plantations, edges and logged forest in the Western Ghats, India. Anim Conserv 5:201–216
Remsen JV, Good DA (1996) Misuse of data from mist-net captures to assess relative abundance in bird populations. Auk 113:381–398
Renjifo LM (1999) Composition changes in a subandean avifauna after long-term forest fragmentation. Conserv Biol 13:1124–1139
Renjifo LM (2001) Effect of natural and anthropogenic landscape matrices on the abundance of subandean bird species. Ecol Appl 11:14–31
Rhoades CC, Coleman DC (1999) Nitrogen mineralization and nitrification following land conversion in montane Ecuador. Soil Biol Biochem 31:1347–1354
Ridgely RS, Greenfield PJ (2001) The birds of Ecuador. Cornell University Press, Ithaca, NY
Schieck J (1997) Biased detection of bird vocalizations affects comparisons of bird abundance among forested habitats. Condor 99:179–190

Sekercioglu CH (2002) Effects of forestry practices on vegetation structure and bird community of Kibale National Park, Uganda. Biol Conserv 107:229–240
Sekercioglu CH, Ehrlich PR, Daily GC, Aygen D, Goehring D, Sandi RF (2002) Disappearance of insectivorous birds from tropical forest fragments. Proc Natl Acad Sci USA 99:263–267
Stattersfield AJ, Crosby MJ, Long AJ, Wege DC (1998) Endemic bird areas of the world: priorities for biodiversity conservation. BirdLife International, Cambridge
Stouffer PC, Bierregaard RO (1995a) Effects of forest fragmentation on understory hummingbirds in Amazonian Brazil. Conserv Biol 9:1085–1094
Stouffer PC, Bierregaard RO (1995b) Use of Amazonian forest fragments by understory insectivorous birds. Ecology 76:2429–2445
ter Braak CJF, Smilauer P (2002) Canoco for Windows. Biometris – Plant Research International, Wageningen
Thiollay JM (1992) Influence of selective logging on bird species-diversity in a Guianan rain-forest. Conserv Biol 6:47–63
Thiollay JM (1995) The role of traditional agroforests in the conservation of rain-forest bird diversity in Sumatra. Conserv Biol 9:335–353
Thiollay JM (1997) Disturbance, selective logging and bird diversity: a neotropical forest study. Biodivers Conserv 6:1155–1173
Thiollay JM (1999) Responses of an avian community to rain forest degradation. Biodivers Conserv 8:513–534
Toral E, Feinsinger P, Crump ML (2002) Frogs and a cloud-forest edge in Ecuador. Conserv Biol 16:735–744
van Rensburg BJ, Chown SL, van Jaarsveld AS, McGeoch MA (2000) Spatial variation and biogeography of sand forest avian assemblages in South Africa. J Biogeogr 27:1385–1401
Walther BA, Martin JL (2001) Species richness estimation of bird communities: how to control for sampling effort? Ibis 143:413–419
Watson JEM, Whittaker RJ, Dawson TP (2004) Habitat structure and proximity to forest edge affect the abundance and distribution of forest-dependent birds in tropical coastal forests of southeastern Madagascar. Biol Conserv 120:311–327
Watson JEM, Whittaker RJ, Freudenberger D (2005) Bird community responses to habitat fragmentation: how consistent are they across landscapes? J Biogeogr 32:1353–1370
Welford MR (2000) The importance of early successional habitats to rare, restricted-range, and endangered birds in the Ecuadorian Andes. Bird Conserv Int 10:351–359
Whitman AA, Hagan JM, Brokaw NVL (1997) A comparison of two bird survey techniques used in a subtropical forest. Condor 99:955–965
Willis KJ, Gillson L, Brncic TM (2004) How ‘virgin’ is virgin rainforest? Science 304:402–403

Biodivers Conserv (2007) 16:1161–1173
DOI 10.1007/s10531-006-9096-8

ORIGINAL PAPER

Human and natural impacts on forests along lower Tana river, Kenya: implications towards conservation and management of endemic primate species and their habitat

Nancy Nthenya Moinde-Fockler · Nicholas Otienoh Oguge · Genesio Mugambi Karere · Daniel Otina · Mbaruk Abdalla Suleman

Received: 9 August 2005 / Accepted: 3 July 2006 / Published online: 27 October 2006

Abstract Seventy-three forest patches were assessed to determine the effects of human and natural impact on native forests along the Lower Tana River flood plains in Kenya between January and March 2001. Seventeen of these forests were within the Tana River Primate National Reserve (TRPNR) while 56 were outside the protected area. Cultivation and dyke construction had the most devastating human impact, which involved partial or complete forest clearing resulting in further fragmentation of forest patches [Suleman MA, Wahungu GM, Mouria PK, Karere GM, Oguge N, Moinde NN (2001) Tana River primate census and forest evaluation. A report to Kenya Wildlife Services]. Natural impacts were either die back or flooding, which appeared to cause progressive degradation of forest structure and biodiversity. Overall, forest area in the Lower Tana significantly reduced by 34.5% ($P < 0.001$) over a 21-year period. Forest loss was greater outside the reserve (38%) than inside (29.2%) reiterating the significant role played by this protected area in habitat and species conservation. Continued forest loss increases extinction risks for the endemic primate species the Tana River Red Colobus (*Procolobus rufomitratus*) and the Crested Mangabey sub-species (*Cercocebus galeritus galeritus*). Initiation of community conservation programmes outside the reserve and introduction of sustainable micro-economic projects were recommended to enhance sustainable livelihoods and the environment.

N. N. Moinde-Fockler (✉) · G. M. Karere · M. A. Suleman
Department of Ecology & Conservation, Institute of Primate Research, National Museums of Kenya, P.O. Box 24481, Karen, Nairobi, Kenya
e-mail: nmoinde@eden.rutgers.edu

N. O. Oguge
Department of Zoology, Kenyatta University, P.O. Box 43844, Nairobi, Kenya

D. Otina
Department of Botany, Nairobi University, P.O. Box 30197, Nairobi, Kenya

Keywords Forest destruction · Human and Natural impacts · Tana River · Red Colobus · Crested Mangabey · Community conservation

Introduction

There are major concerns towards the loss of biodiversity, particularly in tropical forests around the equator where these hotspots are concentrated (Myers et al. 2000; Beck et al. 2002). Deforestation of tropical forests not only jeopardizes biological diversity but also climate systems of the world (Myers 1989; Schwartzman et al. 2000). In addition to high species diversity and endemism, tropical forests are also home to rural communities in need of economic sustainability. Conservation of tropical forest is thus one of the greatest human challenges involving a delicate balance between complex-fragile ecosystems, and impoverished populations. Consequently, shifting cultivation remains the biggest threat to tropical forests (Myers 1987) and has exacerbated the natural fragmentation of landscapes affecting whole ecosystems and biotas (Bender et al. 1998).

The lower Tana riverine forests are unique because they support a high diversity of plants and animals species that exist in a semi-arid environment, which has an annual rainfall of ≤400 mm and show floristic similarities to the western and coastal evergreen forests (Marsh 1976). Of great importance, they provide remaining habitats for two endangered primates: (1) the Tana River Red Colobus (*Procolobus rufomitratus*), and (2) the Tana River Crested Mangabey (*Cercocebus galeritus galeritus*). Five other primate species are among faunal and flora taxa represented here (Suleman et al. 2001). This ecosystem is, however, under severe threat due to intense shifting cultivation practiced by the Pokomo people. These forests have been subjected to increasing destruction of forest cover due to clear cutting, burning and slashing mainly for agriculture as well as forest deterioration due to harvesting and utilization of different forest products (Decker 1994; Medley 1993).Currently, the ecosystem is highly fragmented and exists as isolated patches of various sizes (Karere et al. 2004). One of the direct effects of forest loss since the 1960s has been the notable decline in the two endangered primate populations (Marsh 1986; Homewood 1975; Decker 1994). As a management measure, a 169 km^2 area—the Tana River Primate National Reserve (TRPNR)—was set up in 1976 to protect the two endangered primates species (Marsh 1976). Conservation of these primates and their habitat has since been of high priority nationally and internationally (IUCN 1996).

Aside from the human effects, natural impacts have also had an enormous role in influencing the conditions of the forest here due to dependence on the river seepage for tree survival. The Tana riverine forest ecosystem is highly dynamic being maintained by a balance between forest patches dying off and regeneration driven by regular natural shifts in the course of the river. The Tana River has changed its course several times (Andrews et al. 1975; Butynski and Mwangi 1994a). This is evident by the presence of old river channels, ox-bow lakes and remnant forests around the flood plain due to seasonal flooding regimes (Hughes 1984). According to Hughes (1990), it is evident that the forest patches are ground water dependent and the frequency and duration of flooding of the Tana flood plain affects the distribution and composition of the forests along the lower Tana region. This dynamic nature of the Tana River has consequently led to drying of trees due to either lack of water (natural die-back) or flooding.

We examined the impacts of human activities and natural causes on forest patches in and out of the TRPNR. Herein, we discuss the implication of these impacts on conservation and management of the red colobus and crested mangabey and their habitat.

Methods

Study site

There are currently more than 80 forests distributed in scattered patches on both side of the Tana River varying in sizes along the lower Tana Region (Butynski and Mwangi 1994). The TRPNR, contains 27 of these forest patches, and straddles the lower Tana River and is located entirely within the lower Tana River flood plain (1°50′ S, 40°10′ E) (Fig. 1). The TRPNR, has only 9.5 km^2 out of its total area under forest cover (Medley 1990) and the rest of the reserve is covered by shrubs and grass. The Tana Delta Irrigation project (TDIP), situated south of the reserve is a large rice irrigation project administered by the Tana and Athi River Development Authority (TARDA) and is located at the northern end of the Tana Delta near Garsen (Fig. 1). The plan of the project is to eventually encompass 160 km^2 and currently covers areas that include 21 of the riverine forest patches in the Lower Tana Region. Despite the low human density in the Tana district, there are pockets of high human population concentrating along the Lower Tana River. The Pokomo are the dominant tribe along the River (Kenya Wildlife Service 1996) and are sedentary agriculturists who cultivate land within the flood plain. They practice flood recession and riverbank farming around the along the Tana River, which provides the only source of land in the region that is suitable for arable agriculture because this farming system depends both on floodwater to irrigate their crops, and on the depositions of fertile sediments that the floods bring (IUCN 2003). This form of shifting cultivation along the Lower Tana River, unlike other cultivation systems in tropical wet environments, is largely dictated by the availability of floods, where establishments of farms and their permanence are dictated by soil fertility. The Orma, Somali and Wardei are exclusively pastrolists and make use of in the dry semi-arid areas above the flood plain.

Data collection

Human and natural impact

Data was collected on human and natural impacts while concurrently conducting a primate census between January 2001 and March 2001. The forest fragments that were surveyed varied in size from approximately 500 ha to less than 2 ha. A pre-orientation workshop was held in which all participants familiarized themselves with data collecting techniques and in order to minimize inter-observer variations (Suleman et al. 2001; Karere et al. 2004). The evaluating team was divided into several observer groups consisting of two members each. The name of the forest, reference number, date and names of observers were entered onto the data sheets.

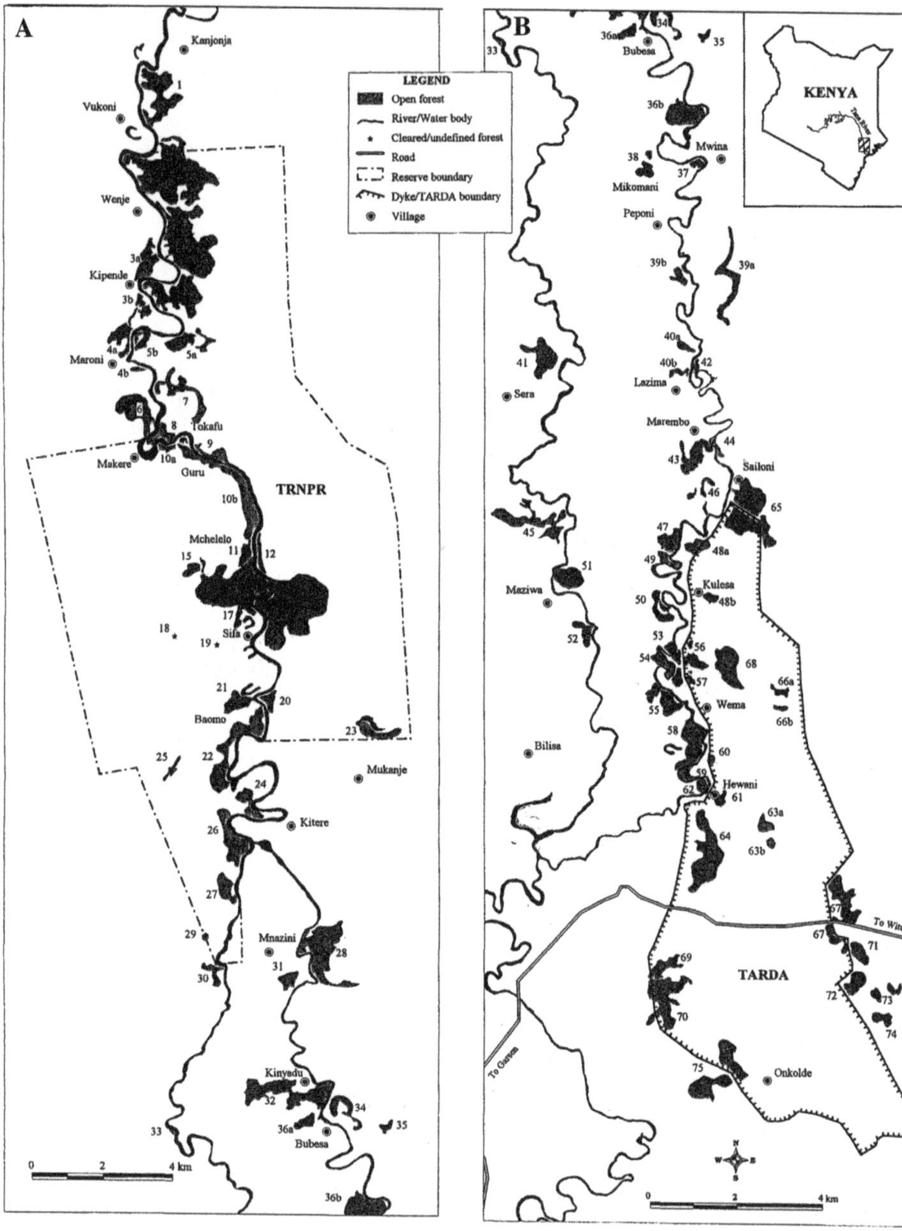

Fig. 1 Distribution of forest patches along the Lower Tana River from Nkanjonja to Onkolde

The observer groups walked parallel to each other (approximately 50–100 m apart, depending on the shape and size of the forest) along pre-determined routes in the forest with the aid of compasses. The observers identified and examined human activities, natural impacts and their frequency of occurrence. Human activities and natural impacts were categorized as follows:

(1) Resource utilization is defined as human practices that do not necessarily result in partial/complete forest cover removal but resulted into deterioration of forest stature. These activities included:

Tree harvesting, which included cutting plant parts for various human utilizations such as thatching, wine tapping, constructions of animal traps and sometimes firewood collection. Thatching and wine tapping involved the chopping off the crowns of trees and tapping of the sap, respectively. Animal trapping involved the use of snares. Firewood collection involved gathering dried twigs and to a lesser extent cutting young stems and branches.

Honey harvesting, which involved digging a hole on a tree stem where bees had a natural hive or cutting the whole tree to harvest the honey.

Logging, which includes cutting trees for construction of canoes, beehives, furniture, building materials and charcoal burning. Charcoal burning involved burning of felled logs under earth mounds from various tree species.

(2) Land use practices are defined as human activities that resulted to partial or complete removal of forest canopy cover. These were identified as follows:

Cultivation entailed the complete or partial clearances of areas of forest for agriculture through slash and burn techniques, which affected all species. This practice sometimes also causes fragmentation of the affected forest patch.

Dyke construction for rice irrigation by the Tana Athi Development Agency (TARDA) which generally destroyed natural vegetation across 50–60 m wide swaths resulting in losses of forest area and further fragmentation of affected forest patches.

(3) Natural Impacts are as a result of excess flooding and natural die back resulting in progressive degradation of forest structure and biodiversity and eventual loss of forest cover. Indicators of natural impacts included:

Excess flooding made evident by swampy forest conditions caused by very heavy rain such as the El Ninõ Southern Oscillation (ENSO) that occurred in 1998, causing the river water to overflow its banks and the excess water remains stagnant for a long period of time in the adjacent forests. This caused the tree roots to suffocate due to lack of aeration and consequently resulted to senescence.

Natural dieback made evident by the drying up of canopy trees and fallen trees due to river dynamism resulting to insufficient ground water seepage to forest adjacent to old river courses.

Forest status

Data obtained from human and natural impact evaluation was used to provide overall assessment of the status of forests surveyed. Each observer group recorded levels of forest disturbance, based on the frequency and effects of human activities and natural impacts on a forest. Disturbance levels were categorized as detailed by Muoria et al. (2002) from level 1 to 4 as follows:

Level 1: Little or no destruction. More specifically little or no human resources utilization and no land use practices and natural impacts observed. Otherwise forest could be pristine.

Level 2: Moderate destruction. Human resource utilization are being observed at a higher frequency as compared to level 1 but less frequently observed than in level 3.

Level 3: Extensive human disturbances and natural impacts. Higher frequency of resource utilization, partial clearing of forest cover due to difference land use impacts such as cultivation and dykes or complete or partial flooding or dieback resulting to overall degradation of forest structure and biodiversity.

Level 4: The highest scale of destruction where larger portions or all of the forest area had been cleared. Clearing of all or large portion of forest area for cultivation or the combinative impact of cultivation and natural dieback or cultivation and flooding resulting to high portions or complete loss of forest area.

Observer groups derived the overall disturbance level in each forest from the average of the disturbance indices recorded. Therefore, the disturbance level that was assigned to each forest was an overall qualitative and accumulated assessment of all human activities and/or natural impacts indicators that had been observed.

Changes in forests sizes

Satellite imagery for the year 2000 and 1979 topographic maps of the study area were used as sources of land cover information and were digitized using MapInfo Version 5.5 (MapInfo Corporation 1985–1999) to obtain forest sizes. Differences in forest sizes between the 2 years were used to determine changes in size of the forests.

Results

Anthropogenic activities in the forest patches along the Tana River

The main human activities observed were logging, tree harvesting and cultivation (Table 1).

Logging was observed in 69 forests and accounted for 39% of human activities, tree harvesting in 45 (25%), and cultivation in 43 (24%). Honey harvesting and dyke constructions were observed in 16 (9%) and 6 (3%) forests visited, respectively. Where observed, cultivation and dyke construction had the most devastating effects on forest cover due to partial or complete vegetation clearance. The most affected species due to tree harvesting were *Borassus aethiopuim*, *Phoenix reclinata* and *Hyphaene compressa*. While the most preferred tree species for construction of canoes and beehives were *Diospyros kabuyeana, Ficus sycomorus, Mimusops fruticosa* and *Mangifera indica*. Bee keeping appeared more sustainable than honey harvesting because although the hives are constructed from a felled tree, the hive can be used for a long period of time while harvesting of honey from standing natural tree hives looked very destructive. Furniture was constructed from *Spyrostachys venenifera*, while building materials were obtained largely from *Phoenix reclinata*.

Table 1 Frequencies and proportional occurrence of categorized human activities in 73 forest patches along the Lower Tana River basin

Activities	Frequency	%
Logging	69	39
Tree harvesting	45	25
Cultivation	43	24
Honey harvesting	16	9
Dyke construction	6	3
Total	179	100

Eight forest patches, severely impacted by cultivation alone were Nkanjonja (no. 1), Wenje Complex (nos. 2a–c), Baomo East (no. 20), Baomo North (no. 21), Baomo South (no. 22), Lazima East (no. 42), Hewani East 1 (no. 59) and Hewani West 2 (no. 2) (Fig. 1). Four forests affected by dyke construction alone were Kulesa East 1 (no. 48a), Wema East 1 (no. 56), Hewani East 2 (no. 60) and Mitapan 2 (no. 70) (Fig. 1). Three forests, Hewani East 1, Hewani East 3 and Hewani West 2 (numbers 59, 61 and 62, respectively) were heavily impacted by human activities and yet satellite imagery indicated an increase in area by 8.2% to 161 ha.

Excess flooding in forest patches along the Tana River

Six forest patches affected by excess flooding included Kipendi 1 (3a), Kipendi 2 (3b), Maroni West 1 (4a) and Maroni West 2 (4b), and are all along channel 2, the current river course (Fig. 1).

Natural die back in forest patches along the lower Tana River

Along the old river channel (channel 1), only one forest patch, Maziwa North (forest no. 51), was affected by natural die back (Fig. 1). Four other forest patches affected by dieback are near the current river channel (Fig. 1). They include Wema East 4 (no. 68), Hewani South 1a (no. 63a), Hewani South 2 (no. 64) and Bvumbwe South 2 (no. 66b) (Fig. 1).

Forests impacted by both natural impacts and human activities along the lower Tana River

Forests impacted by both natural dieback and cultivation were Matalani South (no. 33), Sera (no. 41), Giritu woodlands (no. 45) and Maziwa South (no. 52) (Fig. 1). Flooding and cultivation impacted only Maroni East 1 (5a) and Maroni East 2 (no. 5b) (Fig. 1). Two forest patches impacted by both dieback and dyke constructions were Bvumbwe North (no. 65) and Lango La Simba (67a) (Fig. 1).

Forest status

Out of the 73 forest fragments evaluated, 28 had little or no disturbance while 21 were heavily disturbed. Of the heavily impacted forests, six were in the reserve and 15 outside the protected area (Table 2).

Causes of forest area loss in forest patches along the Tana River

Natural dieback alone impacted on Maziwa North (no. 51) and Hewani South 1a (no. 63a) resulting to area loss of 28.9% and 22.4%, respectively (Table 3). Of the forest patches impacted by flooding alone, Maroni West 2 (no. 24b) and Kipendi 2 (no. 3b) had the highest habitat loss of 85% and 57.1%, respectively (Table 2). Baomo East (no. 20) and Nkanjonja (no. 1) forests were most affected through cultivation with losses of 80.7% and 50.1%, respectively. Construction of irrigation dykes greatly impacted on Kulesa East 1 (no. 48a) resulting in loss of 71.7% of forested area (Table 3).

Table 2 Intensity of destruction of forests and their current areas in and out of the Tana River National Primate Reserve using a scale of 1–4

Destruction levels	Forests in reserve			Forests out of reserve		
	Numbers	Area (ha)	%	Numbers	Area (ha)	%
1	6	469.1	35	22	744.1	27
2	5	486.8	29	12	421.4	21
3	0	0.0	0	7	362.3	13
4	6	595.0	35	15	509.1	39
Total	17	1550.9	100	56	2036.9	100

Scale 1, little or no destruction; scale 2, moderate levels of destruction; scale 3, extensive human destruction with no section of forest completely cleared; scale 4, highest levels of destruction with sections of the forest completely cleared

Four forest patches were affected by a combination of cultivation and natural diebacks. Matalani South (no. 33), Sera (no. 41), Giritu woodlands (no. 52) and Maziwa South (no. 45) forests had area loss of 99, 75.8, 71.1, and 46.8%, respectively (Table 3). Maroni East 1 (no. 5a) and Maroni East 2 (no. 5b) were affected by both cultivation and flooding resulting to a total forest area loss of 50% each (Table 3).

Table 3 Changes in forest sizes and their respective causes in the lower Tana River between 1979 and 2000

Forest patch	No.	Area (ha)		Change in area (%)	Factors leading to loss of area			
		1979	2000		Cult.	Dyke	D/back	Flooding
Nkanjonja	1	168.8	84.2	50.1	+			
Wenje complex	2a–c	683.6	534	21.9	+			
Kipendi 1	3a	55.9	37.4	33.1				+
Kipendi 2	3b	34.5	14.8	57.1				+
Maroni West 1	4a	69.1	30.9	55.3				+
Maroni West 2	4b	27.4	4.1	85.0				+
Maroni East 1,2	5a–b	133.6	54.3	59.4	+			+
Baomo East	20	73.7	14.2	80.7	+			
Baomo North	21	46.1	30.2	34.5	+			
Baomo South	22	261.4	99	62.1	+			
Matalani South	33	240.3	2.3	99.0	+		+	
Sera	41	204.1	59	71.1	+		+	
Lazima East	42	15.5	8.9	42.6	+			
Giritu	45	327.5	79.2	75.8	+		+	
Kulesa East 1	48a	68.1	19.3	71.7		+		
Maziwa North	51	61.2	43.5	28.9			+	
Maziwa South	52	40	21.3	46.8	+		+	
Wema East 1	56	30	28.1	6.3		+		
Hewani East 2	60	7.9	4.2	46.7		+		
Hewani South 1	63a	20.1	15.6	22.4			+	
Bvumbe North	65	260.6	136.5	47.6		+	+	
Lango la Simba	67a	86.4	79.2	8.3		+	+	
Mitapani 2	70	105.3	76.7	27.2		+		
Total		3021.1	1476.9	48.9				

Four factors driving change were identified as cultivation (cult.), dyke construction (dyke), natural dieback (D/back) or flooding. Forests affected by the various factors are specifically indicated by a positive (+) sign

Overall forest area reduced significantly from 5,439 ha to 3,564 ha ($t = 3.807$, $n = 76$, $P < 0.001$) accounting for a 34% decrease between 1979 and 2000 (Table 4). The loss of forest area outside the reserve was 38% from 3,283 ha to 2,037 ha ($t = 2.929$; $n = 57$; $P < 0.005$) and 28.1% from 2,156 ha to 1,551 ha ($t = 2.522$; $n = 21$; $P < 0.02$) in the protected area (Table 4).

Discussion

Human activities and natural impacts on the riverine forests along the lower Tana River region

Our study has shown that through shifting cultivation, dyke constructions, flooding and dieback, human activities, and natural impacts have had a devastating effect on the status of the lower Tana riverine forests leading to loss and increased fragmentation of unique habitats. Anthropogenic activities in the forests persist in the form of slash-burn agriculture, selective logging and several other deleterious uses of forests (Table 1). The resultant change in forest structure, especially removal of large canopy tree species, is of great concern in the conservation of the endangered primates, the Tana red colobus (*Procolobus rufomitratus*) and the Tana crested Mangabey (*Cercocebus galeritus galeritus*). Both species are dependent on gallery forests for food and sleeping groves (Homewood 1976; Marsh 1981; Wahungu 1998; Suleman et al. 2001). Shifting cultivation combined with some natural impacts contributed to a total loss of 1,208 ha or 78.2% of affected forests that were ground "truthed." Dyke construction resulted to a total loss of 210.6 ha or 13.6% of the affected forests that were ground "truthed." The most impacted forest, Matalani south (no. 33), was affected by both cultivation and natural dieback with loss of 99% of area and only 2.3 ha is left standing from 240.3 ha in 1979. Hewani East 1 (59), Hewani West 2 (62) and Hewani East 3 (61) forest patches are reflected on satellite images as indigenous forest patches but upon ground "truthing" it became evident that these patches have mostly been cleared for cultivation and have been replaced by exotic tree species. These three examples provide evidence of the importance of ground "truthing" instead of only using satellite imagery to determine the true status of forests in affected areas.

Our study recorded a loss of 34.5% of total forest area between 1979 and 2000 (Table 4). The loss outside the Tana River National Primate Reserve (1,246 ha) was significantly ($P < 0.005$, $t = 2.929$) larger than loss within the reserve (629 ha) reiterating the significant role played by this protected area in habitat and species

Table 4 Comparison of forest area reduction in and out of the reserve and overall forest area reduction along the Lower Tana River between 1979 and 2000

Location	Number of forests inspected	Area (ha)			% change	
		1979	2000	Loss of area (ha)		
Reserve	21	2156	1527	629	29.2	$P < 0.020$
Outside	57	3283	2037	1246	38.0	$P < 0.005$
Total	76	5439	3564	1875	34.5	$P < 0.001$

conservation. The loss of 29.2% of forest area within the protected area in 21 years is nonetheless of major concern and suggests a loss rate of 29 ha per year. With only 1,527 ha of forest remaining within this reserve, these habitats constitute a biodiversity in risk of extinction within approximately the next five decades. The rate of loss of forest habitats outside the reserve, at 59 ha per year, is twice that of the protected area and the remaining 2,037 ha may disappear in approximately in three and a half decades unless urgent conservation programmes are put in place. Twenty-eight forest patches have experienced the highest levels of destruction (Level 3 & 4) (Table 2). This has accounted for 1,348 ha of forest loss and has, of consequence, severely impacted primate habitats. This equals a loss of 24.8% from the total forested area that existed in 1979 and should be the focus of immediate conservation effort.

The impact of flooding and natural dieback on the forests in the lower Tana River region is enormous. Changes caused by both dieback and flooding in the Lower Tana riverine forests do not necessary immediately remove forest cover, instead they are more likely to cause progressive degradation of forest structure and biodiversity. In the long-term, this progressive degradation leads to partial or complete loss of forest cover. Thus, these impacts share the quality of being difficult to perceive by satellite imagery and are difficult to evaluate without monitoring by ground truthing (Dale et al. 1994). It would be important to note that a forests like Hewani South 2 and Wema East 4 (forest nos. 64 and 68, respectively) where natural dieback occurred, appear to have increased, while in reality the interior of the forest have been affected by tree dieback.

One important aspect that was not evaluated during this study was the loss of mature forest due to bank erosion. This type of evaluation would necessitate long-term monitoring of these potential sites, which was beyond the scope of this research study. Future studies should incorporate the impact of bank erosion and evaluating its role as a natural impact on the forests.

As a whole, the combinative impact of cultivation and natural dieback or cultivation and flooding has resulted in the highest percentage forest area loss in the Lower Tana Region. Both human and natural impacts are responsible for changes in forest cover and forest stature. As this study has indicated, areas that have experienced significant area loss due to the Tana River dynamism could be significantly related to changes in human activities, which further complicates current and potential conservation and management strategies in and out of the reserve.

Effects of forest degradation, destruction and fragmentation on the endangered primate population along the Lower Tana Region

Human exploitation of forest resources can involve rapid, non-sustainable harvesting of particular species (Gentry and Vásquez 1988) while flooding and natural dieback can result in a progressive degradation of forest structure and biodiversity that leaves behind standing but biologically and economically depleted forests. The riverine habitats on the lower Tana River are highly vulnerable to perturbations due to the Tana River dynamism and the continual human overexploitation. An ever-increasing human population continually exacerbates this problem. Many of the tree species that are important to the endangered primates are also vital to the local communities for construction of canoes, poles and other wood products (Marsh 1981; Medley 1990; Kahumbu 1992). This competition for diminishing resources is likely to result in a reduction in the carrying capacity for the endangered primates in the lower Tana region riverine forests (Kahumbu and Davies 1993).

According to the recent primate census (Karere et al. 2004), about 50% of both the red colobus and the crested mangabeys were found outside the protected area. The riverine forests within the protected area represents only 24% of the forest ecosystem and may thus be inadequate to provide resources to stem the current decline in endangered primate populations. However, the importance of the unprotected forest patches situated outside the reserve for the survival of both endangered primate species cannot be overemphasized. The survival of these species depends on the future management and conservation of the majority of forest patches that are situated out of the reserve. The fact that the greatest area of forest loss was outside the reserve implies the immediate need to initiate conservation programmes outside the protected area. That the Tana red colobus, over the last 7 years, have experienced a 15% loss in population outside the reserve (Suleman et al. 2001; Karere et al. 2004) exemplifies these urgent needs. Previous studies conducted in the lower Tana have shown the impact of forest destruction on the red colobus (Marsh 1978; Decker 1989; Mbora and Meikle 2004) and the crested mangabey (Kinnaird 1990; Homewood 1975) populations. Forest destruction can result in declining primate populations (Myers 1987; Gillespie et al. 1999) and in extreme situations, extinction (Yongzu et al. 1989; Boinski 1994). Since these forest patches do not fall under the management of the organization that manages the parks and reserves in Kenya, i.e., Kenya Wildlife Service (KWS), management and conservation strategies should directly involve the participation of the local communities.

A prominent issue that should seriously be addressed in the Lower Tana region is the effect of forest fragmentation on the two endangered primate species because primates have specific responses to fragmentation making them valuable candidates for examining its effect (Estrada and Coates-Estrada 1996; Tutin et al. 1997). Forest fragmentation not only isolates floral and faunal population but it also impedes gene flow between forest patches (Marsh et al. 1987). A study that was conducted on the effect of fragmentation on the Tana river red colobus (Mbora and Miekle 2004) suggests that this primate species may prefer more disturbed forests. However, according to Suleman et al. (2001), the number of primate groups in any given forest along the Tana River was significantly correlated with forest area; suggesting that both endangered primate species along at the Lower Tana region are prone to forest loss and fragmentation. Many primate studies at the Lower Tana region appear to focus on the impact of human activities on these endangered species. This study demonstrates that the effects of natural impacts are just as important, and therefore, future studies should not only examine the long-term effects natural impacts on the endangered primate species, but also study the combinative effects of both natural and human impacts on these species.

Conservation of the riverine forests in the Lower Tana region and the local communities

According to Butynski and Mwangi (1994b), local people living, in the vicinity of the Tana riverine forests are aware of the direct benefits they receive from the remaining forests. The Pokomo, have traditional laws and norms governing land use that determine who can clear land for cultivation and how much may be cleared (CARE-International-Kenya 1992) known as the "Wakijo" (Bunger 1979; Decker 1989). It is not known, however, to what extent these laws actively protect the forest resources or control exploitation. Decker (1989) explains that the traditional forest management was conservative and proposed that their indifference towards forest degradation is more recent phenomenon caused by the displacement of the traditional management by the current protectionist management of the TRPNR. This was an observation that continued to be noted during the duration of this study. The extent to how far these traditional conservation laws are still practiced and their potential effectiveness on the sustainability of the existing forest patches should be examined to reinforce the present management strategies. This type of information is vital to provide guidelines that would assist in strengthening the already existing efforts, if any, of the local communities. The importance of the local communities' full participation in actively and sustainably managing these forest patches is the only way these remaining forests can continue to persist as well as sustain their rich biodiversity.

Acknowledgements We wish to express our gratitude to the World Bank Tana GEF Project that funded this work and the Kenya Wildlife Service. The Senior Warden in Tana River Primate National Reserve, Barasa Otunga and field assistants from Mchelelo Research camp who provided invaluable assistance in logistics and data collection, respectively. Many individuals and organizations within Tana River District assisted in different ways and we are extremely grateful for making it possible for us to accomplish the objectives of the study. We would also like to thank Dennis Milewa for his assistance in the production of the map of the study area and Erick Fockler for his critical editing and constructive comments.

References

Andrews P, Groves CP, Horne JFM (1975) Ecology of the Lower Tana River flood plain (Kenya). J East Afr Nat Hist Soc Nat Mus 152:1–31

Beck J, Schulze CH, Linsenmair KE, Fieldler K (2002) From forest to farmland: diversity of geometrid moths along two habitat gradients on Borneo. J Trop Ecol 18:33–51

Bender DJ, Contreras TA, Fahrig L (1998) Habitat loss and population decline: a meta-analysis of the patch size effect. Ecology 79(2):517–533

Boinski S (1994) Costa Rican squirrel monkey: waltzing towards extinction. Amer J Primatol 33:196–197

Bunger RL Jr (1979) Islamization among the upper Pokomo. Syracuse University Foreign and Comparative Studies Program, Syracuse, New York

Butynski TM, Mwangi G (1994a) Conservation status and distribution of the Tana River red colobus and crested mangabey. Report for Zoo Atlanta, Zoo Atlanta, Kenya Wildlife Services, National Museums of Kenya, Institute Primate Research and East African Wildlife Society

Butynski TM, Mwangi G (1994b) Importance of forest outside the Tana River primate reserve to the conservation of Tana River colobus and crested mangabey. Report for Kenya Wildlife Services and Institute Primate Research

CARE International-Kenya (1992) Conservation of the lower Tana forest and the Tana River Primate national reserve: phase 1 socio-economic survey. Unpublished report

Dale VH, Pearson SM, Offerman HL, O'Niel RV (1994) Relating patterns of land-use change to faunal biodiversity in the Central Amazon. Conserv Biol 8:1027–1086

Decker BS (1989) Effects of habitat disturbance on the behavioural ecology and demographics of the Tana River red colobus (*Colobus badius rufomitratus*). Ph.D. Dissertation, Emory University

Decker BS (1994) Effects of habitat disturbance on the behavioral ecology and demographics of the Tana Red Colobus (*Colobus badius rufomtratus*). Int J Primatol 15(4):703–736

Estrada A, Coates-Estrada R (1996) Tropical rainforest fragmentation and wild population of primates at los Tuxlas, Mexico. Int J Primatol 17(7):759–783

Gentry AH, Vásquez R (1988) Where have all the Ceibas gone? A case history of mismanagement of tropical forest resource. Forest Ecol Manage 23:73–76

Gillespie TP, Balcomb SR, Champmann CA, Skorupa JP, Struhsaker TT (1999) Long-term effects of logging on African primate communities: a 28-year comparison from Kibale National Park, Uganda. Amer J Primatol 49:57

Homewood KM (1975) Can the Tana mangabey survive? Oryx 13:53–59

Homewood KM (1976) The ecology and behaviour of the Tana Mangabey. Ph.D. Dissertation, University Collage, London, UK

Hughes FMR (1984) A comment on the impact of development shemes on the floodplain forests of the Tana River Kenya. Geogr J 150(2):230–244

Hughes FMR (1990) The influence of flooding regimes on forest distribution and composition in the Tana River floodplain, Kenya. J Appl Ecol 27:475–491

IUCN (1996) IUCN red list of threatened animals. IUCN, Grand, Switzerland

IUCN (2003) Tana River, Kenya: integrating downstream values into hydropower planning. Case studies in wetland valuation # 6

Kahumbu P (1992) The sustainability of fig tree *Ficus sycomrous*: harvesting for Canoes in a Kenyan Reserve. MSc Thesis, University of Florida, USA

Kahumbu P, Davies G (1993) Tana River primate national reserve: primate census. E Afr Nat Hist Soc Bull 23:35–44

Karere GM, Oguge NO, Kirathe J, Muoria PK, Moinde NN, Suleman MA (2004) Poulation size and distribution primates in the Lower Tana River Forest, Kenya. Int J Primatol 25(2):351–365

Kinnaird MF (1990) Behavioral ecology and demographic responses to habitat change by the Tana River crested mangabey (*Cercocebus galeritus galeritus*). Ph.D. Thesis, University of Florida, Gainesville

Kenya Wildlife Service (1996) Land use survey and census of communities directly dependent on Tana River Primate National Reserve. Report to World Bank for the Tana GEF project

MapInfo Corporation (1985–1999) MapInfo Version 5.5

Marsh CW (1976) A management plan for the Tana River game reserve. Report to the Kenya Department of Wildlife Conservation and Management

Marsh CW (1978) Ecology and social organization of the Tana River Red Colobus (*Colobus badius rufomitratus*). Ph.D. Dissertation, University of Bristol

Marsh CW (1981) Diet choice among Red Colobus (*Colobus badius rufomtratus*) to the Tana River, Kenya. Folia Primatol 35:147–216

Marsh CW (1986) A resurvey of Tana River primates and their habitat. Primate Conserv 7:353–371

Marsh CW, Johns AD, Ayres JM (1987) Effects of habitat disturbance on rain forest primates. In: Marsh CW, Meittermeier RA (eds) Primate conservation in the tropical rain forest. Alan R. Liss, NY, pp 83–107

Mbora DNM, Meikle DB (2004) Forest fragmentation and the distribution, abundance and conservation of the Tana river red colobus (*Procolobus rufomtitratus*). Biol Conserv 118:67–77

Medley KE (1990) Forest ecology and conservation in the Tana River National Primate National Reserve, Kenya. Ph.D. Dissertation, Michigan State University

Medley KE (1993) Extractive forest resources of the Tana River National Primate Reserve, Kenya. Econ Bot 47(2):171–183

Muoria PK, Karere GM, Moinde NN, Suleman MA (2002) Primate census and habitat evaluation in the Tana delta region, Kenya. Afr J Ecol 40:1–7

Myers N (1987) Trends in the destruction of rain forest. In: Marsh C, Mittermeir RA (eds) Primate conservation in the tropical rain forests. Monogr Primatol 9:3–22

Myers N (1989) Deforestation rates in tropical countries and their climate implications. Friends of the Earth, London

Myers N, Mittermeier RA, Mittermeier CG, Da Fonseca G, Kent J (2000) Biodiversity hotspots for conservation priorities. Nature 403:853–858

Schwartzman S, Moreira A, Nepstad D (2000). Rethinking tropical forest conservation: perils in parks. Conserv Biol 14(5):1351–1357

Suleman MA, Wahungu GM, Mouria PK, Karere GM, Oguge N, Moinde NN (2001) Tana River primate census and forest evaluation. A report to Kenya Wildlife Services

Tutin CEG, White LJ, Mackanga-Amissandzou A (1997) The use of rainforest mammals of natural forest fragments in the equatorial African savanna. Conserv Biol 11:1190–1203

Wahungu GM (1998) Diet and habitat overlap in two sympatric primate species, the Tana crested mangabey *Cercocebus galeritus* and yellow baboon Papio cynocephalus. Afr J of Ecol 36:159–173

Yongzu Z, Guogiang Q, Yonglei L, Southwick C (1989) Extinction of rhesus monkeys (*Macaca mulatta*) in Xinglung, North China. Int J Primatol 10:375–381

Biodivers Conserv (2007) 16:1175–1182
DOI 10.1007/s10531-006-9099-5

ORIGINAL PAPER

Fragmentation causes rarity in common marmosets in the Atlantic forest of northeastern Brazil

Antonio R. Mendes Pontes · Iran C. Normande · Amaro C. A. Fernandes · Patrícia F. Rosas Ribeiro · Marina L. Soares

Received: 20 September 2005 / Accepted: 5 July 2006 / Published online: 27 October 2006

Abstract The purpose of this study was to investigate the population status of the common marmosets, *Callithrix jacchus*, in one of their main habitats, the northeastern Atlantic forest of Brazil, where only 2% of its original area remains as small and isolated forest fragments, to determine long-term viable populations. The study was carried out in 15 forest fragments, measuring from 3,478.3 ha (the largest) to 6 ha, which were invariably subjected to high human impact. Line transect surveys were carried out between January 2002 and December 2004, along transects measuring from 350 to 4,000 m, between 0500 h and 0900 h. Common marmosets were registered in 73% (n = 11) of the forest fragments, were not recorded in the largest one, the Coimbra Forest, and group sizes varied from 1 to 4 individuals. A negative significant relationship was detected between the size of the fragment and the number of sightings of common marmosets. It is shown that the number of groups of common marmosets dwelling in this highly impacted landscape is nowadays lower than necessary for long-term survival. Additionally, smaller fragments having more groups suggests that they live in total association with humans, which assure their subsistence through the supply of introduced and exotic foods.

Keywords Common marmosets · Surveys · Northeastern Atlantic forest of Brazil · Fragmentation · Rareness

A. R. M. Pontes (✉) · I. C. Normande · P. F. R. Ribeiro · M. L. Soares
Depto. Zoologia, Universidade Federal de Pernambuco, CCB, R. Prof. Moraes Rego, 1235, Cidade Universitária, Recife, PE, CEP: 50740-620, Brazil
e-mail: rossano@ufpe.br

A. C. A. Fernandes
Instituto Brasileiro de Meio Ambiente-IBAMA, Av. 17 de Agosto, 1057, Casa Forte, Recife, PE, CEP: 52.060-590, Brazil

Introduction

Callithrix jacchus Linnaeus, 1758 (Callithrichidae, Primate), the common marmosets, is endemic to the Northeastern Atlantic forest of Brazil (Coimbra-Filho 1984). They inhabit a myriad of vegetation types, from evergreen lowland rainforest, to humid and xerophytic semi-deciduous forest, dry thorn scrub, or caatinga (Hershkovitz 1977; Mendes Pontes and Monteiro da Cruz 1995), mangroves (Mendes Pontes 1989), dunes and sandy vegetation (Monteiro da Cruz 1998), where they prefer secondary vegetation (Rylands and Faria 1993; Mendes Pontes and Monteiro da Cruz 1995; Mendes Pontes and Soares 2005) and forest border (Mendes Pontes and Soares 2005). Introduced in the South and South-eastern Brazil they have been highly successful, reproducing considerably, which turned them into pests (Cunha 2005; Lima et al. 2005).

The Brazilian Atlantic forest is one of the world's priorities for conservation (Myers et al. 2000) due to its richness and diversity (Fonseca 1985; Myers et al. 2000). Only ca. 7.5% of its original area remains, which is considered the great tragedy of the last century (Dean 1995). The Northeastern part of the Atlantic forest, located north of the São Francisco River, one of the main habitats of the common marmosets, was identified as an important endemism center in South America, and defined as the Pernambuco Endemism Center (hereafter CEPE) (Silva and Casteletti 2003). It is, however, undergoing a serious biodiversity crisis, which made this sector of the Northeastern Atlantic forest by far the most threatened (Silva and Tabarelli 2001).

Most of the CEPE has been destroyed due to agriculture practices (Silva and Tabarelli 2000, 2001), which created a non-forested matrix of grazing fields, and sugar-cane plantations. The "protected areas" are small and unprotected (Dias et al. 1990), and subsistence hunting is widespread (Almeida et al. 1995). Most of what has remained (less than 2%) comprises an archipelago of thousands of forest fragments, where around 48% are smaller than 10 ha, only 7% are larger than 100 ha (Viana et al. 1997), and the only large fragment measures around 3,400 ha, which has caused profound changes in the ecology and behavior of the common marmosets (Roda and Mendes Pontes 1998; Jordani 2003; Melo et al. 2003).

Most of the medium-sized and large mammals of the CEPE have been extirpated from many areas, or are under threat (Fernandes 2003; Normande 2004; Rosas-Ribeiro 2004). Common marmosets are recognized as very common and adaptable, being able to feed on considerable amounts of plant exudates when no other sources of food are available (Stevenson and Rylands 1988; Scanlon et al. 1991; Mendes Pontes and Soares 2005). Thus, we aimed to assess their population status in this region where they should be one of a few mammals to remain in long-term viable populations.

Methods

Study area

The study was carried out in 15 forest fragments of the CEPE, as follows: small: (<100 ha), medium-sized: (100–1,000 ha), and large: (>1,000 ha). They were located in the three best preserved forest archipelagos: *Gurjaú Ecological Reserve*

(1,077.10 ha) (8°10′00″–8°15′00″S and 35°02′30″–35°05′00″W), with fragments: Cuxiu, São Braz, Xangô, Café; *Frei Caneca Estate* (630,42 ha) (08°42′37″S and 35°50′01″W), with fragments: Quengo, Fervedouro, Ageró, Espelho, *and Serra Grande Estate* (an archipelago of 8,000 ha) (35°52′12″S and 9°00′00″W), with fragments: Coimbra, Cachoeira, Capoeirão, Bom Jesus, Aquidabã, as well as in two completely isolated urban fragments, in the cities of Camaragibe, fragment Vale das Águas, and Igarassú, fragment Charles Darwin (Fig. 1). Fragments measured from 3,400 ha (the largest remnant) to 6 ha (Table 1), and were invariably disturbed by selective logging, hunting, burning, and were crossed by paths used by humans and domestic animals.

Surveys

The surveys were carried out between January 2002 and December 2004 through the Line Transect method (Buckland et al. 1993), as in Mendes Pontes (2004). One transect 1 m wide was established randomly in each of the fragments in the study area, which measured between 350 and 4,000 m, depending on fragment size and shape. Walks were carried out between 0500 h and 0900 h at a speed of 1 km/h.

Whenever an individual or group of common marmosets was sighted we recorded location in the trail, azimuth, day and time of the sighting, group size, distance from animal/group to the trail (NRC 1981; Buckland et al. 1993; Mendes Pontes 2004), and total km walked by the observer each day. Due to not enough sightings being

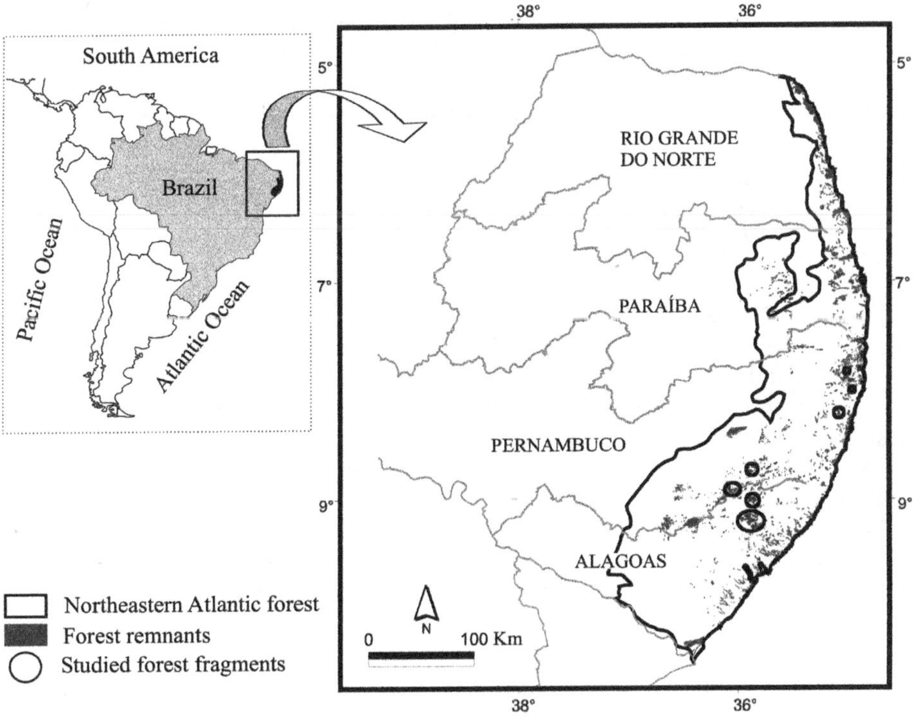

Fig. 1 The location of the Pernambuco Endemism Center (CEPE), in the Northeastern Atlantic forest of Brazil, with studied fragments

Table 1 The abundance of common marmosets in the forest fragments in Pernambuco Endemism Center, North-eastern Atlantic forest of Brazil

Site	Area (ha) of forest fragment	Transect(m)	ha(%) surveyed of the forest fragment	Km walked	No. of sightings	Mean group size
Coimbra	3,478.3	4,000	20 (0.6%)	329.75	0	–
Quengo	500.0	1,900	3.2 (0.6%)	8	1	2
Fervedouro	300.0	1,000	0.6 (0.2%)	7	2	3
Cachoeira	270.9	3,000	15 (5.5%)	172.2	0	–
Capoeirão	122.0	1,500	7.5 (6.1%)	162	0	–
Cuxiu	118.4	900	2.2 (1.8%)	10	4	4
Charles Darwin	60.0	1,500	4.5 (7.5%)	20.5	4	2.5
Ageró	50.0	500	2.8 (5.6%)	7.5	4	1.5
Espelho	50.0	1,000	5.6 (11.2%)	7	1	1
Bom Jesus	41.5	500	2.5 (6%)	7	4	2
São Braz	37.1	700	4.3 (11.6%)	10	12	2.25
Aquidabã	23.9	830	4.1 (17.1%)	11	0	–
Xangô	8.9	500	2.5 (28%)	10	11	1.73
Vale das Aguas	8.0	350	0.8 (10%)	5	3	2
Café	6.8	450	2.2 (32.3%)	10	5	2.6

recorded to allow the calculation of densities as in Mendes Pontes (2004), or sighting rates, as in Chiarello (1999), we analyzed number of sightings only.

For the calculation of area surveyed we used the maximum effective strip width recorded in each forest fragment. For those fragments where only one sighting was recorded, this was used, and when no sightings were recorded, we used the maximum effective strip width recorded in that forest archipelago. Mean group size was the summation of the total number of individuals registered in each group of a given forest fragment divided by the total number of groups, and when only one sighting was available, this was used. After having normalized the data we tested the relationship between the size of the fragment and the number of sightings through the Pearson correlation coefficient (Zar 1996).

Results

Common marmosets were registered (n = 51) in the CEPE in 73% (n = 11) of the forest fragments studied (Table 1). They were not seen in the largest fragment (Coimbra), and only seen once in the second largest one (Quengo). In Serra Grande Estate they were sighted in only one of the fragments (n = 4). In Gurjaú Ecological Reserve and Frei Caneca Estate they were present in all the fragments (n = 32 and n = 8, respectively). In the isolated fragments of Camaragibe and Igarassú they were sighted in 3 and 4 events, respectively.

Mean group sizes ranged from 1 to 4 individuals per group. A negative significant relationship was detected between the size of the fragment and the number of sightings of common marmosets registered (r = –0.584; df = 13; P = 0.002) (Fig. 2).

Discussion

It is known that the richness and abundance of mammals will decline as a function of the human impact upon the forest fragment, and that it will decline as the fragment

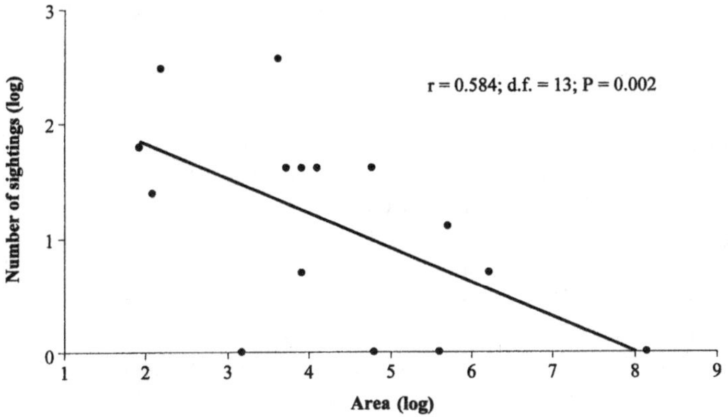

Fig. 2 The relationship between the number of sightings of groups of common marmosets and the size of the forest fragments in the CEPE

becomes smaller (Glanz 1990; Chiarello 1999; Fernandes 2003), reaching a less diverse state (MacArthur and Wilson 1967; Harris 1984). In the case of the CEPE, however, where the level of destruction has left only less than 2% of the original forest, with fragments measuring 10 ha or less (Viana et al. 1997), the future of any of the formerly occurring species is uncertain (Silva and Tabarelli 2000, 2001), even in the case of the highly adaptable ones, such as the common marmosets.

It has been shown that the serious environment disturbances in the CEPE has led the common marmosets to the limit of the carrying capacity of the environment, significantly increasing competition for food and territory, as well as exacerbating their aggressive behavior (Alonso and Langguth 1989; Roda and Mendes Pontes 1998).

The need to adapt to this highly fragmented saturated landscape has further caused the great instability of the groups, forcing them to stray from their modal breeding system, which has resulted in intra- and inter-group infanticide, and ultimately cannibalism, which appears to be a counterstrategy to survive in this limiting environment (Roda and Mendes Pontes 1998; Melo et al. 2003).

Mendes Pontes and Soares (2005) have shown that not withstanding the fact that common marmosets prefer secondary forest and forest border (Rylands and Faria 1993; Mendes Pontes and Monteiro da Cruz 1995) their survival is only assured by the presence of exotic fruit trees introduced by humans in orchards, backyards, and within surrounding forest fragments, the location of these resources contributing of their dynamics.

In this study we show that the number of common marmosets dwelling in this highly impacted landscape is considerably low, much lower than registered in the other few studies available (Jordani 2003; Silva 2002; Stevenson and Rylands 1988, in seasonally-dry forest), in which the number of individuals can reach up to 2 groups per hectare, or 24 individuals per square kilometers. In the only large fragment in this study, the Coimbra forest (c.a. 3,400 ha), marmosets were not recorded.

This suggests an ongoing process of local extinction, or at least, a steady decline in their population, which will certainly affect the ecological processes in which they play a role. Since the line transect method has successfully been used in various other ecosystems with highly acceptable results (Mendes Pontes 1997, 1999, 2004;

Mendes Pontes et al. 2006, *in press*), we do not think that this should be an artifact of the method.

Group sizes were considerably smaller than in the other studies in the CEPE, where they ranged from 5 to 15 individuals (Mendes Pontes and Soares 2005; Mendes Pontes and Monteiro da Cruz 1995; Digby and Barreto 1993), and also smaller than in those studies in seasonally-dry forests, where they ranged from 4 to 13 (Hubrecht 1984, 1985; Monteiro da Cruz 1998).

Although a large preserved control fragment does not exist in the CEPE to determine whether this lower abundance in larger fragments is a natural feature of the species or not, studies in South and South-eastern Brazil with introduced populations show that they have been extremely successful in fragments of various size classes, including remnants much larger than the largest of the CEPE, even competing with highly endangered native species (Lima et al. 2005).

Furthermore, we conjecture that the number of groups is higher the smaller is the fragment. The smaller fragments are those located more closely to, or form a continuum with, human assemblages, having consequently gone through more significant modifications, including the introduction of exotic food trees (as the example of Mendes Pontes and Soares 2005, personal observation). In this novel niche they live in total association with people and have their subsistence assured through the supply of exotic foods.

Finally, we show that the extremely high rate of destruction and modification of the forest remnants of the CEPE has led even the commonest of the remaining mammal species, the common marmosets, to rareness in its original habitat, the typical Atlantic forest of northeastern Brazil.

Acknowledgements We would like to thank M. Tabarelli and J. M. C. da Silva for crucial logistics and financial assistance through Conservation International – CI and Centro de Pesquisas Ambientais do Nordeste, Brazil – CEPAN. The Fundação O Boticário de Proteção à Natureza, Fundação Botânica Margaret Mee, The British Ecological Society – UK, Conselho Nacional de Pesquisas – CNPQ, Brazil, The Ministério do Meio Ambiente – Projeto Probio, Brazil, kindly financed this study. M. Tabarelli, N. Schiel, C. Knogge and L. Marsh made important suggestions on the last draft. We are greatly indebted to the owners and heads of Serra Grande, Frei Caneca, and Gurjaú for their permission to carry out this study in their properties, and to J. Alves and J. C. Baker for their assistance during fieldwork.

References

Almeida RT, Pimentel DS, Silva EMC (1995) The red-handed howling monkey in the State of Pernambuco, Northeast Brazil. Neotrop Primates 3:174–176

Alonso C, Langguth A (1989) Ecologia e comportamento de *Callithrix Jacchus* (Primates: Callitrichidae) numa ilha de floresta atlântica. Rev Nordestina Biol 6(2):105–137

Buckland ST, Anderson DR, Burnham KP, Laake JL (1993) Distance sampling. Estimating abundance of biological populations. Chapman and Hall, London, UK

Chiarello AD (1999) Effects of fragmentation of the Atlantic forest on mammal communities in south-eastern Brazil. Biol Conserv 89:71–82

Coimbra-Filho AF (1984) Situação atual dos calitriquídeos que ocorrem no Brasil (Callitrichidae, Primates). In: Mello MT (ed) A Primatologia no Brasil. University of Brasília, Brasília Brazil, pp 15–33

Cunha AA (2005) Estratificação vertical do macaco-prego (*Cebus* sp.) e do sagüi de tufos brancos *(Callithrix jacchus)* no maciço da Tijuca, Rio de Janeiro, RJ, Brasil. In: Bicca-Marques JC (ed) XI Congresso Brasileiro de Primatologia: Desafios para a Conservação em Paisagens Fragmentadas, pp. 92, Porto Alegre, Rio Grande do Sul, 13–18 February 2005.

Dean W (1995) With the broadax and firebrand, the destruction of the Brazilian Atlantic forest. University of California Press, Berkeley, USA
Dias IS, Câmara IG, Lino CF (1990) Workshop mata Atlântica: problemas, diretrizes e estratégias de conservação. Fundação SOS Mata Atlântica, São Paulo, Brazil
Digby LJ, Barreto CE (1993) Social-organization in a wild population of *Callithrix-jacchus*.1. Group composition and dynamics. Folia Primatol 61(3):123–134
Fernandes ACA (2003) Censo de mamíferos em alguns fragmentos de floresta atlântica no Nordeste do Brasil. Master's Thesis, Universidade Federal de Pernambuco, Pernambuco, Brazil.
Fonseca GAB (1985) The vanishing Brazilian Atlantic forest. Biol Conserv 34:17–34
Glanz WE (1990) Neotropical mammal densities: how unusual is the community in Barro Colorado island, Panamá? In: Gentry AH (ed) Four neotropical rainforests. Yale University Press, New Haven, Connecticut, USA, pp 287–313
Harris LD (1984) The fragmented forest, island biogeography theory and the preservation of biotic diversity. University of Chicago Press, Chicago, Illinois, USA
Hershkovitz P (1977) Living new world monkeys (Platyrrhini) with an introduction to primates. University of Chicago Press, Chicago, Illinois, USA
Hubrecht RC (1984) Field observations on group size and composition of the common marmoset (*Callithrix jacchus jacchus*), at Tapacurá, Brazil. Primates 25:13–21
Hubrecht RC (1985) Home-range size and use and territorial behaviour in the common marmoset, *Callithrix jacchus jacchus*, at the Tapacurá field station, Recife, Brazil. Int J Primatol 6:533–550
Jordani RA (2003) Densidades do sagüi comum, *Callithrix jacchus* (Callithrichidae, Primates) em fragmentos de floresta Atlântica no Nordeste do Brasil. Undergraduate Thesis, Universidade Federal de Pernambuco, Pernambuco, Brazil.
Lima CAS, Ruiz-Miranda CR, Coelho AS, Faria GV, Morais MM, Beck BB (2005) Interações ecológicas entre micos-leões-dourados (*Leontopithecus rosalia*), sagüis de tufo branco (*Callithrix jacchus*) e tufo preto (*Callithrix penicillata*) em fragmentos de mata Atlântica. In: Bicca-Marques JC (ed) XI Congresso Brasileiro de Primatologia: Desafios para a Conservação em Paisagens Fragmentadas, pp. 119, Porto Alegre, Rio Grande do Sul, 13–18 February 2005.
MacArthur RH, Wilson EO (1967) The theory of island biogeography. Princeton, University Press, Princeton USA
Melo L, Mendes Pontes AR, Monteiro da Cruz MAO (2003) Infanticide and cannibalism in wild common marmoset. Folia Primatol 74:48–50
Mendes Pontes AR (1989) Influência dos fatores ambientais e comportamentais na dinâmica de um grupo selvagem de *Callithrix jacchus* em Dois Irmãos, Recife, Brasil. Undergraduate Thesis, Universidade Federal de Pernambuco, Pernambuco, Brasil.
Mendes Pontes AR, Monteiro da Cruz MAO (1995) Home range, intergroup transfers, and reproductive status of common marmosets, *Callithrix jacchus*, in a Forest fragment in North-eastern Brazil. Primates 36(3):335–347
Mendes Pontes AR (1997) Habitat partitioning among primates in Maracá island, Roraima, Northern Brazilian Amazonia. Int J Primatol 18(2):131–157
Mendes Pontes AR (1999). Environmental determinants of primate abundance in Maracá island, Roraima, Brazilian Amazonia. J Zool (Lond) 247:189–199
Mendes Pontes AR (2004) Ecology of a community of mammals in a seasonally dry forest in Roraima, Brazilian Amazon. Mammal Biol 3:1–18
Mendes Pontes AR, Soares ML (2005) Sleeping sites of common marmosets (*Callithrix jacchus*) in defaunated urban forest fragments: an strategy to maximize food intake. J Zool (Lond) 266:1–9
Mendes Pontes AR, Peres PHA, Normande IC, Brazil CM (2006) Mamíferos. In: Porto KC, Almeida-Cortez JS, Tabarelli M (eds) Diversidade Biológica e Conservação da Floresta Atlântica ao Norte do São Francisco. Ministério do Meio Ambiente – Série Biodiversidade, Brasília Brazil, pp 10–50
Mendes Pontes AR, Chivers DJ, Lee PC *In Press*. The effect of biomass on assemblages of large mammals in a seasonally-dry forest in the Brazilian Amazonia. J Zool (Lond)
Monteiro da Cruz MAO (1998) Dinâmica reprodutiva de uma população de sagüis-do-nordeste (*Callithix jacchus*) na Estação Ecológica do Tapacurá, PE. Ph.D. Thesis, Universidade de São Paulo, São Paulo, Brazil
Myers N, Mittermeier RA, Fonseca GAB, Kent J (2000) Biodiversity hotspots for conservation priorities. Nature 403:853–858
National Research Council (NRC) (1981) Techniques for the study of primate population ecology. National Academy of Sciences. National Academic Press, Washington, DC, USA

Normande IC 2004. Impacto da fragmentação da floresta Atlântica sobre a comunidade de mamíferos de médio e grande porte do Centro de Endemismo Pernambuco. Undergraduate Thesis, Universidade Federal de Pernambuco, Pernambuco, Brazil

Roda SA, Mendes Pontes AR (1998) Polygyny and infanticide in common marmosets in a fragment of the Atlantic forest of Brazil. Folia Primatol 69:372–376

Rosas-Ribeiro PF (2004) Mastofauna de médio e grande porte da Reserva Biológica de Saltinho, Pernambuco. Undergraduate Thesis, Universidade Federal de Pernambuco, Pernambuco, Brazil

Rylands AB, Faria DS (1993) Habitats, feeding ecology, and home range size in the genus *Callithrix*. In: Rylands AB (ed) Marmosets and Tamarins systematics, behavior, and ecology. Oxford University Press, Oxford, UK, pp 262–272

Scanlon CE, Monteiro da Cruz MAO, Rylands AB (1991) Exploração de exsudatos vegetais pelo sagüi-comum, *Callithrix jacchus*. In: Rylands AB, Bernardes AT (eds) A Primatologia no Brasil – 3. Fundação Biodiversitas, Belo Horizonte, Minas Gerais, Brazil, pp 197–205

Silva VA (2002) Uma abordagem Bayesiana para estimativa da densidade populacional de animais de vida livre. M.Phil. Thesis, Universidade Federal Rural de Pernambuco, Pernambuco, Brazil

Silva JMC, Casteletti CHM (2003) Status of the biodiversity of the Atlantic forest of Brazil. In: Galindo-Leal C, Câmara IG (eds) The Atlantic forest of South America: biodiversity status, threats, and outlook. CABS and Island Press, Washington, DC, USA, pp 43–59

Silva JMC, Tabarelli M (2000) Tree species impoverishment and the future flora of the Atlantic forest of Northeast Brazil. Nature 404:72–74

Silva JMC, Tabarelli M (2001) The future of the Atlantic forest in Northeastern Brazil. Conserv Biol 15(4):819–820

Stevenson MF, Rylands AB (1988) The Marmosets genus. In: Mittermeier RA, Rylands AB, Coimbra-Filho AF, Fonseca GAB (eds) Ecology and behavior of neotropical primates. World Wildlife Foundation, Washington, DC, USA, pp 131–223

Viana VM, Tabanez AJ, Batista JL (1997) Dynamics and restoration of forest fragments in the Brazilian Atlantic moist forest. In: Laurence WF, Bierregaard RO (eds) Tropical forest remnants: ecology, management, and conservation of fragmented communities. The University of Chicago Press, Chicago, Illinois, USA, pp 351–365

Zar JH (1996) Biostatistical analysis. Prentice-Hall, London, UK

Biodivers Conserv (2007) 16:1183–1195
DOI 10.1007/s10531-006-9101-2

ORIGINAL PAPER

Distribution, abundance, and habitat use of introduced *Boa constrictor* threatening the native biota of Cozumel Island, Mexico

Irene Romero-Nájera · Alfredo D. Cuarón · Cristopher González-Baca

Received: 18 July 2005 / Accepted: 3 July 2006 / Published online: 27 October 2006

Abstract Species introductions are among the most pervasive types of disturbance, seriously affecting island biota and ecosystem dynamics. The management of introduced generalist species, which may live in a wide range of environmental conditions, can be particularly difficult and is a major challenge for the conservation of native insular species. *Boa constrictor* was introduced onto Cozumel Island, Mexico, in 1971. The introduction of this generalist predator has affected negatively the native species (many of them endemic to the island) on which the boa feed. It is important to determine temporal variation in boa abundance, the areas of the island in which boas live, and the vegetation types they use in order to develop management strategies to reduce boa pressure on the native biota. We used nocturnal road transect sampling and occasional boa encounters during field work, to estimate boa distribution, abundance and habitat use, taking into account its spatiotemporal patterns on Cozumel Island. This study confirms that *Boa constrictor* is well established, widespread, and abundant on the island. Our results show that boas are distributed throughout Cozumel, in all vegetation types and geographical regions. Overall, there were 0.11 ± 0.03 boa/10 km road transect. There were no significant spatiotemporal differences in boa activity (time of day) and abundance (monthly, seasonally, by vegetation types or regions of the island). According to the habitat use

I. Romero-Nájera · C. González-Baca
Centro de Investigaciones en Ecosistemas, Universidad Nacional Autónoma de México, Apartado postal 27-3, Santa María, Morelia, Michoacán 58089, México
e-mail: iromero@oikos.unam.mx

A. D. Cuarón
Departamento de Etología y Fauna Silvestre, Facultad de Medicina Veterinaria y Zootecnia, Universidad Nacional Autónoma de México, Ciudad Universitaria, México, D.F. 04510, México

A. D. Cuarón (✉)
Durrell Wildlife Conservation Trust, Reforma Agraria 400-8, Fraccionamiento El Pueblito, Col. San José del Cerrito, Morelia, Michoacán 58341, México
e-mail: cuaron@gmail.com

C. González-Baca
e-mail: chrisgo@oikos.unam.mx

analysis, there were, however, fewer boas than expected in the subdeciduous tropical forest we sampled and in the central-northern region of the island, which coincide with areas inhabited by humans. There were more boas in areas uninhabited by humans, and there was a tendency towards a greater proportion of dead boas in inhabited areas and live boas in uninhabited areas. Cozumel boas are habitat generalists, which are affected by human induced mortality in inhabited areas. There is a vast area uninhabited by humans, with natural vegetation, on the island where boas have suitable habitats available for their continuous existence on Cozumel. This situation, and the adaptability of the boa, makes the control or eradication of this introduced species a critical conservation challenge.

Keywords Abundance · *Boa constrictor* · Cozumel Island · Habitat use · Introduced species · Mexico

Introduction

Species distribution, abundance and habitat use can vary temporally and spatially (Andrewartha and Birch 1974). Abundance can fluctuate in time and differ regionally in response to the species life history, habitat characteristics and availability, variation in resource availability, as well as natural and human disturbance (Seigel et al. 1987; Arthur et al. 1996; Cuarón 2000a). Vegetation, or more generally land-cover, is a good indicator of wildlife habitat (Cuarón 2000b), as this variable summarizes other environmental parameters which are important for species, such as climatic (rainfall, temperature), geographic (latitude, altitude), geologic, edaphic and other local characteristics (Rzedowski 1981; Barbour et al. 1987). In addition, wildlife abundance and habitat use are related to microhabitat characteristics such as food and shelter availability (Seigel et al. 1987).

Species introductions are among the most pervasive types of disturbance (Sakai et al. 2001). Introduced species may have negative effects on ecosystems, and geographical or ecological islands are particularly vulnerable to them. For example, the introduction of the brown treesnake (*Boiga irregularis* Merrem) on Guam caused the extinction of most native vertebrates on the island (Fritts and Rodda 1998). Due to the consequences that the introduction of an exotic species can bring, it is necessary to understand some of its biological characteristics such as its local distribution, abundance, and habitat use in order to implement informed management programs. The management of introduced generalist species, which may live in a wide range of environmental conditions, can be particularly difficult and is a major challenge for the conservation of native insular biota.

Boa constrictor L. was introduced onto Cozumel Island, Mexico, in 1971 (Martínez-Morales and Cuarón 1999). Natural populations of this snake are widely distributed in the Neotropics, from northern Mexico to Argentina (Greene 1997). The boa is a prolific viviparous serpent and a generalist predator (Sironi et al. 2000; Bertona and Chiaraviglio 2003; Chiaraviglio et al. 2003; González-Baca 2006). Because of commercial exploitation and habitat loss, *Boa constrictor* is considered to be threatened in Mexico and other countries (Secretaría de Medio Ambiente y Recursos Naturales 2002; Chiaraviglio et al. 2003). Even though the boa is a popular snake with a large natural distribution, the species has not been systematically studied in the wild until recently (Martínez-Morales and Cuarón 1999; Sironi et al.

2000; Chiaraviglio et al. 2001; Bertona and Chiaraviglio 2003; Chiaraviglio et al. 2003; Romero-Nájera 2004; González-Baca 2006). Most publications on the species are based on anecdotic information, literature and museum research, or focused on captive individuals (e.g., Stone and Holtzman 1996; Greene 1997; Kennedy and Kennedy 2001; Wang et al. 2001; Correa-Sánchez and Godínez-Cano 2002; Boback 2003). With the exception of a study on boa habitat use in Argentina (Attademo et al. 2004), there are no published studies on the abundance and habitat use by this species.

Boa constrictor is now well established and widespread on Cozumel Island (Martínez-Morales and Cuarón 1999). The fact that it is a prolific generalist predator, and that it has no natural predators on Cozumel, may have facilitated its establishment (Martínez-Morales and Cuarón 1999). Since there are no major predators on the island that prey on native vertebrates, *Boa constrictor* has become a formidable keystone predator on Cozumel, inducing top-down effects, interfering on the whole insular system dynamics and causing the decline of Cozumel native species, many of them endemic to the island (Martínez-Morales 1999; Martínez-Morales and Cuarón 1999; Cuarón et al. 2004).

Here we present information on the local distribution, abundance and habitat use of *Boa constrictor* on Cozumel Island, taking into account its spatiotemporal patterns. We considered different land-cover types (as a proxy for macrohabitat types) and areas with different degrees of human presence. We also considered temporal variation in boa activity during the nighttime and in boa abundance by month, and season of the year. This information will help identify the particular areas of Cozumel where boas are more abundant and what vegetation types are related to their presence in order to determine informed management actions towards the eradication or control of this predator with the aim of promoting the conservation of Cozumel native biota. The information will also be useful for other areas where there is an interest in using or recovering *Boa constrictor* populations.

Materials and methods

Cozumel Island, Mexico, is located 17.5 km off the Yucatán Peninsula (20° 20′N, 87° 00′E; 20° 30′N, 86° 50′E) in the Caribbean Sea. It is an oceanic island of coralline origin of approximately 490 km^2. The most common vegetation types are subdeciduous tropical forest, subdeciduous low tropical forest, mangroves, coastal dune vegetation and palm forest (Téllez et al. 1989).

In order to delimitate and quantify land-cover types on Cozumel, we interpreted black and white aerial photographs (taken on 9 February 2000; scale 1:75,000). The interpretation was verified in the field and incorporated into a Geographical Information System (GIS). With this information we generated a base map with all Cozumel vegetation types and roads (Fig. 1). For this purpose we used the program ILWIS version 3.0 (Koolhoven and Wind 2001).

From July 2001 to November 2002, we collected data to estimate boa distribution and abundance on Cozumel. For distribution we used nocturnal road transect sampling (see below) and occasional encounters during other times of the study. For abundance we only gathered information by means of nocturnal road transects by car (40 km/h). Our road transect surveys were restricted at night because of logistical constraints. These nocturnal road transects were made along the main paved road on the island. We accumulated a total sampling effort of 1902 km of nocturnal road transects. We recorded each boa (dead or alive) observed during these transects. All boa sightings were registered with a Global Positioning System (GPS) and were incorporated into the GIS. For analyses we segmented the road in five parts. These segments correspond to different geographical regions of the island, are associated to different vegetation types, and have different degrees of human presence (Fig. 1, Table 1).

We calculated an activity index by time of day. For this purpose, we divided the number of individuals observed by the sampling effort in minutes within an hour interval (ind/h). We used the number of individuals sighted every 10 km as an abundance index (ind/10 km). In all cases, we reported the index mean ± one standard error. We calculated this abundance index for different time periods during the study (month, season of the year) and for each vegetation type and geographical region of the island (Table 1). We considered November–April as the dry season and May–October as the rainy season.

We also considered the overall, dead, and live boa abundance in inhabited areas (those with a permanent human presence) and in uninhabited areas (those with limited human presence; Table 1). There were houses, ranches, hotels and other constructions for tourism in inhabited areas, while in uninhabited areas there were only isolated restaurants (one every 5–10 km).

In order to estimate boa habitat use, we used Bonferroni intervals which consider nonparametric tests for observed and expected data in each habitat (Neu et al. 1974). We conducted this test through the HABUSE program (Byers et al. 1984). We ran one analysis by vegetation types and another for the geographical regions of the island.

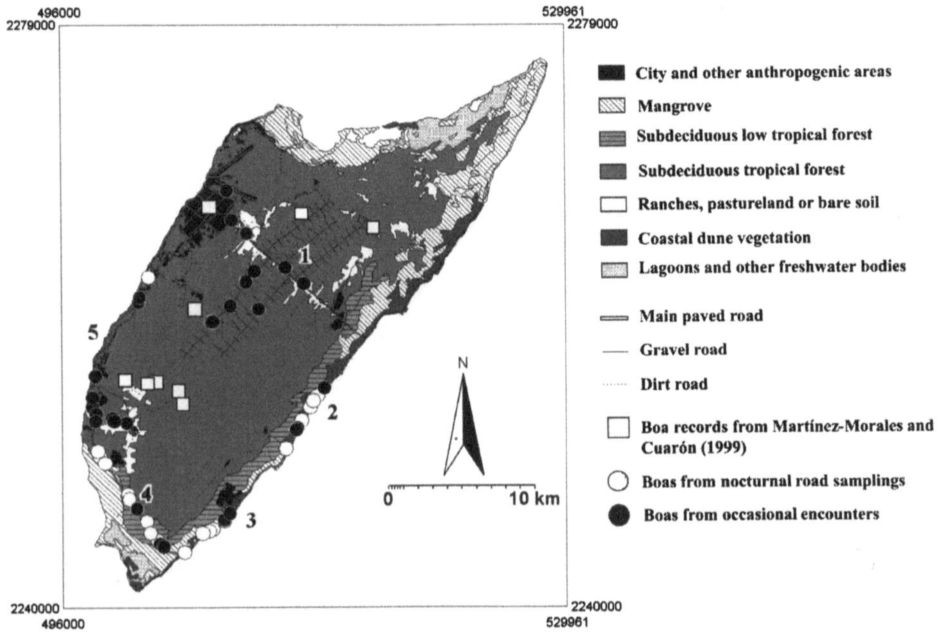

Fig. 1 Cozumel land-cover types, roads and boa records from 1994–1995 (Martínez-Morales and Cuarón 1999), and from this study (occasional encounters and nocturnal road transect samplings). Numbers indicate road segments representing each geographical region mentioned in Table 1. Coordinates are UTM units

Table 1 Characteristics of the five segments along the main paved road where nocturnal road transects were sampled on Cozumel Island. STF = Subdeciduous tropical forest, CDV = Coastal dune vegetation, SLTF = Subdeciduous low tropical forest

Road segment	Length (km)	Geographical region	Vegetation type	Human influence
1	10	Central-northern	STF	Inhabited area
2	10	Eastern	CDV	Uninhabited area
3	10	South-eastern	CDV	Uninhabited area
4	10	Southern	SLTF	Uninhabited area
5	14	South-western	STF	Inhabited area

We used two-tailed nonparametric tests throughout. We used Kruskal–Wallis tests in order to distinguish differences in boa activity by time of day and in boa abundance between months, seasons of the year, vegetation types and geographical regions. We used Mann–Whitney tests to distinguish differences in boa abundance between inhabited and uninhabited areas, and between dead and alive individuals. We also used a contingency table, using the log-likelihood ratio test (G-test), to verify factor interactions (inhabited-uninhabited and dead-alive).

Results

Boa distribution on Cozumel

We found boas ($n = 49$) in all areas of the island where we had access (Fig. 1). The map depicts boas from occasional encounters and from nocturnal road transects (Fig. 1). We registered boas in all land-cover types and in areas with different degrees of human presence. We were not able to visit the northeastern and central-southern regions of Cozumel because of the difficult access to those portions of the island, so we have no boa records for those regions.

Temporal patterns of boa abundance

During nocturnal road transects we found active boas from 19:00 to 24:00 h. Boa activity appeared to be more intensive from 22:01 to 23:00 h (Fig. 2), however, there were no significant differences in boa activity by time of day (Fig. 2: Kruskal–Wallis $H = 6.94$, df = 7, $p = 0.43$). As expected, there was a positive relation between the sampling effort and the number of active individuals registered by time of day ($r = 0.88$, $n = 8$, $p = 0.01$). Thus, the likelihood of observing active boas is similar during the night and it is highly dependent on the sampling effort. These data were gathered exclusively during evening sampling; however, this does not mean that boas are only active during nighttime. In fact, we observed active boas during the daytime in other times of this study.

Overall, we obtained a boa abundance index of 0.11 ± 0.03 boa/10 km. Even if boa abundance fluctuated temporally, there were large variances and we found no significant differences in boa abundance by month (Fig. 3a: Kruskal–Wallis $H = 8.09$, df = 9, $p = 0.52$) or season of the year (Fig. 3b: Kruskal–Wallis $H = 2.61$, df = 3, $p = 0.45$).

Spatial patterns of boa abundance and habitat use

There were no significant differences in boa abundance between the three main vegetation types on the island (Fig. 4a: Kruskal–Wallis $H = 3.05$, df = 2, $p = 0.22$). Nevertheless, according to the habitat use analysis, boas used differentially the three vegetation types ($\chi^2 = 7.214$, df = 2, $p = 0.02$). Taking vegetation availability into account, there were significantly less boas than expected in the subdeciduous tropical forest (Table 2), which suggests that boas avoid or do not prefer this vegetation type on Cozumel (but see discussion).

There were no significant differences in boa abundance by geographical region of the island (Fig. 4b: Kruskal–Wallis $H = 7.47$, df = 4, $p = 0.11$). We did not register any boas in the central-northern region during transect sampling, but we did observe boas in that region at other times of the study (Fig. 1). Although there was a marginal difference in the way boas used the five geographical regions of Cozumel Island ($\chi^2 = 8.762$, df = 4, $p = 0.06$), there were significantly less boas than expected in the central-northern region (Table 3), suggesting that boas avoid or do not prefer this geographical region of Cozumel (but see discussion). Boas used the other four geographical regions of the island we surveyed according to what was expected (Table 3).

Variation in boa abundance in areas with different degree of human presence

We registered boas in both, inhabited and uninhabited areas (Fig. 1). Overall, we registered higher boa abundance in uninhabited areas (0.15 ± 0.04 ind/10 km) than in inhabited areas (0.04 ± 0.02 ind/10 km), and this difference was marginally significant (Mann–Whitney $U = 2805$, $p = 0.06$). We also compared the abundance of both, dead and live boas, in areas with different degrees of human presence. In inhabited areas, the abundance of dead boas tended to be higher than the abundance of live boas, whereas in uninhabited areas the opposite was true (Fig. 5). There were no significant differences (Log-likelihood ratio test $G = 2.4$, df = 1, $p = 0.12$) that could be explained by the interaction of these factors (dead-alive, inhabited-uninhabited).

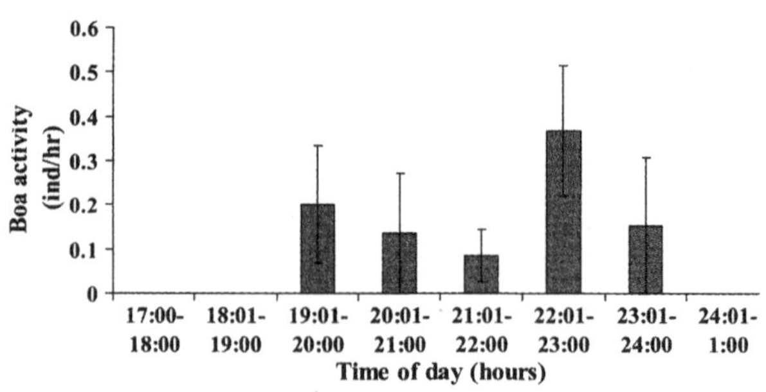

Fig. 2 Boa activity (mean ± standard error) by time of day

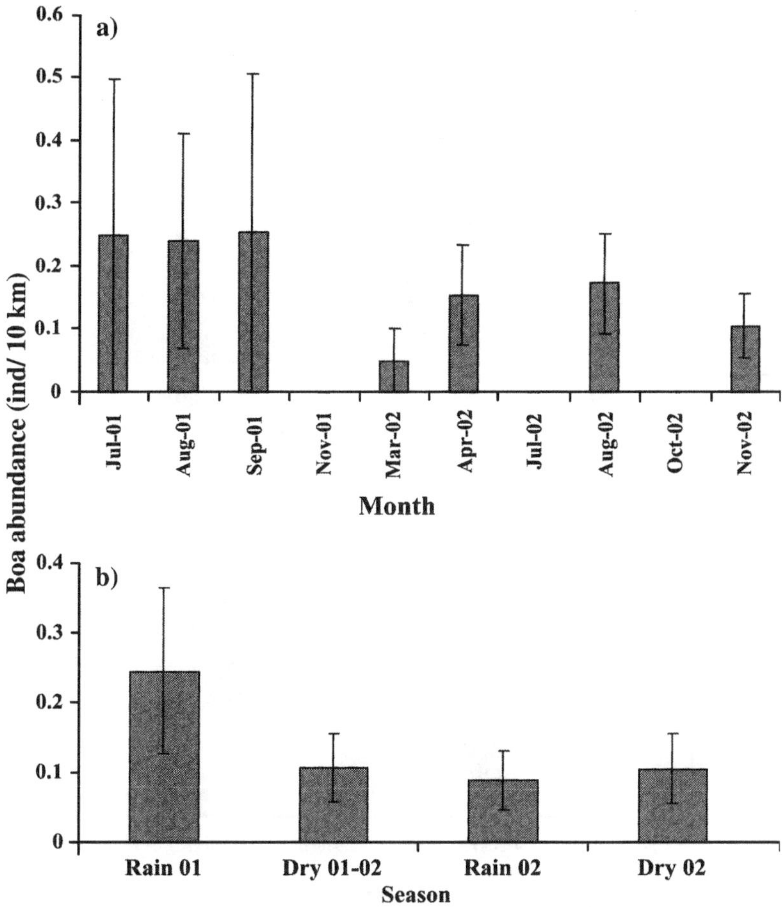

Fig. 3 Boa abundance (mean ± standard error) in 2001 and 2002 by (a) month, and (b) season in both years surveyed. Rain 01 = rainy season of 2001, Dry 01–02 = dry season from 2001 to 2002, Rain 02 = rainy season of 2002 and Dry 02 = dry season of 2002

Discussion

Boa distribution on Cozumel

In this investigation we corroborated the findings of Martínez-Morales and Cuarón (1999), which suggested that the boa has a wide distribution on Cozumel. The boa locality records reported by Martínez-Morales and Cuarón (1999), and those reported herein complement each other, providing a near complete coverage of the island (except for inaccessible areas, where we did not work; Fig. 1). The evidence gathered by nocturnal road transect sampling (this study), by diurnal transect sampling by foot (Martínez-Morales and Cuarón, 1999) and by occasional encounters (both studies) show that boas are present throughout the island in all land-cover types, not only in the vegetation adjacent to the road. In this study we did not consider the urban area, although we have seen boas in the city.

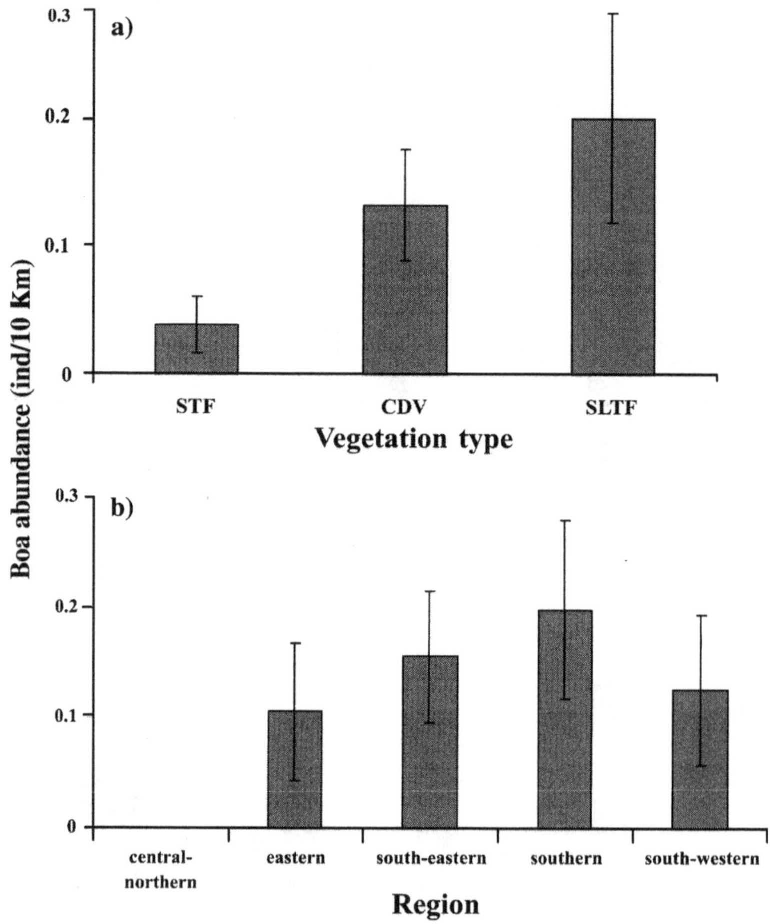

Fig. 4 Boa abundance (mean ± standard error) by (**a**) vegetation type (STF = Subdeciduous tropical forest, CDV = Coastal dune vegetation and SLTF = Subdeciduous low tropical forest), and (**b**) geographical region of Cozumel Island

Table 2 Bonferroni intervals between observed and expected boa habitat use by vegetation type on Cozumel Island

Vegetation type	Observed	Interval	Expected
Subdeciduous tropical forest	0.143	0.000–0.326	0.400*
Subdeciduous low tropical forest	0.381	0.127–0.635	0.200
Coastal dune vegetation	0.476	0.215–0.737	0.400

The asterisk represents the vegetation type where there were significant differences ($p < 0.05$); thus, the number of expected individuals is not within the Bonferroni interval estimated for that particular vegetation type

The local distribution of a species may be determined mainly by prey and shelter availability (Andrewartha and Birch 1974; Huey et al. 1989; Madsen and Shine 1996; Jepsen et al. 2002; Brito 2003; Rutherford and Gregory 2003). This serpent, however, is a generalist predator (Greene 1997; Sironi et al. 2000; González-Baca 2006),

Table 3 Bonferroni intervals between observed and expected boa habitat use by geographical region of Cozumel Island

Geographical region	Observed	Interval	Expected
Centra-northern	0.000	0.000–0.006	0.200*
Eastern	0.190	0.000–0.411	0.200
South-eastern	0.286	0.032–0.540	0.200
Southern	0.381	0.108–0.654	0.200
South-western	0.143	0.000–0.340	0.200

The asterisk represents the geographical region where there were significant differences ($p < 0.05$); thus, the number of expected individuals is not within the Bonferroni interval estimated for that particular region

and potential sites for shelter are abundant on Cozumel, so boas are capable of finding food and establishing themselves throughout the island.

Temporal patterns of abundance

We did not find a peak of activity at any specific time interval during nocturnal road transect sampling. Boas have an important nocturnal activity but we did not systematically survey during daytime, therefore, we can not estimate the intensity of boa activity during nighttime versus daytime. According to Martínez-Morales and Cuarón (1999), and our personal observations during occasional encounters in daytime, we know that boas are also diurnally active on Cozumel. In Argentina, boas also show diurnal activity (Chiaraviglio et al. 2003).

Martinez-Morales and Cuarón (1999) reported 1.8 individuals/100 km (0.18 individuals/10 km) during 1994–1995, which is higher than our overall estimate. This estimate was based on diurnal line transect sampling along footpaths in subdeciduous tropical forest in the central part of the island. The differences in methods do not allow a direct comparison of boa abundance between the two studies, but both estimates are within the same order of magnitude. When our abundance estimates are considered for particular months, seasons, vegetation types, or geographical regions of Cozumel, there are some which are higher or lower than the estimate for the mid-1990s. Apparently there are no boa abundance estimates for elsewhere. Nevertheless, the available evidence strongly suggests that boa abundance is higher in Cozumel than in mainland Mexico (Barreto-Oble 2000; A.D. Cuarón, personal

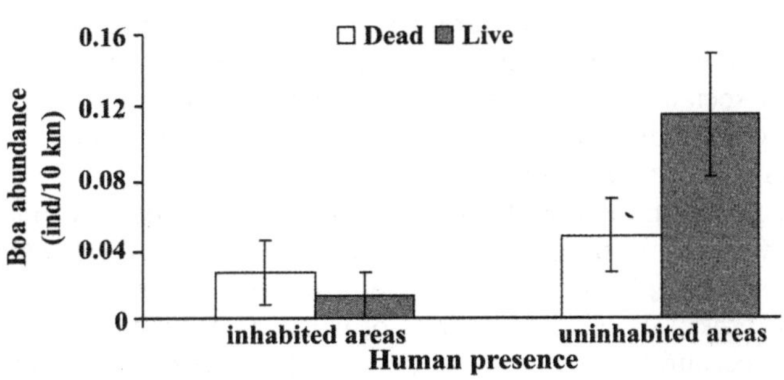

Fig. 5 Live and dead boa abundance (mean ± standard error) in areas with different degree of human presence

observation; V.H. Reynoso, personal communication; I. Suazo, personal communication). Studies of boas in the Mesoamerican mainland are necessary for comparative purposes, and to gain a better understanding of boa population dynamics and the effect of this species on the local biota.

We observed boas all year round. However, there seems to be a declining trend in boa populations from 2001 to 2002 (Fig. 3). Since there were large variances in boa abundance (monthly and seasonally), however, it is anticipated to affirm such a thing and long-term monitoring is needed. On Cozumel, local people claim that boas are most active or abundant during the rainy season. Our data do not support that contention as boa abundance was statistically similar throughout the year. In Argentinean xerophytic forests, boas are active year round, but are more active during the dry season (Chiaraviglio et al. 2003).

It has been documented that climate variation throughout the year directly affects species abundance and local distribution (Andrewartha and Birch 1974; Dalrymple et al. 1991; Secor 1994; Arthur et al. 1996; Greene 1997; Osborne et al. 2001; Zug et al. 2001). In some cases, there is an indirect effect when prey move as a consequence of phenological changes in the vegetation, resulting in predators relocating in response to these movements (Andrewartha and Birch 1974; Madsen and Shine 1996; Jepsen et al. 2002; Brito 2003). It is possible that these factors do not significantly affect boas on Cozumel because they are generalist predators (González-Baca 2006).

Spatial patterns of abundance and habitat use, and human presence

It has been reported that in other ecosystems resource availability and microclimate conditions change according to vegetation type (Schmida and Wilson 1985; Collins and Good 1987; Collins 1990; Schnitzler and Borlea 1998; Romero-Nájera 2000). Boa abundance did not vary significantly with respect to vegetation types on Cozumel. When analyzing boa habitat use by vegetation type, however, there were fewer individuals than expected in the subdeciduous tropical forest, which is the most extensive vegetation type on the island (Fig. 1). When considering boa habitat use by geographical region, there were less boas than expected in the central-northern region of the island, which corresponds precisely to the subdeciduous tropical forest. The abundance estimate of Martínez-Morales and Cuarón (1999) was in subdeciduous tropical forest in areas with no or little anthropogenic influence, which indicates that boas are also abundant in that vegetation type. Thus, on Cozumel, boa abundance seems to be unrelated to vegetation types or geographical regions, except in terms of the different degree of human presence in those areas. Both the subdeciduous tropical forest we sampled and the central-northern region of the island are associated to ranches, houses, and minor tourism infrastructure, where people have a great influence on boa mortality. This is likely to be the main cause of the lower numbers of boas found in this region.

Habitat use is determined by the addition of several factors which interact among them, such as vegetation, sampling period, prey distribution, fresh water and shelter availability (Arthur et al. 1996; Mysterud and Ims 1998; Jepsen et al, 2002) and also anthropogenic pressures (Cuarón 2000a, b; Cuarón et al. 2004). However, preference towards a certain habitat does not necessarily indicate its quality (Arthur et al. 1996). Differences on habitat selection patterns might reflect effects on sex, age or population density (Arthur et al. 1996; Bonnet et al. 1999; Castoe 2002). Habitat use will also depend on habitat availability; if a certain habitat is preferred but is not available, then another one will be used, even though it is not the one preferred (Arthur et al. 1996). Our data indicate that boas on Cozumel are habitat generalists, which use all vegetation types and geographical regions of the island. Human induced mortality resulted in fewer boas than expected in the inhabited subdeciduous tropical forest of the north-central region of the island.

There was higher boa abundance in uninhabited areas than in inhabited areas. Among the most important factors that may determine boa abundance in inhabited areas is human induced mortality. Intentional killing of serpents occur very often because of the negative feelings they inspire on people, which are a mixture of fear, repulsion and hate (Bonnet et al. 1999). On Cozumel, boas are generally perceived to be noxious and responsible for the drastic decline of wildlife populations on the island (Martínez-Morales and Cuarón 1999; Navarro-Ramírez 2005). On the other hand, vehicle traffic on roads has a negative effect on animals because of the risk of being run over by cars when moving from one site to another (Forman and Alexander 1998; Bonnet et al. 1999). The most frequent causes of death were intentional killings (usually with machetes) and roadkills. This suggests that people are controlling the boa population in inhabited areas. This control is spontaneous and is not the product of an organized management program. From the perspective of a control and eradication program for introduced boas, the fact that people can limit boa populations is encouraging. However, there is a vast and inaccessible territory of natural, relatively undisturbed, and uninhabited areas on Cozumel Island (Fig. 1), where control and the possible eradication of the boa population is a major conservation challenge. An organized and well planed management program for boas on Cozumel is necessary, and this study attempts to be a contribution in that direction.

Caution must be exercised when attempting to control boas on Cozumel, as this species is threatened in the mainland: a conservation paradox. Public education in order to avoid confusion between the control of boas on Cozumel (where they are having a negative impact on the native biota; Martínez-Morales 1999; Martínez-Morales and Cuarón 1999; Cuarón et al. 2004), and the illegal use of boas in the mainland, is of paramount importance.

Acknowledgements We are grateful to Jimena Cuarón, Iris Fortes, Denise García-Vasco, Miguel A. Martínez Morales, and Guadalupe Navarro Ramírez for assistance during fieldwork. Jesús Benavides, Rafael Chacón, Héctor González, Raúl López Osorio, Jorge Ortega Joaquín, and Ernesto Vivas facilitated work on Cozumel. We thank Andrés García, Aurelio Ramírez-Bautista, and Victor Hugo Reynoso for useful comments. Heberto Ferreira Medina, Jesús Fuentes Junco, Erna López Granados, Manuel Mendoza Cantú, Vicente Salinas Melgoza and Alberto Valencia García provided computer and GIS support. Funding was provided by CONACYT (project 33635-V to ADC) and Fondo Sectorial de Investigación Ambiental SEMARNAT-CONACYT (project SEMARNAT-2002-001-0571/A1 to ADC), and scholarships from CONACYT and DGEP-UNAM to IRN.

References

Andrewartha HG, Birch LC (1974) The Distribution and Abundance of Animals. University of Chicago Press, Chicago

Arthur SM, Manly BFJ, McDonald LL, Garner GW (1996) Assessing habitat selection when availability changes. Ecology 77:215–227

Attademo A, Bertona M, Kozykariski M, Chiaraviglio M (2004). Uso del hábitat por *Boa constrictor occidentalis* (Serpentes: Boidae) durante la estación seca en Córdoba, Argentina. Cuadernos de Herpetología 18:33–41

Barbour MG, Burk JH, Pitts WD (1987) Terrestrial Plant Ecology. Benjamin Cummings, New York

Barreto-Oble D (2000) Análisis ecológico y distribucional de los anfibios y reptiles de la región de Nizanda, Istmo de Tehuantepec, Oaxaca. Advanced BSc (Licenciatura) thesis, Universidad Nacional Autónoma de México, México, D.F.

Bertona M, Chiaraviglio M (2003) Reproductive biology, mating aggregations and sexual dimorphism of the Argentine boa constrictor (*Boa constrictor occidentalis*). J Herpetol 37:510–516

Boback SM (2003) Body size evolution in snakes: evidence from island populations. Copeia 1:81–94

Bonnet X, Naulleau G, Shine R (1999) The dangers of leaving home: dispersal and mortality in snakes. Biol Conserv 89:39–50

Brito JC (2003) Seasonal variation in movements, home range, and habitat use by male *Vipera latastei* in northern Portugal. J Herpetol 37:155–160

Byers CR, Steinhorst RK, Krausman PR (1984) Clarification of a technique for analysis of utilization-availability data. J Wildlife Manage 48:1050–1053

Castoe TA (2002) Microhabitat selection in *Porthidium nasatum* (serpents: viperidae) in Costa Rica, with comments on ontogenetic variation. Herpetol Rev 33:174–175

Collins SL (1990) Habitat relationships and survivorships of tree seedlings in hemlock-hardwood forest. Can J Bot 68:790–797

Collins SL, Good RE (1987) The seedling regeneration niche: habitat structure of tree seedlings in an oak-pine forest. Oikos 48:89–98

Correa-Sánchez F, Godínez-Cano E (2002) Reproducción de *Boa constrictor imperator* (Serpentes:Boidae) en cautiverio. Boletín de la Sociedad Herpetológica Mexicana 10:1–6

Cuarón AD (2000a) A global perspective on habitat disturbance and tropical rainforest mammals. Conserv Biol 14:1574–1579

Cuarón AD (2000b) Effects of land-cover changes on mammals in a neotropical region: a modeling approach. Conserv Biol 14:1676–1692

Cuarón AD, Martínez-Morales MA, McFadden KW, Valenzuela D, Gompper ME (2004) The status of dwarf carnivores on Cozumel Island, Mexico. Biodivers Conserv 13:317–331

Chiaraviglio M, Bertona M, Cardozo G, Rivera P, Lucino S (2001) Atributos de la historia de vida de la lampalagua (*Boa constrictor occidentalis*) en dos Departamentos de la Provincia de Córdoba. Boletín de la Academia Nacional de Ciencias 66:55–60

Chiaraviglio M, Bertona M, Sironi M, Lucino S (2003) Intrapopulation variation in life history traits of *Boa constrictor occidentalis* in Argentina. Amphibia-Reptilia 24:65–74

Dalrymple GH, Steiner TM, Nodell RJ, Bernardino FS (1991) Seasonal activity of the snakes of Long Pine Key, Everglades National Park. Copeia 2:294–302

Forman RTT, Alexander LE (1998) Roads and their major ecological effects. Annu Rev Ecol Syst 29:207–231

Fritts TH, Rodda GH (1998) The role of introduced species in the degradation of island ecosystems: a case history of Guam. Annu Rev Ecol Syst 29:113–140

González-Baca CA (2006) Ecología de forrajeo de *Boa constrictor*. Un depredador introducido a la Isla Cozumel. MSc thesis, Universidad Nacional Autónoma de México, México D.F.

Greene HW (1997) Snakes: the evolution of mystery in nature. University California Press, Hong Kong

Huey RB, Peterson CR, Arnold SJ, Poter WP (1989) Hot rocks and not so hot rocks: retreat-site selection by garter snakes and its thermal consequences. Ecology 70:931–944

Jepsen JU, Eide NE, Prestrud P, Jacobsen LB (2002) The importance of prey distribution in habitat use by arctic foxes (*Alopex lagopus*). Can J Zoolog 80:418–429

Kennedy M, Kennedy M (2001) Breeding the Colombian boa. Reptile Amphib Hobbyist 7:30–36

Koolhoven W, Wind J (2001) ILWIS 3.0 Academic. ITC, Unit RSG

Madsen T, Shine R (1996) Seasonal migration of predators and prey—a study of phytons and rats in tropical Australia. Ecology 77:149–156

Martínez-Morales MA (1999) Conservation status and habitat preferences of the Cozumel Curassow. Condor 101:11–20

Martínez-Morales MA, Cuarón AD (1999) *Boa constrictor*, an introduced predator threatening the endemic fauna on Cozumel Island, Mexico. Biodivers Conserv 8:957–963

Mysterud A, Ims RA (1998) Functional responses in habitat use: availability influences relative use in trade-off situations. Ecology 79:1435–1441

Navarro-Ramírez MG (2005) Conocimientos y percepciones sobre la fauna por los habitantes de la Isla Cozumel. Advanced BSc (Licenciatura) thesis, Universidad de Guadalajara, Zapopan, Jalisco

Neu CW, Byers CR, Peek JM (1974) A technique for analysis of utilization-availability data. J Wildlife Manage 38:541–545

Osborne PE, Alonso JC, Bryant RG (2001) Modelling landscape-scale habitat use using GIS and remote sensing: a case study with great bustards. J Appl Ecol 38:458–471

Romero-Nájera I (2000) Estructura y condiciones microambientales en bosques perturbados de Los Altos de Chiapas, México. Advanced BSc (Licenciatura) thesis, Universidad Nacional Autónoma de México, México, D.F.

Romero-Nájera I (2004) Distribución, abundancia y uso de hábitat de *Boa constrictor* introducida a la Isla Cozumel. MSc thesis, Universidad Nacional Autónoma de México, México, D.F.

Rutherford PL, Gregory PT (2003) Habitat use and movement patterns of northern alligator lizards (*Elgaria coerulea*) and western skinks (*Eumeces skiltonianus*) in southeastern British Columbia. J Herpetol 37:98–106

Rzedowski J (1981) Vegetación de México. Limusa, México, D.F.

Sakai AK, Allendorf FW, Holt JS, Lodge DM, Molofsky J, With DA, Baughman S, Cabin RJ, Cohen JE, Ellstrand NC, McCauley DE, O'Neil P, Parker IM, Thompson JN, Welleri SG (2001) The population biology of invasive species. Annu Rev Ecol Syst 32:305–332

Schmida A, Wilson MV (1985) Biological determinants of species diversity. J Biogeogr 12:1–20

Schnitzler A, Borlea F (1998) Lessons from natural forests as keys for sustainable management and improvement of naturalness in managed broad leaved forests. Forest Ecol Manag 109:293–303

Secor SM (1994) Ecological significance of movements and activity range for the sidewinder, *Crotalus cerastes.* Copeia 3:631–645

Secretaría de Medio Ambiente y Recursos Naturales. 2002. Norma Oficial Mexicana NOM-059-ECOL-2001. Protección ambiental—Especies nativas de México de flora y fauna silvestres—Categorías de riesgo y especificaciones para su inclusión, exclusión o cambio—Lista de especies en riesgo. Diario Oficial de la Federación (segunda sección; 6 de marzo), México, D.F., pp 95–190

Seigel RA, Collins JT, Novak SS (1987) Snakes: ecology and evolutionary biology. McGraw-Hill, New York

Sironi M, Chiaraviglio M, Cervantes R, Bertona M, Rio M (2000) Dietary habits of *Boa constrictor occidentalis*, in the Cordoba Province, Argentina. Amphibia-Reptilia 21:226–232

Stone A, Holtzman DA (1996) Feeding responses in young boa constrictors are mediated by vomeronasal system. Anim Behav 52:949–955

Téllez O, Cabrera EF, Linares E, Bye R (1989) Las Plantas de Cozumel. Universidad Nacional Autónoma de México, México, D.F.

Wang T, Taylor EW, Andrade D, Abe AS (2001) Autonomic control of heart rate during forced activity and digestion in the snake *Boa constrictor*. J Exp Biol 204:3553–3560

Zug GR, Vitt LJ, Caldwell JP (2001) Herpetology: an introductory biology of amphibians and reptiles. Academic Press, New York

Biodivers Conserv (2007) 16:1197–1204
DOI 10.1007/s10531-006-9105-y

ORIGINAL PAPER

Camera-trap studies of maned wolf density in the Cerrado and the Pantanal of Brazil

Mogens Trolle · Andrew J. Noss · Edson De S. Lima · Julio C. Dalponte

Received: 21 December 2005 / Accepted: 7 August 2006 / Published online: 27 October 2006

Abstract The maned wolf (*Chrysocyon brachyurus*) is threatened by large-scale habitat loss, in particular due to conversion to agricultural land. This is the first published study on maned wolf density and the first test of individual identification from camera-trap photographs. We present results from two Brazilian regions: the Cerrado and the Pantanal. Using capture–recapture analysis of camera-trap data, we estimated densities per 100 square kilometers of 3.64 ± 0.77 individuals at the Cerrado site and 1.56 ± 0.77 individuals at the Pantanal site. Parallel radio-telemetry studies at the Pantanal site showed that maned wolves occupied home ranges of 39–58 km^2 (mean = 50.3 ± 7.67 km^2). Our study in the Cerrado took place in a private farm with a mixture of agricultural land and native habitats, representative of the majority of the present-day Cerrado. Whereas many other mammalian species have suffered in the region, our results show that the maned wolf may cope better with this highly fragmented landscape than one might have feared. Finally, the paper briefly compares maned wolf density with density of puma (*Puma concolor*) in the Pantanal site.

Keywords Brazil · Camera trapping · Cerrado · *Chrysocyon* · Density · Maned wolf · Pantanal · Puma

M. Trolle
Mammal Department, Zoological Museum, University of Copenhagen, Copenhagen, Denmark

A. J. Noss
Wildlife Conservation Society, 4424 NW 13th Street, A2, Gainesville, FL 32609, USA
e-mail: anoss@wcs.org

E. D. S. Lima · J. C. Dalponte
Departamento de Ciências Biológicas - UNEMAT, Campus de Nova Xavantina, 78690-000 Mato Grosso, Nova Xavantina, MT, Brazil
e-mail: edsolima@hotmail.com
e-mail: jc.dalponte@uol.com.br

Present Address:
M. Trolle (✉)
Pile Allé 19A 2tv, 2000 Frederiksberg, Denmark
e-mail: mail@mogenstrolle.dk

Introduction

The savannah-adapted maned wolf (*Chrysocyon brachyurus*) is one of the keystone conservation species of the Brazilian savannahs. The species is threatened by large-scale habitat loss, in particular due to conversion to agricultural land (Fonseca et al. 1994). Around 80% of the original Cerrado vegetation have been lost (Myers et al. 2000), and only about 1.5% of the region are protected (Ratter et al. 1997). The IUCN Canid Specialist Group concludes that population surveys are needed, and that investigations of the suitability of agricultural land as maned wolf habitat are essential (Rodden et al. 2004). This paper addresses both of these aspects.

To the best of our knowledge, no study of maned wolf density has been published before. In order to estimate density of this nocturnal and rarely observed canid we used camera-trapping data combined with capture–recapture analysis. This relatively new method has already proven efficient for a number of elusive tropical mammal species (Karanth 1995; Karanth and Nichols 1998; Noss et al. 2003, 2004; Trolle and Kéry 2003, 2005; Maffei et al. 2004; Silver et al. 2004; Sanderson and Trolle 2005).

This paper presents results from two Brazilian study sites: one in the Cerrado and one in the Pantanal. Our Cerrado study area was a private farm consisting of a mosaic of agricultural land and native habitats, allowing us to assess how the maned wolf is able to cope with the highly fragmented landscape, representative of the majority of the present-day Cerrado region. For the Pantanal site, we were able to compare the camera-trapping results with radio-telemetry data for maned wolf. In addition, we briefly present activity patterns of the maned wolf. Finally, we compare density of maned wolf in the Pantanal site with that of another dominant large carnivore, the puma (*Puma concolor*).

Study areas

The Cerrado study area was the Fazenda Cauaia ranch (headquarters at 19° 28.98′, 44° 01.01′W), Minas Gerais state, approximately 50 km north of Belo Horizonte. The ranch covers 1,760 ha, with 50% cattle pasture and corn plantations and 50% native (although not unaltered) habitats. The main natural habitats were cerrado woodland (‘cerradão’) (characterized by gnarled, thick-barked trees), dry forest related to limestone outcrops (Brina 1998), and marshes. The original cerradão vegetation has been almost entirely eradicated in the study area; only one island of this habitat remained, however, small woodland patches occurred in many places. All natural habitats were used extensively by the 1,300 head of cattle, which undoubtedly had a substantial impact on the vegetation.

The Pantanal study area (field stations at 16° 42.66′, 56° 01.65′W and 16° 41.20′, 56° 10.49′W) was the 106,000 ha private reserve Estância Ecológica SESC Pantanal, between the Rio Cuiabá river and the tributary Rio São Lourenço, north-eastern Pantanal wetlands, Mato Grosso. It consists of gallery forest, semi-deciduous forest with the understorey dominated by acurí palms (*Scheelea phalerata*), Cerrado woodland, scrubland, and seasonally inundated grassland. Typical of the Pantanal (Trolle 2003), the study area was affected considerably by ranching-related activities such as dry-season fires to promote introduced pasture up until the cattle were removed when the reserve was established in 1998.

Materials and methods

Field methodology

Passive TrailMaster camera-trap equipment was used for the surveys (Goodson and Associates, Inc., Lenexa, Kansas) (Trolle 2003). At both sites, traps covered all major habitats and were set up along either car tracks or animal trails.

In the SESC Pantanal reserve, a camera-trapping area of approximately 54 km^2 was chosen in the north-eastern corner of the reserve. In each of four sub-areas, 14 trapping stations were camera trapped for nine consecutive nights. Trap density was 1 trap/km^2. Two cameras per trap were used (for more details see Trolle and Kéry 2005). At the Cerrado study area, data were obtained during a general survey of large and medium-sized mammals. We used 15 camera traps with single cameras. The methodology outlined in Trolle and Kéry (2003) was followed. Trappers' lure (Pro's Choice, Carman's Superior Animal Lures) was applied to attract carnivores. During the study period, 45 points were camera trapped, covering a minimum convex polygon area of 16 km^2.

In the SESC Pantanal area five adult maned wolves (3 females and 2 males) were captured in wood traps, made according to the specifications supplied by Dietz (1984), fitted with radio collars (Wildlife Materials, Inc.) and monitored for 5–12 months in 2002.

Data analysis

The maned wolf photos were examined to determine characteristics that could serve to distinguish individuals: black markings on legs and face, white tip of tail, tail shape, hair patterns on the flanks, and body structure. We took care to account for the differences in the observed features resulting from differences in camera angle, body position, and lighting conditions. Based on the number of "captures" and "recaptures" during the survey, it is possible to estimate population abundance using the closed population models of the program CAPTURE (White et al. 1978; Rexstad and Burnham 1991). We assumed that the maned wolf population was closed during the 25–36-day survey periods. CAPTURE provides estimators for seven models that make different assumptions about sources of variation of detection probability, and recommends the model that best fits the data.

To estimate population density, we divided the abundance estimate from the recommended model by the effective sample area that includes a circular buffer around each camera-trap site. We applied two alternative buffers: (1) half the mean maximum distance moved (HMMDM) among multiple captures of individual maned wolves during the survey period (Wilson and Anderson 1985), and (2) the full MMDM (Parmenter et al. 2003). For the Pantanal site, we conducted the same density analysis for puma as described above for the maned wolf (and in Kelly et al. 2004). We used the time information recorded on all camera-trap photographs to evaluate activity patterns for maned wolves. To determine home ranges from radio-telemetry data we used the Minimum Convex Polygon method (Mohr 1947).

Results

At the Cerrado site we obtained 93 maned wolf photographs (31 separate observations), and at the Pantanal site 15 photographs (10 separate observations). Maned wolves were very interested in the lure used in the Cerrado study, often remaining at the traps for extended periods (2–9 min), photographing both flanks with the single camera. At each site we identified four individuals (Table 1).

In the Cerrado case we considered each night to be an independent sampling occasion and grouped multiple observations of a single individual on the same night. This resulted in a matrix of 18 captures (individual × trap-night) over the 25-day survey period. In the Pantanal case, we pooled the data across the four sub-areas, grouping the first day of trapping in each sub-area to constitute day 1, the second day of trapping in each sub-area to constitute day 2, etc., for a total of 9 sampling occasions. CAPTURE recommended the M(th) model as the one that best fits the Cerrado data, combining heterogeneous capture probability with a differential response over time among individuals. In the case of the Pantanal data, Capture recommended the M(o) model, assuming a constant capture probability for all individuals.

The mean maximum distance moved (MMDM) for the maned wolf individuals was 3.7 km at the Cerrado site and 4.5 km at the Pantanal site. We estimate conservatively that population density is 1.56 ± 0.83/100 km^2 at the Pantanal site (Table 2). Population density is likely to be higher at the Cerrado site, 3.64 ± 0.77/100 km^2, but this estimate is only tentative until a systematic survey covering a wider area can be conducted at this site. Maned wolf activity recorded was strictly crepuscular and nocturnal, with a peak between 18.00 and 03.00. Home ranges for the five maned wolves radio collared in the SESC Pantanal area based on 31–201 locations covered 39–58 km^2 (mean = 50.3 ± 7.67 km^2).

In the Pantanal site we identified three individual pumas from 10 photographs obtained. In one adult male the tail has no black tuft at the end, while faint spotting is also visible on the flanks. A second adult male has a full black tail tuft, as well as a radio-collar. The third individual is a juvenile animal with heavy spotting on its

Table 1 Features used for distinguishing individual maned wolves (a—Cerrado site; b—Pantanal site)

Ind.	ID
(a)	
1	Thin tail with a markedly white tip, and a diagonal marking in the hair pattern on the right flank
2	Grey spots on the face/muzzle, grey/black extending far up on front and hind legs, and a relatively bushy tail
3	Dark lines or parts in the hair where the head joins the neck and where the neck joins the body, a diagonal marking in the hair pattern on the left flank, a tail that is in between the shapes of the tails for individuals 1 and 2, and distinct patterns of black on the legs in comparison with the other individuals
4	Relatively little black on its legs, and a wider/fuller body
(b)	
1	Thin tail with no white, diagonal marking in the hair pattern on the left flank
2	Tail slightly fuller, white tip, kinked
3	Tail fuller, white extends 1/3 of its length from the tip, black on hindlimbs extends above heel
4	Tail fuller, white extends 1/2 of its length from the tip, diagonal marking in the hair pattern on left flank distinct from individual 1

Table 2 Estimation of maned wolf population size and density (a—Cerrado site; b—Pantanal site)

Method	Buffer (km)	Area (km^2)	*N* (SE)	*D* (SE)
Panel a[a]				
HMMDM	1.85	51	4 (0.61)	7.84 (1.47)
MMDM	3.70	110	4 (0.61)	3.64 (0.77)
Panel b[b]				
HMMDM	2.26	134	4 (1.13)	2.99 (2.04)
MMDM	4.53	257	4 (1.13)	1.56 (0.83)

[a] *N* is the number of independent maned wolves in the study area according to the M(th) model

[b] *D* is the estimated number of independent maned wolves per 100 km^2 . N is the number of independent maned wolves in the study area according to the M(o) model

flanks. For the two adult males we recorded MMDMs of 6.5 and 2.5 km, respectively. Density estimates are given in Table 3. Photographic records were concentrated between 17.30–21.15 and 02.30–06.45, overlapping with maned wolf activity patterns.

Discussion

Biological considerations

Camera trapping is a promising new method for conducting density studies of the maned wolf, with no studies of maned wolf density published before. However, various radio-telemetry studies of home ranges have been conducted. For the Cerrado region, the following home-range sizes have been published: 21.7–30.0 km^2 (average 25.2 ± 4.4 km^2; $n = 3$ pairs) in Serra da Canastra National Park (Dietz 1984); 15.6–104.9 km^2 (average 57.0 ± 34.3 km^2; $n = 5$) in Águas Emendadas Ecological Station (Rodrigues 2002); 4.7–79.5 km^2 (average 49.0 ± 31.8 km^2; $n = 5$) in Emas National Park (Silveira 1999), and 75 ± 34.6 km^2 ($n = 3$) in São Paulo (Carvalho and Vasconcelos 1995). In the SESC Pantanal area, home range sizes were 39–58 km^2 (mean = 50.3 ± 7.67 km^2; $n = 5$).

Maned wolves appear to be facultatively monogamous and territories are defended against adjacent pairs, although they may overlap at the boundaries (Rodrigues 2002; Rodden et al. 2004). In the SESC Pantanal area, radio-tracked individuals that were assumed to be mated pairs shared 50–90% of their observed ranges, whereas females with adjacent territories shared only 10–20% and males of adjacent territories only 1% of their ranges.

Table 3 Estimation of puma population size and density at the Pantanal site

Method	Buffer (km)	Area (km^2)	*N* (SE)	*D* (SE)
HMMDM	2.24	133	4 (1.35)	3.01 (1.38)
MMDM	4.48	254	4 (1.35)	1.57 (0.64)

D is the estimated number of independent pumas per 100 km^2 . *N* is the number of independent pumas in the study area according to the M(h) model, the model which best fits the data according to the Capture program

The four individuals identified at each of our study sites may thus represent two pairs whose territories overlap at the edges, or may include subadults and/or floater individuals moving along the boundaries of occupied territories (Dietz 1984). These variables considered, it is difficult to directly compare the home range sizes given above with the densities found in our study areas. Nevertheless, it seems that our results (3.64 ± 0.77 and 1.56 ± 0.77/100 km^2, respectively) reflect a higher abundance of maned wolf than expected, given the altered state of the study areas. However, our density estimates also reinforce the conclusions that large suitable areas are required to maintain viable populations of maned wolves.

The general mammal survey we conducted in the Fazenda Cauaia of the Cerrado region revealed that many of the larger mammal species have either disappeared altogether or become rare (M. Trolle, personal observation). However, the maned wolf seems to be one of the positive exceptions. The maned wolf is an omnivore whose diet consists primarily of fruit and small mammals such as rodents (Rodden et al. 2004). Although the human-induced fragmentation and disappearance of natural habitats has undoubtedly had a negative effect on the small mammal biodiversity, the overall *biomass* of small mammalian prey may still be high, dominated by a few species that are able to exploit the agricultural land. This could be one major factor contributing to the continued well being of the maned wolf. In addition to food, the maned wolves require shelter for their dens, such as shrubs and tall grass (Rodden et al. 2004), and plenty of suitable hideouts are still found in the study area.

In conclusion, it appears that the maned wolf may in fact be able to adapt well to a mixture of agricultural land and natural habitats. Further studies in unprotected areas of the Cerrado, both of maned wolf density and the food availability for the species, are needed to investigate whether the very interesting indications of this study are representative for the Cerrado region as a whole. If this was indeed the case, then the conservation status of maned wolves may be better than previously thought.

Pumas and maned wolves were the two large carnivores recorded in the Pantanal study site (jaguars are known to occur in the SESC area, but seem to prefer the wetter western parts of the reserve). The preliminary data presented here show both spatial and temporal overlap between the two species, and further research is needed to explain how these two sympatric carnivores co-habit. Pumas in turn present population densities significantly lower than those recorded for Chaco and Chiquitano dry forest sites (3–7/100 km^2) in neighboring Bolivia where jaguars are present at densities of 1.5–5/100 km^2 (Kelly et al. 2004).

Methodological considerations

The camera-trapping study at the Cerrado site was designed as a species inventory, not a systematic density study of maned wolves. As a result, the area covered by the camera traps, 16 km^2, was less than the average home range of the species. Noss and Maffei (2005) recommend that the area covered by camera traps be at least 4 times the size of the average home range of the species in question. Even the Pantanal survey covered only 54 km^2, insufficient since home ranges for maned wolves inhabiting this area average 50 km^2. To ensure that our density estimate is conservative, therefore, we have used the largest recommended buffer and the minimum abundance estimate (equivalent in both cases to the number of individuals identified).

The MMDM is used in the camera-trapping surveys as a proxy for home range diameter, and is used to estimate effective survey area in two ways: Wilson and Anderson (1985) recommend applying a buffer of HMMDM around trap sites, whereas Parmenter et al. (2003) recommend applying a buffer equal to MMDM. The smaller the buffer, the smaller the survey area and the higher the estimated population density. The full MMDM of 3.7 km for the Cerrado site coincides with the mean home-range radius (as opposed to the diameter) from the radio-telemetry studies at other Cerrado sites reported above: 3.68 km ± 0.75. In turn, assuming the home ranges described by the telemetry study at SESC Pantanal are circular, then average radius of 3.99 km (± 0.31) coincides roughly with the full MMDM described above of 4.53 km (± 1.03). We therefore recommend using the full MMDM as a buffer, which in these cases corresponds to home range radius. A camera trap survey covering a wider area at each of the sites would serve to confirm the density estimates presented here.

Acknowledgements The Pantanal camera-trap study was a collaboration between Zoological Museum, University of Copenhagen, Museu Nacional, Universidade Federal do Rio de Janeiro, and SESC Pantanal. The Cerrado project was a collaboration between the Zoological Museum, University of Copenhagen and Laboratório de Estudos Evolutivos Humanos, Instituto de Biociências, Universidade de São Paulo. The project was made possible by the generous support of SESC Pantanal, WWF-Denmark/Novo Nordisk, Dronning Margrethe og Prins Henriks Fond, Beckett-Fonden, Torben og Alice Frimodts Fond, Inge og Skjold Burnes Fond, Ingeniør K.A. Rohde og hustrus legat, Fonden Kjebi, and Dansk Naturhistorisk Forening. Many thanks are due to João Alves de Oliveira, Luiz Flamarion de Oliveira, José Luiz Cordeiro, Walter Alves Neves, Renato Kipnis, Marcos Cesar Bissaro Junior, Helbert Medeiros Prado, and José Hein for logistical support as well as to Ullas Karanth, Luis Beethoven Piló, Jon Fjeldså, Hans J. Baagøe, Mogens Andersen, Abdi Hedayat, Michael Sterll, Birgitte Holten, and Rosa Leny Cuéllar.

References

Brina AE (1998) Aspectos da dinâmica da vegetação associada a afloramentos calcários na APA Carste de Lagoa Santa, MG. Master Thesis, Universidade Federal de Minas Gerais, Brazil

Carvalho CT, Vasconcellos LEM (1995) Disease, food and reproduction of the maned wolf – *Chrysocyon brachyurus* (Illiger) (Carnivora, Canidae) in southeast Brazil. Rev Bras Zool 12:627–640

Dietz JM (1984) Ecology and social organization of the maned wolf (*Chrysocyon brachyurus*). Smithsonian Contrib Zool 392:1–51

Fonseca GAB, Rylands AB, Costa CMR, Machado RB, Leite YR (eds) (1994) Livro Vermelho dos Mamíferos Ameaçados de Extinção. Fundação Biodiversitas, Belo Horizonte Brazil

Karanth KU (1995) Estimating tiger *Panthera tigris* populations from camera trap data using capture–recapture models. Biol Conserv 71:333–338

Karanth KU, Nichols JD (1998) Estimation of tiger densities in India using photographic captures and recaptures. Ecology 79:2852–2862

Kelly MJ, Noss AJ, Camblos HB, Rumiz DI (2004) Sympatric pumas and jaguars: data from camera trapping in Bolivia and Belize. 18th Annual Meeting of the Society for Conservation Biology, July 30-August 2, 2004, New York, USA

Maffei L, Cuéllar E, Noss A (2004) One thousand jaguars (*Panthera onca*) in Bolivia's Chaco? Camera trapping in the Kaa-Iya National Park. J Zool 262:295–304

Mohr CO (1947) Table of equivalent populations of North American mammals. Am Midl Nat 37:223–249

Myers N, Mittermeier RA, Mittermeier CG, Fonseca GAB, Kent J (2000) Biodiversity hotspots for conservation priorities. Nature 403:853–858

Noss AJ, Maffei L (2005) How small is too small? Camera trap survey areas and density estimates for ocelot in the Bolivian Chaco. Society for Conservation Biology Annual Meeting 2005, 15–19 July, 2005, Brasilia, Brazil

Noss AJ, Peña R, Rumiz DI (2004) Camera trapping *Priodontes maximus* in the dry forests of Santa Cruz, Bolivia. Endangered Spec Update 21:43–52

Noss AJ, Cuéllar RL, Barrientos J, Maffei L, Cuéllar E, Arispe R, Rumiz DI, Rivero K (2003) A camera trapping and radio telemetry study of *Tapirus terrestris* in Bolivian dry forests. Tapir Conserv 12:24–32

Parmenter RR, Yates TL, Anderson DR, Burnham KP, Dunnum JL, Franklin AB, Friggens MT, Lubow BC, Miller M, Olson GS, Parmenter CA, Pollard J, Rexstad E, Shenk TM, Stanley TR, White GC (2003) Small-mammal density estimation: a field comparison of grid-based vs. web-based density estimators. Ecol Monogr 73:1–26

Ratter JA, Riveiro JF, Bridgewater S (1997) The Brazilian cerrado vegetation and threats to its biodiversity. Ann Bot 80:223–230

Rexstad E, Burnham KP (1991) Users Guide for Interactive Program CAPTURE. Colorado Cooperative Fish & Wildlife Research Unit, Colorado State University, Fort Collins, Colorado.

Rodden M, Rodrigues F, Bestelmeyer S (2004) Maned wolf. In: Sillero-Zubiri C, Hoffmann M, Macdonald DW (eds), Canids: foxes, wolves, jackals and dogs – 2004 Status Survey and Conservation Action Plan. IUCN/SSC Canid Specialist Group, Gland Switzerland, pp 38–43

Rodrigues FHG (2002) Biologia e conservação do lobo-guará na Estação Ecológica de Águas Emendadas, DF. Ph.D. Thesis, Universidade Estadual de Campinas, Brazil

Sanderson JG, Trolle M (2005) Monitoring elusive mammals. Am Sci 93:148–155

Silveira L (1999) Ecologia e conservação dos mamíferos carnívoros do Parque Nacional das Emas, Goiás. Master thesis, Universidade Federal de Goiás, Brazil

Silver CS, Ostro LET, Marsh LK, Maffei L, Noss AJ, Kelly M, Wallace RB, Gomez H, Ayala G (2004) The use of camera traps for estimating jaguar *Panthera onca* abundance and density using capture/recapture analysis. Oryx 38:148–154

Trolle M (2003) Mammal survey in the southeastern Pantanal, Brazil. Biodivers Conserv 12:823–836

Trolle M, Kéry M (2003) Ocelot density estimation in the Pantanal using capture–recapture analysis of camera-trapping data. J Mammal 84:607–614

Trolle M, Kéry M (2005) Camera-trap study of ocelot and other secretive mammals in the northern Pantanal. Mammalia 69:405–412

White GC, Burnham KP, Otis DL, Anderson DR (1978) User's Manual for Program CAPTURE. Utah State University Press, Logan Utah USA

Wilson KR, Anderson DR (1985) Evaluation of two density estimators of small mammal populatin size. J Mammal 66:13–21

Biodivers Conserv (2007) 16:1205–1211
DOI 10.1007/s10531-006-9106-x

ORIGINAL PAPER

Mammal survey at a ranch of the Brazilian Cerrado

Mogens Trolle · Marcos Cesar Bissaro · Helbert Medeiros Prado

Received: 21 December 2005 / Accepted: 7 August 2006 / Published online: 27 October 2006

Abstract The wildlife of the Brazilian Cerrado is threatened by large-scale habitat loss, in particular due to conversion to agricultural land. It is essential to study how the mammal fauna copes with the highly fragmented, human-influenced, non-protected landscape. The paper presents the results of a survey of the large to medium-sized mammals of a typical cattle ranch with a mixture of human-created and natural vegetation types. We recorded 18 species. Surprisingly, several species were found to still thrive in the area, however, many species are rare or have become extinct. We conclude the paper with comments relevant for the conservation of mammals in the Cerrado as a whole.

Keywords Brazil · Camera trapping · Cerrado · Conservation · Lagoa Santa · Mammal survey

Introduction

Being the second largest biome of South America, the Cerrado of central Brazil harbors a rich mammal fauna, yet, the region lingers unstudied (Eisenberg and Redford 1999; Fonseca et al. 1999). Around 80% of the original Cerrado vegetation have been lost (Myers et al. 2000). Protected areas account for only 1.5% of the total

M. Trolle (✉)
Mammal Department, Zoological Museum, University of Copenhagen, Copenhagen, Denmark
e-mail: mail@mogenstrolle.dk

M. C. Bissaro · H. M. Prado
Laboratório de Estudos Humanos, Departamento de Genética e Biologia Evolutiva, Instituto de Biociências, Universidade de São Paulo, São Paulo, Brazil

Present Address:
M. Trolle
Pile Allé 19A 2tv, 2000 Frederiksberg, Denmark

area of the Cerrado, and pasture and cash crops are dominating the region (Ratter et al. 1997; Fonseca et al. 1999).

Mammal research in the region has typically focused on protected areas. However, in order to assess the conservation status of wildlife of the Cerrado and plan adequate conservation strategies, it is essential to study how the mammal fauna copes with the highly fragmented, human-influenced, non-protected landscape. The objective of this investigation was to inventory the large- to medium-sized mammals (i.e., species that could be identified without capture) in a typical cattle ranch with a mixture of human-created and natural vegetation types.

The study area was situated in the Lagoa Santa limestone region of the Rio das Velhas Valley. The region is famous for its fossil-rich caves originally excavated by the Danish naturalist Peter Wilhelm Lund from 1835–1844 (Rubbioli 2002). Lund (1842) published a list of extant mammals of Rio das Velhas Valley (a list brought up to date by Paula Couto (1950) and Leite and Costa (2002)). A number of recent publications have dealt with the small-mammal fauna of the region (Hershkovitz 1990, 1992; Voss and Myers 1991; Voss 1993; Musser et al. 1998). In contrast, no thorough field study of the large- to medium-sized mammals has been published since the time of Lund.

We conclude the paper with comments relevant for the conservation of mammals in the study region as well as the Cerrado as a whole.

Study area

The study area was the ranch Fazenda Cauaia (field station at 19°28.98′S, 44°01.01′W), situated in the Minas Gerais state approximately 50 km north of Belo Horizonte. The ranch covers about 1,760 ha, out of which approximately half consists of cattle pasture and corn plantations and half of native (although not unaltered) habitats.

The main natural habitats were Cerrado woodland ('Cerradão') (characterized by gnarled, thick-barked trees), secondary dry forest related to limestone outcrops (Brina 1998), and marshes. The original Cerradão vegetation has been almost entirely eradicated in the study area; only one island of this habitat remained, however, small woodland patches occurred in many places. All natural habitats were used extensively by the cattle, and the stock of around 1,300 head of cattle of the farm undoubtedly has a great impact on the vegetation.

Excavations in the Fazenda Cauaia have shown, that hunting has occurred in the area for at least 8,000 years (R. Kipnis, pers. comm.), a fact that undoubtedly affects the detectability of many mammal species. Recreational hunting by ranch hands has been normal practice in the region previously; however, the amount of hunting has diminished significantly within recent years. The focal mammalian species have been deer, armadillos, capybara (*Hydrochaeris hydrochaeris*), and paca (*Cuniculus paca*).

The study area lies within the tropics and is characterized by two well-defined seasons: a warm rainy season (typically October–March) and a colder dry season (April–September). The average annual precipitation is 1,350 mm (Brina 1998).

Materials and methods

The diverse fauna of large to medium-sized mammals of the Cerrado contains a number of categories that each requires a targeted inventory technique, including a high percentage of rare and secretive terrestrial species of various habitat preferences, arboreal/diurnal species, arboreal/nocturnal species, and semiaquatic species. We applied the following methods: camera trapping, diurnal and nocturnal walked transects, intensive track surveys (in particular along water), observations of additional animal signs such as burrows and feces, and interviews.

Camera trapping has proved to be highly efficient at recording elusive, terrestrial Neotropical mammals (Trolle 2003a, b, 2004; Trolle and Kéry 2003, 2005; Maffei et al. 2004; Silver et al. 2004; Trolle and Emmons 2004; Sanderson and Trolle 2005). We used 15 camera traps with passive infrared sensors (TrailMaster model TM550; Goodson and Associates, Inc., Lenexa, Kansas). The methodology outlined in Trolle (2003a) was followed. Trappers' lure (Pro's Choice, Carman's Superior Animal Lures) was applied to attract carnivores. During the study period, 45 points were camera trapped, evenly distributed throughout the study area in all major habitats. Traps were placed mainly by trails (78%) or dirt roads (18%).

A system of nine trails was established in the forests of the study area and used for the daily surveys. The positions of primate observations were recorded by GPS, allowing an assessment of minimum number of groups present in the study area.

Results

We recorded 18 species of large to medium-sized mammals and locals reported seven additional species (Table 1). A summary of records by species and method is given in Table 2.

We recorded a minimum of nine groups of *Callithrix penicillata*, two groups of *Alouatta fusca*, and two groups of *Cebus apella.*

Discussion

Our results allows an assessment of the fauna of large to medium-sized mammal species in the study area. Although a longer-term investigation is needed, these preliminary observations are highly valuable given the fact that the Cerrado, and in particular the human-influenced areas making up the majority of the region, is so little known.

We found one species to be exceptionally common: *Chrysocyon brachyurus*. Several species were apparently relatively common: *Didelphis albiventris, Cabassous* (at least in Cerrado vegetation), *Dasypus novemcinctus*, *Euphractus sexcinctus, Callithrix penicillata, Cerdocyon thous, Procyon cancrivorus, Leopardus pardalis, Puma concolor, Hydrochaeris hydrochaeris*, and *Sylvilagus brasiliensis*. On the other hand, a number of species seem to occur at relatively low densities: *Myrmecophaga tridactyla, Alouatta fusca, Cebus apella, Nasua nasua, Panthera onca, Mazama*, and *Cuniculus paca.*

Table 1 List of mammals of the Fazenda Cauaia area

Didelphidae	*Didelphis albiventris*	White-eared opossum
Dasypodidae	*Cabassous* sp.	Naked-tailed armadillo
	Dasyprocta novemcinctus	Nine-banded long-nosed armadillo
	Dasypus septemcinctus[a]	Seven-banded long-nosed armadillo
	Euphractus sexcinctus	Six-banded armadillo
Myrmecophagidae	*Myrmecophaga tridactyla*	Giant anteater
	Tamandua tetradactyla	Southern tamandua
Callitrichidae	*Callithrix penicillata*	Tufted-ear marmoset
Cebidae	*Alouatta fusca*	Brown howler monkey
	Cebus apella	Brown capuchin monkey
Canidae	*Cerdocyon thous*	Crab-eating fox
	Chrysocyon brachyurus	Maned wolf
Procyonidae	*Nasua nasua*	South American coati
	Procyon cancrivorus	Crab-eating raccoon
Mustelidae	*Eira barbara*	Tayra
	Galictis sp.	Grison
	Lontra longicaudis	Neotropical otter
Felidae	*Herpailurus yaguarondi*	Jaguarundi
	Leopardus pardalis	Ocelot
	Panthera onca[b]	Jaguar
	Puma concolor	Puma
Cervidae	*Mazama* sp.[c] (cf. *americana*)	Brocket deer
Hydrochaeridae	*Hydrochaeris hydrochaeris*	Capybara
Cuniculidae	*Cuniculus paca*[b]	Paca
Leporidae	*Sylvilagus brasiliensis*	Brazilian rabbit

[a] Locals describe a small species of armadillo in the area, distinguishable from *D. novemcinctus*. This is most likely *D. septemcinctus* (Leite and Costa 2002)

[b] Reported from the neighboring ranch, Fazenda Jaguara

[c] The locals recognize only one type of deer, and one ranch hand who worked for 20 years in the Fazenda Cauaia described the species as reddish, which would be *M. americana*. However, this needs to be confirmed. The one photo we obtained of *Mazama* did not allow species identification

Various species from the list of extant mammals in the region by Lund (1842) were not found in the study area. Of the species that, given the diversity of habitats of the study area, can be presumed to have occurred in the Fazenda Cauaia previously, it seems safe to conclude that the following have become extinct: *Tapirus terrestris*, *Pecari tajacu*, *Tayassu pecari*, *Blastocerus dichotomus*, *Ozotoceros bezoarticus*, *Dasyprocta azarae* and, possibly, *Conepatus semistriatus*, *Pteronura brasiliensis*, one of the *Mazama* species (most likely *gouazoubira*), and *Sciurus aestuans*. Besides, the greater rhea (*Rhea americana*), that according to locals used to be common in the ranch, is extinct.

More camera-trap studies and nocturnal censuses are needed to determine the presence/absence status of *Pseudalopex vetulus*, *Leopardus wiedii*, *Leopardus tigrinus*, and *Coendou/Sphiggurus*.

Conservation considerations

Scientists affiliated with the federal university as well as environmental institutions have claimed that there would be no interesting mammal fauna left in the Fazenda Cauaia region, and this is a widespread opinion (José Hein, pers. comm.;

Table 2 Summary of records at Fazenda Cauaia by species and method

	Camera trapping	Walked census	Miscellaneous	Total	Interviews
Didelphis albiventris	1	2	0	3	–
Cabassous sp.	1	0	Many burrows	1 (+ burrows)	–
Dasyprocta novemcinctus	3	0	Many burrows	3 (+ burrows)	–
Dasypus septemcinctus[1]	0	0	0	0	Yes
Euphractus sexcinctus	0	2	Many burrows	2 (+ burrows)	–
Myrmecophaga tridactyla	0	0	0	0	Yes
Tamandua tetradactyla	1	0	Tracks	1 (+ tracks)	–
Callithrix penicillata	0	17	0	17	–
Alouatta fusca	0	3	Heard only once	4	–
Cebus apella	0	4	0	4	–
Cerdocyon thous	17	0	Tracks	17 (+ tracks)	–
Chrysocyon brachyurus	45	1	Many feces	46 (+ feces)	–
Nasua nasua	0	0	0	0	Yes
Procyon cancrivorus	6	1	Many tracks	7 (+ tracks)	–
Eira barbara	2	0	0	2	–
Galictis sp.	1	0	0	1	–
Lontra longicaudis	0	0	0	0	Yes
Herpailurus yaguarondi	0	0	0	0	Yes
Leopardus pardalis	8	0	0	8	–
Panthera onca	0	0	0	0	Yes
Puma concolor	9	0	0	9	–
Mazama sp.	1	0	0	1	Yes
Hydrochaeris hydrochaeris	3	2	Many tracks and feces	5 (+ tracks/feces)	–
Cuniculus paca	0	0	0	0	Yes
Sylvilagus brasiliensis	10	1	0	11	–

Walter A. Neves, pers. comm.). Our study has proved that many important mammal species survive in the area. However, there is a great need for concern: many species have become extinct or are rare (and some are quite possibly in the process of disappearing). The fragmentation of the natural vegetation is no doubt the major factor contributing to this, although hunting may also have played a part for certain species (e.g., the peccaries).

Of utmost importance when setting conservation priorities for the Cerrado region as a whole is to protect the true Cerrado habitats, i.e., cerrado woodland (*cerradão*) and native savannas/grasslands (*cerrado/campos*). All people we interviewed shared the view, that the remaining forest patches are the major conservation concern, based on the conception that this is where the wildlife is likely to hide. This is reflected by the fact, that both Fazenda Cauaia and the neighboring Fazenda Jaguara have protected their forests, but not the more open areas. However, all of the forest species of the Valle Rio das Velhas region are widely distributed throughout tropical South America and also occur in the Atlantic forest and/or the Amazon rainforest (Emmons and Feer 1997; Eisenberg and Redford 1999). In a conservation sense, the most valuable species of the Cerrado are: (1) species depending on the savanna landscape such as *Chrysocyon brachyurus*, *Cerdocyon thous*, *Pseudalopex vetulus*, *Ozotoceros bezoarticus*, and *Euphractus sexcinctus*, and (2) species that may attain particularly high densities in the Cerrado, e.g., *Myrmecophaga tridactyla*, *Priodontes maximus*, *Cabassous*, and *Mazama gouazoubira*.

In the study area, only one patch of Cerrado woodland was left (Cerrado woodland had been cleared in several places within the last 15 years), and this is representative for the Lagoa Santa region as a whole. The landscape is dominated by cattle pasture on cleared land interspersed with forest patches on the limestone outcrops, and barely any typical cerrado vegetation remains.

In conclusion, we have found that there is indeed something valuable left to protect in the Lagoa Santa region. However, if the mammal fauna is to be preserved, given the high degree of anthropological impact on the environment, there is a great need for efficient conservation measures, in particular focusing on the original Cerrado vegetation.

Acknowledgements The project was a collaboration between the Zoological Museum, University of Copenhagen, Denmark and Laboratório de Estudos Evolutivos Humanos, Instituto de Biociências, Universidade de São Paulo, Brazil. Many thanks to Walter Alves Neves and Renato Kipnis as well as José Hein and the employees of Fazenda Cauaia for supplying the logistics making the project possible. MT greatly appreciates funding from Dronning Margrethe og Prins Henriks Fond, Torben og Alice Frimodts Fond, Dansk Naturhistorisk Forening and Fonden Kjebi. Thanks are also due to Luis Beethoven Piló, Jon Fjeldså, Hans J. Baagøe, Mogens Andersen, Abdi Hedayat, Michael Sterll, Birgitte Holten, Ane Elisa Brina, and Yuri L.R. Leite.

References

Brina AE (1998) Aspectos da dinâmica da vegetação associada a afloramentos calcários na APA Carste de Lagoa Santa, MG. Master thesis, Instituto Ciências Biológicas, Universidade Federal de Minas Gerais, Minas Gerais

Eisenberg JF, Redford KH (1999) Mammals of the Neotropics, vol 3. the central neotropics: Ecuador, Peru, Bolivia, Brazil. University of Chicago Press, Chicago

Emmons LH, Feer F (1997) Neotropical rainforest mammals: a field guide, 2nd edn. University of Chicago Press, Chicago

da Fonseca GAB, Herrmann G, Leite YLR (1999) Macrogeography of Brazilian mammals. In: Eisenberg JF, Redford KH (eds) Mammals of the Neotropics, vol 3. The central neotropics: Ecuador, Peru, Bolivia, Brazil. University of Chicago Press, Chicago, pp 549–563

Hershkovitz P (1990) The Brazilian rodent genus *Thalpomys* (Sigmodontinae, Cricetidae) with a description of a new species. J Nat Hist 24:763–783

Hershkovitz P (1992) The South American gracile mouse opossums, genus *Gracilinanus* Gardner and Creighton, 1989 (Marmosidae, Marsupialia): a taxonomic review with notes on general morphology and relationships. Fieldiana Zool ns 70:1–56

Leite YLR, Costa LP (2002) Peter Wilhelm Lund e a fauna de mamíferos do Vale do Rio das Velhas. In: Rubbioli EL (ed) Bicentenário do nascimento de Peter Wilhelm Lund, pp 32–41 O Caste 14 (special edition)

Lund PW (1842) Blik paa Brasiliens Dyreverden för sidste Jordomvaeltning. Fjerde Afhandling: Fortsaettelse af Pattedyrene. Det Kongelige Danske Videnskabernes Selskabs Naturvidenskabelige og Mathematiske Afhandlinger 9:137–208

Maffei L, Cuéllar E, Noss A (2004) One thousand jaguars (*Panthera onca*) in Bolivia's Chaco? Camera trapping in the Kaa-Iya National Park. J Zool 262:295–304

Musser GG, Carleton MD, Brothers EM, Gardner AI (1998) Systematic studies of oryzomyine rodents (Muridae, Sigmodontinae): diagnosis and distributions of species formerly assigned to *Oryzomys "capito"*. Bull Am Museum of Nat Hist 236:1–376

Myers N, Mittermeier RA, Mittermeier CG, Fonseca GAB, Kent J (2000) Biodiversity hotspots for conservation priorities. Nature 403:853–858

de Paula Couto C (1950) Lista atualizada dos mamíferos fósseis e atuais do Vale do Rio das Velhas, segundo os trabalhos de Peter Wilhelm Lund e Herluf Winge. In: Lund, PW (ed) Memórias sobre a Paleontologia Brasileira (revistas e comentadas por Carlos de Paula Couto). Instituto Nacional do Livro, Rio de Janeiro, pp 537–548

Ratter JA, Riveiro JF, Bridgewater S (1997) The Brazilian cerrado vegetation and threats to its biodiversity. Ann Bot 80:223–230

Rubbioli EL (ed) (2002) Bicentenário do nascimento de Peter Wilhelm Lund. O Caste 14 (special edition)

Sanderson JG, Trolle M (2005) Monitoring elusive mammals. Am Sci 93:148–155

Silver CS, Ostro LET, Marsh LK, Maffei L, Noss AJ, Kelly M, Wallace RB, Gomez H, Ayala G (2004) The use of camera traps for estimating jaguar *Panthera onca* abundance and density using capture/recapture analysis. Oryx 38:148–154

Trolle M (2003a) Mammal survey in the southeastern Pantanal, Brazil. Biodivers Conserv 12:823–836

Trolle M (2003b) Mammal survey in the Rio Jauaperí region, Rio Negro Basin, the Amazon, Brazil. Mammalia 67:75–83

Trolle M (2004) Flashed in the forest. BBC Wildlife Mag 22:30–37

Trolle M, Emmons LH (2004) A record of dwarf brocket from lowland Madre de Dios, Peru. Deer Specialist Group Newslett 19:2–5

Trolle M, Kéry M (2003) Ocelot density estimation in the Pantanal using capture-recapture analysis of camera-trapping data. J Mammal 84:607–614

Trolle M, Kéry M (2005) Camera-trap study of ocelot and other secretive mammals in the northern Pantanal. Mammalia 69:405–412

Voss RS (1993) A revision of the Brazilian muroid rodent genus *Delomys* with remarks on "Thomasomyine" characters. Am Mus Novitates 3073:1–44

Voss RS, Myers P (1991) *Pseudoryzomys simplex* and the significance of Lund's collections from the caves of Lagoa Santa, Brazil. Bull Am Mus Nat Hist 206:414–432

Biodivers Conserv (2007) 16:1213–1230
DOI 10.1007/s10531-006-9114-x

ORIGINAL PAPER

Surveying carnivores at large spatial scales: a comparison of four broad-applied methods

José Miguel Barea-Azcón · Emilio Virgós · Elena Ballesteros-Duperón · Marcos Moleón · Manuel Chirosa

Received: 3 October 2005 / Accepted: 22 May 2006 / Published online: 3 September 2006

Abstract Reliable methods to estimate species richness are very important to managers and conservationists because they provide key data to make the right decisions in conservation programmes. In the case of carnivore mammals, traditional methods, such as direct count censuses, are not useful since these animals are usually scarce, elusive and nocturnal. Difficulties in carnivore sampling are compounded when monitoring programmes are developed at large spatial scales, where high economic costs and field efforts are necessary to achieve reliable richness or abundance estimates. These problems have highlighted the need to find more effective carnivore survey methods, especially in regions with high rates of landscape change, such as the Mediterranean basin. The present study, performed in a typical Mediterranean area, was the first in Europe to test simultaneously the relative efficiencies of four broad-applied sampling methods to detect carnivore species at large spatial scales. Sign surveys based on scat detection, scent stations, camera-trapping and live-trapping were investigated. We compared efficiencies using biological parameters and by considering both the logistic and economic costs of each method. Overall, scent stations and sign surveys were the most efficient methods both in economic and logistic terms. In addition, the use of scent stations may be necessary to detect

J. M. Barea-Azcón (✉) · E. Ballesteros-Duperón · M. Moleón
Empresa de Gestión Medio Ambiental, Consejería de Medio Ambiente, Junta de Andalucía, C/ Marqués de la Ensenada, 4, 1° C and D, E-18004 Granada, Spain
e-mail: jbarea@egmasa.es

E. Virgós
Departamento de Matemáticas, Física Aplicada y Ciencias de la Naturaleza, Área de Biodiversidad y Conservación, Escuela Superior de Ciencias Experimentales y Tecnología, C/ Tulipán s/n, E-28933 Móstoles, Madrid, Spain

M. Moleón
Departamento de Biología Animal y Ecología, Facultad de Ciencias, Universidad de Granada, E-18071 Granada, Spain

M. Chirosa
Centro de Investigación y Formación Agraria (CIFA), Junta de Andalucía, Camino de Purchil, s/n, 2027, E-18080 Granada, Spain

species rarely detected by scats. Detailed and extensive training programmes for conducting sign surveys and scent stations may overcome perceived problems thus enhancing the widespread use of both methods. Our results are applicable not only to other Mediterranean areas, but also to other habitats and regions of the world. More research into the suitability of these and other methods in relation to different landscapes, seasons and species is required.

Keywords Carnivore conservation · Economic costs · Mediterranean region · Large-scale monitoring · Relative abundance · Species richness

Introduction

Managers and conservationists need reliable methods to estimate species richness in communities and/or population size of target species in order to make the right decisions in conservation practices (Caughley and Sinclair 1994; Sadlier et al. 2004). This fact has been emphasised by an accelerated lost of biodiversity in recent decades which has reinforced the urgent need for monitoring programmes and the necessity of studies relating habitat and species occurrence and abundance worldwide.

Carnivores are a group where worldwide conservation or management problems have been identified (Ginsberg and Macdonald 1990; Zielinski and Kucera 1995; Gese 2001; Wilson and Delahay 2001). Their position in the top of the food webs and their large potential impact on different human activities (game, agriculture, etc..., Saunders et al. 1995; Reynolds. and Tapper 1996; Moore et al. 1999) make this group especially significant in conservation and management actions. However, because carnivores are usually scarce, elusive and nocturnal, they are very difficult to count by traditional methods such as transects based on direct observation (e.g. spotlight or line-transects, Heydon et al. 2000; Sharp et al. 2001) or capture-recapture estimates based on trapping records (Tuyttens et al. 1999). These problems are especially relevant when the study needs to consider large spatial scales (Zielinski and Kucera 1995; Zielinski and Stauffer 1996; Webbon et al. 2004), where high economic costs and field efforts are necessary to achieve a reliable estimate (Smallwood and Fitzhugh 1995; Sadlier et al. 2004). However, conservation biology is largely related to large spatial scales problem (Edwards et al. 1994) and we need robust, efficient and low cost sampling methods to survey species at these scales. Several authors, mainly in the USA (Linhart and Knowlton 1975; Smallwood and Fitzhugh 1995; Zielinski and Kucera 1995; Zielinski and Stauffer 1996), but also in the United Kingdom (Tuyttens et al. 2001; Wilson and Delahay 2001; Sadlier et al. 2004; Webbon et al. 2004), have emphasised the use of alternative sampling methods to obtain density estimates or, at least, relative abundance indexes of different carnivore species at large spatial scales. These methods were also proposed under the assumption of their usefulness in detecting species under a variety of density ranges (Sadlier et al. 2004).

The most classical method to detect the presence of carnivores and estimate relative abundance is the use of field signs (scats, footprints, dens). Under the so-called 'sign surveys' (Clevenger 1993) include many different survey methodologies, from snow tracking (Pullianen 1981; Zielinski and Kucera 1995) to scat counting along paths (Cavallini 1994; Virgós et al. 2000; Virgós 2001; Webbon et al. 2004), den enumeration (Cresswell et al. 1989; Wilson et al. 1997) or spoor counts

(Smallwood and Fitzhugh 1995; Staender 1998). Most of these methods have not been tested in relation to true population density, but recent research has indicated a close fit between sign abundance and density for several carnivore species (Staender 1998; Sharp et al. 2001; Tuyttens et al. 2001; Sadlier et al. 2004; Webbon et al. 2004), despite early criticisms about accuracy (Kruuk et al. 1986; Norton 1990). As a consequence, several researchers recommended the use of sign surveys as the main element in national monitoring programmes or large-scale studies (Macdonald et al. 1998; Toms et al. 1999; Baker et al. 2002; Sadlier et al. 2004; Webbon et al. 2004; but see Birks et al. 2004).

Scent station and track plate methodologies, are derived from sign surveys and are routinely used in the USA to estimate population trends or as an index of relative abundance of different species (Roughton and Sweeny 1982; Conner et al. 1983; Sargeant et al. 1998). Overall, evidence has indicated good correlation between abundance indexes derived from this method and population densities (Sargeant et al. 1998). Silveira et al. (2003) found this the best method to detect species richness and relative abundance in a tropical area. However, contrary to its large use in North America and some regions of south America (Travaini et al. 2003; Silveira et al. 2003), the use of scent stations or similar methodologies has been limited elsewhere (but see Travaini et al. 1996).

Both sign surveys and scent station methods showed good efficiencies in carnivore detection rates and low economic costs, but their use may be limited because they require people with good skills in tracking or identification (Burnham et al. 1980; Smallwood and Fitzhugh 1995, Birks et al. 2004). Macdonald et al. (1998) and Sadlier et al. (2004) recognised this problem but they suggested that a well-designed training programme may overcome most of the problems with scat or track differentiation.

Another traditional method to obtain an index of abundance in carnivores is through trapping records (Brand and Keith 1979; Tapper 1992). Few studies, however, have controlled bag numbers by trapping effort which may invalidate the abundance estimates (McDonald and Harris 1999). Nevertheless, the method may be useful when trapping effort is taken into account, as in some of the carnivore survey protocols in Spain (Guzmán et al. 2003), and also in the some surveys with small carnivores in United Kingdom (Birks and Kitchener 1999; McDonald and Harris 1999) or with badgers (Tuyttens et al. 1999). Data from these studies may be recorded as number of trapped individuals by trap/nights (McDonald and Harris 1999), or in small-scale studies as an estimate of density from capture-recapture records (Otis et al. 1978). The latter method, is not useful at a large-scale because capture rates are very low (Tuyttens et al. 1999; Baker et al. 2001), and hence estimates show large confidence intervals which preclude a reasonable density estimation (Rexstad and Burnham 1991). In addition, large areas could be sampled only with a high cost and effort, which can rarely may be achieved by national conservation agencies or researchers.

Finally, more recent sampling methods have included camera-trapping, using a similar approach to traditional trapping, but substituting traps or other capture devices with a 35-mm camera system (Mace et al. 1994; Zielinski and Kucera 1995; Karanth and Nichols 1998). The method has been applied with apparent success for several species, although the relationship with true density has not been tested so far.

Unfortunately, the number of studies and the number of species and environmental conditions where efficiency of the methods in terms of species detection, time constraints and economic costs has been tested remains very low (Bull et al. 1992; Zielinski and Kucera 1995; Foresman and Pearson 1998; Silveira et al. 2003). Results are partially contradictory, which may reflect different species response to methods and/or changes in the efficiency of methods under different environmental conditions (e.g. tropics vs. large woods in the USA, Foresman and Pearson 1998; Silveira et al. 2003). All authors agree that more research about the performance of different methods in other regions, species or conditions is needed.

In Europe, comparison of different methods to estimate abundance of carnivores is restricted to the UK, and to a small diversity of species, mainly badgers and red foxes (reviewed in Wilson and Delahay 2001; Sadlier et al. 2004). Data are absent from continental Europe, and specifically the Mediterranean region, despite the large diversity of habitats, species and land use regimes in this area. Moreover, the Mediterranean region has considerable conservation value for carnivores because most of the species reach the highest densities in this area. However, despite the high conserved status of this region as compared to UK or central Europe, most of the countries involved (Spain, Portugal, Greece) have accelerated their economic development as a consequence of EU funding, which has had a strong impact on nature conservation (de Juana 2004; Díaz et al. 1997). Monitoring of carnivores in this region is thus very important, and must be prioritised in order to guarantee conservation in a progressively changing landscape. This requires more research on the suitability of different sampling methods in this region.

The aim here is to test the performance of four broadly applied methods to detect carnivore species at large spatial scales: sign surveys based on scat detection, scent stations, camera-trapping and live-trapping, in a typical Mediterranean landscape. We compared these methods with the view that large-scale surveys must optimise precision and repeatability of the results, yet have low cost and resource demands. The usefulness of the different methods were considered by comparing their suitability to detect the different species and then estimate carnivore community composition (species richness), and latency to first detection (sensu Foresman and Pearson 1998) of particular species and by considering both logistics and costs of each method.

Methods

Study area

The present study was conducted in Sierra Harana (Subbetic mountains, Granada province, SE Spain, Fig. 1.) during the winter season of 2004. The study area encompassed 5,000 ha and comprised dense forest formations (old Allepo pine *Pinus halepensis* reforestations but also some Holm oak *Quercus ilex* forest) which covered 38% of the total area, dense scrub formations (16.2%), scattered scrub formations (22.6%) and a mixture of soil-stone-pasture and scrublands (18%). The scrublands were mainly composed of Small-flowered gorse (*Ulex parviflorus*), Rosemary (*Rosmarinus officinalis*) and Prickly juniper (*Juniperus oxycedrus*). In addition to natural vegetative formations, 10% of the area was cultivated (mainly olive orchards but also cereal croplands).

Annual mean temperature and rainfall for the study area was 14°C and 620 mm respectively, i.e. a typical continental Mediterranean climate. The human population density inside the study area was very low and restricted to some dispersed cottages. Road density was also low, averaging 0.128 km/km^2. Six carnivore species (Red fox *Vulpes vulpes*, Eurasian badger *Meles meles*, Stone marten *Martes foina*, Weasel *Mustela nivalis*, wildcat *Felis silvestris* and Genet *Genetta genetta*) were present in the study area (Barea-Azcón et al. 2005). These from the typical carnivore community in most of Mediterranean Spain. Weasels have been excluded from this study because of their small size compared to the other five, medium-sized carnivores.

Sampling protocol

We selected a sample unit of 2.5 km^2 (hereafter plots). This area was recommended by Zielinski and Kucera (1995) to survey medium-sized carnivore communities because it assures that the species with the largest home range will have the least chance of being overlooked in a survey area. A total of eight plots in this mountain area were sampled. These plots covered about the 30% of the mountain range and held all the representative habitats of the study area. Plots were sampled at short time periods to avoid or mitigate fluctuations in the composition of the carnivore community among them. To coincide with a low resource availability season in southern Spain, which may enhance visitation to bait devices, we selected the period between December and January. This period also coincided with the mating season of many species (Macdonald and Barrett 1993) increasing the probability of attraction by scent bait. Each time, we simultaneously sampled two plots with camera traps, scent stations and box-traps to mimic a representative sampling protocol for a large-scale survey. We also performed a sign survey transect in each plot. The distribution of survey routes and devices in each sampling plot is depicted in the Fig. 2.

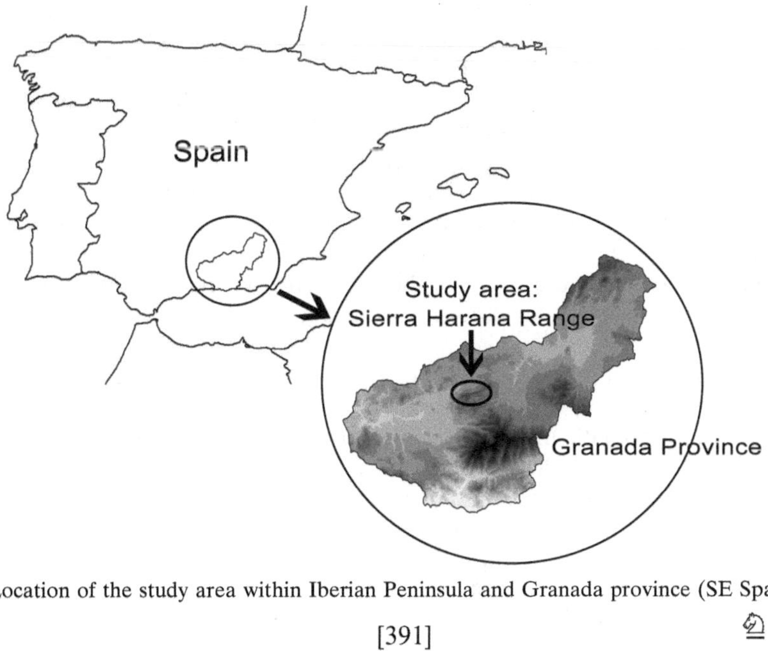

Fig. 1 Location of the study area within Iberian Peninsula and Granada province (SE Spain)

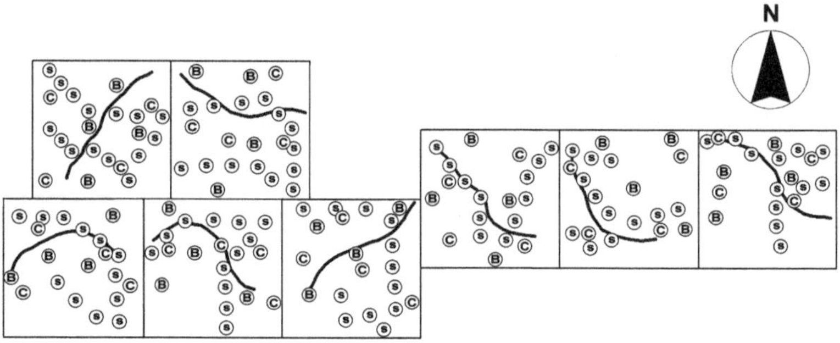

Fig. 2 Sample units and placement of each kind of device. B: box traps; C: camera traps; S: scent station; *Black line*: sing survey transect

Sign surveys

One sign survey transect measuring 3 km was carried out in every plot, following a common procedure for sampling carnivores in large scale surveys (Clevenger 1993; Wilson and Delahay 2001; Virgós et al. 2000; Virgós 2001). Each transect was divided into 250 m intervals (12 segments) following previous protocols (Clevenger 1993; Virgós et al. 2000; Lozano et al. 2003). At every station we searched for faeces of the studied species and we recorded the presence/absence of each species in every segment. Faeces of the different species were identified according to shape, size, odour and location. When signs showed an ambiguous identity they were not used. We did not consider footprints because their presence or absence is dependent on substrate, which invalidates between-plot comparisons.

Scent stations lines

Twelve scent stations were placed in every plot for two weeks (168 scent stations/night) following the protocols recommended by Zielinski and Kucera (1995). Each scent station consisted of a 0.9-m diameter circle of thin and smoothed earth with a centrally placed cork bark (0.5 m above the ground) impregnated with a synthetic attractant (Conner et al. 1983) (an specific carnivore lure; Armería Alvarez, s.l.). Scent stations were placed along unpaved roads at 400 m intervals and all habitat types and sections of each plot were included. All the stations were checked every 3.5 days. At each check, we noted the visit of different carnivore species from tracks and faeces, and we removed the signs once recorded and repaired the bait. There can be no confusion between the tracks of the five carnivore species inhabiting the study area, but when signs were unclear, they were not recorded.

Camera trapping with a live bait

We placed four automatic line-triggered (Canon Prima) camera systems in each plot for two weeks (56 traps/night). The devices were distributed within the main habitat types. The cameras were placed inside a wooden-made protective structure, to allow it to operate in rain. Despite the fact that the sampling occurred during the coldest season of the year, there were no snowfalls during this period. The cameras worked correctly even during the coldest days. Cameras were baited with a live pigeon similar to the method used to detect carnivores in other Spanish regions (Guzmán et al. 2003). Each camera was revisited every 3.5 days. Cameras were triggered by the pressure caused by the animal weight over a plate system connected to the camera. We followed the sampling protocols of Zielinski and Kucera (1995).

Box-traps with a live bait

Four box traps (0.5 × 0.8 × 1.3 m) were placed in every plot for two weeks (56 traps/night). As for the other devices, the box traps were placed within as many habitat types of the plot as possible whilst trying to cover the different sections of each sampled plot. The box traps were baited with live pigeons in a similar way to previous trapping campaigns for carnivores. Each box-trap was checked every morning in order to minimise possible injuries in captured individuals. All the traps were covered with scrub branches to avoid the capture of raptors.

Cost evaluation

We also estimated the economic costs of every method tested. This evaluation was made based on the average Spanish prices i.e. 200€ per set of camera equip (including camera system, pigeon, pigeon box and food and water supply for pigeon) 2€ per scent station (including earth, synthetic attractive and cork bark) 100€ per box-trap equip (including box trap, pigeon and pigeon food and water), 35€ per day car rental for a 4 × 4 vehicle and fuel and 60€ per day for the services of a qualified technical expert. We also have included the costs associated with the technical formation of surveyors. Sign surveys and scent station monitoring require the training in track and scat identification which is a very important element of monitoring efficiency (Davison et al. 2002; Sadlier et al. 2004). Training was performed by an expert surveyor during 10 days and with a cost of 120€ per day. One day of training includes the cost of the trainer (60€) and the cost of the technician (60€). For camera and box-trapping using live bait, training was focused in placement selection of the different devices, handling and safety of catched carnivores and pigeons used as a bait. Training was performed by an expert with the same cost of 120€ per day during 5 days (also including the cost of the trainer and of the technician).

Data analyses

We compared differences in the number of species (response variable) detected by different methods by using one-way ANOVA with the different methods sampled as the fixed factor. Prior to analysis we checked for normality and homoscedasticity of variables. The relative efficiency of the different methods in detecting each species (number of plots where a species is detected/total number of plots where a species was recorded) was evaluated by means of a χ^2 test applied to a 2 × 4 contingency table. Differences between methods in the latency to first detection (LTD) were analysed by means of Friedman ANOVA by ranks because data did not fit normality assumptions. In this analysis, species were used combined as replicates and the three methods tested (camera trapping, scent station lines and box-trapping) as the fixed factor. Sign surveys were not included because the survey was conducted only on one day.

In all statistical tests we used two-tailed probabilities and the typical $P < 0.05$ as a criterion of significance. Statistica 99 software was used for all statistical analyses.

Results

Number of species detected in the sampled plots

The number of species detected (species richness) differs significantly among the four methods considered (ANOVA: $F_{3\ 28} = 5.97$, $P < 0.01$). Post-hoc Duncan tests (not shown) revealed that differences were largely due to the higher values of sign surveys and scent stations as compared to camera-trapping and box-trapping (Fig. 3). Sign surveys and scent stations were not statistically different, nor were differences detected in richness values when camera-trapping and box trapping methods (all $P's > 0.5$) were compared. The highest richness values were found for sign surveys (that also showed narrow variability values), followed by scent stations and box trapping, with the lowest recorded values for camera-trapping (Fig. 3). A relatively large variability was observed in both scent station lines and box trapping methods (Fig. 3).

The largest values of richness were recorded in sign surveys in three of the sampled plots, while scent stations yielded the highest values in two cases. Camera-trapping and box-trapping did not yield the highest richness values in any of the sampled plots. The largest values of species richness were recorded for sign surveys in three of the sampled plots, while scent stations yielded the highest values in one plot. In other two plots the highest value of richness was recorded both for sign surveys and scent stations. In the remainder two plots, the largest value of species richness was obtained in one case both for scent stations and box-trapping and in another case both for sign surveys and camera trapping (Table 1).

We also tested the relative efficiency of each method in each plot. First, we recorded the total species richness in each plot by summing the records of individual species by each method (Table 2). This information was needed to evaluate the values of efficiency rate computed. We detected a significant difference between them (ANOVA: $F_{3\ 28} = 4.02$, $P = 0.02$). Post-hoc Duncan tests (not shown) indicated that, as in the case of species richness, the differences were due to higher efficiencies recorded in sign surveys and scent stations than in camera-trapping and box trapping ($0.03 < P > 0.01$, Fig. 4). Sign surveys and scent stations did not differ in efficiency ($P = 0.95$) and the same was observed between camera-trapping and box-trapping ($P = 0.88$). In contrast to the number of species recorded, the largest mean values of relative efficiencies were found in scent station lines followed by sign surveys, but the latter showed a lower variability (Fig. 4). Camera-trapping and box-trapping yielded the lowest values of relative efficiency (Fig. 4) as we also observed for number of species recorded. Despite the large mean values for the scent station method, we observed that both sign surveys and scent stations yielded the largest efficiency scores in two plots each (Table 2). In the additional four plots we observed a more complicated pattern than in the case of species richness. The largest value was recorded both for sign surveys and scent stations in one plot, while in another plot the largest values were observed in box-trapping and scent station lines. In the two remaining plots, the highest values were shown both by camera-trapping and scent stations in one plot and for all sampling methods in the other (Table 2).

Efficiency in species detection

We tested the relative efficiency of the different methods in detecting each particular species in each sampled plot. Three species were relatively rare in our study area

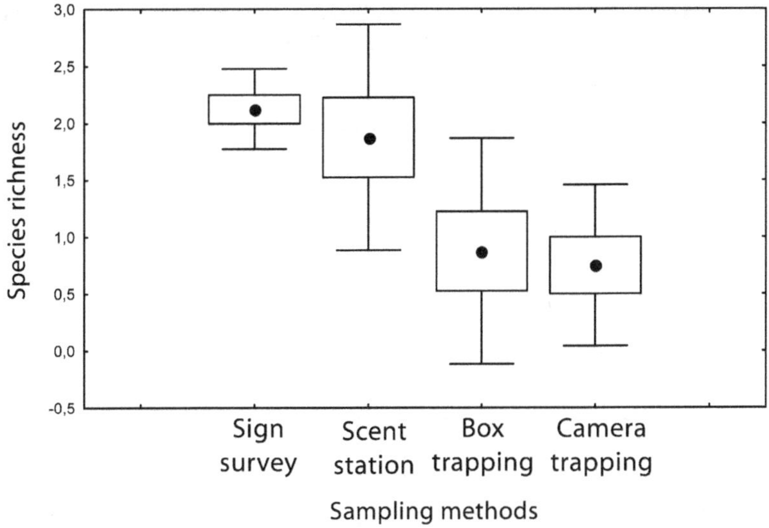

Fig. 3 Means (*black points*), standard errors (*boxes*) and standard deviations (*whiskers*) of species richness for each one of the sampling methods tested

(badger, wildcat and genet), with these species being present in only two or three plots. Red foxes and stone martens were recorded in all plots (Table 3). This information is needed to interpret efficiency data because occupancy patterns affected our computed efficiency rate index.

For detection of Red foxes, we observed a significant difference between sampling methods ($\chi^2 = 13.5$, 3 df, $P < 0.01$). The best method was sign surveys, with a very low efficiency for the remainder methods (Table 3). In the case of stone martens, we also found a significant difference between the methods ($\chi^2 = 9.5$, 3 df., $P = 0.02$). The highest score in efficiency rate was found for scent station methodology, although the value for sign surveys was also very high (Table 3). It was not possible to calculate the chi-square statistic for differences between methods for the other species due to their low occupancy rates. In badgers, both sign surveys and scent stations yielded the highest values. In the case of wildcats, both sign surveys and scent stations had low efficiency for detecting the species, with a relatively larger value both for camera-trapping and box-trapping. For genets, the most efficient

Table 1 Number of species detected (richness) for the different sampled methods in each plot

Plots	Sign surveys	Scent stations	Box-trapping	Camera-trapping	Total richness
1	3	3	0	1	4
2	2	3	3	0	3
3	2	1	1	2	4
4	2	1	0	0	2
5	2	3	0	0	3
6	2	1	1	1	3
7	2	2	1	1	3
8	2	1	1	1	2
Total richness	4	5	4	3	

Table 2 Richness and relative efficiency (number of species detected by a single method/number of species recorded in each plot by combining all sampling methods) of the different methods in each sampled plot

Plots	Sign surveys	Scent stations	Box-trapping	Camera-trapping	Richness
1	0.75	0.75	0	0.25	4
2	0.66	1	1	0	3
3	0.5	0.25	0.25	0.5	4
4	1	0.5	0	0	2
5	0.66	1	0	0	3
6	0.66	0.33	0.33	0.33	3
7	0.33	0.66	0.33	0.66	3
8	0.5	0.5	0.5	0.5	2

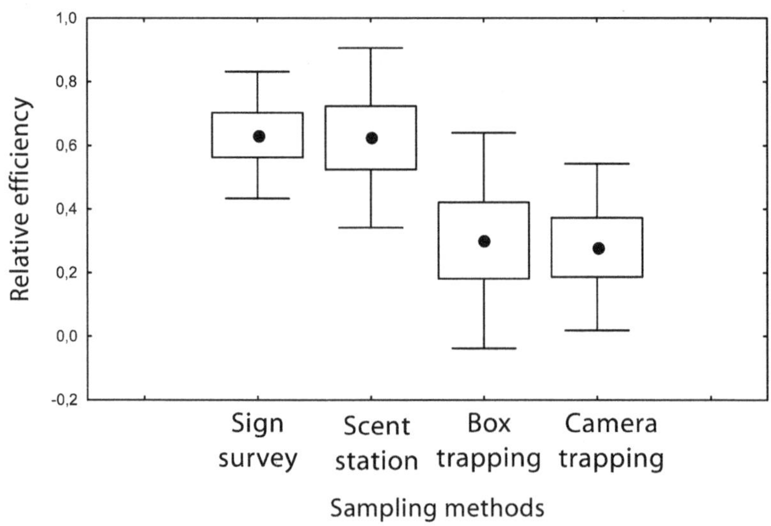

Fig. 4 Means (*black points*), standard errors (*boxes*) and standard deviations (*whiskers*) of the values of relative efficiency for each one of the sampling methods tested

method was scent station lines, followed by box-trapping; this species was not detected by sign surveys or camera trapping (Table 3). It is important to note that the values of 0.66 for efficiency in the case of wildcats and genets were influenced by low occupancy patterns (only three plots).

Latency to first detection

The LTD did not differ significantly among the three methods tested (Friedman ANOVA by ranks: $\chi^2 = 0.73$; 2 df., $P = 0.69$). Though values did not significantly differ between methods, it is clear that for a same method there was a large difference in the time elapsed until the first detection between species (Table 4).

For all species combined, camera-trapping yielded the lowest first detection time (mean value: 6.5 ± 1.14 days), followed by scent-station lines (6.67 ± 1.75 days) and box trapping (8.5 ± 0.79 days). For the Red fox, box trapping yielded the lowest latency time, while scent stations showed the largest values (Table 4). In contrast, for

Table 3 Efficiency value (between brackets) for each species (number of plots where a species was detected/total number of plots where a species was recorded)

Species	Sign surveys	Scent stations	Box-trapping	Camera-trapping
Vulpes vulpes	8/8 (1.0)	2/8 (0.25)	2/8 (0.25)	2/8 (0.25)
Martes foina	6/8 (0.75)	8/8 (1)	2/8 (0.25)	2/8 (0.25)
Meles meles	2/2 (1)	2/2 (1)	0/2 (0)	0/2 (0)
Felis silvestris	1/3 (0.33)	1/3 (0.33)	2/3 (0.66)	2/3 (0.66)
Genetta genetta	0/3 (0)	2/3 (0.66)	1/3 (0.33)	0/3 (0)
Mean efficiency	0.62	0.65	0.3	0.23

We also give the original rate to indicate the current value of occupancy rate (number of plots with the presence of a particular species) for each species

stone martens the lowest LTDs was obtained with scent station lines and the highest with box-trapping, and an identical result was observed for wildcats (Table 4). The genet showed high LTDs and it was not recorded in camera-trapping devices. The lowest LTD was observed with box-trapping methodology and the largest with scent stations. Badgers were not detected by camera-trapping or box-trapping, yielding a relatively larger value for scent station lines (Table 4). We did not perform statistical tests for each species because of the low sample sizes obtained.

Although, these LTDs can be used as a comparative tool between the three methods, they are minimum estimates because in some of the sampled plots we did not detect some of the species despite the fact that they were present (see above).

Economic evaluation for each method

Taking into account that we managed two plots simultaneously, we can estimate the cost for the camera trapping method amounted to 4,680€, including two technical experts, eight camera sets and four visits per two plots (one for collocation, two for revision and the last for revision and removal) and training. For the scent station method we estimated a total price of 3,872€, including two technical experts 24 scent stations per pair of plots and the maintenance costs (four visits per two plots) and training. For box trapping, we estimated a total amount of 3,880€, including the equipment and its maintenance (as for to the other two methods, we employed two technical experts and four visit days per each pair of plots) and training. Only two technical experts were needed for the sign surveys transect and one car per day. In one day it was possible to complete four plots, thus the total cost for this method was 1,510€ to complete the survey of eight plots and considering the 10-day period of training.

Table 4 Mean LTD values (in days) for the five studied species in each one sampled method considered (except, sign surveys, see Methods)

Species	Scent stations	Box-trapping	Camera-trapping
Vulpes vulpes	10.5	7	9.5
Martes foina	3.94	7	5
Meles meles	8.75	Not recorded	Not recorded
Felis silvestris	3.5	7.5	4.5
Genetta genetta	10.5	9	Not recorded

Because we only tested four methods, correlations can not be computed, but we showed the correlation values of the costs of the different methods with the species richness (Table 1) and the mean efficiency for species detection (Table 3) to illustrate this topic. Species richness and computed costs were highly negatively correlated ($r_s = -0.63$), and a stronger negative correlation was found between economic costs and mean efficiency in species detection ($r_s = -0.80$).

Summary of results

We ranked the methods according to the three indices of suitability used: species richness, mean efficiency in species detection and economic costs (Table 5). Scent stations were ranked as the superior method for the four mentioned sampling methods, whereas sign surveys was ranked highest for economic costs. Overall, scent stations and sign surveys were the preferred methods on this basis.

Discussion

The data in this study are the first to simultaneously test the relative efficiencies of several commonly used sampling methods to detect carnivore species in Europe. We compared efficiencies not only by biological parameters but also by economic costs, a key point in any management practice on a large spatial scale. Carnivores are elusive and show relatively low densities as compared to other vertebrate groups (Zielinski and Kucera 1995; Gros et al. 1996; Gese 2001). Several census methods have been proposed but we know little about their relative efficiencies (Zielinski and Kucera 1995; Gros et al. 1996; Foresman and Pearson 1998; Staender 1998; Silveira et al. 2003). However, managers need this information to plan any monitoring programme (Zielinski and Kucera 1995; Gros et al. 1996; Sadlier et al. 2004). In Spain and other European countries, several methods have been used to monitor abundance and population trends, but our results indicated that only a combination of several of these methods allowed us to obtain a true representation of the carnivore community in a location or region. Despite this fact, the different methods showed different suitabilities or efficiencies that may be used when planning monitoring programmes on a large spatial scale.

Overall, scent stations and sign surveys were clearly superior to methods based on attraction of species to a live prey. The number of species detected and the relative efficiencies were clearly lower for camera-trapping and box-trapping (using a pigeon as bait) than for scent stations and sign surveys. These bait methods have been used

Table 5 Ranking of the four tested sampling methods based on the following variables: number of species detection (richness), mean efficiency in species detection and economic costs

Sampling methods	Species richness	Mean efficiency	Economic costs
Sign surveys	2	2	1
Scent stations	1	1	2
Box trapping	3	3	3
Camera trapping	4	4	4

commonly in several carnivore projects in Spain (Guzmán et al. 2003). The use of live pigeons or other potential prey was only justified from our data for wildcats, as this was the only case where this method was superior to sign and scent stations surveys. This method has been mainly used to detect the highly threatened Iberian lynx (*Lynx pardinus*), where it has been shown to be superior to others (Garrote, com. pers.). In wildcats, we suggest that only this method gives confidence for detecting this species in a sampling plot, at least in areas of low density. For the rest of the species we investigated, a combination of sign surveys and scent stations was the preferred methodology both for sampling efficiency and economics. The inclusion of the latter factor is very important when large-scale surveys are planned because money and time are usually the main constraints (see Zielinski and Kucera 1995; Foresman and Pearson 1998). Sign surveys and scent stations are cheaper than camera and box-trapping (even if a lure instead a life prey was used as attractant). An eight-plot survey using both scent stations and sign surveys could be covered by two people in less than one month, and for a cost of 2,300€, which is a lower cost than either camera or box trapping. In addition, where a live prey is used as bait, ethic considerations also need to be considered. Any method that allows the estimation of abundance or presence of the species without intrusions is clearly preferred (i.e. non-invasive methods, Bekoff and Jamieson 1996).

Recent studies have recommended a broader use of sign surveys as a fast and accurate method for monitoring programmes (Gros et al. 1996; Sadlier et al. 2004; Webbon et al. 2004; but see Silveira et al. 2003). More information is available about the suitability of scent stations for undertaking large-scale surveys (Linhart and Knowlton 1975; Conner et al. 1983; Sargeant et al. 1998; Travaini et al. 1996), especially in the USA, but its use in Europe is less well documented (but see Travaini et al. 1996). Our results, advocate a more wide use of scent stations to monitor carnivore populations in Europe. This method was the only one able to detect the five species in our study area as a whole, although it failed to detect some species in particular plots. However, scent stations may be not preferred to camera-trapping when surveyors are not well-trained on track identification (Silveira et al. 2003), but this may be overcome by training programmes before any large-scale survey project (Sadlier et al. 2004; Webbon et al. 2004). In addition, the use of scent stations may be necessary to detect species rarely located by searching for scats on tracks or paths, as was the case for the genets and to a lesser extent the badgers (Virgós et al. 2000; Virgós 2001). In our study area, genets were not detected in sign surveys, but they were detected relatively well using a commercial lure in scent station lines. Other researchers have indicated that genets can be effectively monitored using commercial lures in combination with remote cameras (Torre 2003), but they are also well detected using scent stations, which are much cheaper than camera-trapping. Hence, we advocate using a combination of commercial lures and scent stations to detect genets.

Badgers were also attracted effectively to commercial lures attached to scent lines, a result that confirms previous results in an other area of southern Spain (Travaini et al. 1996), although they were also detected well in the sign surveys. We were able to detect latrines by following paths and tracks as was found in the United Kingdom, where latrine surveys following linear structures in the landscape is known to be a powerful monitoring technique (Tuyttens et al. 2001; Hutchings et al. 2002; Sadlier et al. 2004). The detection of latrines on paths and tracks in Mediterranean environments is more difficult as Mediterranean badgers have a different marking behaviour, and tend to lay faeces in latrines around the sett rather than at the edge of territories or signalling profitable patches (Pigozzi 1990; Revilla and Palomares 2002; own unpubl. data). However, an ongoing study about the usefulness of latrine surveys in different Mediterranean habitats has shown that they may be located in relatively large numbers on narrow paths and tracks (Mangas et al., in prep.), and is in agreement with our results in this area. Sign surveys may also be useful for badgers, although a combination with scent stations will increase the confidence of the results in large-scale surveys. More research in other low density areas may be useful to ascertain which method is superior for surveying badgers on a large spatial scale.

In the case of stone martens, scent stations were the best method, although sign surveys were also effective. In contrast, for red foxes, sign surveys were the most efficient method, as in a recent, detailed study in the United Kingdom, where it was shown that faecal counts were related to fox density (Baker et al. 2002; Webbon et al. 2004; but see Beltrán et al. 1991). Although our study was not aimed to test relationships between true density and indirect counts, it supports the suitability and low-cost of red fox monitoring on large spatial scales using faeces recorded in tracks and paths. Because the red fox is one of the first target species in predator control and game management in most of Europe (Tapper 1992; Harris and Saunders 1993; Reynolds and Tapper 1996; Virgós and Travaini 2005) the combined data from these studies represents good news for managers that use this easy and efficient way to quickly monitor red foxes. This then allows decisions to be taken based on real data rather than subjective ratings of abundance or trends.

In the case of wildcats, more research is needed on the use of other alternative commercial lures; cat urine may be a good candidate as shown for Iberian lynx sampling (Guzmán et al. 2003). Sign surveys have been used for wildcats (Lozano et al. 2003) reliant on the large marking behaviour of this species in paths and tracks (Corbett 1979). Our results did not support the suitability of sign surveys for this species, although it is important to note that the area showed a low wildcat density (0.17 ind/km^2, Ballesteros-Duperón et al., in prep.), and it may require higher effort to detect the species, for example by using a repeated sign survey on the same paths. More research is needed in areas of different wildcat density.

Several researchers have criticised the use of scat searching as a key sampling method to record carnivore presence or abundance on a large scale (Kruuk et al. 1986; Messenger et al. 2000; Birks et al. 2004). Scat surveys may be invalidated by a combination of non-random sampling, errors in species identification and seasonal or habitat-related differences in detectability or decay rates (Birks et al. 2004). Davison et al. (2002) have shown that even expert scat surveyors may mis-identify a large proportion of faeces, especially in low density areas (Messenger et al. 2000). As in the case of footprints in scent stations, the wide use of sign surveys based on scats needs well-designed training programmes (see also Sadlier et al. 2004). Trial pre-survey projects may also be needed where scat identification are compared with molecular tagging (Davison et al. 2002). Despite the problems derived from scat identification, the method remains attractive for large-scale surveys due to the low cost and faster sampling of large areas. More research is needed about the implications of non-random sampling or about the effects of seasonal changes in detectability or density effects on probability of false negative recordings (see Andelt and Andelt 1984; Walsh and White 1999).

Conservation implications

Carnivore monitoring and conservation requires suitable and efficient sampling protocols. Our results clearly show that large-scale surveys can be undertaken without the use of sophisticated and costly devices in Mediterranean landscapes. Our results could be also applied in other habitats and regions of the world.

We have demonstrated that scent stations and sign surveys based on scat detection are good methods to detect carnivore species and then evaluating species richness. Managers and conservationists are thus advantaged in that these two sampling methods are the cheapest among all the methods compared here. The negative relationship between the efficiency and the economic cost is good news for managers, since it encourages monitoring programs to be carried out on a large spatial scale. Unlike the monitoring of many other animal taxa, survey programs of carnivores (which normally have low population densities) are usually planned over large areas. Consequently, the results from this study are key to many conservation activities.

Unfortunately there have been few attempts to estimate the efficiency of different methods at different environmental situations and for different taxa (Zielinski and Kucera 1995; Foresman and Pearson 1998; Silveira et al. 2003). We advocate a strong research effort in this area, but other key research topic for the future is to address the reasons behind detection differences between methods (e.g. behaviour of different species to the same lure, behaviour differences to the same trap). Efforts should be made to test the relative efficiencies and accuracy of each method when applied to a known population size of each species (see Smith et al. 1994).

Acknowledgements We are very grateful to José M. Gil-Sánchez, Manuel Otero, Rogelio López, Francisco Martín, José F. Sánchez-Clemot and Juan Bellido for their assistance during the field work. Miguel Á. Simón, Ignacio Maldonado, Rafael Cadenas and Ramón Suárez for their logistical assistance. We also thank the people of the Delegación Provincia de Granada of the Consejería de Medio Ambiente (Junta de Andalucía), Egmasa (Empresa de Gestión Medio Ambiental. Consejería de Medio Ambiente. Junta de Andalucía) and the Endangered Species Recovery Centre "El Blanqueo". We thank Prof. Peter Jeffries, University of Kent, UK, for improving the English editing of this manuscript. We also thank the useful comments and improvements of an anonymous referee. This work has been supported by the project "Carnivore mammals in the Granada province: distribution, status and management", conducted by the Consejería de Medio Ambiente (Junta de Andalucía).

References

Andelt WF, Andelt SH (1984) Diet bias in scat deposition-rate surveys of coyote density. Wildlife Soc Bull 12:74–77

Baker PJ, Harris S, Robertson CPJ, Saunders G, White PCL (2001) Differences in the capture rate of cage-trapped red foxes *Vulpes vulpes* and an evaluation of rabies control measures in Britain. J Appl Ecol 38:823–835

Baker PJ, Harris S, Webbon CC (2002) Effect of British hunting ban on fox numbers. Nature 419:34

Barea-Azcón JM, Ballesteros-Duperón E, Moleón M, Gil-Sánchez JM, Virgós E, Chirosa M (2005) Distribución de los mamíferos carnívoros en la provincia de Granada. Acta Granat 3:43–53

Bekoff M, Jamieson D (1996) Ethics and the study of carnivores: doing science while respecting animals. In: Gittleman JL (ed) Carnivore behaviour, ecology and evolution, vol 2. Cornell University Press, Ithaca, London, pp 15–45

Beltrán JF, Delibes M, Rau JR (1991) Methods of censusing red fox (*Vulpes vulpes*) populations. Hystrix 3:199–214
Birks J, Kitchener AC (1999) The distribution and status of the Polecat *Mustela putorius* in Britain in the 1990s. The Vincent Wildlife Trust, London
Birks J, Messenger J, Braithwaite T, Davison A, Brookes R, Strachan C (2004) Are scat surveys a reliable method for assessing distribution and population status of pine martens? In: Harrison DJ, Fuller AK, Proulx G (eds) Marten and fishers in human-altered environments. Springer-Verlag, New York, pp 235–252
Brand CJ, Keith LB (1979) Lynx demography during a snowshoe hare decline in Alberta. J Wildlife Manage 43:827–849
Bull EL, Holthausen RS, Bright LR (1992) Comparison of three techniques to monitor marten. Wildlife Soc Bull 20:406–410
Burnham KP, Anderson DR, Laake JL (1980) Estimation of density from line transect sampling of biological populations. Wildlife Monogr 72:1–202
Caughley G, Sinclair ARE (1994) Wildlife ecology and management. Blackwell Science Publications, Oxford
Cavallini P (1994) Faeces count as an index of fox abundance. Acta Theriol 39:417–424
Clevenger AP (1993) Sign surveys as an important tool in carnivore conservation research and management programmes. In: Seminar on the management of small populations of threatened mammals. Council of Europe, Sofia, pp 44–54
Conner MC, Labisky RF, Progulske DR (1983) Scent station indices as measures of population abundance for bobcats, racoons, grey foxes and opossums. Wildlife Soc Bull 11:146–152
Corbett LK (1979) Feeding ecology and social organization of wildcats (*Felis silvestris*) and domestics cats (*Felis catus*) in Scotland. PhD. Thesis, University of Aberdeen
Cresswell P, Harris S, Bunce RGH, Jefferies D (1989) The badger, *Meles meles* in Britain: present status and future population changes. Biol J Linn Soc 38:91–101
Davison A, Birks J, Brookes RC, Braithwaite AC, Messenger JE (2002) On the origin of faeces: morphological versus molecular methods for surveying rare carnivores from their scats. J Zool Lond 257:141–143
Díaz M, Campos P, Pulido FJ (1997) The Spanish dehesas: a diversity in land-use and wildlife. In: Pain DJ, Pienkowski MW (eds) Farming and birds in Europe. Academic Press, San Diego, pp 178–209
Edwards PJ, May RM, Webb NR (1994) Large-scale ecology and conservation biology. British Ecological Society. Blackwell Science, Oxford
Foresman KR, Pearson DE (1998) Comparison of proposed survey procedures for detection of forest carnivores. J Wildlife Manage 62:1217–1226
Gese EM (2001) Monitoring of terrestrial carnivore populations. In: Gittleman JL, Funk SM, Macdonald DW, Wayne KR (eds) Carnivore conservation. Cambridge University Press, Ithaca, New York, pp 372–396
Ginsberg JR, Macdonald DW (1990) Foxes, wolves, jackals and dogs: an action plan for the conservation of canids. IUCN, Gland, Switzerland
Gros PM, Kelly MJ, Caro TM (1996) Estimating carnivore densities for conservation purposes: indirect methods compared to baseline demographic data. Oikos 77:197–206
Guzmán JN, García FJ, Garrote G, Pérez de Ayala R, Iglesias MC (2003) Censo diagnostico de las poblaciones de lince ibérico en España 2000–2003. Boletín de Programas de Conservación de Especies Amenazadas y del Inventario de Biodiversidad 5. Dirección General de Conservación de la Naturaleza, Madrid
Harris S, Saunders G (1993) The control of canid populations. Symp Zool Soc Lond 65:441–464
Heydon MJ, Reynolds JC, Short MJ (2000) Variation in abundance of foxes (*Vulpes vulpes*) between three regions of rural Britain, in relation to landscape and other variables. J Zool Lond 251:253–264
Hutchings MR, Service KM, Harris S (2002) Is population density correlated with faecal and urine scent marking in European badgers (*Meles meles*) in the UK? Mammal Biol 67:286–293
de Juana E (2004) Cambios en el estado de conservación de las aves en España, años 1954 a 2004. Ardeola 51:19–50
Karanth KU, Nichols JD (1998) Estimation of tiger densities in India using photographic captures and recaptures. Ecology 79:2852–2862
Kruuk H, Conroy JWH, Glimmerveen U, Ouwerkerk EJ (1986) The use of spraints to survey populations of otters *Lutra lutra*. Biol Conserv 35:87–94

Linhart SB, Knowlton FF (1975) Determining the relative abundance of coyotes by scent station lines. Wildlife Soc Bull 3:119–124

Lozano J, Virgós E, Malo AF, López-Huertas D, Casanovas JG (2003) Importance of scrub-pastureland mosaics for wild-living cats occurrence in a Mediterranean area: implications for the conservation of the wildcat (*Felis silvestris*). Biodivers Conserv 12:921–935

Mace RD, Minta SC, Manley TL, Aune KA (1994) Estimating grizzly bear population size using camera sightings. Wildlife Soc Bull 22:74–83

Macdonald DW, Barrett P (1993) Mammals of Britain and Europe. Harper Collins, London

McDonald RA, Harris S (1999) The use of trapping records to monitor populations of stoats *Mustela erminea* and weasels *Mustela nivalis*: the importance of trapping effort. J Appl Ecol 36:679–688

Macdonald DW, Mace G, Rushton S (1998) Proposals for future monitoring of British mammals. Department of the Environment, Transport and The Regions, London

Messenger JE, Birks J, Jefferies DJ (2000) Monitoring the very rare: pine marten populations. In: Griffiths HI (ed) Mustelids in a modern world. Management and conservation aspects of small carnivore human interactions. Backhuys, Leiden, Netherlands, pp 217–230

Moore N, Whiterow A, Kelly P, Garthwaite D, Bishop J, Langton S, Cheeseman C (1999) Survey of badger *Meles meles* damage to agriculture in England and Wales. J Appl Ecol 36:974–988

Norton PM (1990) How many leopards? A criticism of Martin and de Meulenaer's population estimates for Africa. S Afr J Sci 86:218–220

Otis DL, Burnham KP, White GC, Anderson DR (1978) Statistical inference from capture data on closed animal populations. Wildlife Monogr 62:1–135

Pigozzi G (1990) Latrine use and the function of territoriality in the European badger, *Meles meles*, in a Mediterranean coastal area. Anim Behav 39:1000–1002

Pullianen E (1981) A transect survey of small land carnivores and red fox populations on a subartic fell in Finnish forest Lapland over 13 winters. Ann Zool Fenn 18:270–278

Revilla E, Palomares F (2002) Spatial organization, group living and ecological correlates in low-density populations of Eurasian badgers, *Meles meles*. J Anim Ecol 71:497–512

Rexstad E, Burnham KP (1991) User's guide for interactive program CAPTURE. Abundance estimation of closed animal populations. Colorado State University, Fort Collins, CO

Reynolds JC, Tapper SC (1996) Control of mammalian predators in game management and conservation. Mammal Rev 26:127–156

Roughton RD, Sweeny MW (1982) Refinements in scent-station methodology for assessing trends in carnivore populations. J Wildlife Manage 46:217–229

Sadlier LMJ, Webbon CC, Baker PJ, Harris S (2004) Methods of monitoring red foxes *Vulpes vulpes* and badgers *Meles meles*: are field signs the answer? Mammal Rev 34:75–98

Sargeant GA, Johnson DH, Berg WE (1998) Interpreting carnivore scent-station surveys. J Wildlife Manage 62:1235–1245

Saunders G, McIlroy J, Berghout M, Kay B, Gifford E, Perry R, van de Ven R (1995) The effects of induced sterility on the territorial behaviour and survival of foxes. J Appl Ecol 39(1):56–66

Sharp A, Norton M, Marks A, Holmes K (2001) An evaluation of two indices of red fox (*Vulpes vulpes*) abundance in an arid environment. Wildlife Res 28:419–424

Silveira L, Jácomo ATA, Alexandre J, Diniz-Filho F (2003) Camera trap, line transect census and track surveys: a comparative evaluation. Biol Conserv 114:351–355

Smallwood KS, Fitzhugh EL (1995) A track count for estimating mountain lion *Felis concolor californica* population trend. Biol Conserv 71:251–259

Smith WP, Borden DL, Endres KM (1994) Scent station visits as an index to abundance of raccons – an experimental manipulation. J Mammal 75:637–647

Staender PE (1998) Spoor counts as indices of large carnivore populations: the relationship between spoor frequency, sampling effort and true density. J Appl Ecol 35:378–385

Tapper S (1992) Game heritage: an ecological review from shooting and gamekeeping records. The Game Conservancy Trust, Fordingbridge

Toms MP, Siriwardena GP, Greenwood JJD (1999) Developing a mammal monitoring programme for the UK. British Trust for Ornithology, Thetford, Norfolk

Torre I (2003) Distribution, population dynamics and habitat selection of small mammals in Mediterranean environments: the role of climate, vegetation structure and predation risk. PhD. Thesis, Universidad de Barcelona

Travaini A, Laffite R, Delibes M (1996) Determining the relative abundance of European red foxes by scent station methodology. Wildlife Soc Bull 24:500–504

Travaini A, Zapata SC, Zoratti C, Soria G, Escobar F, Aguilera G, Collavino P (2003) Diseño de un programa de seguimiento de poblaciones de cánidos silvestres en ambientes esteparios de la Patagonia, Argentina. Acta Zool Mex 90:1–14

Tuyttens FAM, Macdonald DW, Swait S, Cheeseman CL (1999) Estimating population size of Eurasian badgers (*Meles meles*) using mark-recapture and mark-resight data. J Mammal 80:950–960

Tuyttens FAM, Long B, Fawcett T, Skinner A, Brown JA, Cheeseman CL, Roddam AW, Macdonald DW (2001) Estimating group size and population density of Eurasian badgers *Meles meles* by quantifying latrine use. J Appl Ecol 38:1114–1121

Virgós E (2001) Relative value of riparian woodlands in landscapes with different forest cover for the medium-sized Iberian carnivores. Biodivers Conserv 10:1039–1049

Virgós E, Travaini A (2005) Relationship between small game hunting and carnivore diversity in central Spain. Biodivers Conserv 14:3475–3486

Virgós E, Recio MR, Cortés Y (2000) Stone Marten (*Martes foina*) use of different landscape types in the mountains of central Spain. Z Säugetierk 65:375–379

Walsh PD, White LJT (1999) What will take to monitor forest elephant populations? Conserv Biol 13:1194–1202

Webbon C, Baker PJ, Harris S (2004) Faecal counting for monitoring changes in red fox numbers in rural Britain. J Appl Ecol 41:768–779

Wilson GJ, Delahay RJ (2001) A review of the methods to estimate the abundance of terrestrial carnivores using field signs and observation. Wildlife Res 28:151–164

Wilson G, Harris S, McLaren G (1997) Changes in the British badger population 1988 to 1997. People's Trust for Endangered Species, London

Zielinski WJ, Kucera TE (1995) American marten, fisher, lynx and wolverine: survey methods for their detection. General Technical Report PSW-157. US Department of Agriculture Forest Service, Pacific Southwest Research Station, Berkeley, CA, USA

Zielinski WJ, Stauffer HB (1996) Monitoring *Martes* populations in California: survey design and power analysis. Ecol Appl 6:1254–1267

Biodivers Conserv (2007) 16:1231–1244
DOI 10.1007/s10531-006-9127-5

ORIGINAL PAPER

Habitat and landscape factors associated with neotropical waterbird occurrence and richness in wetland fragments

Demetrio Luis Guadagnin · Leonardo Maltchik

Received: 27 January 2006 / Accepted: 12 September 2006 / Published online: 27 October 2006

Abstract Habitat destruction alters the structure and spatial arrangement of remnants and creates a novel matrix, affecting the structure of biological communities. This study evaluated the contribution of patch, class and landscape level attributes of 42 wetland fragments and surrounding landscapes on the richness, abundance and presence of waterbird species. Twelve monthly censuses identified 55 species—ranging from 3 to 40 among fragments, and a total of 84,286 birds—ranging from 15 to 19,322. Wetland areas varied from 0.2 ha to 145.2 ha. The presence of 37 out of 42 species could be predicted from the structure of the fragments and the landscapes. The most important predictors were the wetland area (22 models), the rice field matrix permeability (13 models), the microhabitat richness (7 models), the wetland connectivity (5 models), and the wetland isolation (5 models). Discriminant analysis showed that fragments richer in species than expected from their areas, are found in landscapes with greater connectivity and matrix permeability than the species-poor fragments. The total area of rice fields was not related to the richness, abundance or presence of waterbirds in the wetlands fragments. The results suggest that the maintenance of the connectivity among fragments and the matrix permeability through rice-fields could be managed to foster the waterbird conservation at the landscape scale.

Keywords Waterfowl · Occupancy · Redundancy analysis · Logistic regression · Discriminant analysis · Fragstats · Area · Shape · Contrast weighted edge density · Habitat heterogeneity · Mean nearest neighbour distance · Proximity index · Area weighted gyration index · Correlation length · Rice fields

Introduction

Human activities are major contributors to biodiversity loss and the degradation of biological community structure. As natural habitats are modified by human use, the

D. L. Guadagnin (✉) · L. Maltchik
Lab of Ecology and Conservation of Aquatic Ecosystems, UNISINOS, Av. Unisinos 950, Sao Leopoldo, RS 93022-000, Brazil
e-mail: dlg@unisinos.br

remnants became smaller and more isolated (Sharpe et al. 1981; Wiens 1995) and the total area of habitats available at the landscape scale is reduced (Fahrig 2002; Villard et al. 1999). Changes in remnant shape are known to cause edge effects, impacting several ecological processes (Saunders et al. 1991; van Rensburg et al. 2000). The novel matrix of modified habitats reduces the permeability to animal movements (Johnson et al. 1992; Johnson and Boeriijst 2002), but remains hospitable for some species, complementing or supplementing their resource needs (Brotons et al. 2003; Norton et al. 2000).

Island biogeography, first proposed to explain variations in island species richness as a function of their area and isolation (MacArthur and Wilson 1967), has been successfully applied to the analysis of terrestrial fragments and is now an important tool in conservation planning (Diamond 1976). Area loss, or fragment size, have been shown to be a key factor for species richness (Lomolino 2000), but may simply be a confounding factor, the true cause been the covarying increase in microhabitat richness or a simple random placement effect (Coleman 1981). Moreover, contrary to real islands, area and isolation are related factors in the fragmentation process (Sharpe et al. 1981; Wiens 1995).

While the effects of forest loss and fragmentation on terrestrial birds has been extensively studied, the impact of wetland loss on waterbird assemblages has not (Brown and Dinsmore 1986; Fairbairn and Dinsmore 2001; Riffell et al. 2001). Waterbirds are strictly dependent upon wetlands, which tend to occur naturally as islands in a matrix of upland habitats. Therefore, it is reasonable to expect waterbird assemblage structure to conform to island biodiversity rules.

The State of Rio Grande do Sul has vast wetlands (Maltchik et al. 2003) and rich waterbird assemblages (Guadagnin et al. 2005). Rice cultivation is a major economic activity in the state (Gomes and Magalhães Júnior 2004) and a major factor of wetland loss. It is known that several waterbird species use irrigated rice fields—man-made wetlands. Several studies compared assemblages in natural and man-made wetlands (Czech and Parsons 2002; Erwin 2002; Tourenq et al. 2001), but the role of the matrix of rice fields as a connecting factor in the landscape has not yet been studied.

This study explores the relationship of the structure of neotropical waterbird assemblages with the structure of wetland fragments (area, shape and microhabitat richness), the configuration of the wetland mosaic (isolation, proximity, fragmentation, connectivity and habitat availability), the matrix permeability (wetland/upland and wetland/rice field contrast weighted edge density) and the total area of rice fields. We first analysed the contribution of different fragment and landscape attributes as predictors of the bird species occurrence. We then considered the effect of wetland area on waterbird species richness. Finally, excluding the confounding effect of the fragment area, we assessed the difference between species-rich and species-poor fragments in terms of their spatial structure and landscape context. Our aim was to evaluate the possibilities of improving biodiversity conservation in the agricultural landscape, finding key spatial attributes for management.

Methods

Study area and design

The coastal zone of Rio Grande do Sul is among the most important regions of South America for waterbirds. Apart from its outstanding species richness, it has important stop-over sites for migratory species in the southern cone (Guadagnin et al. 2005; Scott and Carbonell 1986). One main land use dominates the region—a matrix of rice fields and drained meadows in a roughly 4 year rotation. Scattered in the matrix there are remnants of wetlands and native forests, natural lagoons, reservoirs and *Pinus* and *Eucalyptus* plantations. Larger wetland fragments are used as water reservoirs for rice plantations and as sites for recreational or illegal fishing and hunting. Both large and small fragments are used for watering cattle. The climate is subtropical, with mean annual temperature of 19°C and total annual rainfall of 1200 mm evenly distributed over the year.

We carried out a complete aerial photographic survey of the 71,300 ha (30°56′–30°22′S 50°58′–50°22′W; Fig. 1) study area, producing 250 oblique photographs. A total of 212 wetland fragments were identified and 50 were randomly selected for the study. Eight fragments were excluded because they were drained or dry during field sampling or because of the poor quality of the vertical aerial photographs (see below). The final set comprised 42 fragments.

Waterbird census

Monthly censuses were done during 2003, changing the sequence of site visits for each census. Total bird counts were done during the daylight hours. Passerines and some secretive Rallidae were not considered because the counting method was not suitable for them—some rallid species can only be detected using play-back equipment. Data from the 12 censuses were pooled for the analyses, providing an accurate estimate of richness, composition and abundance of waterbirds at each site. (Guadagnin et al. 2005) describes the waterbird assemblage and counting methods in detail.

Wetland and landscape spatial attributes

We used FRAGSTATS 3.3 (MacGarigal and Marks 1995) to calculate patch, class and landscape metrics (Table 1). Patch structure was measured from vertical

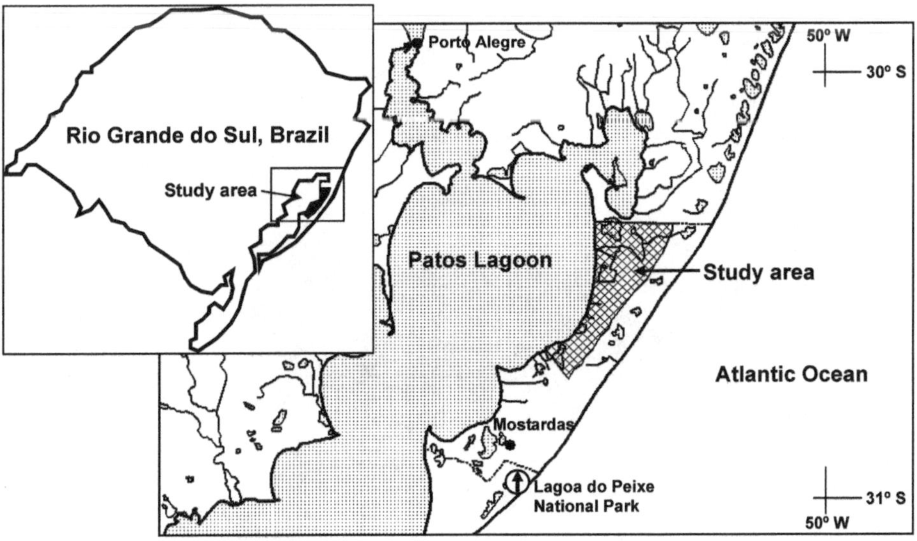

Fig. 1 Study area (dashed) in Coastal zone of Rio Grande do Sul, Brazil

Table 1 Spatial attributes of wetlands and the landscape surrounding 42 wetland fragments in the coastal zone of Rio Grande do Sul, Brazil. Metrics were calculated in program Fragstats 3.3 (MacGarigal and Marks 1995) and follow its syntax

Level	Attribute	Metric	Mean	Min.	Max.	CV
Patch	Area (ha)	AREA	16.28	0.19	145.24	155.19
	Shape index	SHAPE	1.81	1.08	3.12	28.99
	Microhabitat richness (N)	PR	3.98	2	7	42.84
Class	Habitat availability in the landscape— total wetland area in 5 × 5 km (ha)	CA-wet	323.15	159.94	1078.94	55.02
	Area of rice fields in 5 × 5 km (ha)	CA-rice	1168.95	498.44	1612.19	24.57
	Structural connectivity—Area weighted radius of gyration (m)	GYRATE_AM	354.92	151.25	1243.96	65.61
	Fragmentation—Interspersion and juxtaposition index (%)	IJI	60.56	34.77	74.34	16.47
	Isolation—Mean euclidian nearest neighbour distance (m)	ENN_MN	122.39	96.98	151.26	10.17
	Isolation—Proximity index	PROX_MN	111.41	16.31	1356.65	197.17
Landscape	Matrix permeability—Contrast weighted edge density (m/ha)	CWED	43.95	19.76	56.05	17.76

photographs taken from 1900–2000 m a.s.l. with a 35 mm equipment. Each photograph covered a land surface of 3 × 4 km (1:12.3 in the negative). The photographs were scanned with a resolution of 500 dpi and exported to the SIG IDRISI 3.2 to classify and quantify the area of the different types of macrophyte banks, as previously recognized in the field (Fig. 2). At the patch level we measured the fragment area, shape and microhabitat richness. The number of macrophyte cover types is a straightforward measure of microhabitat richness that avoids the use of indirect

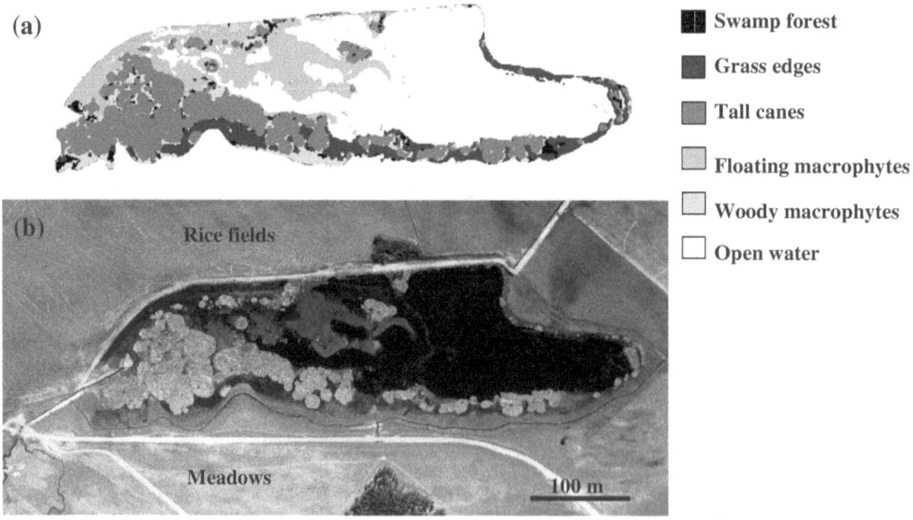

Fig. 2 (**a**) Example of the classification and mapping of the microhabitat structure of a wetland fragment and (**b**) original photograph showing the vegetation cover and surrounding matrix

measures of heterogeneity, as commonly applied in forest studies (Ricklefs and Lovette 1999; Tews et al. 2004).

Class and landscape structure were measured from LandSat^{TM} 7 windows of 5 × 5 km, centered in each fragment. We classified the image in the software SPRING 4.0 (INPE 2002) using 10 different cover classes previously recognized in the field. The final classification grouped the original classes into three—upland, rice fields and wetlands (including open water bodies) (Fig. 3). At the class level we considered (1) the total area of wetlands in the landscape—a straightforward measure of habitat availability; (2) the mean Euclidian distance between all neighbouring wetlands, and (3) the proximity index as measures of isolation; (4) the index of interspersion and juxtaposition to evaluate the degree of fragmentation; (5) the area weighted radius of gyration as a measure of structural connectivity; and (6) the total area of rice fields in the landscape, as a measure of supplementary habitat availability. The index of interspersion and juxtaposition measures the pattern of dispersion of fragments in the landscape and is not directly affected by the area and number of fragments. The mean nearest neighbour distance takes into account only the distance between fragments, whereas the proximity index weights the area of the neighbours. The radius of gyration, also known as correlation length, measures the mean distance an individual can move in the landscape, from a random starting point, without leaving the mosaic of wetland habitats (including fragments, channels and other water bodies).

Fragmentation, isolation, proximity and connectivity are related concepts, but report different aspects of habitat structure and configuration. Fragmentation refers explicitly to the degree of habitat subdivision and the resulting spatial arrangement of the remnants, independent of their areas (Fahrig and Merriam 1994; Mcgarigal and Marks 1995). Isolation and proximity derive from the island approach and refer

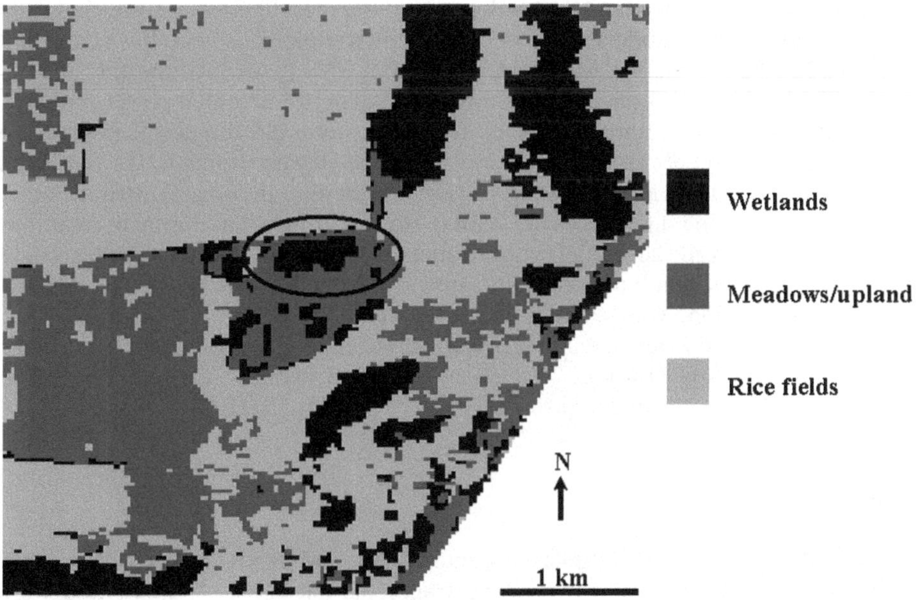

Fig. 3 Landscape pattern around the fragment on Fig. 2 (circled) from a LandSat^{TM} 7 window

explicitly only to the distance between patches of habitat, independent of matrix effects (Gustafson and Parker 1992). Connectivity, on other hand, refers to the capacity of the landscape to allow for individual movement, and is a spatially and species-specific concept (With et al. 1999), whose measure and application is debatable (Moilanen and Hanski 2001; Tischendorf and Fahrig 2001).

Connectivity can be either an structural measure at the class level, evaluating the total distance an organism can move through a specific kind of habitat, or a functional measure at the landscape level, taking into account the matrix's permeability to the individual movement (Mcgarigal and Marks 1995). At the landscape level we measured the contrast weighted edge density as functional measure of connectivity. A weight of 0.5 was set for the contrast between rice fields and wetland fragments and zero for all other cases, in order to evaluate the role of the matrix of rice fields as an element of permeability.

Analyses

We employed Mantel tests to determine the independence of species richness, species composition (log-transformed) and spatial attributes (centralized and normalized) in relation to geographical proximity (Legendre and Legendre 1998). The index of similarity of Jaccard was employed as resemblance measure in the case of richness and composition and Euclidian distance in the case of spatial metrics. The Mantel tests were carried out in the program MULTIV 2.0 (Pillar 2000). We analyzed the relationship among the spatial metrics and the species richness and abundance of individuals through Pearson correlations on log-transformed variables.

We employed logistic regressions to evaluate the effect of the spatial attributes at patch, class and landscape levels on the presence of waterbird species. Only species occurring in at least three fragments were included. First, all models with a single explanatory variable (metric) were run. Variables with significance level above 80% were retained for multiple models. Models with the largest likelihood were chosen as the best model. Exact models were used when the sample sizes were low (King and Ryan 2002). Logistic models were implemented in SAS (SAS Institute Inc. 2003).

We fitted the species–area relationship using the log-transformed power function. Alternative models (not shown) yielded poorer fits than did the power model. We then classified fragments as species-poor and species-rich according to the predicted richness, and used Discriminant Analysis (DA) to identify the spatial attributes that best accounted for the differences between these two groups. Discriminant functions were built from an initial model with all metrics, dropping them successively and testing for different combinations to maximize the group's discrimination and the canonical correlation, with significance level above 95%. We ran two analyses, representing two arbitrary deviation levels—discriminate fragments with residuals equal or greater than 10% and 20% of the expected richness. Correlations, curve fitting, and DA's were carried out in SPSS 10 (SPSS Inc. 2003).

Results

The spatial structure of the studied fragments and landscapes is relatively homogenous (Table 1). In general, the wetland fragments had a simple shape and were immersed in a landscape with a high degree of fragmentation, low isolation and high contrast with the surrounding matrix (low permeability). The metrics with the largest range were fragment area and proximity (CV greater than 100%). Microhabitat richness, structural connectivity and total area of wetlands in the landscape showed intermediate levels of variation (CV ~ 50%). The matrix was dominated by rice plantations, which cover from 20.0% to 64.4% (mean of 46.8%) of the landscape surrounding the wetland remnants.

We recorded 84,286 birds, belonging to 55 species (Table 2). The fragment with the highest species number had 78% of the total richness. More than 75% of the birds were recorded in nine fragments (upper percentile 20), while only 5% were cumulatively recorded in the lower percentile 50.

The attributes were correlated within levels, but not between levels (fragments × landscapes; Table 3). Large fragments tended to show a higher microhabitat richness and more complex shapes. Landscapes with greater availability of wetlands tend to show greater structural connectivity and proximity, and lower matrix permeability and total area of rice fields. The isolation and the degree of fragmentation did not correlate with other metrics. The species richness and the abundance of individuals correlated with each other and with the wetland area, the wetland shape, the microhabitat richness and the matrix permeability.

The fragments were spatially autocorrelated according to the metrics ($R = 0.128$; $P = 0.001$) and species composition ($R = 0.127$; $P = 0.002$), but not according to species richness ($R = -0.018$; $P = 0.62$). Due to low R values, no attempt was made to correct the degrees of freedom. These autocorrelations may have affected marginally some parameter estimates, but the low magnitude of this confounding factor does not invalidate the conclusions or justify corrections for geographical effects.

The structure and configuration of the fragments and landscape could be used to predict the presence of 37 out of 42 waterbird species (Table 4). Five metrics appeared frequently in the models. The presence of species varied positively with area (22 models), microhabitat richness (seven models) and structural connectivity (five models), and negatively with the matrix permeability (13 models). Isolation related negatively with the presence of four species, but positively with the presence of *Plegadis chihi*. Significant models were not found for only five species—*Syrigma sibilatrix, Ciconia maguari, Anas versicolor, Sterna superciliaris* and *Ceryle torquata*. The total area of rice fields in the landscape was not among the best predictors of waterbird richness and presence.

Waterbird richness varied positively with the fragment area ($F_{[1,40]} = 70.2$, $R^2 = 0.63$). Out of the 42 fragments, 15 showed species richness at least 10% greater than expected (corresponding to one species in the smallest fragments) and 17 showed richness lower than expected. Respectively, 10 and 11 fragments deviated 20% from expected (two species in the smallest fragments). The best functions discriminating species-poor and species-rich fragments employed two metrics—for both levels of deviation from the expected richness by the species–area relationship,

Table 2 Structure of the waterbird assemblage in 42 wetland fragments in the coastal zone of Rio Grande do Sul, Brazil

	Mean	Min.	Max.	CV	Total
Species richness	17.3	3	40	49.7	55
Abundance	2,006.8	15	19,322	204.4	84,286

Table 3 Pearson correlation coefficients of metrics describing the structure of waterbird assemblages and spatial structure of 42 wetland fragments and surrounding landscape in the coastal zone of Rio Grande do Sul, Brazil. See acronyms in Table 1

Level/attribute		Assemblage	Patch (fragment)			Class (wetlands or rice fields)						Landscape
		Species richness	Area	Shape	Microhabitat richness	Area of rice fields	Wetland availability	Structural connectivity	Fragmentation	Isolation	Proximity	Matrix permeability
Assemblage	Waterbird abundance	0.83**	0.66**	0.38*	0.64**	0.17	0.20	0.30	0.01	0.08	0.30	–0.42**
Landscape	Matrix permeability	–0.47**	–0.29*	–0.23	–0.41**	0.20	–0.59**	–0.63**	–0.10	0.05	–0.64**	
Class (wetlands or rice fields)	Proximity	0.34*	0.14	0.25	0.20	–0.41**	0.69**	0.78**	0.08	0.03		
	Isolation	–0.10	–0.28*	0.19	–0.25	–0.20	0.13	0.15	–0.19			
	Fragmentation	0.00	0.05	–0.10	0.18	0.10	–0.08	–0.03				
	Structural connectivity	0.35*	0.10	0.19	0.02	–0.52**	0.80**					
	Wetland availability	0.20	0.06	0.06	0.06	–0.51**						
	Area of rice fields	0.09	0.20	0.06	0.22							
Patch (fragments)	Microhabitat richness	0.63**	0.70**	0.35*								
	Shape	0.39**	0.32*									
	Area	0.80**										

Significance level (two-tailed): 0.05*; 0.01**

Table 4 Regression coefficients of the best logistic models of the effects of the spatial structure of 42 wetland fragments and surrounding landscape on the presence of 42 waterbird species. For five of these species we show two equally predictive models (names repeated twice). No significant models were found for five species (not shown). See acronyms in Table 1

Species	Freq[a]	AREA	SHAPE	PR	PR_RES	CA	GYRATE_AM	IJI	ENN_ MN	PROX_ MN	CWED
Podiceps major	6	0.14***									
Podilymbus podiceps	21	0.08**							–0.07*		
Rollandia Rolland	4	0.03**					0.005**				
Phalacrocorax brasilianus	19	0.07*								0.02**	
Butorides striatus	12			0.62**							–0.12*
Nycticorax nycticorax	28	0.22***									
Casmerodius albus	38		3.72**								
Ardea cocoi	29										–0.20***
Egretta thula	33	0.22**									
Mycteria Americana	19	0.07*		0.64***							
Plegadis chihi	35			1.85***					0.18**		
Phimosus infuscatus	36			2.09**							
Theristicus caerulensces	9	0.12**				0.01**					
Platalea ajaja	15	0.09**					0.005**				
Chauna torquata	21	0.10**									–0.294***
Dendrocygna viduata	34		2.69**								
Dendrocygna bicolor	13			0.61***							
Dendrocygna bicolor	13										–0.14***
Anãs flavirostris	17		2.02***								
Netta peposaca	4	0.08***									
Callonetta leucophrys	6	0.02*									
Rostrhamus sociabilis	13	0.07*									–0.10*
Rostrhamus sociabilis	13	0.08**									
Circus buffoni	15	0.10**									–0.14**
Aramus guarauna	25										–0.12**
Aramides ypecaha	12								–0.11**		–0.15**
Pardirallus sanguinolentus	9								–0.13**		–0.24***
Pardirallus nigricans	3	0.04*							–0.16*		
Gallinula melanops	13			0.42**							
Gallinula chloropus	31	0.23***									
Fulica leucoptera	11	0.07*									

Table 4 continued

Species	Freq[a]	AREA	SHAPE	PR	PR_RES	CA	GYRATE_AM	IJI	ENN_ MN	PROX_ MN	CWED
Rhynchops niger	3					0.01**					
Jacana jacana	38	0.42**									
Himantopus himantopus	21	0.11**									
Himantopus himantopus	21										–0.17**
Pluvialis dominica	7	0.027**					0.01***				
Charadrius collaris	4										–0.17**
Charadrius collaris	4						0.01***				
Tringa melanoleuca	4										–0.19***
Gallinago paraguaiae	21	0.09**				0.01**					
Larus maculipennis	14			0.93**			0.01**				–0.24**

Significance level (two-tailed): 0.1*; 0.05**; 0.01***

[a]Frequency of occurrence of waterbirds in 42 wetland fragments

we found that species-rich wetland fragments are located in landscapes with greater connectivity and lower matrix permeability (10% deviation: Wilk's Lambda = 0.68; $P = 0.003$; canonical correlation = 0.57; 20% deviation: Wilk's Lambda = 0.7; $P < 0.04$; canonical correlation = 0.55; Fig. 4).

Discussion

This work showed that area is the major factor related with the presence and richness of waterbird species in wetlands fragments. Matrix permeability, wetland connectivity and microhabitat richness were also found to be important contibuting factors (Boecklen 1986; Celada and Bogliani 1993; McAlpine and Eyre 2002). Wetland shape, isolation, and total availability in the landscape contributed less.

The effect of fragment area on species presence and richness is a matter of intense debate (Cao et al. 2002; Oertli et al. 2002). In our study area correlated with several other spatial attributes, as has been frequently found (Riffell et al. 2001; Tsao 2000). This complicates the evaluation of the contribution of each covarying factor. (Guadagnin DL et al. submitted for publication) showed that most of the variation in waterbird species richness in the coastal zone of Rio Grande do Sul is explained by a pure area effect that deviates from a random pattern of individual distribution. In this paper we employed the residuals of the species–area regression as an independent variable to demonstrate that class and landscape attributes account for variations in species richness not related to the area of the fragments.

Connectivity, both structural and functional, was shown to be an important factor in the structure of waterbird assemblages. For highly mobile species, as many waterbirds are, it may reflect the need individuals have to use multiple sites to complement or supplement resource needs (Haig et al. 1998; Plissner et al. 2000; Riffell et al. 2003; Sanzenbacher and Haig 2002). The access to resources available in the landscape depends not only on the distance between them (Dunning et al. 1992), but also on the degree that the physical characteristics of the routes and edges facilitate or impend bird movement (Taylor et al. 1993). Even highly mobile birds

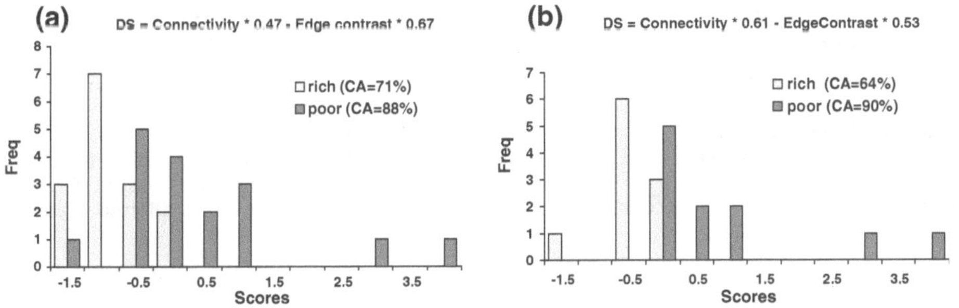

Fig. 4 Histograms of the discriminant scores (DA) of species-poor (grey bars) and species-rich (white bars) wetland fragments according to the spatial metrics of the fragments and the surrounding landscape. The classification criterion was waterbird richness greater or lesser than (**a**) 10% ($n = 32$) and (**b**) 20% ($n = 21$) of the richness predicted by the species-area regression (power model). Structural connectivity through the wetlands was measured by the Area Weighted Radius of Gyration. The Contrast Weighted Edge Density was applied as a measure of the permeability of the matrix of rice fields

may prefer to move over amenable matrices. Rice fields, and the intricate net of irrigation channels and small wetlands might be filling these roles (Day and Colwell 1998; Maeda 2001).

The total area of rice fields was not related to the richness, abundance or presence of waterbirds in the wetland fragments, even though it is known that several species use rice fields (Czech and Parsons 2002; Elphick 2000; Tourenq et al. 2001). The combined findings about the effects of rice fields—important connectivity element, but poor area effect—suggest that the spatial arrangement of rice fields and wetlands may be more important for the waterbirds than the total area of rice fields available at the landscape scale.

Wetland shape, isolation, and total availability in the landscape were not among the most important predictors of waterbird presence or factors that discriminate between species-poor and species-rich sites. (Guadagnin DL et al. submitted for publication) found a small, significant effect of wetland availability in the landscape on the waterbird richness, independent from the fragment's area. This could be related to the small range of values of those metrics among the studied fragments and landscapes and due to their correlation with the most important factors. Patch-centred studies produced similar results (Celada and Bogliani 1993), while landscape-centred studies found more prominent effects of habitat availability (Fairbairn and Dinsmore 2001; Norton et al. 2000; Villard et al. 1999).

Our findings provide insights for the investigation and planning of the agricultural matrix contribution to biodiversity conservation. (Guadagnin et al. 2005) showed that the region is still rich in waterbird species, despite the high loss and fragmentation of natural wetlands, and that the richness, composition and abundance of assemblages found in these fragments is comparable to that found in large, natural reference sites. The matrix permeability and the connectivity provided by the large number of small fragments and channels seem to be key factors for the maintenance of rich waterbird assemblages. While natural wetlands might locally harbour greater richness and abundance of waterbirds than rice fields (Tourenq et al. 2001), the connectivity provided by agricultural wetlands appears to be essential for the maintenance of the existing level of biodiversity (Ovaskainen and Hanski 2002; Sanchez-Zapata et al. 2005). The effect of the matrix of rice fields opens an interesting debate over the role of these man-made wetlands that are the major cause of habitat loss in many regions (Lawler 2001). Amenable matrices, including both small fragments and land uses that complement or supplement resource needs, can improve the capacity for biodiversity conservation outside protected areas (Fischer et al. 2005).

Acknowledgements Marco Aurélio Pizo, Emerson Monteiro Vieira and Paul Fisher kindly revised early versions of the manuscript. This study was founded by the Fundação O Boticário/McArthur Foundation (grant 528/2002-1), CAPES Foundation (grant BEX0431/03-6) and UNISINOS (grants 032/01, 02.00.030/02-0 and 38.00.005/01-1). We thank several landowners that allowed access to wetlands on their properties.

References

Boecklen WJ (1986) Effects of habitat heterogeneity on the species area relationships of forest birds. J Biogeogr 13:59–68

Brotons L, Monkkonen M, Martin JL (2003) Are fragments islands? Landscape context and density–area relationships in boreal forest birds. Am Nat 162:343–357

Brown M, Dinsmore JJ (1986) Implications of marsh size and isolation for marsh bird management. J Wildl Manage 50:392–397
Cao Y, Williams DD, Larsen DP (2002) Comparison of ecological communities: the problem of sample representativeness. Ecol Monogr 72:41–56
Celada C, Bogliani G (1993) Breeding bird communities in fragmented wetlands. Boll Zool 60:73–80
Coleman BD (1981) On random placement and species–area relations. Math Biosci 54:191–215
Czech HA, Parsons KC (2002) Agricultural wetlands and waterbirds: a review. Waterbirds 25:56–65
Day JH, Colwell MA (1998) Waterbird communities in rice fields subjected to different post-harvest treatments. Colonial Waterbirds 21:185–197
Diamond JM (1976) Island biogeography and conservation—strategy and limitations. Science 193:1027–1029
Dunning JB, Danielson BJ, Pulliam HR (1992) Ecological processes that affect populations in complex landscapes. Oikos 65:169–175
Elphick CS (2000) Functional equivalency between rice fields and seminatural wetland habitats. Conserv Biol 14:181–191
Erwin RM (2002) Integrated management of waterbirds: beyond the conventional. Waterbirds 25:5–12
Fahrig L (2002) Effect of habitat fragmentation on the extinction threshold: a synthesis. Ecol Appl 12:346–353
Fahrig L, Merriam G (1994) Conservation of fragmented populations. Conserv Biol 8:50–59
Fairbairn SE, Dinsmore JJ (2001) Local and landscape-level influences on wetland bird communities of the prairie pothole region of Iowa, USA. Wetlands 21:41–47
Fischer J, Fazey I, Briese R, Lindenmayer DB (2005) Making the matrix matter: challenges in Australian grazing landscapes. Biodivers Conserv 14:561–578
Gomes AdS, Magalhães Júnior AMd (2004) Arroz irrigado no Sul do Brasil. Embrapa, Pelotas
Guadagnin DL, Peter AS, Perello LFC, Maltchik L (2005) Spatial and temporal patterns of waterbird assemblages in fragmented wetlands of Southern Brazil. Waterbirds 28:261–272
Gustafson EJ Parker GR (1992) Relationships between landcover proportion and indexes of landscape spatial pattern. Landsc Ecol 7:101–110
Haig SM, Mehlman DW, Oring LW (1998) Avian movements and wetland connectivity in landscape conservation. Conserv Biol 12:749–758
INPE (2002) Tutorial de introdução ao SPRING (Sistema de Processamento de Informações Georreferenciadas) Version Instituto Nacional de Pesquisas Espaciais, São José dos Campos, São Paulo
Johnson AR, Wiens JA, Milne BT, Crist TO (1992) Animal movements and population-dynamics in heterogeneous landscapes. Landsc Ecol 7:63–75
Johnson CR, Boeriijst MC (2002) Selection at the level of the community: the importance of spatial structure. Trends Ecol Evol 17:83–90
King EN, Ryan TP (2002) A preliminary investigation of maximum likelihood logistic regression versus exact logistic regression. Am Stat 56:163–170
Lawler SP (2001) Rice fields as temporary wetlands: a review. Isr J Zool 47:513–528
Legendre P, Legendre L (1998) Numerical ecology. Elsevier, New York
Lomolino MV (2000) Ecology's most general, yet protean pattern: the species–area relationship. J Biogeogr 27:17–26
MacArthur RH, Wilson EO (1967) The theory of island biogeography. Princeton University Press, Princeton
Maeda T (2001) Patterns of bird abundance and habitat use in rice fields of the Kanto Plain, central Japan. Ecol Res 16:569–585
Maltchik L, Schneider E, Becker G, Escobar A (2003) Inventory of wetlands of Rio Grande do Sul (Brazil). Pesquisas Botânica 53: 89–100
McAlpine CA, Eyre TJ (2002) Testing landscape metrics as indicators of habitat loss and fragmentation in continuous eucalypt forests (Queensland, Australia). Landsc Ecol 17:711–728
Mcgarigal K, Marks BJ (1995) Fragstats: spatial pattern analysis program for quantifying landscape structure. Version U.S. Department of Agriculture, Forest Service, Pacific Northwest Research Station, Portland. Gen Tech Rep PNW-GTR-351
Moilanen A, Hanski I (2001) On the use of connectivity measures in spatial ecology. Oikos 95:147–151
Norton MR, Hannon SJ, Schmiegelow FKA (2000) Fragments are not islands: patch vs landscape perspectives on songbird presence and abundance in a harvested boreal forest. Ecography 23:209–223

Oertli B, Joye DA, Castella E, Juge R, Cambin D, Lachavanne JB (2002) Does size matter? The relationship between pond area and biodiversity. Biol Conserv 104:59–70

Ovaskainen O, Hanski I (2002) Transient dynamics in metapopulation response to perturbation. Theor Popul Biol 61:285–295

Pillar VDP (2000) Multiv - Multivariate exploratory analysis, randomization testing and bootstrap resampling: User's Guide. Version 2.0. Universidade Federal do Rio Grande do Sul, Porto Alegre

Plissner JH, Haig SM Oring LW (2000) Postbreeding movements of American Avocets and implications for wetland connectivity in the western Great Basin. Auk 117:290–298

Ricklefs RE, Lovette IJ (1999) The roles of island area per se and habitat diversity in the species-area relationships of four Lesser Antillean faunal groups. J Animal Ecol 68:1142–1160

Riffell SK, Keas BE, Burton TM (2001) Area and habitat relationships of birds in Great Lakes coastal wet meadows. Wetlands 21:492–507

Riffell SK, Keas BE, Burton TM (2003) Birds in North American Great Lakes coastal wet meadows: is landscape context important? Landsc Ecol 18:95–111

Sanchez-Zapata JA, Anadon JD, Carrete M, Gimenez A, Navarro J, Villacorta C, Botella F (2005) Breeding waterbirds in relation to artificial pond attributes: implications for the design of irrigation facilities. Biodivers Conserv 14:1627–1639

Sanzenbacher PM Haig SM (2002) Residency and movement patterns of wintering Dunlin in the Willamette Valley of Oregon. Condor 104:271–280

SAS Institute Inc (2003) Proceedings of the Twenty-Eighth Annual SAS Users Group International Conference. SAS Institute Inc., Cary, NC

Saunders DA, Hobbs RJ, Margules CR (1991) Biological consequences of ecosystem fragmentation—A Review. Conserv Biol 5:18–32

Scott D, Carbonell M (1986) Directorio de los Humedales de la Región Neotropical. International Waterfowl and Wetlands Research Bureau and International Union for the Conservation of Nature, Slimbridge and Gland

Sharpe DM, Stearns FW, Burgess RL, Johnson WC (1981) Spatio-temporal patterns of forest ecosystems in man-dominated landscape. In: Tjallingii SP, de Veers AA (eds). Perspectives in landscape ecology. PUDOC, Wageningen, The Netherlands, pp 109–116

SPSS Inc (2003) SPSS for Windows Release 12.0.1 Standard Version. Version 10.0.1 (27 Oct 1999). SPSS Inc, Chicago IL

Taylor PD, Fahrig L., Henein K., Merriam G (1993) Connectivity Is A Vital Element of Landscape Structure. Oikos 68:571–573

Tews J, Brose U, Grimm V, Tielborger K Wichmann MC, Schwager M, Jeltsch F (2004) Animal species diversity driven by habitat heterogeneity/diversity: the importance of keystone structures. J Biogeogr 31:79–92

Tischendorf L, Fahrig L (2001) On the use of connectivity measures in spatial ecology. A reply. Oikos 95:152–155

Tourenq C, Bennetts RE, Kowalski H, Vialet E, Lucchesi JL, Kayser Y, Isenmann P (2001) Are ricefields a good alternative to natural marshes for waterbird communities in the Camargue, southern France? Biol Conserv 100:335–343

Tsao M (2000) On the random placement hypothesis. Environ Ecol Stat 7:415–422

van Rensburg BJ, McGeoch MA, Matthews W, Chown SL, van Jaarsveld AS (2000) Testing generalities in the shape of patch occupancy frequency distributions. Ecology 81:3163–3177

Villard MA, Trzcinski MK, Merriam G (1999) Fragmentation effects on forest birds: relative influence of woodland cover and configuration on landscape occupancy. Conserv Biol 13:774–783

Wiens JA (1995) Habitat fragmentation - Island V landscape perspectives on bird Conservation. Ibis 137:S97–S104

With KA, Cadaret SJ, Davis C (1999) Movement responses to patch structure in experimental fractal landscpes. Ecology 80:1340–1353

Biodivers Conserv (2007) 16:1245–1260
DOI 10.1007/s10531-006-9124-8

ORIGINAL PAPER

Livestock husbandry as a tool for carnivore conservation in Africa's community rangelands: a case–control study

Rosie Woodroffe · Laurence G. Frank · Peter A. Lindsey · Symon M. K. ole Ranah · Stephanie Romañach

Received: 12 March 2006 / Accepted: 5 September 2006 / Published online: 27 October 2006

Abstract Conflict between people and wildlife is a major issue in both wildlife conservation and rural development. In African rangelands, species such as African wild dogs (*Lycaon pictus*), cheetahs (*Acinonyx jubatus*), lions (*Panthera leo*), leopards (*Panthera pardus*), and spotted hyaenas (*Crocuta crocuta*) may kill livestock and are therefore themselves killed by local pastoralists. Such conflict has led to the extirpation of these species from many areas, and also impacts the livelihoods of local livestock farmers. To investigate the possibilities for coexistence of people, livestock, and large predators in community rangelands, we measured the effectiveness of traditional livestock husbandry in reducing depredation by wild carnivores, using a case–control approach. Different measures were effective against different predator species but, overall, the risk of predator attack by day was lowest for small herds, accompanied by herd dogs as well as human herders, grazing in open habitat. By night, the risk of attack was lowest for herds held in enclosures ('bomas') with dense walls, pierced by few gates, where both men and domestic dogs were present. Unexpectedly, the presence of scarecrows increased the risks of attack on bomas. Our findings suggest that improvements to livestock husbandry can contribute to the conservation and recovery of large carnivores in community rangelands, although other measures such as prey conservation and control of domestic dog diseases are also likely to be necessary for some species.

R. Woodroffe (✉)
Department of Wildlife, Fish & Conservation Biology, University of California, One Shields Ave, Davis, CA 95616, USA
e-mail: rwoodroffe@ucdavis.edu

L. G. Frank · P. A. Lindsey · S. M. K. ole Ranah · S. Romañach · R. Woodroffe
Mpala Research Centre, PO Box 555, Nanyuki, Kenya

L. G. Frank
Museum of Vertebrate Zoology, University of California, Berkeley, CA 94720, USA

Present Address:
P. A. Lindsey · S. Romañach
Save Valley Conservancy, PO Box 47, Birchenough Bridge, Zimbabwe

Keywords African wild dog · African lion · Carnivore conservation · Cheetah · Community conservation · Human--wildlife conflict · Livestock depredation · Livestock husbandry · Predator control · Vertebrate pest management

Introduction

Conflict between people and wildlife is a major issue in both wildlife conservation and rural development. Deliberate killing of animals that are considered pests has driven several species to extinction, and has contributed to the endangerment of many others (Woodroffe et al. 2005b). Despite conservation concern about such lethal control, there is no doubt that some rare and endangered species genuinely threaten human lives and livelihoods (Thirgood et al. 2005). Human–wildlife conflict is a particularly serious issue in the conservation of large mammalian carnivores, which can prey on livestock (and occasionally people), and are therefore frequently killed by farmers or wildlife managers.

Lethal control is often carried out in response to specific attacks on livestock (e.g. Angst 2001; Bangs et al. 2005; Woodroffe and Frank 2005); hence any measure which reduces livestock depredation is also likely to reduce offtake from carnivore populations (Ogada et al. 2003), benefiting both people and wildlife. There is a clear need, therefore, to identify measures which demonstrably reduce livestock depredation by wild carnivores. Such measures may be particularly valuable where populations of highly threatened carnivores live alongside people and livestock, and where local people may benefit from carnivores' presence through tourism or hunting.

There is evidence that various forms of livestock husbandry can effectively reduce livestock depredation by wild carnivores (Breitenmoser et al. 2005). While modern approaches such as electric fencing and radio-activated guards appear useful in North America and Europe (Breitenmoser et al. 2005), they are costly and may be inappropriate for use in less developed areas with little infrastructure. In contrast, Ogada et al. (2003) showed that, on commercial ranches in East Africa, livestock husbandry similar to that practiced for generations by local Masai pastoralists was very effective at reducing conflict between predators and livestock farmers.

Building on Ogada et al.'s (2003) findings from commercial ranches, we conducted a case–control study of livestock depredation on neighbouring community lands. The study aimed to reduce predators' impacts on local people's livelihoods by identifying the most effective husbandry methods, and also to investigate the possibilities for sustainable conservation of African wild dogs (*Lycaon pictus*), cheetahs (*Acinonyx jubatus*), lions (*Panthera leo*), leopards (*Panthera pardus*), and spotted hyaenas (*Crocuta crocuta*) on community lands.

Methods

Study area

This study was carried out between January 2001 and June 2005 in northern Kenya, in Laikipia District (37°2′ E, 0°6′ N), and parts of neighbouring Samburu, Isiolo and Baringo Districts (Figs. 1, 2). The area is mainly semi-arid bush land and savanna, used for subsistence pastoralism and commercial ranching, as well as for tourism and small-scale agriculture. Within the study area, some properties are owned by private individuals, but this study was conducted primarily on land owned or occupied by local communities (Figs. 1, 2). None of the area is formally protected, and livestock occur throughout the region. However, wildlife is abundant in some areas (Khaemba et al. 2001; Mizutani 1999). Populations of leopards, lions, cheetahs and spotted hyaenas have persisted in the region despite extensive human use of the landscape, and wild dogs recolonised Laikipia naturally in 2000 (Frank et al. 2005).

Data collection

Most of the data were gathered by 18 staff employed in partnership with local non-governmental conservation organisations and paid, in whole or in part, by the project. These staff were educated people chosen from, and resident within, local communities, well placed to gather often-sensitive information about depredation. The staff did not collect data over the whole area, and their home locations (shown in Fig. 1) and work schedules were dictated by their other duties: hence the number and distribution of predator attacks that they investigated may not give an accurate picture of the relative depredation rate in different areas or by different predators.

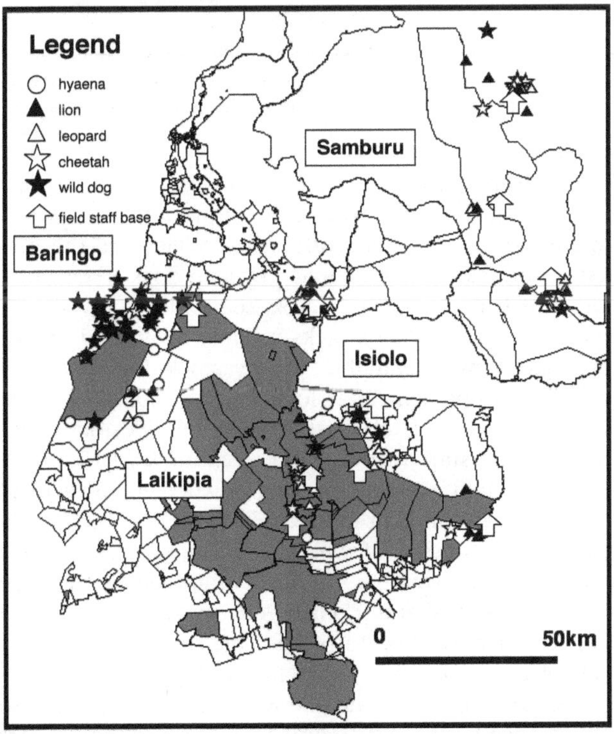

Fig. 1 The distribution within the study area of reported predator attacks on herds grazing by day, along with the approximate home locations of field staff investigating attacks. The number of field staff locations is smaller than the number of field staff because some staff worked in pairs. Commercial ranches are shaded in grey; communities inhabit remaining areas

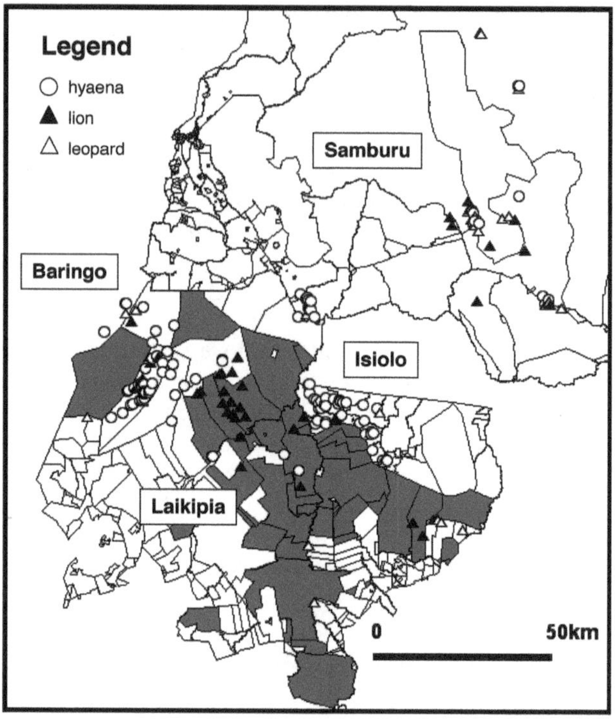

Fig. 2 The distribution within the study area of predator attacks on herds enclosed in bomas at night. Commercial ranches are shaded in grey; communities inhabit remaining areas

In particular two staff were employed in the north-west of the study areas specifically to monitor a pack of wild dogs that were chronic predators of livestock; attacks by this pack were highly localised (Woodroffe et al. 2005a) and the proportion of the investigated livestock attacks that were made by wild dogs is therefore very unlikely to be representative of true proportion across the entire study area.

Field staff investigating reported attacks initially confirmed the veracity of each report; no compensation was available for livestock losses, so farmers had no financial incentive to misrepresent their losses and a separate analysis of predator diet showed that, for wild dogs at least, reports of livestock loss were reliable (Woodroffe et al. 2005a). Staff recorded the location and timing of each attack, whether livestock were killed while out grazing or when enclosed in a boma, and the number and age of livestock killed or injured. For attacks on grazing herds, they also recorded the herder's perception of whether the predator(s) initiated the attack by approaching the herd, or whether the herd stumbled upon a resting predator. Detailed data on livestock husbandry on the day of the attack were then collected for the herd or boma which had been attacked (the case herd) and between one and three nearby control herds or bomas that had not been attacked. Controls were selected on the basis of proximity alone and were not matched to cases in any other way. The median distance between case grazing herds and their controls was 656 m, and the median distance between case and control bomas was 323 m; hence a high proportion of control herds and bomas are likely to have been detectable to the predators that attacked the case herds.

For each grazing herd, observers recorded the number of livestock of different species present, and the number of herders (men, women and children), and domestic dogs, accompanying the herd. The type of terrain grazed was recorded in five categories (flat, rolling hills, single escarpment, hilly, mountainous), and the density of the vegetation in the grazing area was recorded in four categories (open grassland, light bush, dense bush, very dense bush), using reference photographs provided to all observers. For case herds, details of terrain and habitat density were recorded for the site of the attack, as well as for the grazing area in general.

For each boma, observers categorised the type of construction as thornbrush (constructed of whole trees or large branches cut and laid on their sides), solid (constructed of solid poles or stone), wire, or open (an area where livestock are bedded down with no physical barrier between them and the surrounding environment). A small number of bomas were constructed from a combination of solid posts and wire; classifying these as either "solid" or "wire" did not influence overall conclusions. For thornbrush, solid and wire bomas, the number of gates into the boma was also recorded. The height (in cm) of the boma walls was recorded at 10 m intervals around the perimeter of the boma. Likewise, for thornbrush bomas, the thickness of the boma walls was measured at 10 m intervals using a thin pole and a tape-measure. In addition, the density of thornbrush boma walls was estimated every 10 m by measuring the transparency of the wall; transparency was scored by having an assistant hold a 'chequerboard' marked with 100 2 cm × 2 cm squares, shaded alternately black and white, against the inside boma wall at a height of 1 m from the ground, and recording the number of white squares which were more than 50% visible to an observer outside the boma with their eye also 1 m from the ground (hence maximum transparency received a score of 50). Analyses used the minimum height and width, and the maximum transparency, excluding (where appropriate) any site where the predator or livestock had broken through the boma wall. For thornbrush bomas, observers also recorded whether the majority of tree branches were facing the inside or the outside of the boma, and also the time since the boma walls were last repaired.

In addition to gathering information on the construction of the boma itself, observers recorded the number of possible predator deterrents present at the boma on the night of the attack, including the number of people (men, women and children), the number of houses, the number of domestic dogs, the number of guns, the number of fires left burning outside boma walls, and the number of scarecrows (pieces of cloth hung in nearby trees or on boma walls). Details of the numbers of livestock present were also recorded (cattle, sheep and goats, camels and donkeys); for cattle the presence or absence of adult bulls was noted because these animals are the largest and strongest and hence considered most likely to break through boma walls if panicked by a predator attack. Finally, the terrain and habitat density surrounding the boma were recorded using the same categories as for grazing herds.

Sample sizes quoted in descriptions of livestock husbandry and depredation vary slightly due to occasional missing data.

Data analysis

The characteristics of matched case and control herds or bomas were compared using conditional logistic regression, using the program Egret (Cytel Software Corporation; Cambridge, MA, USA). The two primary analyses considered all data on all attacks on either (i) grazing herds, or (ii) bomas, pooling data for all predators and all livestock prey species. Data were pooled in this way because farmers experience the same loss whether a calf is, for example, killed by a lion or a hyaena. Subsequent analyses were restricted to attacks by single species of predator, to investigate the husbandry measures most effective against each. Some predator species apparently specialised on particular livestock prey (e.g. cheetahs killed sheep and goats, but never cattle); hence a nearby control herd might have been ignored by the predator not because of differences in husbandry, but because it did not contain the preferred prey species. Analyses of attacks by specific predator species were therefore restricted to the cases involving only the most common prey species, omitting also any control herds or bomas not including the favoured prey. Hence, for example, analyses for leopards involved 39 attacks on grazing sheep and goat herds (excluding nine attacks on grazing cattle) and 35 attacks on sheep and goats in bomas (excluding one attack on cattle).

Initially, univariate analyses were carried out on each characteristic of livestock husbandry or habitat in turn. Where sample sizes were restricted, the data on habitat density were collapsed into two categories (open grassland + light bush, and dense bush + very dense bush). Data on numbers of domestic dogs were also simplified to presence/absence of domestic dogs. Variables with univariate test results at $P \leq 0.25$ were subsequently entered into multivariate analyses. Final models were selected by including all of the selected variables, and then dropping terms from the model one by one until only significant predictors remained.

Exploratory analyses showed that some husbandry characteristics were significantly intercorrelated. For example, grazing herds accompanied by domestic dogs also had more men herding them ($t_{498} = 4.90$, $P < 0.001$), and the thickness of boma walls was correlated with their transparency ($r = -0.11$, $n = 412$, $P = 0.024$). We investigated the robustness of the final models by analysing their sensitivity to these intercorrelations. In models that included predictor variables that were significantly intercorrelated with other possible predictors, we replaced each predictor in turn with its correlate(s) to determine whether this improved the fit of the model.

The results of case–control analyses are presented as odds ratios (and their associated 95% confidence intervals (CI)). These odds ratios indicate the magnitude of the estimated effects; for example the odds ratio of 0.371 (95% CI 0.195–0.705) associated with domestic dogs' presence in Table 1 indicates that dogs reduce the probability of attack by a factor of about 0.37, i.e. a 63% reduction.

In addition to these case–control comparisons, we also compared the habitat density and terrain at sites where attacks occurred on grazing herds, with those in the

Table 1 Predictors of the probability of attack, by any predator species, on herds grazing during the daytime, based on conditional logistic regression analysis of 147 matched case–control sets

Variable	Odds ratio (95% CI)	χ^2	*df*	*P*
Domestic dogs				
Present vs. absent	0.371 (0.195–0.705)	9.14	1	0.003
Herd size	1.006 (1.002–1.011)	7.27	1	0.007
Habitat density		11.71	3	0.008
Light bush vs. open	1.667 (0.635–4.381)			
Dense bush vs. open	4.059 (1.278–12.894)			
Very dense bush vs. open	25.372 (2.312–278.45)			

general area grazed by that herd on that day. For these analyses, habitat density and terrain were ranked on ordinal scales (from "open" = 1 to "very dense bush" = 4, and "flat" = 1 to "mountainous" = 5) and compared using Wilcoxon paired rank tests.

Results

Characteristics of predator attacks

The timing and nature of attacks on livestock varied among predators. As shown in Fig. 3, hyaenas mainly attacked herds enclosed in bomas at night, cheetahs and wild dogs attacked herds only when they were grazing by day, and leopards and lions attacked both grazing herds and those enclosed in bomas. Daytime attacks on herds

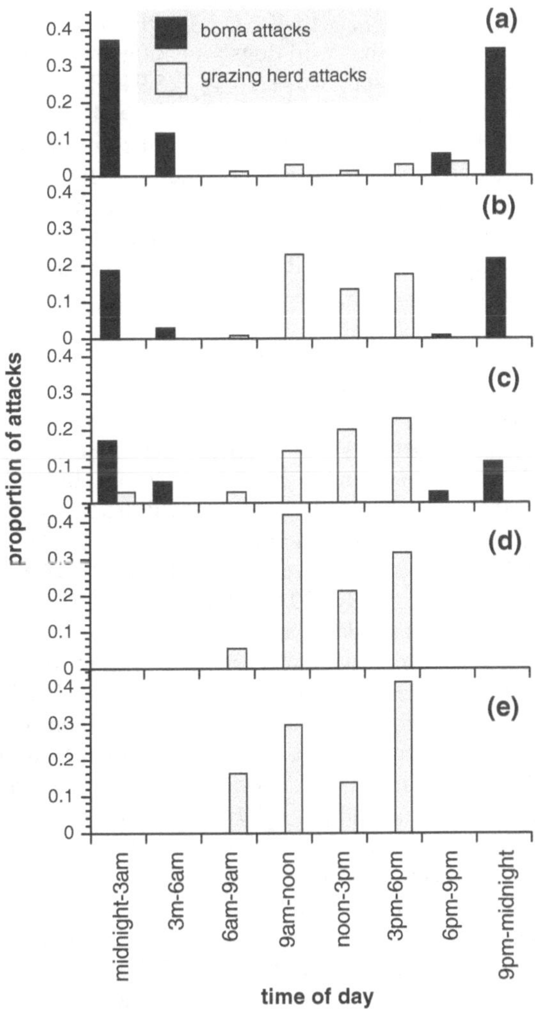

Fig. 3 Timing of predator attacks on livestock which were either enclosed in bomas at night (shaded bars) or grazing by day (open bars), by (**a**) spotted hyaenas (n = 105 attacks); (**b**) leopards (n = 96 attacks); (**c**) lions (n = 35 attacks); (**d**) cheetahs (n = 19 attacks); (**e**) wild dogs (n = 44 attacks)

were perceived to have been initiated by the predator approaching the herd (rather than the herd stumbling upon a resting predator) in 95% of 19 cheetah attacks, 85% of 40 wild dog attacks, 91% of 11 hyaena attacks, 65% of 17 lion attacks and 65% of 49 leopard attacks. When data were combined for the three primarily nocturnal species (leopards, lions and hyaenas) there was a non-significant trend suggesting that a smaller proportion of attacks were initiated by predators in the hot hours in the middle of the day, with predators approaching herds on 75% of 4 attacks occurring between 0600 h and 0900 h, 76% of 25 attacks occurring between 0900 h and 1200 h, 44% of 18 attacks occurring between 1200 h and 1500 h, and 76% of 29 attacks occurring between 1500 h and 1900 h ($\chi^2 = 6.28$, $df = 3$, $P = 0.099$). There was no such trend for the diurnal predators (cheetahs and wild dogs, $\chi^2 = 3.51$, $df = 3$, $P = 0.32$).

The number of livestock killed per attack also varied between predators (Table 2). When attacking herds grazing by day, wild dogs killed more sheep and goats per attack than did leopards, cheetahs or hyaenas (Kruskal–Wallis test, excluding two lion attacks, $H_{44,36,17,10} = 21.00$, $P < 0.001$), and also injured more animals ($H_{44,36,17,10} = 18.46$, $P < 0.001$). In contrast, there was a near-significant trend suggesting that wild dogs attacking grazing herds killed fewer cattle per attack than did lions or leopards ($H_{15,8,7} = 5.94$, $P = 0.051$), though there was no effect on the number of cattle injured ($H_{15,8,7} = 4.31$, $P = 0.116$). When attacking herds enclosed in bomas at night, hyaenas killed more sheep and goats per attack than did leopards (one lion attack excluded, Mann–Whitney U test: $U_{86,44} = 2264.5$, $P = 0.030$), though there was no difference in the number of sheep and goats injured ($U_{86,44} = 1987.5$, $P = 0.471$), nor was there any difference between hyaenas and lions in the number of cattle killed (one leopard attack excluded, $U_{16,8} = 76$, $P = 0.376$) or injured ($U_{16,8} = 69.5$, $P = 0.634$) per attack on livestock inside bomas.

Characteristics of livestock husbandry

Ninety-seven percent of 502 grazing herds (cases and controls combined) for which data were available were accompanied by herders. The average grazing herd was accompanied by 2.1 ± 1.9 SD herders (range 0–19): 0.6 ± 1.0 men, 0.1 ± 0.4 women, and 1.3 ± 1.2 children. Thirty-seven percent of 505 herds included cattle, 76% included sheep or goats, 9% included donkeys and 3% included camels. The average herd comprised 82.6 ± 73.4 SD individual animals (range 2–400): 11.8 ± 29.6 adult cattle, 2.0 ± 4.4 calves, 68.3 ± 75.4 sheep and goats, 0.3 ± 1.4 donkeys, and 0.7 ± 3.9 camels. Twenty-four percent of 502 grazing herds were accompanied by one or more domestic dogs, with an average of 1.3 ± 0.5 (range 1–3) dogs per accompanied herd.

Of 491 bomas examined (cases and controls combined), 446 (91%) were constructed of thornbrush, 27 (5%) were built from solid poles or stone, 12 (2%) were made of wire, and six (1%) were "open" bomas with no physical wall. The average thornbrush boma had 2.4 ± 1.7 SD gates (range 1–12). Cattle were present at 62% of 477 bomas with 54.1 ± 76.6 SD (range 1–603) adult cattle and 9.5 ± 12.4 SD (range 0–83) calves present. Seventy percent of 89 cattle herds for which data were available included one or more adult bulls. Sheep or goats were present at 86% of 477 bomas with 144.1 ± 189.9 (range 4–1300) individuals present. The average boma was occupied by 11.3 ± 8.2 SD people: 2.3 ± 2.0 men, 2.6 ± 2.2 women, and 6.4 ± 5.5 children. Domestic dogs were present at 71% of 484 bomas, with an average of

Table 2 Numbers of livestock killed and injured on predator attacks

Predator	Cattle			Sheep & goats			Camels			Donkeys		
	n	Killed	Injured	*n*	Killed	Injured	*n*	Killed	Injured	*n*	Killed	Injured
Leopard												
Herd	8	1.9 ± 1.5 (1–5)	0.1 ± 0.4 (0–1)	45	2.1 ± 2.3 (0–12)	0.2 ± 0.6 (0–2)	0	–	–	1	2 ± 0	0
Boma	1	1	0	44	1.4 ± 1.2 (1–16)	0.2 ± 0.4 (0–2)	0	–	–	0	–	–
Lion												
Herd	15	1.3 ± 1.1 (0–5)	0.3 ± 0.6 (0–2)	3	1.7 ± 0.6 (1–2)	0.3 ± 0.6 (0–1)	4	1 ± 0 (1–1)	0.5 ± 0.6 (0–1)	0	–	–
Boma	16	1.3 ± 0.4 (1–2)	0.2 ± 0.4 (0–1)	1	1	1	0	–	–	1	1	0
Cheetah												
Herd	0	–	–	19	1.3 ± 0.7 (1–4)	0.1 ± 0.3 (0–1)	0	–	–	0	–	–
Boma	0	–	–	0	–	–	0	–	–	0	–	–
Hyaena												
Herd	2	1.0 ± 0 (1–1)	0	11	1.6 ± 1.9 (0–7)	0.3 ± 0.5 (0–1)	0	–	–	0	–	–
Boma	8	1.0 ± 0.8 (0–2)	0.4 ± 0.7 (0–2)	86	2.5 ± 4.0 (0–33)	0.5 ± 3.0 (0–28)	0	–	–	2	2.5 ± 2.1 (1–4)	0 ± 0
Wild dog												
Herd	7	0.7 ± 0.5 (0–1)	0.7 ± 0.8 (0–2)	38	4.1 ± 3.9 (0–16)	1.0 ± 1.2 (0–4)	0	–	–	0	–	–
Boma	0	–	–	0	–	–	0	–	–	0	–	–

For each predator species, prey species, and attack type (on herds grazing by day, or enclosed in bomas at night), *n* denotes the sample size of attacks for which full data were available, and other columns give the numbers of animals killed or injured (as mean ± SD (range))

2.0 ± 1.3 dogs per boma. Fires were left burning outside 22% of 481 bomas, and scarecrows were present at 44% of 483 bomas with an average of 2.4 ± 1.5 scarecrows per boma. Only 6% of 483 bomas reported having a gun.

Predictors of predator attacks on herds grazing during the daytime

Across all species of predator and livestock prey, the probability of depredation was lowest for smaller herds, that were accompanied by one or more domestic dogs, and grazed in more open habitat (Table 1). Sensitivity analyses showed that the fit of the model was reduced by substituting the number of herdsmen for the presence of herd dogs, with which dogs' presence was intercorrelated (overall model deviance 292.7, $df = 5$, compared with 280.8, $df = 5$).

Analysis of attacks by different predator species on their most common domestic prey revealed marked variation in which husbandry features were most effective. Results for leopards were similar to those for all predators (perhaps because leopard attacks were most common, Table 2); multivariate analysis ($n = 39$ matched sets) revealed that leopard attacks were more likely for sheep and goat herds that were unaccompanied by herd dogs (odds ratio 0.091, 95% CI 0.02–0.45, $\chi^2 = 8.63$, $df = 1$, $P = 0.003$), and those that grazed in more dense habitat (open & light bush vs. dense & very dense bush; odds ratio 7.33, 95% CI 1.22–43.94, $\chi^2 = 4.76$, $df = 1$, $P = 0.029$). There were no statistically significant predictors of lion attacks, although an association with dense habitat approached significance (odds ratio 6.75, 95% CI 0.77–59.35, $\chi^2 = 2.97$, $df = 1$, $P = 0.085$, $n = 22$ matched sets). There were likewise no significant predictors of cheetah attacks (for which only limited data were available), although a non-significant trend suggested that attacks might be more common in sheep and goat herds unaccompanied by domestic dogs (odds ratio 0.28, 95% CI 0.06–1.38, $\chi^2 = 2.44$, $df = 1$, $P = 0.118$, $n = 17$ matched sets). Attacks by wild dogs were the only ones for which effects of herder characteristics were detectable; wild dogs were more likely to attack sheep and goat herds accompanied by larger numbers of child herders (odds ratio 2.125, 95% CI 1.003–4.50, $\chi^2 = 3.87$, $df = 1$, $P = 0.049$, $n = 35$ matched sets), as well as those grazing in less open habitat (open & light bush vs. dense & very dense bush; odds ratio 3.95, 95% CI 1.15–13.61, $\chi^2 = 4.73$, $df = 1$, $P = 0.030$). In contrast with the findings for other species, there was a non-significant trend suggesting that sheep and goat herds accompanied by domestic dogs might be *more* likely to experience attacks by wild dogs (odds ratio after adjusting for effects of habitat and child herders 4.01 95% CI 0.86–18.62, $\chi^2 = 3.15$, $df = 1$, $P = 0.076$). The sample size for hyaena attacks on sheep and goats was very small (only 11 matched sets) and there were no significant predictors of attack risk. Sensitivity analyses revealed no evidence that any of the effects detected were generated by intercorrelations among possible explanatory variables.

Pairwise comparisons of the habitat density recorded at attack sites, and in overall areas grazed, revealed that leopard attacks tended to occur when herds entered dense bush (Wilcoxon signed rank test $P = 0.034$), as did lion attacks (Wilcoxon signed rank test $P = 0.034$). There were no such significant effects of habitat density for attacks by cheetahs, hyaenas or wild dogs (Wilcoxon signed rank tests, P values all >0.15), nor were there effects of terrain for any species of predator (P values all >0.25).

Predictors of predator attacks on herds enclosed in bomas at night

When data were pooled for all predator species, the type of boma construction (thornbrush, solid, wire or open) did not influence the probability of attack ($\chi^2 = 5.07$, $df = 3$, $P = 0.167$, $n = 244$ matched sets). Considering only thornbrush bomas, the most common type, Table 3 shows that the probability of attack was influenced by features of boma construction (the transparency of the boma wall, and the number of gates) as well as by deterrents present at the boma (men, and domestic dogs). Unexpectedly, the number of scarecrows, a widely-used deterrent, was associated with an increased risk of attack. Sensitivity analyses revealed no evidence that these effects were generated by intercorrelations among possible explanatory variables.

As for daytime attacks on herds, multivariate analyses revealed marked variation in the husbandry characteristics most effective at preventing boma attacks by different predator species on their most common domestic prey. There was no significant effect of the type of boma construction for any species, so analyses considered only thornbrush bomas, the most common type. The only significant predictor of leopard attacks on sheep and goat herds was the number of scarecrows present at the boma (odds ratio 1.54, 95% CI 1.001–2.36, $\chi^2 = 3.87$, $df = 1$, $P = 0.049$, $n = 35$ matched sets). The only significant predictor of lion attacks on cattle herds was the total number of people present at the boma, with attacks more likely where human activity was lower (odds ratio 0.84, 95% CI 0.71–0.99, $\chi^2 = 4.21$, $df = 1$, $P = 0.040$, $n = 31$ matched sets). The probability of hyaena attack on sheep and goat herds was greatest at bomas with thin walls (odds ratio 0.99, 95% CI 0.97–1.00, $\chi^2 = 5.25$, $df = 1$, $P = 0.022$, $n = 95$ matched sets), with larger numbers of gates (odds ratio 1.39, 95% CI 1.03–1.87, $\chi^2 = 4.61$, $df = 1$, $P = 0.032$), and where domestic dogs were absent (odds ratio 0.30, 95% CI 0.13–0.72, $\chi^2 = 7.38$, $df = 1$, $P = 0.007$). Sensitivity analysis suggested that, in this analysis, the transparency of the boma walls could be substituted for their thickness (odds ratio 1.10, 95% CI 1.03–1.17, $\chi^2 = 9.14$, $df = 1$, $P = 0.003$). In this case, however, the effect of number of gates became non-significant (odds ratio after adjusting for boma transparency and domestic dogs 1.25, 95% CI 0.93–1.67, $\chi^2 = 2.21$, $df = 1$, $P = 0.137$) and, after dropping this covariate, the overall model had (non-significantly) greater deviance than the original model incorporating significant effects of wall thickness, gates and domestic dogs (difference in deviance = 1.81, $df = 1$, $P = 0.178$). There was no evidence that effects detected for leopards or lions were influenced by intercorrelations among possible explanatory variables.

Table 3 Predictors of the probability of attack, by any predator species, on herds enclosed in thornbrush bomas at night, based on conditional logistic regression analysis of 178 matched case–control sets

Variable	Odds Ratio (95% CI)	χ^2	*df*	*P*
Number of boma gates	1.402 (1.119–1.756)	8.63	1	0.003
Transparency of boma wall	1.039 (1.005–1.075)	4.94	1	0.026
Number of scarecrows	1.451 (1.157–1.820)	10.34	1	0.001
Domestic dogs				
present vs. absent	0.412 (0.223–0.759)	8.11	1	0.004
Number of men	0.764 (0.615–0.949)	5.9	1	0.015

Discussion

Our results confirm that, in our study area, predator attacks on livestock are, to some extent at least, preventable without resorting to lethal control. This raises the possibility that at least some of the carnivore species may be sustainably conserved on community lands in other areas. Various aspects of livestock husbandry, including boma design, herding practices, and deterrents, were effective in reducing the probability of predator attack. Since killing of wild carnivores is associated with livestock predation on both commercial (Ogada et al. 2003) and community (Romañach et al. in review) lands, such measures almost certainly reduced the level of lethal control carried out. These methods are therefore likely to be valuable not only in reducing predators' impacts on local people's livelihoods, but also in conserving wild carnivores, both locally and more widely.

Our study area was characterised by very attentive livestock husbandry (e.g. 97% of herds were accompanied by herders), in comparison with other parts of Africa, western Europe, North and South America, where traditional livestock husbandry practices have typically been abandoned (Breitenmoser et al. 2005). There are probably multiple reasons for the retention of this traditional husbandry, but one key factor appears to be the widespread risk of stock theft which commits farmers to guard their herds carefully even in the absence of predators (Frank et al. 2005).

As mentioned briefly in the *Methods* section, the distribution of investigated attacks shown in Figs. 1 and 2 does not necessarily reflect the actual distribution of attacks, since observers did not cover the whole area, were not able to investigate all attacks, and because one pair of observers was employed specifically to collect data in an area suffering chronic livestock depredation by wild dogs. Nevertheless, Figs. 1 and 2 suggest a non-random distribution of attacks by different predators. Wild dog attacks appear concentrated in the north-western part of the study area; reasons for this are discussed in detail elsewhere (Woodroffe et al. 2005a) but probably reflect severe depletion of wild prey in this area. Figures 1 and 2 also indicate that most of the investigated lion attacks occurred on or near commercial ranches, even though observers were working primarily on community land. This probably reflects lions' marked preference for commercial ranches (Frank et al. 2005), perhaps because wild prey are more abundant there (Khaemba et al. 2001).

The numbers of attacks investigated likewise give only a very approximate picture of each predator species' impact on local people's livelihoods. This impact will be influenced not only by the frequency of attacks, but also by the numbers of animals killed or injured on each attack. The latter varies substantially between species, with wild dogs particularly likely to kill multiple animals when attacking grazing herds, and hyaenas particularly likely to make multiple kills inside bomas (Table 1). Since our analyses did not take into account the age, sex, or reproductive state of each animal killed, we did not quantify the impact of depredation on local systems of livestock production (Mizutani et al. 2005).

Our findings are broadly comparable with those of previous studies (Kruuk 1980; Ogada et al. 2003), showing that herding practices and boma construction can greatly reduce the probabilities of livestock loss. However, our case–control approach allowed us to more precisely quantify the relative impacts of different husbandry factors and should be helpful in indicating improvements that could reduce livestock losses to predators. For example, analyses indicate that having a domestic dog accompanying a herd reduces the risk of attack by 63% (Table 1), yet only 24% of herds were so accompanied even though dogs were present at 71% of bomas. Likewise, we found that each gate in a boma wall increases the likelihood of attack by 40% (Table 3), yet the average boma had 2.4 gates. These sorts of findings are helpful because they allows farmers (and conservation managers) to make informed decisions about optimal livestock husbandry, taking into account, for example, that a more densely constructed boma wall will provide more effective protection for livestock (and hence lead to fewer predators being killed), but will also be more labour-intensive to construct (and involve greater offtake from slowly-growing native trees, Okello et al. 2001).

While there is considerable local knowledge and experience concerning effective livestock husbandry, the quantitative approach taken in this study revealed two results that were somewhat surprising, namely that attack probability was higher for (i) bomas with more scarecrows; and (ii) herds accompanied by more child herders. Concurrent findings in line with expectations—for example that predation events were more likely to occur in dense bush, or that domestic dogs help to deter attacks—suggest that these counter-intuitive findings should not be dismissed as anomalous. The detrimental effect of scarecrows was particularly marked for leopards although, after adjusting for the significant effects reported above, non-significant trends for lions ($P = 0.137$) and hyaenas ($P = 0.127$) were in the same direction. This unintended effect of scarecrows might perhaps reflect attraction to a visual cue (cloth moving in the wind)—although similar moving flags have been shown to deter wolves from attacking livestock (Musiani et al. 2003). While the detrimental effect of multiple child herders was unforeseen, community members had an immediate explanation for this findings: lone herd boys are likely to be more attentive to their flocks than those with companions to talk or play with.

Our results demonstrate clear benefits of having domestic dogs both at bomas and accompanying grazing herds (Tables 1, 3). It should be noted that the dogs involved are not selectively bred for guarding livestock, nor do they actively chase predators away; their main role is to alert people to predators' presence. It is perhaps not surprising that domestic dogs reduce livestock depredation, and, by extension, reduce the numbers of predators killed by livestock farmers. This finding is important, however, because domestic dogs also have negative consequences for predator conservation: they carry infectious diseases that have the potential to substantially reduce the viability of wild carnivore populations (Cleaveland and Dye 1995; Laurenson et al. 1998; Roelke-Parker et al. 1996), and are also widely used for hunting wild prey. While, from a disease perspective, control of domestic dog populations might be desirable (Laurenson et al. 2004), our findings suggest that benefits to carnivore conservation achieved by reduced disease risks might be countered by increased retaliatory killing associated with livestock depredation. This reinforces the idea that—outside protected areas at least—vaccination (rather than control) of domestic dog populations is a more sustainable solution to disease threats for wild carnivores (Laurenson et al. 2004). These benefits of domestic dogs are widely recognised by local people: interviews conducted at 562 bomas revealed that 49% of interviewees wanted more dogs at their boma, compared with 42% wanting to keep the same number and just 9% wanting fewer (R. Woodroffe unpubl. data). The possibly (though not statistically significantly) detrimental effects of domestic dogs on livestock predation by wild dogs are perhaps unsurprising; there are several anecdotal reports of wild dogs being attracted to domestic dogs (R. Woodroffe unpubl. data; Butynski 1974).

The different husbandry practices that appear effective against different predator species influence the extent to which our overall findings may be generally applicable. Since attacks by leopards and hyaenas dominated our dataset, effects on these species will have strong effects on the overall results presented in Tables 1 and 3; hence findings for the particular species may be more appropriate for applying to other areas with different predator problems. For example, in the majority of community areas studied, the principal predators of livestock were leopards and hyaenas (Figs. 1, 2), suggesting that the most effective husbandry approaches would involve ensuring that domestic dogs accompany grazing herds as well as staying at bomas at night, building bomas with thick walls and the minimum number of gates (ideally reinforced to further reduce losses), and removing scarecrows from bomas. In the north-west of our study area, however, wild dogs were the principal predator of livestock. Here, therefore, the most important improvements to livestock husbandry would be to avoid assigning herding duties to groups of children, and perhaps also to discourage domestic dogs from accompanying herds. These differences illustrate how local conditions need to be taken into account in extrapolating our findings to other areas where predator communities may differ.

In conclusion, our study indicates that relatively small improvements to livestock husbandry have the capacity to substantially reduce depredation on livestock by wild carnivores and, hence, to reduce both predator impacts on rural people's livelihoods and lethal control of species of conservation concern. Similar approaches would presumably also be effective in other parts of Africa, or for similar species elsewhere. However, whether such approaches would be adopted in other areas depends on multiple factors including local herding traditions, access to materials to construct bomas, the availability of education and alternative employment for potential herders, and opportunities for generating income from predator conservation (through tourism or trophy hunting) as well as from livestock farming. Moreover, preventing wild carnivores from killing livestock will only contribute to their conservation if alternative, wild, prey are available to them. For species such as wild dogs and cheetahs, which can survive in areas of comparatively low prey density (Durant 1998; Mills and Gorman 1997), the maintenance of traditional livestock husbandry (perhaps combined with control of domestic dog diseases) may be sufficient to allow populations to persist in pastoralist community lands (Woodroffe et al. 2005a). In contrast, evidence suggests that lions may require higher densities of wild prey (Frank et al. 2005), indicating that improvements to livestock husbandry will not allow lion populations to recover in community lands in the absence of measures to promote conservation of their prey.

Acknowledgements We would like to thanks Leonard Chenguli, Barnabas Ekaran, Nteere Gitonga, Martin Kahindi, Morias Kisio, Daniel Lekilit, Peter Lekitasharan, Ntirango Lekodei, Lkekui Lelesit, Leiyer Lempusia, John Katarin Lenanyankera, Saidimu Lengarawet, Lekoria Lengolooni, John Lesepe, Sinyati Lesowapir, Sammy Letepetaa, Julius Leturuka, Daniel Lomoe, Jackson Lorkitama, David Masere, Peter Mengich, Benson Minyarit, Joseph Musiany, Symon Ntunyoi, Julius ole Silikan and Andrew Stein for assistance in the field. This study was funded by Lincoln Park Zoo, the Wildlife Conservation Society, the African Wildlife Foundation and St Louis Zoo. This project is carried out under permission from the Kenya Ministry of Education, Science and Technology, through a collaboration with the Kenya Wildlife Service.

References

Angst C (2001) Procedure to selectively remove stock raiding lynx in Switzerland. Carnivore Damage Prevent News, 4:8

Bangs EE, Fontaine JA, Jimenez MD, Meier TJ, Bradley EH, Niemeyer CC, Smith DW, Mack CM, Asher V, Oakleaf JK (2005) Managing wolf–human conflict in the northwestern United States. In: Woodroffe R, Thirgood S, Rabinowitz AR (eds) People and wildlife: conflict or coexistence? Cambridge University Press, Cambridge, pp 340–356

Breitenmoser U, Angst C, Landry J-M, Breitenmoser-Würsten C, Linnell JDC, Weber J-M (2005) Non-lethal techniques for reducing depredation. In: Woodroffe R, Thirgood S, Rabinowitz AR (eds) People and wildlife: conflict or coexistence? Cambridge University Press, Cambridge, pp 49–71

Butynski TM (1974) An encounter between African wild dog and domestic dog. East Afr Wildl J 12:243

Cleaveland S, Dye C (1995) Maintenance of a microparasite infecting several host species: rabies in the Serengeti. Parasitology 111:S33–S47

Durant SM (1998) Competition refuges and coexistence: an example from Serengeti carnivores. J Anim Ecol 67:370–386

Frank LG, Woodroffe R, Ogada MO (2005) People and predators in Laikipia District, Kenya. In: Woodroffe R, Thirgood S, Rabinowitz AR (eds) People and wildlife – conflict or coexistence? Cambridge University Press, Cambridge, pp 286–304

Khaemba WM, Stein A, Rasch D, De Leeuw J, Georgiadis N (2001) Empirically simulated study to compare and validate sampling methods used in aerial surveys of wildlife populations. Afr J Ecol 39:374–382

Kruuk H (1980) The effects of large carnivores on livestock and animal husbandry in Marsabit District, Kenya, Rep. No. E-4. UNESCO, Nairobi

Laurenson MK, Sillero-Zubiri C, Thompson H, Shiferaw F, Thirgood S, Malcolm J (1998) Disease as a threat to endangered species: Ethiopian wolves, domestic dogs and canine pathogens. Anim Conserv 1:273–280

Laurenson MK, Cleaveland S, Artois M, Woodroffe R (2004) Canids and disease. In: Sillero-Zubiri C, Macdonald DW (eds) Wild canids: status survey and conservation action plan. IUCN, Gland, pp 246–256

Mills MGL, Gorman ML (1997) Factors affecting the density and distribution of wild dogs in the Kruger National Park. Conserv Biol 11:1397–1406

Mizutani F (1999) Biomass density of wild and domestic herbivores and carrying capacity on a working ranch in Laikipia District, Kenya. Afr J Ecol 37:226–240

Mizutani F, Muthiani E, Kristjanson P, Recke H (2005) Impact and value of wildlife in pastoral livestock production systems in Kenya: possibilities for healthy ecosystem conservation and livestock development for the poor. In: Osofsky SA, Cleaveland S, Karesh WB, Kock MD, Nyhus PJ, Starr L, Yang A (eds) Conservation and development interventions at the wildlife/livestock interface: implications for wildlife, livestock and human health. IUCN, Gland, Switzerland

Musiani M, Mamo C, Boitani L, Callaghan C, Gates CC, Mattei L, Visalberghi E, Breck S, Volpi G (2003) Wolf depredation trends and the use of fladry barriers to protect livestock in western North America. Conserv Biol 17:1538–1547

Ogada MO, Woodroffe R, Oguge N, Frank LG (2003) Limiting depredation by African carnivores: the role of livestock husbandry. Conserv Biol 17:1521–1530

Okello BD, O'Connor TG, Young TP (2001) Growth, biomass estimates, and charcoal production of *Acacia drepanolobium* in Laikipia, Kenya. For Ecol Manage 142:143–153

Roelke-Parker ME, Munson L, Packer C, Kock R, Cleaveland S, Carpenter M, O'Brien SJ, Popischil A, Hofmann-Lehmann R, Lutz H, Mwamengele GLM, Mgasa MN, Machange GA, Summers BA, Appel MJG (1996) A canine distemper virus epidemic in Serengeti lions (*Panthera leo*). Nature 379:441–445

Romañach SS, Lindsey PA, Woodroffe R. Determinants of attitudes toward predators in central Kenya, and suggestions for increasing tolerance in a livestock dominated landscape (in review)

Thirgood S, Woodroffe R, Rabinowitz AR (2005) The impact of human–wildlife conflict on human lives and livelihoods. In: Woodroffe R, Thirgood S, Rabinowitz AR (eds) People and wildlife: conflict or coexistence? Cambridge University Press, Cambridge, pp 13–26

Woodroffe R, Frank LG (2005) Lethal control of African lions (*Panthera leo*): local and regional population impacts. Anim Conserv 8:91–98

Woodroffe R, Lindsey PA, Romañach SS, Stein A, ole Ranah SMK (2005a) Livestock predation by endangered African wild dogs (*Lycaon pictus*) in northern Kenya. Biol Conserv 124:225–234

Woodroffe R, Thirgood S, Rabinowitz AR (2005b) The impact of human–wildlife conflict on natural systems. In: Woodroffe R, Thirgood S, Rabinowitz AR (eds) People and wildlife: conflict or coexistence? Cambridge UniversityPress, Cambridge, pp 1–12

Biodiversity and Conservation (2007) 16:1261–1281
DOI 10.1007/s10531-006-6735-z

Comparison of funding and demand for the conservation of the charismatic koala with those for the critically endangered wombat *Lasiorhinus krefftii*

CLEM TISDELL* and HEMANATH SWARNA NANTHA
*School of Economics, The University of Queensland, Brisbane 4072, Australia; *Author for correspondence (e-mail: c.tisdell@economics.uq.ezdu.au; phone: +61-7-33656570; fax: +61-7-33657299)*

Received 3 May 2005; accepted in revised form 5 January 2006

Key words: Charismatic fauna, Conservation demand, Conservation funding, Contingent valuation, Endangerment, Koala, *Lasiorhinus krefftii*, Northern hairy-nosed wombat, *Phascolarctos cinereus*

Abstract. This study contrasts the actual conservation spending and the Australian public's demand for conservation funding for two Australian mammal species, the koala and the northern hairy-nosed wombat. It involves a survey of 204 members of the Australian public. Willingness to fund conservation action to protect the northern hairy-nosed wombat was found to be higher than that for the koala despite the koala's immense popularity. The critically endangered status of the northern-hairy nosed wombat and the more secure conservation status of the koala is a factor likely to have influenced the comparative willingness-to-pay decisions. Actual annual conservation expenditure for both species is lower than the estimated aggregate willingness-to-pay for their conservation. Furthermore, conservation funding for the koala is much more than that for the northern hairy-nosed wombat even though the estimated public willingness-to-pay (demand) for funding koala conservation was less than for this wombat species. Reasons for this are suggested. They may also help to explain misalignment between demand for conservation funding of other species involving differences in charisma and endangerment.

Abbreviation: WTP – Willingness to pay

Introduction

There is evidence that charismatic wildlife species are likely to obtain more public funding for efforts to conserve them than less charismatic ones if both are equally endangered, or even if the latter are more endangered (Metrick and Weitzman 1996, 1998; Naidoo and Adamowicz 2001). Nonetheless, it is still unclear how closely public funding of conservation efforts for different species reflects the public's comparative support for those efforts. It is, for example, possible that this actual comparative public support does not closely reflect public demand because pure public goods or mixed goods[1] are involved, and complex non-market mechanisms, such as political mechanisms, determine actual allocations. There may be 'excessive' public allocation of funds for conservation efforts for charismatic species compared to what the public actually demands. The purpose of this article is to explore this possibility, taking the charismatic koala *Phascolartos cinerus* and the critically endangered northern hairy-nosed wombat *Lasiorhinus krefftii* as comparative cases.

The public's stated willingness to donate funds to support efforts for the conservation of wildlife species appears to depend mainly on the extent of the public's knowledge of the species, their likeability and the perceived degree to which they are endangered (*cf.* Samples et al. 1986; Tkac 1998). The likeability of the species may depend on factors such as their charismatic nature, size, whether they are human-like and so on (Metrick and Weitzman 1996). Lorenz (1970), Gould (1980), Kellert (1980), Plous (1993) and Gunnthorsdottir (2001) indicate that more human-like or physically attractive species are liked more and regarded with greater affection than those that are not so, and are likely to receive more public support.

The koala is charismatic, has a face without a muzzle, eyes placed forward rather than on the side of its head and frequently assumes an upright posture in trees. It is, therefore, considered to be quite humanoid (Lee and Martin 1988; Martin and Handasyde 1999). While the wombat has some charisma and humanoid features, these are less pronounced than in the case of the koala. Therefore, given the above theories, one might expect koalas to be liked more than wombats. Whereas the northern hairy-nosed wombat is critically endangered (IUCN 2004), the koala is considered to be at low risk of extinction; the IUCN (2004) classifies the koala as being at "least concern/near threatened". If likeability happened to be the dominant influence on the public's support for funding the conservation of species, we would expect the public to give greater support to the funding of the conservation of the koala than to the hairy-nosed wombat. On the other hand, if endangerment is the dominant influence, we would expect the reverse preference.

In these circumstances, does the stated willingness of the Queensland public to contribute funds for conservation of the koala exceed that for the northern hairy-nosed wombat or is the opposite the case? To what extent does the actual distribution of funds for conserving koalas and northern hairy-nosed wombats accord with the comparative stated willingness of members of the Queensland public to contribute funds to aid their conservation? Both these questions are considered here. In addition, possible reasons for a discrepancy, should there be a discrepancy, will be considered.

---

[1]A pure public good can be described as a good that is nonexclusive (it is impossible to exclude everyone/anyone from enjoying the good) and nonrival (the enjoyment of the good by one person does not diminish another's enjoyment of the good) (*cf.* Tisdell 1982, p. 406; 2005, Ch. 3). The impossibility of or impracticality of exclusion means that the good cannot be marketed. In the case of mixed goods, partial exclusion may be possible. So limited marketing of these can occur.

Experimental surveys are used to gather evidence on the likeability of the focal species, the stated willingness of survey participants to contribute funds for the conservation of species and the impacts of information provision on these factors. Information provision increased the knowledge of participants about the focal species, including their conservation status.

After briefly providing some general background on the focal species, the methodology is outlined. The willingness to pay (demand) results from the experimental survey are then reported. This is followed by an estimate of actual Queensland's expenditures on the conservation of koalas and hairy-nosed wombats and these are compared with the estimated willingness-to-pay figures. Discrepancies are observed and then discussed and possible limitations of the study are outlined.

Brief observations on the focal species

The koala is a tufty-eared, pale-grey/grey-brown marsupial that inhabits sclerophyll forests and woodlands from north Queensland down to New South Wales and Victoria (Martin and Handasyde 1995; Menkhorst and Knight 2004). It is arboreal and dependent primarily on eucalyptus leaves for food (Hindell and Lee 1990). This species is being threatened in some localities by the recession of its habitat due to land clearing (Maxwell et al. 1996; Melzer et al. 2000, p. 623). Overall, it is considered to have a low risk of extinction (IUCN 2004). Regionally, the conservation status of the koala in Queensland is 'vulnerable' in the south-east but 'common' elsewhere (Queensland Environmental Protection Agency 2003, p. 6), and the Victorian koala population is more secure than the New South Wales population, which is classified as vulnerable Jackson et al. 2003, p. 148).

Generations of Australian children have grown up developing an affinity for the koala as a result of their exposure to classic children's storybooks such as *Magic Pudding* (Lindsay 1936) and *Blinky Bill* (Wall 1939). They have also been exposed to the koala in children's television programmes such as Hanna Barbera's cartoon "Kwicky Koala Show". Due to their popularity, koalas are thought to be useful means for creating awareness and educating the public about wildlife conservation (Finnie 1990). They have recently been used in posters (displayed in the waiting rooms of doctors' surgeries) to encourage Australians to take precaution against sun exposure. The koala is a one of the world's most widely recognised and loved mammals (Cork et al. 2000). Koalas draw a great number of foreign tourists to Australia annually (Hundloe and Hamilton 1997). Tourists are able to see and interact with this creature in many wildlife parks such as the Featherdale Wildlife Park in Sydney and the Lone Pine Koala Sanctuary in Brisbane. They are also held in numerous zoos throughout the world (Jackson 2001; Lees and Johnson 2002). The koala is a prominent symbol used in tourism advertisements, magazine advertisements, on billboards and in promotions of products and services by businesses such as

Qantas Airlines (Martin and Handasyde 1999, p. 2) and is frequently the mascot of Australian teams at the Olympics (Phillips 1990, p. 6).

The northern hairy-nosed wombat is the largest and rarest of the three wombat species in Australia. The more abundant species are the common wombat *Vombatus ursinus*, found in south-eastern Australia, and the southern hairy-nosed wombat *Lasiorhinus latifrons*, occurring in South Australia and south-eastern Western Australia (Triggs 1996; Jackson 2003). The northern hairy-nosed wombat is a rotund, burrowing, herbivorous mammal with silky fur, pointy ears and a square muzzle (Horsup 1999; Menkhorst and Knight 2004). It is very secretive, nocturnal (McDonald and Norris 2001; Woodford 2002) and spends most of its time underground (Horsup 1999).

Historically, it used to inhabit native perennial grassland and open woodland areas in the semi-arid zone encompassing Queensland and New South Wales, but due to overgrazing by cattle and drought, its range declined and at present the species is limited to a 300-hectare area in Epping Forest National Park, located inland in central Queensland (between Townsville and Rockhampton) (Horsup 1999). Its population size was estimated to be 113 (Banks et al. 2003) but has recently dropped to 90 due to dingo predation and drought (Australian Geographic 2005; Alan Horsup, personal communication, 19th April 2005). It is classified as 'critically endangered' (IUCN 2004). The major threat to its existing population is its small size and single location, which makes it vulnerable to disease, to environmental changes and disturbances and to inbreeding (Horsup 1999).

Cultural factors have helped to endear the wombat to many Australians. The wombat in general has featured in popular children's books like the *Magic Pudding* and others (Lindsay 1936; Trinca and Argent 1987; French and Whatley 2003), and is the lead character in the Australian Broadcasting Corporation's (2005) radio serialisation of *The Muddle-Headed Wombat* stories by Ruth Park (1962). The wombat also features in adult literature, such as in the poems of Ogden Nash (Nash 1954), and in the drawings and verses of the 18th century Italian painter and poet, Dante Gabriel Rossetti (Archer 1965; McGann 2000).

Unlike the koala, the northern hairy-nosed wombat cannot be seen in zoos and wildlife parks nor do the public have access to it in the wild. At Epping Forest National Park, entry to its habitat is restricted to those trying to ensure its survival. This implies that currently the total economic value[2] of the two species consist of different components.

---

[2]An individual's total gain in wellbeing obtained from a change in policy is usually measured, in this case, by the individual's willingness to pay for that change (or to avoid it) (Bateman et al. 2002, p. 28). This willingness to pay, once aggregated across all individuals in a society, is often used to measure the total economic value of that change to society. Total economic value of an environmental good is sometimes expressed as follows: Total economic value = Use values + non-use values = [Direct (consumptive) use value + indirect (non-consumptive) use value + option value] + [existence value + bequest value + altruistic value] (see Pearce and Moran 1994, p. 12; see also Tisdell 2005, Ch. 3).

The total economic value of the northern hairy-nosed wombat consists only of its non-use (or passive) economic values, such as existence and bequest values. It currently has no use value but this could be a potential economic value in the future. Presently, it can be classified as a pure public good.[3] On the other hand, the koala has economic use as well as non-use values because Australians value its continuing existence in the wild. The current economic use value of the koala is non-consumptive and consists of viewing and photographing it in the wild, in zoos and in private wildlife parks. In the latter case, individuals are often photographed (for a fee) holding a koala. Koalas in private wildlife parks and zoos can be classified as virtually private economic goods because those who do not pay can be excluded from the establishment. Some rivalry occurs in use, e.g., only one person can hold a koala at a time for a photograph and crowding can reduce viewing opportunities. Because the use of the koala involves private good plus public good components, it can be regarded as a mixed economic good. In the wild, the koala is also a common property resource in many situations. For example, access to most national parks and state protected areas in Queensland where koalas may be seen is free, but use of these resources is subject to state (communal) regulations. Because the components of the economic value of these two species differ, this may (as taken up in the discussion section) affect the comparative amount of funds actually allocated for their conservation.

Methods

Survey methodology and questions

This study was conducted using two serial questionnaires, Survey I and Survey II. Drafts of these were pre-tested on a group of university students and were subsequently modified for greater clarity. A stratified sample of the public in Brisbane was obtained for the surveys by means of 1500 invitations letterbox-dropped in varied suburbs to acquire a representative socio-economic sample of the local population. The circulars informed potential respondents that the surveys would be about the use of Australia's tropical resources and that participants would be offered Aus$20, a presentation, refreshments and an opportunity to win Aus$200 in a draw. The precise aims of the study were withheld to avoid selection bias. After screening respondents to match the adult age (18 years old or more) and gender distribution of the Brisbane population, 204 participants were selected for the survey. Participants were divided into five groups of about equal size for the survey sessions held during the working week and at the weekend. This allowed participants flexibility in choosing a convenient time and helped to increase the representativeness of the sample.

[3]See Note 1.

At the beginning of the survey sessions, participants filled out questionnaire Survey I. This provided information on the participants' socio-economic background and their attitude towards the use and conservation of a number of Australian animal species of which the set of mammals is the only relevant one here. The koala and the northern hairy-nosed wombat were included in the mammal set. Once the forms in Survey I were completed, they were collected.

The nine species in the mammal set were selected to include species that all occur in tropical Australia (some of which it was assumed would be well known to Brisbane residents and some poorly known) and to include a mixture of species with varied conservation status. The selected species were the (1) dugong *Dugong dugon*, (2) eastern pebble-mound mouse *Pseudomys patrius*, (3) koala *Phascolarctos cinereus*, (4) mahogany glider *Petaurus gracilis*, (5) northern bettong *Bettongia tropica*, (6) northern hairy-nosed wombat *Lasiorhinus krefftii*, (7) northern quoll *Dasyurus hallucatus*, (8) red kangaroo *Macropus rufus* and (9) tree kangaroo *Dendrolagus lumholtzi*.

Participants were asked to self-rate their knowledge of all the species ("very good", "good", "poor", "non-existent"); how much they liked or disliked a species ("strongly like", "like", "dislike", "strongly dislike" or "uncertain of feelings towards species"); and whether they favoured the survival each of the species or not ("yes", "no" or "indifferent"). While these measures are subjective, they are able to rank comparatively species and highlight relative changes between Survey I and Survey II for the focal species. They act as ordinal indicators.

This procedure was followed by the presentation of willingness-to-pay questions. Three different indicators of willingness-to-pay for conservation of the focal species were used: (i) stated willingness to contribute money out of the participants' own pocket for funds (using the single-bid method) to conserve the koala and the northern hairy-nosed wombat, and reasons for the amounts stated if the amount is different between the species; (ii) the allocation of a windfall of Aus$1000 for conservation between nine mammal species; and (iii) comparative willingness to contribute a windfall of Aus$1000 for conservation between two wombat species, the northern hairy-nosed wombat and the southern (hairy-nosed) wombat and reasons for the allocations stated.

The exact questions corresponding to the above three willingness-to-pay indicators are stated in the results section. All involve variations on the contingent valuation method (*cf.* Champ et al. 2003, pp. 101–103). The first estimate involves a single-bid approach rather than iterative bidding. Bishop and Heberlein (1990) suggest that the single-bid approach tends to underestimate willingness to pay and may not be as accurate as iterative bidding. However, it is a less costly method than iterative bidding to apply and causes less fatigue to respondents. Since respondents were asked many questions in our survey, this was an important consideration in our choice.[4]

[4]Furthermore, perfect accuracy of estimates is not always needed for ideal policy choices nor even for perfect rationality Baumol and Quandt 1964; Tisdell 1996, Chs. 2 3). Even in the absence of perfect knowledge, rational conclusions can sometimes be drawn as will transpire in this case.

It is, however, well known that stated willingness to pay estimates are subject to several limitations. These include strategic bias, informational bias, hypothetical bias, and instrument bias. Strategic bias occurs, for example, when individuals deliberately inflate (or possibly deflate) their stated willingness to pay for, say, conservation of particular species or all species so as to influence conservation policy in their favour. The provision of information by interviewers could be slanted or defective and cause bias. The outcome may also be subject to income effects; persons on higher income are able to pay more to satisfy their preferences.

The two fixed-pie questions involving the allocation of windfall gains between the species provide further evidence about the relative support of respondents for funding conservation efforts of the different species. The formulation means that possible income effects on the willingness to pay are lessened and strategic bias is reduced (*cf.* Samples et al. 1986, p. 309; Gunnthorsdottir 2001, p. 207). The income influence is eliminated by the fixed pie method because all obtain the windfall gain which can only be allocated to funding conservation of the species listed. Strategic bias is reduced because the total funds for conservation are fixed and in allocating more funds to one species, the respondent must reduce allocations to others, that is must make trade-offs.

Upon completing Survey I and handing in their survey forms, participants were given a tea break. Participants then attended a presentation by the Queensland Museum's senior Curator of Vertebrates about Australia's tropical wildlife. After his presentation, participants were each given a booklet containing information on each of the focal species of the study. The information consisted of descriptions and coloured photographs of the species, their geographic distributions, life histories and conservation status. The information provided for each species was approximately of the same amount, and was kept brief and descriptive. The participants were requested to take the booklet home and read it before filling out the second questionnaire, Survey II, and returning this questionnaire in the postage pre-paid envelopes provided. Survey II contained the same set of questions as those described earlier for Survey I. The purpose of Survey II was to quantify possible changes resulting from information provision in participants' attitudes towards the species and their willingness to pay to conserve them compared to Survey I.[5]

Data analysis of survey data

The percentage of participants who stated their knowledge of the koala and the northern hairy-nosed wombat was 'very good' or 'good' was calculated in both surveys. Indices of how much a species is liked (likeability) were constructed by

[5]Note that completed Survey I forms were collected prior to participants being given forms for Survey II. This was to help ensure that responses were given independently.

assigning weights to each attribute of the Likert scale (2 for "strongly like", 1 for "like", 0 for "uncertain of feelings towards species", −1 for "dislike" and −2 "for strongly dislike") and by averaging the input of all participants for each species. The percentages of participants who answered "yes" to the question of whether they favoured the survival of the species were also calculated.

The single-bid willingness-to-pay (WTP) values given by each participant for both the koala and the northern hairy-nosed wombat were averaged. Care was taken with zero bids. The mean WTP values obtained for both species in Survey I and Survey II were compared using the Wilcoxon test for paired samples (Zar 1999). This statistical procedure was also used to compare the change in the mean WTP for each species between surveys. This non-parametric test was used because the distribution of the WTP values from the sample population appears to be non-normal. The mean allocation of funds for the koala and the northern hairy-nosed wombat (obtained from the allocation exercise involving all the mammal species in the survey) were also compared using the Wilcoxon test. The same statistical test was also applied in comparing the mean allocations of funds between the two wombat species (the northern hairy-nosed wombat and the southern wombat). Reasons given by participants for their stated single-bid WTP and their allocation of funds were reviewed.

The aggregate willingness to pay for the conservation of the koala and the northern hairy-nosed wombat in Queensland was then calculated as described by Bateman et al. (2002, Ch. 9) and as performed by Bateman et al. 2000) and Tisdell et al. (2005). This involved extrapolating survey participants' mean WTP for the koala and the wombat to Queensland's adult population. A necessary assumption for use of this method is that the sample is reasonably representative of the population. Our sample was chosen so as to be reasonably representative of Brisbane's population so we provide estimates of aggregate willingness to pay for this population. The wider the population considered, e.g., in this case the entire Australian population, the less accurate is the aggregate WTP likely to be due to the 'distance decay' effect (Bateman et al. 2002, p. 333).

Method of estimating actual expenditure on conservation of koalas and northern hairy-nosed wombats

Estimates of the willingness of individuals to pay for programs for the conservation of the koala and the northern hairy-nosed wombat are compared with estimates of actual payments for these programs in Queensland. Estimates for the koala are minimum values but for the northern hairy-nosed wombat are close to actual values. They are obtained by using secondary data, information provided by personal communications and various assumptions outlined later.

Results

Knowledge, likeability and support for survival

In Survey I, more than three-quarters of participants stated that they have very good or good knowledge of the koala whereas only a third of the participants claimed to have similar levels of knowledge about the northern hairy-nosed wombat (Table 1). After information provision, the number of participants who stated that they have very good or good knowledge of the northern hairy-nosed wombat increased by 70% but this still accounted for only a little more than half of all participants. The koala remained the better-known species by far. Participants also, on average, found the koala more likeable than the wombat in both surveys (Table 1). In Survey I, 98% and 96.1% of participants were strongly in favour of the survival of the koala and northern hairy-nosed wombat, respectively. In Survey II, it was 96.6% and 95.6% respectively.

Stated willingness to contribute to conservation funding from own funds to conserve the northern hairy-nosed wombat and the koala

The following questions were asked:

> "Assume that the government or a conservation body is raising money to save the northern hairy-nosed wombat. What is the maximum amount you would be willing to pay per week to conduct research, protect and conserve (such as by buying land) this species for the next ten years? Aus$ .... a week
>
> Now assume that instead of the northern hairy-nosed wombat, funds are being raised to conduct research, protect and conserve the koalas. What is the maximum amount you would be willing to pay per week for the conservation of koalas for the next ten years? Aus$ .... a week"

Results indicate that participants were willing to pay more to conserve the northern hairy-nosed wombat on average than to conserve the koala in both surveys (before and after participants were provided information about the species) despite the fact that the koala is liked more than the northern hairy-

Table 1. Percentages of participants stating that they have 'very good' or 'good' levels of knowledge of the species[a] and their estimated likeability indices.

	Knowledge level (% very good or good)		Likeability indices	
	Northern hairy-nosed wombat	Koala	Northern hairy-nosed wombat	Koala
Survey I	33	79	1.27	1.53
Survey II	56	83	1.20	1.42

[a]The remainder have poor or no knowledge of the species.

nosed wombat (Table 2). The differences in the mean WTP between the species are statistically significant at the 99% confidence level in both surveys. The rise in mean WTP of participants for the northern hairy-nosed wombat between surveys is large and statistically significant whereas the change in the mean WTP for the koala is small and not statistically significant. While for most participants the WTP amounts were equal for both species, the proportion of participants willing to pay relatively more for the wombat than for the koala was large and increased in Survey II (Table 3).

The allocations received by the koala and the northern hairy-nosed wombat from a hypothetical fund to conserve the nine mammals species studied

In order to obtain further evidence about whether the participants' demand for conservation efforts to support the northern hairy-nosed wombat are greater than for the koala, the following proposition was put to the participants:

> "Suppose that you are given Aus$1,000. This time you can only donate it to organisations in Australia to help conserve mammals in Australia,

Table 2. The northern hairy-nosed wombat and the koala: mean stated WTP for their conservation per week.

	Mean WTP ($)		Significance of difference between species (W, p)
	Northern hairy-nosed wombat	Koala	
Survey I (n = 185)	1.73	1.40	1409, <0.01**
Survey II (n = 192)	1.94	1.45	3512, <0.01**
Significance of difference between surveys (W, p)	−1208, 0.02*	454, 0.43	

Statistical significance: **99% confidence level, *95% confidence level.

Table 3. Distributions in the sample of WTP for the conservation of northern hairy-nosed wombat and the koala.

	Survey I No. (and as a % of all participants)	Survey II No. (and as a % of all participants)
Participants who gave equally to both species	117 (63.2%)	98 (51.0%)
Participants who gave more for the wombat	55 (29.7%)	83 (43.2%)
Participants who gave more for the koala	13 (7.0%)	12 (6.3%)
Participants who did not answer	19 (10.3%)	12 (6.3%)
Total responding participants	185	192

including marsupials, in the list below. What percentage of it would you allocate for the conservation of each of the mammals listed below? Your total should add up to 100%."

In Survey I, while the mean allocation for the koala was less than for the northern hairy-nosed wombat, the difference was not statistically significant

Animals (Mammals)	(%)
Tree kangaroos	
Red kangaroos	
Koalas	
Mahogany gliders (similar to the squirrel glider)	
Dugongs (a sea cow, not related to seals or whales)	
Northern quolls (a native marsupial cat)	
Northern bettongs (a small kangaroo-like marsupial)	
Northern hairy-nosed wombats (two related species are found in southern Australia)	
Eastern pebble-mound mice (four related species are found in Australia)	100

(Table 4). In Survey II, however, there was a statistically significant rise (at the 95% confidence level) in the mean allocation for this wombat, whereas the mean allocation for the koala fell significantly (at the 99% confidence level). The difference between the mean allocation of funds for the koala and the wombat in Survey II is statistically significant at the 99% confidence level. The change is probably mainly a result of respondents obtaining more information about the conservation status of each of these species. Note that if all nine mammal species in the survey received equal allocation, then each would receive an average 11.1% of the funds. In Survey II, the allocation for the northern hairy-nosed wombat is 29% higher and the allocation for the koala is 16% lower than that 11.1% average.

Table 4. Mean stated percentage allocations of funds for conserving nine mammal species: percentage allocated to the northern hairy-nosed wombat and the koala.

	Mean allocation (%)		Significance of difference between species (W, p)
	Northern hairy-nosed wombat	Koala	
Survey I (n = 159)	13.0	12.4	7, 0.99
Survey II (n = 166)	14.3	9.3	2892, <0.01**
Significance of difference between surveys (W, p)	−1379, 0.03*	3106, <0.01**	

Statistical significance: **99% confidence level, *95% confidence level.

The northern hairy-nosed wombat and the southern wombat: percentage allocation and distribution of funds for their conservation

In order to check whether support for programs to conserve the northern hairy-nosed wombat was not merely a consequence of it being a wombat, participants were asked the following:

> "You are given Aus $1,000 that you can allocate to conserve the northern hairy-nosed wombat or the southern wombat (this species is common). What percentage would you allocate to each?
>
> Northern hairy-nosed wombat ....%
>
> Southern wombat ....%"

When asked to allocate funds between these two, morphologically similar wombat species, the results shown in Table 5 were obtained. The northern hairy-nosed wombat was on average allocated a significantly greater proportion of the funds than the southern wombat in both surveys. Following provision of information about the species, the mean allocation for the northern hairy-nosed wombat rose significantly in Survey II at the expense of the allocation for the southern wombat.

More than two-thirds of the participants gave greater allocation of the funds to the northern hairy-nosed wombat, less than a quarter allocated equally between the species, and none allocated more funds to the southern wombat in Survey II (Table 6). Hence, support for conservation efforts for the northern hairy-nosed wombat significantly exceeds that for the southern wombat which is not under threat of extinction at present.

Aggregate WTP for projects to conserve the koala and the northern hairy-nosed wombat and actual expenditure on these

Table 7 provides point estimates of the aggregate (annual) WTP of Brisbane adult residents, the Queensland adult population and the Australian adult population for conservation projects for the koala and the northern hairy-nosed wombat. These are obtained by the benefit transfer method Bateman et al. 2002). This is also sometimes called the simple transferring point estimate approach and has been used, for example, by Hadker et al. (1997), Loomis and Ekstrand (1998) and Loomis et al. (2000). Table 7 is based on the results in Table 2, times 52 (weeks in a year) and multiplied in each instance by the estimates of the Australian Bureau of Statistics (2002, 2004a, b) of the adult populations involved. Note that these figures are annual sums and are based upon the stated willingness of respondents to contribute weekly to the conservation of the species for the next 10 years. The results from Survey I correspond to the situation in which the public is provided with no extra information about the species beyond what they already have whereas the aggregates corresponding to Survey II assume that the public is provided with the extra information given to survey participants in this experiment.

Table 5. Northern hairy-nosed wombat and southern wombat: average percentage allocation of funds for conservation of the species in Survey I and Survey II.

	Mean allocation (%)		Significance of difference between species (W, p)
	Northern hairy-nosed wombat	Southern wombat	
Survey I (n = 177)	70.2	29.2	5585, <0.01*
Survey II (n = 189)	78.1	22.0	10010, <0.01*
Significance of difference between surveys (W, p)	−3223, <0.01*	3051, <0.01*	

*Significant at the 99% confidence level.

Table 6. Distribution of allocation of funds for conservation of the northern hairy-nosed wombat and southern wombat.

	Survey I No. (and as a percentage of all participants)	Survey II No. (and as a percentage of all participants)
Participants who allocated more for the n. wombat	105 (51.5%)	142 (69.6%)
Participants who allocated equally	70 (34.3%)	47 (23.0%)
Participants who allocated more for the s. wombat	2 (1.0%)	0 (0.0%)
Participants who did not answer	27 (13.2%)	15 (7.4%)
Total responding participants	177	189

Table 7. Estimates in millions of Australian dollars of aggregate annual willingness to contribute funds for conservation projects for the koala and the northern hairy-nosed wombat by different constituencies based on aggregated benefits in Table 2 times 52.

Population, survey and species	Brisbane (in millions)	Queensland (in millions)	Australia (in millions)
Adult population size (millions)	1.20	2.92	15.31
Aggregate WTP ($) (millions)			
Survey I			
Koala	87.36	212.68	1114.36
Northern hairy-nosed wombat	108.16	262.60	1378.00
Survey II			
Koala	90.48	219.96	1154.40
Northern hairy-nosed wombat	121.16	292.76	1544.40

Sources of population estimates: Australian Bureau of Statistics (2002, 2004a, b). Adult population defined as those 18 years old and older.

The estimates are likely to be of decreasing accuracy the larger the constituencies. Those for Brisbane should be the most accurate and those for Australia the least. If one wants to reflect the actual demand of the public for those conservation programs, results from Survey I are probably the most appropriate. However, if one is concerned about the demand of a better informed public, then those for Survey II are more relevant. A better informed public increases its relative demand for projects to conserve the more threatened species, namely the northern hairy-nosed wombat.

Estimating actual expenditure on each of the two focal species is difficult, especially for the koala. This is so because data from the Australian Koala Foundation on its allocations, for example, are very limited. Nevertheless, it is possible to obtain a rough minimum estimate for Queensland for the koala and a relatively precise estimate for the northern hairy-nosed wombat. These estimates and their bases are given in Table 8.

Comparing Tables 7 and 8, a large discrepancy is apparent between (i) the estimated demand of adult Queenslanders for conservation funding of the koala and the northern hairy-nosed wombat and (ii) the comparative allocation of funds for conservation of these species. Whereas the Queensland

Table 8. Estimated recent (approximately 2003–2004) annual expenditure on conservation projects for the koala and the northern hairy-nosed wombat in Queensland in Australian dollars.

Funding source	Koala ($)	Northern hairy-nosed wombat ($)
Queensland state government	≈700,000[a]	149,625[b]
Public donations/funds from NGOs	83,333[c]	51,377[d]
Research grants	130,000[e]	106,800[f]
Funds for community-level conservation initiatives	13,467[g]	0
Total annual expenditure	926,800	307,802

[a]See Queensland Environmental Protection Agency (2003, p. 14).
[b]Financial year 2003–2004 (Alan Horsup, Queensland Parks and Wildlife Service, Coordinator of the northern hairy-nosed wombat recovery program, pers. comm.).
[c]Average annual contribution of the Australian Koala Foundation till end of 2004 would be approximately Aus$250,000 (the organisation claims to have allocated Aus$2 million to koala research and conservation projects since 1986) (Australian Koala Foundation, undated). This figure is for funds raised for koalas Australia-wide, not for Queensland alone. Therefore, funds potentially allocated annually for each of the three states where koalas mainly occur would be, on average, a third of the Aus$250,000 sum. Note that there are many sources of funds from the public and from non-governmental institutions (local and overseas). The San Diego Zoo in the United States, for example, loans its koalas to other zoos and part of the funds obtained from these loans are donated to koala habitat conservation in Australia (Zoological Society of San Diego 2005).
[d]From donations, financial year 2003–2004 (Alan Horsup, personal communication).
[e]Average annual funds from Rio Tinto Coal Australia to The University of Queensland's Koala Study Program for a three-year research and management program in central Queensland to help conserve koalas (The University of Queensland 2005).
[f]From the Federal Government, financial year 2003–2004 (Alan Horsup, personal communication).
[g]Average annual grants from The National Heritage Trust's ENVIROFUND (Commonwealth Government) for community-based local conservation projects. Over the past 3 years, funding for community koala conservation projects in Queensland totalled Aus$40,402.

public's demand for conservation spending on the northern hairy-nosed wombat exceeds that for the koala, actual annual spending in Queensland on conserving the northern hairy-nosed wombat is a third of that spent on the koala. A significant imbalance is present. Overall, conservation spending on each of these species falls far short of what appears to be demanded by the Queensland public. There may, however, be some strategic bias present. The WTP figures may overstate what respondents are really willing to pay. Nonetheless, the results in Table 4 (the case in which respondents allocate a fixed sum of money for the conservation of species) also supports the view that demand for expenditure on conserving the northern hairy-nosed wombat at least equals that for expenditure on projects to conserve the koala. In fact, when respondents are better informed about the conservation status of the species, demand for conservation projects for the former exceeds that for the koala. Note that the amount of funding koala conservation receives exceeds that apparent from the Queensland data. Both New South Wales (NSW National Parks and Wildlife Service 2003) and Victoria have projects to conserve the koala too but not the northern hairy-nosed wombat because it does not occur in these states.

Discussion and conclusions

The koala, a well-known iconic species, is at present not as threatened as the less well-known northern hairy-nosed wombat which is critically endangered. Our sample of Brisbane respondents said that they liked both species but the koala was liked most on average. This may be because the koala is more humanoid in appearance than the northern hairy-nosed wombat and has been given greater cultural coverage in Australian than the wombat. Nevertheless, our Brisbane respondents stated that they are willing to contribute more towards projects to conserve the northern hairy-nosed wombat than the koala, particularly so when they were better informed about the relative conservation status of the species. Aggregate actual funding for the conservation of both species is also lower than the aggregate WTP for their conservation even if only the Brisbane population is considered.

Differences in the respondents' stated likeability of the species did not appear to be the major influence on the stated willingness of respondents to contribute funds for projects to conserve these species. The major influence seems in this case to be differences in respondents' perception of the degree of endangerment of the species. In turn, this reflects the degree of urgency of conservation actions demanded (*cf.* Bandara and Tisdell, 2005). While likeability plays a role in influencing the public's demand for projects to conserve species, the degree of endangerment of the species appears to be the over-riding influence in many cases, as in this case (*cf.* Tkac 1998; Tisdell et al. in press). On the whole, this appears to be a rational approach to biodiversity conservation, although there are signs that some individuals will support projects to conserve a species when it is known that the species cannot be saved (Samples et al. 1986; DeKay and McClelland 1996), which does not seem rational unless there is uncertainty about whether the species can be saved. Nevertheless, this action can reflect a deeply held moral commitment.

Despite the above, public demand for conservation projects for different wildlife species is not always well reflected in actual aggregate expenditures on these projects because wildlife species are often pure public economic goods or mixed economic goods. This leaves scope for free-riding and political influences on the funding of wildlife conservation. In this case study, the northern hairy-nosed wombat is a pure public good whereas the koala is a mixed economic good. Actual aggregate funding in Queensland for koala conservation was found to be greater than that for the northern hairy-nosed wombat even though public demand is more in favour of this wombat.

Reasons for this imbalance could be the following:

(1) The koala has private and public good components whereas the northern hairy-nosed wombat is a pure public good. Private beneficiaries from the existence of the koala, such as wildlife parks and zoos, and tourism bodies that indirectly benefit from its existence may have an incentive to lobby governments to contribute to its conservation. As can be seen from Table 8, government funds account for the lion's share of funds for the conservation of these focal species. No private appropriation of economic benefits results from the existence of the northern hairy-nosed wombat.

(2) This point is partly related to (1). The koala is regarded as a significant international tourism attraction for Australia (Hundloe and Hamilton 1997). Hundloe and Hamilton (1997) estimated that the contribution of koalas to the revenue of the Australian tourism industry in 1996 was Aus$1.1 billion. By contrast, the northern hairy-nosed wombat would make no contribution to the Australian tourism industry. This industry, would, therefore, have an incentive to lobby in favour of government funding for conservation of the koala rather than the northern hairy-nosed wombat.

(3) Wildlife parks and zoos that utilise koalas may be anxious to purchase 'moral respectability' by donating to koala conservation, informing visitors of this and also providing an opportunity for visitors to contribute. This may help to counteract critics who object to koalas being kept in captivity, and who especially oppose koalas being handled by the public.

(4) The fact that the koala is more widespread and better known than the northern hairy-nosed wombat may also favour conservation support for the former.

(5) The koala has some well established NGOs, such as the Australian Koala Foundation, to campaign for its conservation whereas this is not so for the northern hairy-nosed wombat. This may indicate that NGOs can more easily obtain funds to support conservation initiatives for the koala.

The theory of public goods indicates that these goods are likely be unsupplied or undersupplied compared with the demand for them and this is also likely to be so for a mixed good that contains a large public good component. The results reported in Tables 7 and 8 accord with this theory. In Queensland, government policy has failed to compensate for the undersupply of the conservation effort for the two focal species compared to the estimated demand for this effort. However, whether demand is as high as estimated is unclear. For example, strategic bias may be present; respondents may have exaggerated their willingness to pay. Nevertheless, the gap between conservation support and estimated demand is very large. Hence, gross overstatement would be required for the above mentioned result not to hold. (6) As pointed out by one of the referees, it is also possible that the Queensland government believes that it is acting in accordance with public preferences. However, not having surveyed public opinion, it has no way of knowing if it is right or wrong.

Furthermore, it is necessary to consider limitations of the methods used. The results are based on a sample. While care was taken in the selection of the sample, bias is always possible. It would be desirable to draw further and larger samples to see if the results are consistent with those reported here. The type of indicators used to take account of knowledge and likeability of the species are more qualitative or ordinal in nature than cardinal but this suffices for comparative purposes. Further refinement of these measures would be useful. In addition, contingent valuation methods are subject to several limitations of which strategic bias and embedding are just a couple (Bateman et al. 2002, Ch. 8).

However, for the purpose of this particular exercise, exact valuations are unnecessary – considerable variations in the valuation estimates would still be consistent with the main observation that *relative* funding for koala conservation far exceeds that for the northern hairy-nosed wombat even though public demand does not support such a large disparity. According to Buchanan (1996), the koala is in a privileged conservation position – funding for its conservation exceeds that of most of the other 210 mammal, bird, amphibian and reptile species listed in the Australian Federal Government's *Endangered Species Protection Act 1992*. But this may not reflect the demand or preferences of the general public according to the results reported here. It is even less likely to accord with preferences of some ecologists, for example, May (2002), who expressed concern about preferences for the conservation of 'cute' species. He says too much conservation effort is aimed at what the "heart engages", the "furries and featheries" and charismatic megafauna.[6] Similar types of imbalances in

[6]Given this, Clark and May (2002) think that such bias could may not be so bad after all: public support for charismatic species like the koala could have "trickle-down benefits for less charismatic species". Other scientists argue that the conservation of the koala can confer protection to a larger number of naturally co-occurring species (the umbrella species concept) (*cf.*[6] Caro 2003; Roberge and Angelstam 2004). However, Stein et al. (2002) argue that giving excessive focus to charismatic organisms to raise public support and conservation funds leads to the underrepresentation of a vast majority of species that are at a much greater risk of extinction and that require more conservation attention. Also, conservation based purely on the umbrella species approach, especially if based on just a single species, raises questions such as whether the resulting protected area fulfils the principles of comprehensiveness, adequacy and representativeness of species and ecosystems.

conservation funding to that observed here may also occur for other species for similar reasons to those suggested in this case. Other cases are worthy of investigation.

Acknowledgement

We gratefully acknowledge the support of the Australian Research Council Discovery grant for this research on the economics of conserving Australia's tropical wildlife. We thank Clevo Wilson for his involvement in the project. We also thank Dr. Alan Horsup and Dr. Frank Carrick for providing some information useful in preparing this article.

References

Archer M. 1965. Rossetti and the Wombat. Apollo, March.

Australian Broadcasting Corporation. 2005. The Muddle-Headed Wombat. Available from: http://shop.abc.net.au/browse/product.asp?productid = 236970 [Accessed 18 February 2005].

Australian Bureau of Statistics. 2002. 2001 Census Basic Community Profile and Snapshot: 305 Brisbane (Statistical Division). Available from: http://www.abs.gov.au/ausstats/abs%40census.nsf/ddc9b4f92657325cca256c3e000bdbaf/6714103d8b354507ca256bbf0000dbf9!OpenDocument [Accessed 18 April 2005].

Australian Bureau of Statistics. 2004a. Population by age and sex, Australian states and territories: Table 3. Estimated Resident Population By Single Year Of Age, Queensland (June 2004). Available from: http://www.abs.gov.au/Ausstats/abs%40.nsf/lookupresponses/08eb60bce5c600 17ca25688d00098cc0?opendocument [Accessed 18 April 2005].

Australian Bureau of Statistics. 2004b. Population by age and sex, Australian states and territories: Table 9. Estimated Resident Population By Single Year Of Age, Australia (June 2004). Available from: http://www.abs.gov.au/Ausstats/abs%40.nsf/lookupresponses/31cb0c9f795059d1ca25688 d00099381?opendocument [Accessed 18 April 2005].

Australian Geographic. 2005. Wombat numbers drop. Australian Geographic 78: 12.

Australian Koala Foundation. Undated. Profile - AKF. Available from: http://www.savethekoala.com/mediaprofakf.html [Accessed 15 February 2005].

Bandara R. and Tisdell C.A. 2005. Changing abundance of elephants and willingness to pay for their conservation. J. Environ. Manage. 76: 47–59.

Banks S.C., Hoyle S.D., Horsup A., Sunnucks P. and Taylor A.C. 2003. Demographic monitoring of an entire species (the northern hairy-nosed wombat, *Lasiorhinus krefftii*) by genetic analysis of non-invasively collected material. Anim. Conserv. 6: 101–107.

Bateman I.J., Carson R., Day B., Hanemann M., Hanley N., Hett T., Jones-Lee M., Loomes G., Mourato S., Özdemiroglu E., Pearce D.W., Sugden R. and Swanson J. 2002. Economic Valuation with Stated Preference Techniques: A Manual. Edward Elgar, Cheltenham, UK.

Bateman I.J., Langford I.H., Nishikawa N. and Lake I. 2000. The Axford debate revisited: a case study illustrating different approaches to the aggregation of benefits data. J. Environ. Planning Manage. 43: 291–302.

Baumol W.J. and Quandt R.E. 1964. Rules of thumb and optimally imperfect decisions. Am. Econ. Rev. 54: 23–46.

Bishop R.C. and Heberlein T.A. 1990. The contingent valuation method. In: Johnson R.L. and Johnson G.V. (eds), Economic Valuation of Natural Resources: Issues, Theory and Applications. Westview, Boulder, Colorado, pp. 81–104.

Buchanan R. 1996. Survival of the Cutest. The Politics of being an Endangered Species. The Age, April 27, Extra 1–2.
Caro T.M. 2003. Umbrella species: critique and lessons from East Africa. Animal Conservation 6: 171–181.
Champ P.A., Boyle K.J. and Brown T.C. (eds) 2003. A Primer on Nonmarket Valuation. Kluwer Academic Publishers, Dordrecht, The Netherlands.
Clark J.A. and May R.M. 2002. Taxonomic bias in conservation research. Science 297: 191–192.
Cork S.J., Clark T.W. and Mazur N. 2000. Introduction: an interdisciplinary effort for koala conservation. Conserv. Biol. 14: 606–609.
DeKay M.L. and McClelland G.H. 1996. Probability and utility components of endangered species preservation programs. J. Exp. Psychol. Appl. 2: 60–83.
Finnie T. 1990. The role of zoos in the conservation of koalas. In: Lunney D., Urquhart C.A. and Reed P. (eds), Koala Summit – Managing Koalas in New South Wales. New South Wales National Parks and Wildlife Service, Hurstville, Sydney, pp. 109.
French J. and Whatley B. 2003. Diary of a Wombat. Clarion Books, New York.
Gould S.J. 1980. A Biological Homage to Mickey Mouse. Penguin Books, Hammondsworth, Middlesex, United Kingdom.
Gunnthorsdottir A. 2001. Physical attractiveness of an animal species as a decision factor for its preservation. Anthrozoös 14: 204–216.
Hadker N., Sharma S., David A. and Muraleedharan T.R. 1997. Willingness to pay for Boriuli National Park: evidence form contingent valuation. Ecol. Econ. 21: 105–122.
Hindell M.A. and Lee A.K. 1990. Tree preferences of the koala. In: Lee A.K., Handasyde K.A. and Sanson G.D. (eds), Biology of the Koala. Surrey Beatty & Sons, Sydney, pp. 117–121.
Horsup A. 1999. Recovery Plan for the Northern Hairy-nosed Wombat (*Lasiorhinus krefftii*) 1998–2002. Queensland Parks and Wildlife Service, Brisbane, Queensland.
Hundloe T. and Hamilton C. 1997. Koala and Tourism: An Economic Valuation. Discussion Paper No. 13. The Australia Institute, Lyneham, Australian Capital Territory.
IUCN. 2004. IUCN Red List of Endangered Species. Available from: http://redlist.org [Accessed 10 February 2005].
Jackson S.M. 2001. Koalas. In: Bell C. (ed.), Encyclopedia of the World's Zoo. Fitzroy Dearborn, Chicago, pp. 687–690.
Jackson S. 2003. Wombats. In: Jackson S. (ed.), Australian Mammals: Biology and Captive Management. CSIRO Publishing, Collingwood, Victoria, Australia, pp. 183–204.
Jackson S., Reid K., Spittal D. and Romer L. 2003. Koalas. In: Jackson S. (ed.), Australian Mammals: Biology and Captive Management. CSIRO Publishing, Collingwood, Victoria, Australia, pp. 145–182.
Kellert S.R. 1980. American attitudes toward and knowledge of animals: an update. Int. J. Study Anim. Problems 1. 87–119.
Lee A.K. and Martin R. 1988. The Koala – A Natural History. New South Wales University Press, Sydney.
Lees C. and Johnson K. 2002. Australasian Species Management Program: Regional Census and Plan. 12th ed. Australasian Regional Association of Zoological Parks and Aquaria, Sydney.
Lindsay N. 1936. The Magic Pudding: Being the Adventures of Bunyip Bluegum and his Friends Bill Barnacle and Sam Sawnoff. Farrar & Rinehart, New York.
Loomis J.B. and Ekstrand E. 1998. Alternative approaches for incorporating respondent uncertainty when estimating willingness to pay: the case of the Mexican spotted owl. Ecol. Econ. 27: 29–41.
Loomis J.B., Kent P., Strange L., Fausch K. and Covich A. 2000. Measuring the total economic value of restoring ecosystem service in an impaired river basin from a contingent valuation survey. Ecol. Econ. 33: 103–117.
Lorenz K. 1970. Studies in Animal and Human Behavior, Vol. 2 [translated from German by Robert Martin]. Methuen & Co. Ltd, London.
Martin R. and Handasyde K. 1995. Family Phascolarctidae: koala. In: Strahan R. (ed.), The Mammals of Australia. Reed Books, Chatswood, New South Wales, pp. 195–198.

Martin R. and Handasyde K. 1999. The Koala: Natural History, Conservation and Management. UNSW Press, Sydney.
Maxwell S., Burbidge A.A. and Morris K. 1996. The 1996 Action Plan for Australian Marsupials and Monotremes. Project 500, for the IUCN/SSC Australasian Marsupial and Monotreme Specialist Group of the World Conservation Union/Species Survival Commission. Endangered Species Program, Environment Australia, Canberra.
May R.M. 2002. The future of biological diversity in a crowded world. Curr. Sci. 82: 1325–1331.
McDonald D. and Norris S. (eds) 2001. The New Encyclopedia of Mammals. Oxford, Oxford University Press.
McGann J.J. (ed.) 2000. The Complete Writings and Pictures of Dante Gabriel Rossetti: A Hypermedia Research Archive. Institute for Advanced Technology in the Humanities, University of Virginia, Charlottesville, Virginia. Available from: http://www.rossettiarchive.org/ [Accessed 10 February 2005].
Melzer A., Carrick F., Menkhorst P., Lunney D. and St. John B. 2000. Overview, critical assessment, and conservation implications of koala distribution and abundance. Conserv. Biol. 14: 619–628.
Menkhorst P. and Knight F. 2004. A Field Guide to the Mammals of Australia, 2nd ed. Oxford University Press, South Melbourne, Australia.
Metrick A. and Weitzman M.L. 1996. Patterns of behaviour in endangered species preservation. Land Econ. 72: 1–16.
Metrick A. and Weitzman M.L. 1998. Conflicts and choices in biodiversity preservation. J. Econ. Persp. 12: 21–34.
Naidoo R. and Adamowicz W.L. 2001. Effects of economic prosperity on numbers of threatened species. Conserv. Biol. 15: 1021–1029.
Nash O. 1954. Many Long Years Ago. Dent, London.
NSW National Parks and Wildlife Service 2003. Recovery Plan for the Koala (*Phascolarctos cinereus*) (Draft for Public Comment). NSW National Parks and Wildlife Service, Hurstville, NSW. Available from: http://www.nationalparks.nsw.gov.au/npws.nsf/Content/Koala+-+draft+recovery+plan [Accessed 11 February 2005].
Park R. 1962. The Muddle-headed Wombat. Educational Press, Sydney.
Pearce D.W. and Moran D. 1994. Economic Value of Biodiversity. Earthscan Publications Ltd, London.
Phillips B. 1990. Koalas: The Little Australians We'd All Hate to Lose. Australian Government Publishing Service, Canberra.
Plous S. 1993. Psychological mechanisms in the human use of animals. J. Social Issues 49: 11–52.
Queensland Environmental Protection Agency. 2003. Public Supports Plan to Save Qld's Koalas. *EQ Newsletter*, Issue 21. Available from: http://www.epa.qld.gov.au/about_the_epa/public_reporting/epa_bulletin/ [Accessed 22 February 2005].
Roberge J. and Angelstam P. 2004. Usefulness of the umbrella species concept as a conservation tool. Conserv. Biol. 18: 76–85.
Samples K.C., Dixon J.A. and Gowen M.M. 1986. Information disclosure and endangered species valuation. Land Econ. 62: 306–312.
Stein B.A., Master L.L. and Morse L.E. 2002. Taxonomic bias and vulnerable species. Science 297: 1807.
The University of Queensland. 2005. UQ News Online: State-of-the-art Koala Venture. Available from: http://www.uq.edu.au/news/index.phtml?article=2227 [Accessed 28 February 2005].
Tisdell C.A. 1982. Microeconomics of Markets. John Wiley & Sons, New York.
Tisdell C.A. 1996. Bounded Rationality and Economic Evolution. Edward Elgar, Cheltenham, UK.
Tisdell C.A. 2005. Economics of Environmental Conservation, 2nd ed. Edward Elgar, Cheltenham, UK and Northampton, MA, USA.
Tisdell C., Swarna Nantha H. and Wilson C. (in press). Endangerment and likeability of wildlife species: how important are they for proposed payments for conservation. Ecol. Econ.

Tisdell C., Wilson C. and Swarna Nantha H. 2005. Policies for saving a rare Australian glider: economics and ecology. Biol. Conserv. 123: 237–248.

Tkac J. 1998. The effects of information on willingness-to-pay values of endangered species. Am. J. Agri. Econ. 80: 1214–1220.

Triggs B. 1996. The Wombat: Common Wombats in Australia. University of NSW Press, Sydney.

Trinca R. and Argent K. 1987. One Woolly Wombat. Kane/Miller Book Publishers, Brooklyn, New York.

Wall D. 1939. Complete Adventures of Blinky Bill: Containing Blinky Bill, Blinky Bill grows up, Blinky Bill and Nutsy. Angus and Robertson, Sydney.

Woodford J. 2002. The Secret Life of Wombats. Text Publishing, Melbourne, Australia.

Zar J.H. 1999. Biostatistical Analysis, 4th ed. Prentice Hall, Upper Saddle River, New Jersey.

Zoological Society of San Diego. 2005. Animal Bytes: Koalas. Available from: http://www.sandiegozoo.org/animalbytes/t-koala.html [Accessed 28 February 2005].

Biodivers Conserv (2007) 16:1283–1298
DOI 10.1007/s10531-007-9162-x

ORIGINAL PAPER

Nest site selection in middle and great spotted woodpeckers *Dendrocopos medius* & *D. major*: implications for forest management and conservation

Gilberto Pasinelli

Received: 2 January 2006 / Accepted: 13 February 2007 / Published online: 8 March 2007

Abstract Success of species conservation depends to a large extent on comprehensive management that considers all critical aspects of a species' niche. Many studies have examined habitat factors in relation to occurrence, abundance or foraging behaviour of European woodpecker species, while relatively little is known about nest site selection. I compared habitat structures used for nesting by middle and great spotted woodpeckers *Dendrocopos medius* and *D. major* with available structures in an oak forest in the Swiss lowlands. I first tested if nest trees were randomly selected among available trees by focusing on species, condition and diameter of nest trees, and on the presence of the fruiting body (hereafter sporophore) of polypores (wood-decomposing fungi). Second, I examined if the nesting niches of the two species were differentiated. Both species showed strong preferences for oaks, large trees, dead trees and for trees with sporophores. Nest sites of the two species differed most strongly with respect to the presence of sporophores, cavity age and tree condition, pointing towards interspecific competition for nest sites. Old living or dead trees with sporophores are central components of the nesting niche of middle and great spotted woodpeckers. Conservation plans for the threatened middle spotted woodpecker have so far mostly focused on the needs in terms of distribution and foraging; future conservation strategies and forest management must take into account the preference for dead and decaying trees with sporophores as another vital resource. This will also provide benefits for other woodpecker species as well as for the community of secondary cavity nesters.

Keywords Dead trees · Forestry · Nesting ecology · Niche · *Piciformes* · Resource selection

G. Pasinelli (✉)
University of Zurich, Institute of Zoology, Ecology Group, Winterthurerstrasse 190, 8057 Zurich, Switzerland
e-mail: gpasi@zool.unizh.ch

Introduction

The removal of old and dead trees is considered to be one of the most negative aspects of intensive forestry in remaining woodlands of temperate zones. Intensively managed forests are generally younger and structurally impoverished compared to not or only extensively managed forests (e.g. Bretz Guby and Dobbertin 1996; Scherzinger 1996; Hunter 1999). This decline of tree age and structural diversity has not remained without impact on biodiversity in forest ecosystems. For example, a large proportion of saproxylic (dead-wood dependent) insects are threatened today (e.g. Speight 1989; Mawdsley and Stork 1995), and communities of these insects are generally depauperated in forests harvested since centuries (Okland 1996; Nilsson and Baranowski 1997). Besides a deficit of dead wood, managed forests often lack characteristic structures of unmanaged forests, as for example 'legacy' trees, i.e. old-growth trees having been spared during harvest, or having survived stand-replacing natural disturbances (Franklin 1990 cit. in Mazurek and Zielinski 2004). Such legacy trees have been shown to be important structures for a variety of wildlife species (Mazurek and Zielinski 2004 and studies therein). Overall, 40% of forest wildlife species in Europe are estimated to be threatened or in danger of extinction (INSECTA and Zaric, 1995 cit. in Bretz Guby and Dobbertin 1996).

Primary cavity-nesters (i.e. species that excavate their own breeding cavity) are particularly vulnerable to intensive forestry, despite their ability to create suitable nest sites. Many of these species require specific habitat structures for excavating their nests (e.g. Raphael and White 1984; Harestad and Keisker 1989; Martin and Eadie 1999; Conner et al. 2001; Steeger and Dulisse 2003), but these structures often coincide with those that are most efficiently removed by forestry.

As in many other species, the selection of suitable nest sites is of central importance in the life history of woodpeckers. Despite this, relatively few studies have examined nest site selection in European woodpecker species, and still fewer have addressed use in relation to available habitat structures. A general assumption seems to be that knowing both habitat selection on the landscape level and foraging requirements within habitats is sufficient to protect and promote habitats of endangered species. However, such an approach is insufficient, and many examples exist that show how structures needed for nesting differ from those required for other activities (e.g. red-cockaded woodpecker *Picoides borealis*, Jackson 2003; cavity nesting waterfowl, Eadie and Gauthier 1985; Eadie et al. 1998). Thus, the success of species conservation is contingent upon comprehensive management plans that include detailed information on all critical aspects of a species' ecology.

A particularly striking example for the lack of information on nest site selection is the middle spotted woodpecker *Dendrocopos medius*. As a consequence of habitat destruction through intensive forestry, this species has suffered strong declines in the 20th century, resulting in highly fragmented populations throughout Europe (Pasinelli 2003). The middle spotted woodpecker belongs to the Annex I of the Birds directive of the European Union listing threatened species that require special conservation measures to be taken for their habitats. While many studies exist on habitat selection and foraging behaviour of the middle spotted woodpecker, knowledge on nest site selection in relation to available resources is very limited (reviewed by Pasinelli 2003; but see Kosinski and Winiecki 2004). A similar situation is found in the great spotted woodpecker *Dendrocopos major* (reviewed by Michalek and Miettinen 2003). This species, however, is far more numerous and widespread than the middle spotted woodpecker and not threatened so far.

Here, my first goal is to examine habitat structures used for nesting by middle and great spotted woodpeckers in a central European oak forest. Specifically, I test whether nest trees are a random sub-sample of available trees by focusing on species, condition and diameter of nest trees, and on the presence of fruiting bodies (or sporophores) of polypores (wood-decomposing fungi). This information will then be used to suggest guidelines for forest management.

A second goal is to compare the nest sites of middle spotted and great spotted woodpeckers, which allows insight into niche differentiation in terms of the nesting ecology of these species. The competitive exclusion principle (Gause's principle) states that two species that compete for the same resources cannot stably coexist (Gause 1934). Given that middle and great spotted woodpeckers are morphologically very similar and are often syntopic, many studies have focused on niche differentiation in terms of foraging behaviour and habitat use of these species to explain their coexistence (e.g. Jenni 1983; Lovaty 1985; Török 1990; Michalek 1994; Hertel 2003). Much less is known about niche differentiation with respect to nesting, although this may be at least as important as interspecific differences in foraging and habitat use for explaining coexistence.

Methods

Study site

From 1993–1999, nest site selection of the two woodpecker species was examined in the northeastern part of the Swiss lowlands. The study area (47°37'N, 8°37'E; 120 ha, 380 m a.s.l.) was situated in the Niderholz, an 800-ha forest, 35 km north of Zurich near the Rhine River. The Niderholz is one of the most important breeding areas of the middle spotted woodpecker in Switzerland, hosting 40–60 breeding pairs, although substantial declines have recently been noticed (Bühlmann et al. 2003). The Niderholz consists of an old oak–hornbeam forest managed for centuries as coppice-with-standards (Mittelwald), a forestry practice that results in a two-layered forest with large trees forming the canopy and small trees below the canopy. In the study area, the canopy is dominated by oak *Quercus* sp., with occasional Scots pine *Pinus sylvestris*, Norway spruce *Picea abies*, and Norway maple *Acer platanoides*, whereas the dominant tree species below the canopy are hornbeam *Carpinus betulus* and lime *Tilia* sp. The study area is surrounded by deciduous and coniferous forest of varying tree species composition and age classes, with some further oak–hornbeam forest stands (Pasinelli 2000a).

Over the course of the study, breeding densities of middle and great spotted woodpeckers averaged 0.83 and 2.03 pairs 10 ha^{-1}, respectively, in the study area. Besides the two target species of this study, four other woodpecker species regularly breed in the Niderholz, these being black woodpecker *Dryocopus martius*, grey-faced woodpecker *Picus canus*, green woodpecker *P. viridis* and lesser spotted woodpecker *Dendrocopos minor*.

Nest trees and available habitat

Nests were searched each year from mid-April to mid-June. This time span covers the usual nesting period of both woodpecker species in Central Europe (Michalek and Miettinen 2003; Pasinelli 2003). From 1993–1996, nests of the middle spotted woodpecker were found by following individuals carrying radio transmitters that had been attached to them during a project on the ecology of this species (for details see Pasinelli 2000a, b, 2001; Pasinelli et al. 2001). From 1997–1999, nests were searched by acoustically locating the chirping begging calls of young and/or by following parents carrying food in their beaks. Annually, 7–13 territories were searched for nests. In the great spotted woodpecker, some nests were detected during cavity construction, but most were found in the nestling stage by acoustically locating the loud chirping begging calls of young and/or by following parents carrying food in their beaks. Annually, 18–29 territories were searched for nests in this way.

Some nests may have gone undetected if the brood died before it could be found, particularly during the very wet breeding season of 1995 (cf. Pasinelli 2001). However, daily survival rates of nests and nesting success of both woodpecker species are very high (Glue and Boswell 1994; Smith 1997; Michalek and Miettinen 2003), so that the majority of nests probably survived long enough to be found.

For each active nest cavity, the following parameters were recorded: tree species, condition of the cavity tree (hereafter referred to as tree condition) and of the substrate containing the cavity (substrate condition) as either alive or dead, diameter at breast height (dbh; recorded after the young had left the cavity), substrate diameter at cavity entrance (estimated by comparing with the breadth of the woodpecker's dorsum), type of substrate containing the cavity (substrate type: trunk or limb), position of substrate containing the cavity (substrate position: vertical or inclined), height of cavity entrance above ground (cavity height, estimated visually), cavity age (freshly excavated before breeding season yes or no), presence or absence of one or more fruiting bodies (hereafter referred to as sporophores) of polypores (wood-decomposing fungi, with sporophores on the tree surface ≥3 cm in length or breadth) within 1 m of the cavity entrance (sporophore).

Data on available habitat were sampled in the winter months following breeding attempts on 286 circular plots of 0.03 ha, located at the intersections of an 150 m by 80 m grid placed randomly over the study area. The area sampled corresponded to 7.2% of the total study area. For every tree on each plot, species identity, dbh (≥8 cm), tree condition, and occurrence of one or more sporophores were recorded. These data were used to determine resource availability, which then was compared with nest tree characteristics.

Statistical analysis

In both woodpecker species, cavities are mainly constructed by males, which may live for several years in the breeding territories once selected (Michalek and Miettinen 2003; Pasinelli 2003). Some individuals may thus have contributed more than one observation to the data, but to retain as much information as possible, each nest was considered as one independent observation for two reasons. First, based on data of marked middle spotted woodpeckers, only a small fraction of individuals was present for more than one breeding season (three males and two females out of 26 marked individuals, with two males being present in three, the other individuals in

2 years). Second, intra-individual variation in nest site selection is presumably high (own observation), although detailed studies are lacking. Besides these potential intra-individual dependencies, cavity-related dependencies due to the use of the same cavity or nest tree by the birds in more than 1 year may arise. Of the 196 breeding cavities examined here (see Results), six cavities and 10 nest trees were used more than once by the great spotted woodpecker, the corresponding numbers in the middle spotted woodpecker being 0 cavities and four nest trees. However, because structural attributes related to a cavity do not necessarily remain constant over time, I considered the use of a cavity in a given year as independent from the use of the same cavity in another year. Structural attributes such as tree dbh or substrate diameter increase with time, while other attributes such as the presence of sporophores or cavity age can change between years. Note that results did not change if each tree and cavity was used only once (data not shown).

Following Manly et al. (1993), selection indices were used to see if one resource was preferred, avoided or used as expected based on its availability (see below). Selection indices were calculated by dividing the proportion of a used resource by its proportional availability. A selection index significantly greater than one indicated preference, while a value significantly smaller than one implied avoidance. A selection index was considered to be significantly different from one if the simultaneous 95% Bonferroni confidence interval calculated over all categories did not contain the value one (Manly et al. 1993), i.e. the null hypothesis of use being proportional to availability was rejected. To calculate the proportion of available trees, I only considered those trees with a dbh of at least 20 cm and 15 cm in middle spotted and great spotted woodpecker, respectively, since these are the minimum dbh of nest trees recorded for these two species so far (Michalek and Miettinen 2003; Pasinelli 2003). Consideration of all trees (≥8 cm) would have led to trivial results, since most trees had very small dbh and for that reason were unsuitable for cavity construction. It should be noted that the selection analyses were redone for the great spotted woodpecker by considering only those trees with a dbh of at least 20 cm as available (as in the middle spotted woodpecker), but since all results were identical to using 15 cm as the limit for available trees, I will not consider them further in the result section.

Nest sites of the two woodpecker species were compared with logistic regression (PROC LOGISTIC, SAS Institute Inc 1999–2001), using species as the categorical dependent variable. Categorical explanatory variables were tree condition, substrate condition, substrate type, substrate position, cavity age, and sporophore. Continuous explanatory variables were dbh of cavity tree, substrate diameter and cavity height. Results did not change if cavity height was treated as categorical variable, with height above ground assigned to one of six categories (<5 m, 5–10 m, 10–15 m, 15–20 m, 20–25 m, ≥25 m) (data not shown). Tree species was excluded from this analysis, since the majority of the cavities of both woodpecker species was built in oaks (see Results).

Results of multiple regression analysis may be compromised by strong correlations among the explanatory variables (Chatterjee and Price 1991; Menard 1995). Spearman rank correlations indicated two pairs of highly correlated explanatory variables ($r > 0.7$), these being tree condition with substrate condition ($r_s = 0.80$, $n = 107$) and substrate type with substrate position ($r_s = 0.78$, $n = 107$), respectively. I therefore analysed five models, starting with a global model containing all explanatory variables. Then, reduced models were analysed by including only one member of each of the two strongly related variable pairs along with all the other explanatory variables; by altering which of the highly related variables were included, four such reduced models could be analysed. For each model, I then calculated Akaike's Information Criterion (=AIC, Akaike 1973; Burnham and Anderson 1998) corrected for small sample sizes (AICc, Burnham and Anderson 1998). The model with the lowest AICc value is considered the most parsimonious one, i.e. explaining most of the variance with the fewest parameters (Burnham and Anderson 1998; Anderson et al. 2000; Burnham and Anderson 2001). Akaike model weights were calculated to determine the level of support for each of the five models by the data (Burnham and Anderson 1998). Model weights add up to 1 (by definition), with higher weights indicating better explanatory power. Effect sizes of parameters were calculated with model averaging (Burnham and Anderson 1998; Johnson and Omland 2004). Similarly, model-averaged standard errors (SE) were calculated for parameter estimates following Johnson and Omland (2004). All effect sizes in the results section refer to these weighted averages and SE. Finally, I checked the fit of each model with residual analysis (McCullagh and Nelder 1989).

Results

In total, 57 nest cavities of middle spotted woodpeckers were found, 52 of which had been built in oaks. In the great spotted woodpecker, 139 nest cavities were located, and 137 of them had been constructed in oaks. Given the prevalence of oaks as nest trees, I focused on availability and use of oaks versus all other tree species in the analyses of nest tree species selection.

Characteristics of nest trees

Both woodpecker species significantly preferred oaks as nest trees, while other tree species were avoided, as indicated by the selection indices and their associated confidence intervals (Table 1). Both woodpecker species most often built their nest cavity in living trees, which were much more abundant than dead trees. However, when comparing available and used trees, the selection indices indicated a strong preference for dead trees and an avoidance of living trees in the middle spotted woodpecker. Similarly, the great spotted woodpecker clearly preferred dead trees, but used living ones almost as expected from their availability (Table 1). Trees with large fruit bodies (sporophores) of polypores were very rare in the study area, but such trees were nevertheless strongly selected as nest trees in both woodpecker species.

Close inspection of Table 1 shows that most nest trees used by the middle spotted woodpecker were oaks and that most nest trees had sporophores. Thus, the positive selection of oaks and of trees with sporophores may not be independent of each other, since oaks (not least due to their age) may be more likely than other tree species to have sporophores. However, there was no such association between nest tree species and the presence of sporophores (Fisher's exact test $P > 0.99$, $n = 57$, Table 2). Similarly, in the great spotted woodpecker no significant association between nest tree species and the presence of sporophores was found ($P > 0.52$, $n = 139$, Table 2).

Table 1 Selection of nest trees in middle (msw) and great spotted woodpeckers (gsw)

	Tree species		Tree condition		Sporophore present	
	Oak	Other	Alive	Dead	Yes	No
MSW						
Availability[a]	263	445	704	4	27	681
%	37.15	62.85	99.44	0.56	3.81	96.19
Use	52	5	44	13	52	5
%	91.23	8.77	77.19	22.81	91.23	8.77
Selection index	2.46	0.14	0.78	40.37	23.92	0.09
95% CI	2.23–2.68	0.01–0.27	0.65–0.90	18.32–62.41	21.72–26.12	0.00–0.18
GSW						
Availability[b]	274	753	1020	7	27	1000
%	26.68	73.32	99.32	0.68	2.63	97.37
Use	137	2	130	9	43	96
%	98.56	1.44	93.53	6.47	30.94	69.06
Selection index	3.69	0.02	0.94	9.50	11.77	0.71
95% CI	3.61–3.78	–0.01–0.05	0.90–0.99	2.64–16.36	8.43–15.11	0.62–0.80

See text for calculation of selection indices. The variable 'sporophore present' refers to the occurrence of a sporophore of ≥ 3 cm length or breadth of one or more polyporous fungi

[a] based on trees ≥ 20 cm dbh

[b] based on trees ≥ 15 cm dbh

Table 2 Occurrence of sporophores in relation to species and condition of nest trees in middle (msw) and great spotted woodpeckers (gsw)

Woodpecker species	Sporophore present	Tree species		Tree condition	
		Oaks	Other	Alive	Dead
MSW	Yes	47	5	40	12
	No	5	0	4	1
GSW	Yes	42	1	40	3
	No	95	1	90	6

Along the same line of arguments, the positive selection of dead trees and of trees with sporophores may not be independent of each other, since dead trees may be more likely than living ones to have sporophores. However, I did not find any relation between the presence of sporophores and the condition of nest trees (Fisher's exact tests, middle spotted woodpecker: $P > 0.42$, $n = 57$; great spotted woodpecker: $P > 0.27$, $n = 139$; Table 2).

Nest trees of both woodpecker species had on average significantly larger dbhs than available trees (Mann–Whitney U-tests, $P < 0.001$ in both species, Fig. 1a). Since oaks were the most often used nest trees (see above), I redid the analyses by considering only oaks as used and available trees. Again, both woodpecker species selected on average larger trees for nesting compared to the available trees (Fig. 1b, Mann–Whitney U-tests, $P < 0.001$ in both species).

There was no relation between the presence of sporophores on oaks and dbh (logistic regression, estimate ± SE = 0.013 ± 0.015, $\chi^2_1 = 0.81$, $P > 0.36$, $n = 260$ oaks with dbh ≥ 20 cm; results identical for dbh ≥ 15 cm, data not shown). Because there were many more oaks without ($n = 234$) than with sporophores ($n = 26$), I

Fig. 1 Median diameter at breast height (dbh) of nest trees (filled bars) compared to available trees (empty bars). (**a**) Middle spotted woodpecker, (**b**) great spotted woodpecker; numbers inside bars indicate sample sizes. Bars indicate 25–75% quartile ranges. Trees used for nesting significantly larger than available trees in all cases (Mann–Whitney *U*-tests, $P < 0.001$)

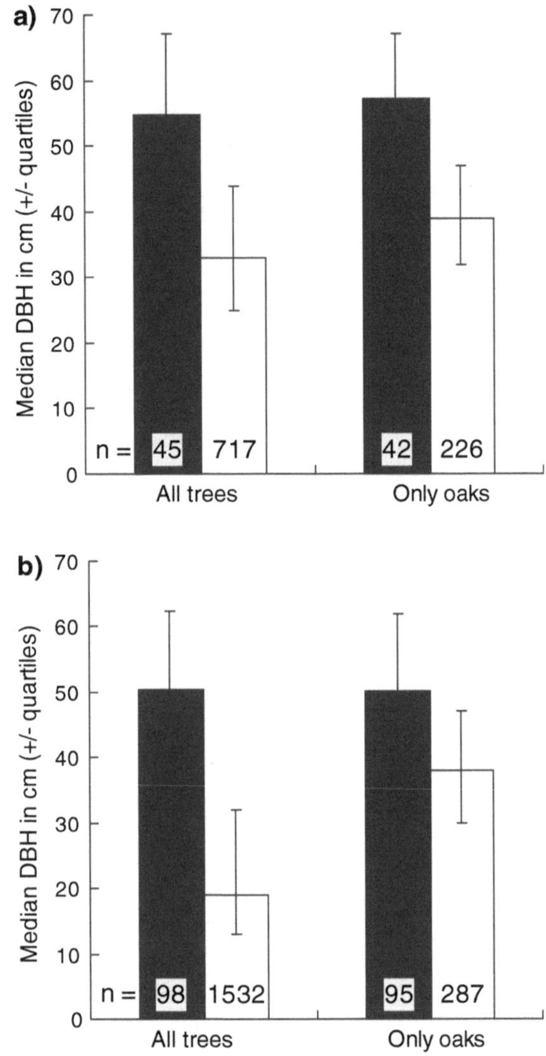

redid the analysis by randomly selecting 26 oaks without sporophores to be used in the logistic regression. Still, I did not find any significant relation between the presence of sporophores and the dbh of an oak (0.002 ± 0.022, $\chi^2_1 = 0.01$, $P > 0.93$, $n = 52$). Thus, the positive selection of large nest trees appeared to be independent of the presence of sporophores.

Comparison of nest sites of middle and great spotted woodpeckers

Of the five models examined (see Methods), two were well supported by the data, as indicated by both ΔAICc values and model weights (Table 3). These well-supported models 1 and 2 included the variable tree condition instead of substrate condition and further included either substrate type or substrate position along with all the other explanatory variables (dbh, substrate diameter, cavity height, cavity age,

Table 3 Differentiation of nest sites between middle and great spotted woodpeckers

Model	Variables in model	K	−2 Log L	AICc	ΔAICc	Weight
1	Dbh, substrate diameter, cavity height, tree condition, substrate type, cavity age, sporophore	8	74.18	−130.89	0.00	0.50
2	Dbh, substrate diameter, cavity height, tree condition, substrate position, cavity age, sporophore	8	73.95	−130.43	0.46	0.40

Modelled was the probability of a nest cavity having been used by the middle spotted woodpecker. Only models with ΔAICc ≤ 2 are shown. *K* = number of parameters estimated (i.e. variables in the model + intercept); −2 Log L = −2 * maximum likelihood estimate, derived from SAS; ΔAICc = difference in AICc to the best model; weight = relative support of a particular model compared to the other models, with higher values indicating better support. *N* = 107 cavities (28 middle spotted, 79 great spotted woodpecker).

sporophore, Table 3). Conversely, models 3 and 4 received little support (ΔAICc: 3.91 and 5.34, Akaike weights: 0.07 and 0.03, respectively). These models did not contain tree condition, but substrate condition in both cases, and either substrate type or substrate position along with the other explanatory variables. Together with model-averaged effect sizes and standard errors (SEs, see below), this suggests tree condition to be more important than substrate condition for differentiating between the nest sites of the two woodpecker species. Hardly any support was found for the global model with all the variables (Model 5, ΔAICc = 10.37, weight = 0.003).

Three variables had effect sizes larger than their SEs when considering model-averaged effect sizes and SEs. This suggests that the probability of a nest cavity having been built and used by the middle spotted woodpecker increased with the presence of sporophores next to the cavity (effect size ± S.E. = 1.339 ± 0.647) and decreased with cavities being old (−1.067 ± 0.565) and trees being alive (−0.794 ± 0.627). In other words, cavities of the middle spotted woodpecker were more often freshly excavated and found close to sporophores in dead trees than cavities of the great spotted woodpecker (Table 3, Fig. 2). Model averaged effect sizes ± SE of the other variables were 0.154 ± 0.505 for substrate diameter, −0.111 ± 0.220 for substrate condition, 0.094 ± 0.290 for cavity height, −0.068 ± 0.513 for substrate type, −0.106 ± 0.435 for substrate position, and −0.004 ± 0.166 for dbh

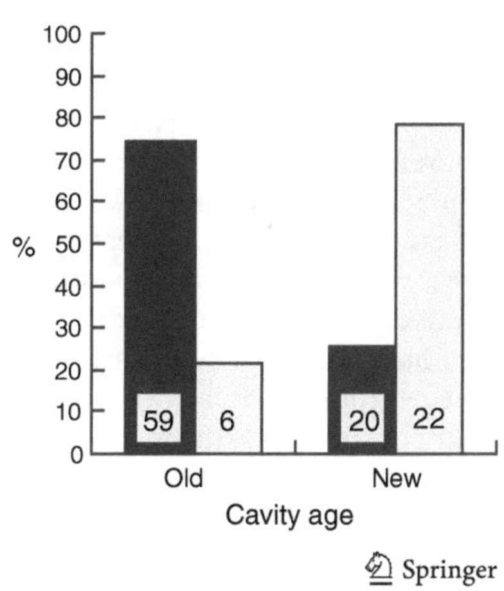

Fig. 2 Nest cavity age in great spotted woodpeckers (filled bars) and middle spotted woodpeckers (empty bars). Numbers within bars indicate sample size

of nest tree. Apparently, these variables contributed little to the separation of the two species' nest sites.

Discussion

Trees used for nesting by middle and great spotted woodpeckers were not a random sub-sample of available trees: both species showed preferences for oaks, for dead trees and for trees with large sporophores of polypores close to the cavity entrance. In addition, nest trees were significantly larger in diameter than available trees.

Middle spotted woodpecker

Very few studies have so far examined nest tree use in relation to availability in the middle spotted woodpecker (Pasinelli 2003). Only one study reported preference for oaks as nest trees in comparison with tree species availability (Mazgajski 1997), while another one found a preference for dead trees, but not for particular tree species (Kosinski and Winiecki 2004). Wesolowski and Tomialojc (1986) suggested selection of cavity tree species to reflect availability. One explanation for the preference of oaks reported here may be that oaks used as nest trees in my study area more often had sporophores than other tree species. However, I did not find such a relation, so the preference for oaks cannot be explained by an increased presence of sporophores. More likely, since the middle spotted woodpecker prefers to nest in large trees, as evidenced by the significantly larger dbh of nest trees compared to available trees in this study, oaks appear to be particularly attractive as nest trees because of their generally large dbh and hence age (also see Fauvel 2001; Kosinski and Winiecki 2004). Presumably, large trees are important resources for nesting in the middle spotted woodpecker, regardless of tree species. The middle spotted woodpecker has been shown to regularly use more than 20 tree species for nesting (Pasinelli 2003) and usually excavates its breeding cavities in large trees, averaging from 39.1 cm to 97.0 cm (Wesolowski 1989; Prill 1991; Mazgajski 1997; Fauvel 2001; Kossenko and Kaygorodova 2003; Kosinski and Winiecki 2004). Although most of these studies did not present data on available trees, these findings support the notion that large trees are important for cavity construction in this woodpecker species, whereas tree species per se is not.

Throughout its range, the middle spotted woodpecker excavates nest cavities in decaying or dead parts of living, decaying or dead trees (Pasinelli 2003), very often close to a sporophore of a polypore (e.g. *Fomes fomentarius*, *Laetiporus sulphureus, Phellinus* sp.), an old cavity, a limb hole or below a dead limb (Pettersson 1984; Dubreuil et al. 1998; Pasinelli 2000a; Fauvel 2001; Spitznagel 2001; Kosinski and Winiecki 2004). Fungus-infected wood is softer than healthy wood, which decreases the effort of excavating a cavity, and this may explain the positive selection for nest trees with sporophores found in this study. Furthermore, neck muscles and bill are relatively weak in the middle spotted woodpecker, hampering extensive excavation into hard, living substrates (Rüger 1972; Jenni 1981). This is also reflected by the species' foraging behaviour consisting mainly of probing into bark crevices rather than pecking for arthropods in the bark (Jenni 1983; Pettersson 1983; Pasinelli and Hegelbach 1997; Pasinelli 2000b). The preference for nest trees with sporophores was not a by-product of the positive selection of old oaks, because there was neither an association between trees with sporophores and tree species nor any relation between dbh of oaks and the presence of sporophores.

Besides trees with sporophores, dead trees were also preferred despite their striking rarity in the study forest (below 0.7% of all trees, Table 1). Dead trees have been reported as nest substrates earlier, although comparisons with availability are lacking (Pasinelli 2003; exception: Kosinski and Winiecki 2004). The soft wood of decaying parts of dead trees likely offer similar conditions for cavity excavation as living trees with polypores, not least because dead trees are probably always infected by fungi (even if no fruit body is visible). The prevalence of sporophores on both living and dead nest trees (Table 2) suggests, however, that condition of trees (alive or dead) used for nesting is of secondary importance with respect to nest site selection. This is supported by the lacking association between nest tree condition and the presence of sporophores (although the generality of this independence should be examined with data including more dead nest trees than in this study). Thus, the middle spotted woodpecker appears to use the sporophores of polypores, rather than tree condition, as a cue for selecting suitable sites for cavity excavation.

Great spotted woodpecker

The great spotted woodpecker is the most flexible *Dendrocopos* species in Europe, both in terms of its foraging and nesting ecology (Michalek and Miettinen 2003). This flexibility is reflected in this study: despite significant preferences for trees with sporophores and dead trees for nesting, great spotted woodpeckers regularly excavated their nests also in trees without sporophores and in living trees (cf. Table 1). Although this does not rule out the possibility that nest sites may have been built in places of rotten wood, my findings agree with previous studies indicating that the great spotted woodpecker is much less dependent 'on sites such as rotten, dead, broken, stump, under bracket fungus as other woodpeckers' (Michalek and Miettinen 2003; but see Smith 1997; Kosinski and Winiecki 2004). In terms of tree species, oaks were positively selected, but as in the middle spotted woodpecker, this selection likely reflects the preference for large trees rather than for oaks per se (Fig. 1). Similarly, Michalek and Miettinen (2003) found no evidence for associations of nest cavities with particular tree species. In contrast to my findings, Smith (1997) did not demonstrate any marked selection for nests to be in trees with large diameter, although the majority of cavities were in trees with 28 cm dbh or above. On the other hand, Smith (1997) found strong selection for dead trees or dead parts of trees as in this study, but depending on tree species. In contrast, Kosinski and Winiecki (2004) found positive selection for oak and alder, but not for dead trees. Overall, these contradictory results suggest that the great spotted woodpecker is much more able to adjust resource use for nesting to local availability than perhaps any other European woodpecker species.

Interspecific comparison

The comparison of nest site characteristics between the two woodpecker species revealed differentiation of their nesting niches mainly in terms of tree condition, the presence of sporophores and cavity age. Middle spotted woodpeckers more frequently nested in freshly excavated cavities close to sporophores of tree fungi in dead trees than did great spotted woodpeckers. In contrast to my study, the middle spotted woodpecker has been found to excavate its cavities higher up in trees (Wesolowski and Tomialojc 1986; Prill 1991; Günther 1993; but see Fauvel 2001), to use substrates with smaller diameters at cavity entrance (Günther 1993) and to use limbs more often (Fauvel 2001) than the great spotted woodpecker. The two woodpecker species differ in some morphological traits important for cavity excavation and cavity location, for example in bill size, neck muscles and body size (Rüger 1972; Jenni 1981), and therefore, some differences in nesting niches are to be expected (also see Kosinski and Winiecki 2004). Further, it is likely that habitat structure of forests differed among studies, which may explain differences in niche differentiation observed (although note that all studies were done in oak-dominated habitats comparable to the one in my study). Nevertheless, regardless of habitat, all studies suggested nesting niches of the two species to be differentiated, which may indicate the presence of interspecific competition for nest sites. If the nest niches were 'fixed' properties of each species, determined primarily by morphology, but not influenced by nest site selection of the other species, we would expect to find similar niche differentiation in all the studies, which is not the case.

Management implications

This study has shown that nest trees of middle and great spotted woodpeckers are large and thus old and show characteristics of trees usually found in primeval forests. Specifically, the presence of sporophores of polypores in living trees indicates rotten wood usually in advanced stage of decay, which dramatically lowers the economic value of the trees. Such trees, along with completely dead trees, are quickly removed by forestry to improve growing conditions for healthy and economically more promising trees, and to prevent infestation of neighbouring trees. Yet, the removal of such economically unimportant trees reduces the ecological quality of forests that are otherwise suitable for the highly specialised middle spotted woodpecker. This may reduce carrying capacities in those forests, because home range size of the middle spotted woodpecker is inversely related to the densities of potential cavity trees (i.e. trees with sporophores, limb holes, or old cavities) and old oaks (Pasinelli 2000a). In terms of management of middle spotted woodpecker habitats, oaks with large dbh and with sporophores as well as dead trees should be retained and can serve both as future cavity trees and as foraging trees. It is important to note though that densities of trees suitable for cavity construction and for foraging can be fairly high, averaging 26 potential cavity trees per ha in my study site (Pasinelli 2000a) and ranging from 15–80 large oaks (potential foraging trees) per ha (Pettersson 1984; Bühlmann and Pasinelli 1996; Coch 1997; Pasinelli 2000a; Michalek et al. 2001).

Retention of currently existing potential cavity trees may not suffice to meet current and future requirements of the middle spotted woodpecker. Because nest cavities are preferentially built in decaying wood, their re-usability is very poor. In fact, not one of the breeding cavities and only four (7%) of 57 nest trees in this study had been used more than once for nesting by the middle spotted woodpecker (own unpublished data). Forest management may therefore be required to actively recruit potential cavity trees, particularly in forests with low availability of such trees. Several methods have been proposed for producing decaying and dead trees, for example girdling (Aulén 1991) or topping of trees (Bull and Partridge 1986; Filip et al. 2004). Girdling may be most promising when applied at a tree height of 5–15 m above ground, where cavities of the middle spotted woodpecker are usually found

(Pasinelli 2003). Inoculating parts of living trees with (naturally occurring) wood-decaying fungi may also hold promise (Parks et al. 1995; but see Filip et al. 2004).

The use of artificial cavities has been a very successful management tool for the red-cockaded woodpecker in the south-eastern United States (Walters et al. 1992; Conner et al. 2001). Unlike the middle spotted woodpecker, however, red-cockaded woodpeckers continue to use the cavities (both natural and artificial ones) for several years (Conner et al. 2001), and thus, providing artificial cavities for middle spotted woodpeckers may be only of temporally limited use.

Increasing the availability of dying and dead trees may also be important to reduce impacts of potential interspecific competition for nest sites between the two *Dendrocopos* species studied. In addition, although the great spotted woodpecker is currently the most numerous and widespread woodpecker species throughout Europe, this study and others (reviewed by Michalek and Miettinen 2003) suggest that the species will benefit as well if old and dead trees are retained. Given the potential role of the great spotted woodpecker as a key stone species in managed forests, increasing the availability of old trees and of dying and dead trees may be the best way to also benefit the whole community of secondary cavity nesters. The importance of old and dead trees for biodiversity and forest ecosystem functioning is generally accepted (e.g. Harmon et al. 1986; Aulén 1991; McComb and Lindenmayer 1999), but the implications of this for forest management and conservation still lags behind.

Acknowledgements I am grateful to the many persons that helped finding woodpecker nests or sampling habitat data, the latter often on cold winter days. The manuscript has benefited from fruitful comments by K. Schiegg, B. Miranda, P. Pechacek, K. Michalek, and J.-M. Thiollay.

References

Akaike H (1973) Information theory and an extension of the maximum likelihood principle. In: Petrov BN, Csáki F (eds), Proc. 2nd Int. Symp. Information Theory. Akadémiai Kiado, Budapest pp 267–281

Anderson DR, Burnham KP, Thomson WL (2000) Null hypothesis testing: problems, prevalence, and an alternative. J Wildl Manage 64:912–923

Aulén G (1991) Increasing insect abundance by killing deciduous trees: a method of improving the food situation for endangered woodpeckers. Holarctic Ecol 14:68–80

Bretz Guby NA, Dobbertin M (1996) Quantitative estimates of coarse woody debris and standing dead trees in selected Swiss forests. Global Ecol Biogeogr 5:327–341

Bühlmann J, Müller W, Pasinelli G, Weggler M (2003) Entwicklung von Bestand und Verbreitung des Mittelspechts *Dendrocopos medius* 1978–2002 im Kanton Zürich: Analyse der Veränderungen und Folgerungen für den Artenschutz. Ornithol Beob 100:343–355

Bühlmann J, Pasinelli G (1996) Beeinflussen kleinflächige Waldnutzung und Wetter die Siedlungsdichte des Mittelspechts *Dendrocopos medius*? Ornithol Beob 93:267–276

Bull EL, Partridge AD (1986) Methods of killing trees for use by cavity nesters. Wildl Soc Bull 14:142–146

Burnham KP, Anderson DR (1998) Model selection and inference. A practical information-theoretic approach. Springer, New York

Burnham KP, Anderson DR (2001) Kullback–Leibler information as a basis for strong inference in ecological studies. Wildlife Res 28:111–119

Chatterjee S, Price B (1991) Regression analysis by example. Wiley, New York

Coch T (1997) Spechte (Gattung *Picoides*) und Strukturmerkmale als Wegweiser einer Eigenart bewahrenden Pflege und Entwicklung ehemaliger Mittelwälder. Ph.D. thesis, Albert-Ludwigs-Universität, Freiburg im Breisgau

Conner RN, Rudolph DC, Walters JR (2001) The red-cockaded woodpecker: surviving in a fire-maintained ecosystem. University of Texas Press, Austin

Dubreuil B, Pautz F, L'Huillier J (1998) Importance du phellin robuste *Phellinus robustus* Karst, champignon parasite du chêne, sur la nidification des oiseaux cavernicoles. Ciconia 22:25–32

Eadie J, Sherman P, Semel B (1998) Conspecific brood parasitism, population dynamics, and the conservation of cavity-nesting birds. In: Caro T (ed) Behavioral ecology and conservation biology. Oxford University Press, New York pp 306–340

Eadie JM, Gauthier G (1985) Prospecting for nest sites by cavity-nesting ducks of the genus *Bucephala*. Condor 87:528–534

Fauvel B (2001) Écologie du Pic mar *Dendrocopos medius* en Champagne (Est France). Alauda 69:87–101

Filip GM, Parks CG, Baker FA, Daniels SE (2004) Artificial inoculation of decay fungi into Douglas-Fir with rifle or shotgun to produce wildlife trees in western Oregon. West J Appl For 19:211–215

Gause GF (1934) The struggle for existence. Williams & Wilkins, Baltimore, MD

Glue DE, Boswell T (1994) Comparative nesting ecology of the three British breeding woodpeckers. British Birds 87:253–269

Günther E (1993) Zur Wahl des Höhlenstandortes von Bunt- und Mittelspecht (*Dendrocopos major* und *D. medius*) im nordöstlichen Harz (Sachsen-Anhalt). Ornithol Jber Mus Heineanum 11:67–73

Harestad AS, Keisker DG (1989) Nest tree use by primary cavity-nesting birds in south central British Columbia. Can J Zool 67:1067–1073

Harmon ME, Franklin JF, Swanson FJ, Sollins P, Gregory SV, Lattin JD, Anderson NH, Cline SP, Aumen NG, Sedell JR, Lienkaemper GW, Cromack KJ, Cummins KW (1986) Ecology of coarse woody debris in temperate ecosystems. Adv Ecol Res 15:133–302

Hertel F (2003) Habitatnutzung und Nahrungserwerb von Mittelspecht *Picoides medius* und Buntspecht *Picoides major* in bewirtschafteten und unbewirtschafteten Buchenwäldern des nordostdeutschen Tieflandes. In: Pechacek P, d'Oleire-Oltmanns W (eds) International woodpecker symposium. Nationalparkverwaltung Berchtesgaden, Berchtesgaden pp 69–80

Hunter ML Jr (1999) Maintaining biodiversity in forest ecosystems. Cambridge University Press, Cambridge UK

Jackson JA (2003) Niche concepts and woodpecker conservation: understanding why some species are endangered. In: Pechacek P, d'Oleire-Oltmanns W (eds) International woodpecker symposium. Nationalparkverwaltung Berchtesgaden, Berchtesgaden pp 87–96

Jenni L (1981) Das Skelettmuskelsystem des Halses von Buntspecht und Mittelspecht (*Dendrocopos major* und *medius*). J Ornithol 122:37–63

Jenni L (1983) Habitatnutzung, Nahrungserwerb und Nahrung von Mittel- und Buntspecht (*Dendrocopos medius* und *D. major*) sowie Bemerkungen zur Verbreitungsgeschichte des Mittelspechts. Ornithol Beob 80:29–57

Johnson JB, Omland KS (2004) Model selection in ecology and evolution. Trends Ecol Evol 19:101–108

Kosinski Z, Winiecki A (2004) Nest-site selection and niche partitioning among the great spotted woodpecker *Dendrocopos major* and middle spotted woodpecker *Dendrocopos medius* in riverine forest of Central Europe. Ornis Fennica 81:145–156

Kossenko SM, Kaygorodova EY (2003) Osobennosti ekologii srednego pyostrogo dyatla (*Dendrocopos medius*) v Desnyanskom Polesye [Ecological features of the Middle Spotted Woodpecker (*Dendrocopos medius*) in the Desna Polesye]. Ornithologia (Moscow) 30:94–103

Lovaty F (1985) Sur l'organisation spatiale du peuplement de pics épeiche, mars et épeichettes nichant dans une vielle futaie de chênes. Le Grand Duc 26: 33–36

Manly B, McDonald L, Thomas D (1993) Resource selection by animals. Chapman & Hall, London

Martin K, Eadie JM (1999) Nest webs: a community-wide approach to the management and conservation of cavity-nesting forest birds. For Ecol Manage 115:243–257

Mawdsley NA, Stork NE (1995) Species extinctions in insects: ecological and biogeographical considerations. In: Harrington R, Stork NE (eds), Insects in a changing environment. Academic Press, London pp 321–369

Mazgajski TD (1997) Zmiany liczebnosci i miejsca gniazdowania dzieciola duzego *Dendrocopos major* i sredniego *D. medius* w rezerwacie "Las Bielanski" w Warszawie [Changes in the numbers and nest sites of the great spotted woodpecker *Dendrocopos major* and the middle spotted woodpecker *D. medius* in the Las Bielanski reserve in Warsaw]. Ochrona Przyrody 54:155–160

Mazurek MJ, Zielinski WJ, (2004) Individual legacy trees influence vertebrate wildlife diversity in commercial forests. For Ecol Manage 193:321–334
McComb W, Lindenmayer D (1999) Dying, dead, and down trees. In: Hunter ML (ed) Maintaining biodiversity in forest ecosystems. Cambridge University Press, Cambridge pp 335–372
McCullagh P, Nelder JA (1989) Generalized linear models. Chapman & Hall, London New York
Menard S (1995) Applied logistic regression analysis. Sage University Paper, Thousand Oaks, California
Michalek KG (1994) Lokomotion und Nahrungserwerb bei Buntspecht (*Picoides major*) und Mittelspecht (*Picoides medius*). Master thesis, Universität Wien, Wien
Michalek KG, Auer JA, Großberger H, Schmalzer A, Winkler H (2001) Die Einflüsse von Lebensraum, Witterung und Waldbewirtschaftung auf die Brutdichte von Bunt- und Mittelspecht (*Picoides major* und *P. medius*) im Wienerwald. Abh. Ber. Mus. Heineanum 5 (Sonderheft): 31–58
Michalek KG, Miettinen J (2003) *Dendrocopos major* Great Spotted Woodpecker. BWP Update 5:101–184
Nilsson SG, Baranowski R (1997) Habitat predictability and occurrence of wood beetles in old-growth beech forests. Ecography 20:491–498
Okland B (1996) Unlogged forests: important sites for preserving the diversity of Mycetophilids (*Diptera*: *Sciaroidea*). Biol Conserv 76:297–310
Parks CG, Bull EL, Filip GM (1995) Using artifical inoculated decay fungi to create wildlife habitat. In: Aguirre-Bravo C, Eskew L, Vilas-Salas AB, Gonzalez-Vicente CE (eds), Partnerships for sustainable forest ecosystem management, USDA Forest Service General Technical Report RM-GTR-266, pp 175–177
Pasinelli G (2000a) Oaks (*Quercus* sp.) and only oaks? Relations between habitat structure and home range size of the middle spotted woodpecker (*Dendrocopos medius*). Biol Conserv 93:227–235
Pasinelli G (2000b) Sexual dimorphism and foraging niche partitioning in the Middle Spotted Woodpecker *Dendrocopos medius*. Ibis 142:635–644
Pasinelli G (2001) Breeding performance of the middle spotted woodpecker *Dendrocopos medius* in relation to weather and territory quality. Ardea 89:353–361
Pasinelli G (2003) *Dendrocopos medius* Middle Spotted Woodpecker. BWP Update 5:49–99
Pasinelli G, Hegelbach J (1997) Characteristics of trees preferred by foraging Middle Spotted Woodpecker *Dendrocopos medius* in northern Switzerland. Ardea 85:203–209
Pasinelli G, Hegelbach J, Reyer H-U (2001) Spacing behavior of the middle spotted woodpecker in Central Europe. J Wildl Manage 65:432–441
Pettersson B (1983) Foraging behaviour of the middle spotted woodpecker *Dendrocopos medius* in Sweden. Holarctic Ecol 6:263–269
Pettersson B (1984) Ecology of an isolated population of the middle spotted woodpecker (*Dendrocopos medius*, L.) in the extinction phase. Ph.D. thesis, Swedish University of Agricultural Sciences, Uppsala
Prill H (1991) Untersuchungen an Spechten und deren Bedeutung für andere höhlenbewohnende Vögel im Naturschutzgebiet Serrahn. Ornith Rundb Mecklbg-Vorpommern 34:52–65
Raphael MG, White M (1984) Use of snags by cavity-nesting birds in the Sierra Nevada. Wildl Monogr 86:1–66
Rüger A (1972) Funktionell-anatomische Untersuchungen an Spechten. Z wiss Zool 84:63–163
SAS Institute Inc 1999–2001. SAS Proprietary software release 8.02, Cary, North Carolina
Scherzinger W (1996) Naturschutz im Wald. Ulmer, Stuttgart
Smith KW (1997) Nest site selection of the Great spotted woodpecker *Dendrocopos major* in two oak woods in southern England and its implications for woodland management. Biol Conserv 80:283–288
Speight MCD (1989) Saproxylic invertebrates and their conservation. Nature and Environment Series No. 42, Strasbourg
Spitznagel A (2001) *Picoides medius* (Linnaeus 1758) Mittelspecht. In: Hölzinger J, Mahler U (eds) Die Vögel Baden-Württembergs. Ulmer, Stuttgart pp 436–464
Steeger C, Dulisse J (2003) Temporal use patterns of woodpecker nest trees in British Columbia. In: Pechacek P, d'Oleire-Oltmanns W (eds) International woodpecker symposium. Nationalparkverwaltung Berchtesgaden, Berchtesgaden pp 157–163
Török J (1990) Resource partitioning among three woodpecker species *Dendrocopos spp* during the breeding season. Holarctic Ecol 13:257–264
Walters JR, Copeyon CK, Carter JH III (1992) Test of the ecological basis of cooperative breeding in red-cockaded woodpeckers. Auk 109:90–97

Wesolowski T (1989) Nest-sites of hole-nesters in a primaeval temperate forest (Bialowieza National Park, Poland). Acta Ornithol 25:321–351
Wesolowski T, Tomialojc L (1986) The breeding ecology of woodpeckers in a temperate primaeval forest—preliminary data. Acta Ornithol 22:1–21

Biodivers Conserv (2007) 16:1299–1320
DOI 10.1007/s10531-006-9068-z

ORIGINAL PAPER

An evaluation of the contribution of cultivated allspice (*Pimenta Dioca*) to vertebrate biodiversity conservation in Nicaragua

David I. King · Martin D. Hernandez-Mayorga · Richard Trubey · Raul Raudales · John H. Rappole

Received: 12 April 2005 / Accepted: 19 May 2006 / Published online: 12 July 2006

Abstract Tropical deforestation has emerged as one of the most important conservation challenges of our time, both because of the high species diversity and rates of endemism of tropical forests, and because of the rapid rate at which this process is proceeding. Recent studies indicate that areas of low-intensity agroforestry have similar levels of vertebrate diversity as some primary habitats, leading some researchers and conservationists to conclude that this type of commodity production could contribute to the conservation of biodiversity. We compared the composition of bird, mammal and herpetofaunal communities in primary forest, secondary forest, and pasture—and within the allspice productive systems that have replaced pasture. We found that mammal species richness was higher in primary forest than all other habitats; however for resident and migrant birds, amphibians and reptiles, species richness was similar between primary forest and the other habitats. Despite similarities in overall numbers of species, there were numerous species that were encountered only in primary habitats. We conclude that the cultivation of allspice in a mixed productive system can offset some of the losses to biodiversity; however it should be complemented by the establishment and maintenance of protected areas to accommodate populations of primary forest specialists that are unable to persist in altered habitats.

D. I. King (✉)
USDA Forest Service Northeastern Research Station, University of Massachusetts, Amherst, MA 01003, USA
e-mail: dking@fs.fed.us

M. D. Hernandez-Mayorga
Martin Domingo Mayorga, Ecólogo, Cooperativa CoopeSiuna, Siuna, Nicaragua

R. Trubey · R. Raudales
Mesoamerican Development Institute, 669 Stevens Street, Lowell, MA 018514519, USA

J. H. Rappole
Smithsonian Conservation and Research Center, 1500 Remount Road, Front Royal, VA 22630, USA

Keywords Agroforestry · Allspice · Amphibians · Birds · Conservation · Mammals · Reptiles · Tropics

Introduction

Tropical deforestation has emerged as one of the most important conservation challenges of our time, both because of the high species diversity and rates of endemism of tropical forests, and because of the rapid rate at which this process is proceeding. Various initiatives have been forwarded to address this problem. Clearly, outright acquisition and strict protection is the most desirable approach from the perspective of biodiversity maintenance, although this solution is not feasible in many areas because it ignores political realities as well as the legitimate needs of the human population who are entitled to economic security and well-being (Kamaljit and Seidler 1998).

Some tropical habitats altered by low intensity cultivation of commercially valuable crops in combination with tree cover, or "agroforestry," support native tropical species. For example, numerous studies have reported that bird species richness in shade coffee is comparable to some natural habitats (Wunderle and Latta 1996; Greenberg et al. 1997a, b; Tejada-Cruz and Sutherland 2004). Similarly, cacao grown under diverse planted shade in Mexico appears to be good habitat for some migrant bird species (Greenberg et al. 2000; Reitsma et al. 2001). Hughes et al. (2002) found that "countryside habitats" were occupied by nearly 40% of native species detected in their sites in Costa Rica. These findings have led some to conclude that these habitats have an important role in augmenting habitat area, connectivity and range conditions represented by reserves (Daily et al. 2003).

Despite these findings, it is becoming clear that the propagation of these habitats cannot be a substitute for a program of rigorous habitat protection (Rappole et al. 2003a, b, Tejada-Cruz and Sutherland 2004). Although species diversity in some agroforestry systems is similar to primary forest, some forest residents are scarce or absent from cultivated habitats (Tejada-Cruz and Sutherland 2004), perhaps because of the absence of resources or substrates needed for Feeding and reproduction (Greenberg et al. 2000; Reitsma et al. 2001). The absence of these primary forest specialists from these managed forests point out the limitations of these systems, and thus the task has shifted to determining the role of altered forests in tropical biodiversity conservation.

For example, it is thought that the promotion of less intensive forms of cultivation might contribute to the conservation of some species, or might be useful as a buffer zone around existing preserves (Moguel and Toledo 1999; Tejada-Cruz and Sutherland 2004). Perhaps with knowledge of the relative value of various countryside habitats, biodiversity can be explicitly integrated into agricultural policy (Hughes et al. 2002; Daily et al. 2003).

In 1999, the Mesoamerican Development Institute (MDI) in cooperation with the Programa a Campesino a Campesino (PCAC), the Nicaraguan Government, and the World Bank/Global Environment Facility (GEF), initiated a program to develop and promote the cultivation of indigenous allspice (*Pimenta dioca*) on previously deforested parcels for the production of essential oils. Allspice, called locally "pimienta," is a native species in that region that occurs as a midstory tree, which is traditionally harvested by felling mature trees. Allspice is currently being cultivated by a cooperative of nearly 100 subsistence farmers (CoopeSiuna) in northern Nicaragua in the buffer zone of the largest remaining stand of tropical rain forest north of the Amazon basin—the BOSAWAS Biosphere Reserve (Smith, 2003). Within the buffer zone, which has been settled by ex-combatants from civil war, forest clearing for subsistence farming has increased along with extensive livestock operations, commercial logging, and mining. The estimated rate of deforestation in this region is 80,000 ha/year, or about 2.1% of remaining forest cover (Global Environment Facility (GEF) 1997). In an effort to develop an alternative to slash and burn agriculture and consequential advance of the agricultural frontier, which threatens the BOSAWAS, a program has been initiated through which indigenous allspice along with banana, cocoa, citrus, and other native shade trees have been introduced as a mixed-productive system on previously deforested parcels. These productive systems are to provide the raw products that will be processed to produce essential oils of allspice, lemongrass, vetiver, and ginger for a growing international market. Through this new agro-industrial activity, it is hoped that the cooperative members can break from the slash and burn cycle of subsistence farming and increase biodiversity within the buffer zone.

Allspice cultivated in mixed productive systems is characterized by a diverse, multistoried habitat similar to other types of crops known to support native biodiversity, such as coffee and cocoa. To determine the extent to which biodiversity conservation can be realized through the cultivation of allspice in a mixed productive system, we conducted an intensive, multi-taxon survey comparing species richness and composition of allspice plantations with that of primary forest, secondary forest and pasture.

Methods

Study area

The study took place in the region around Siuna, Nicaragua (13°40′ N, 85°50′ W) in areas consisting of tropical moist and wet forest between 170 and 600 m in elevation. The landscape in this area is a diverse mosaic of patches of remnant primary forest, degraded or regenerating secondary forest, "agroforestry systems" such as mixed plantings of coffee, cacao, citrus and native overstory trees, pasture and cereal crops. In recent years, allspice, which is native to the region and grows as a midstory tree in its natural state, has been cultivated as the principal plant in agroforestry systems for use in the production of essential oils. The earliest plantings took place in 1999, and new areas are currently being planted.

Vertebrate sampling

We sampled birds, mammals, amphibians and reptiles in several primary forest, secondary forest, allspice and pasture sites. Primary forest areas existed as large (>30 ha) patches of forest that had experienced no timber harvest, and were characterized by large, tall trees and open understory. Secondary forest had most of the original canopy removed, and had shorter, smaller trees and denser understory than primary forest. Allspice agroforestry systems consisted of allspice that had been planted in pasture with other commercially valuable species such as coffee, cacao, and citrus between 3 to 5 years before the commencement of the study. Allspice had shorter, smaller trees than secondary forest, but similar understory structure. Pasture consisted of grazed or recently grazed areas with grass or forb cover with scattered shrubs.

We sampled birds, mammals, reptiles and amphibians from July 2002 to April 2004. Mammals were sampled during the dry season (July–November) 2002 and 2003, and birds were sampled in the rainy season (January–April) 2003 and 2004. Sampling for birds was conducted at 23 sites, mammals at 27 sites, and herpetofauna at 19 sites. Although habitats were sampled in a random sequence within each season, it was not possible to randomize the selection of sites because of limited availability.

Bird species distribution and species composition were sampled using mist nets (Karr 1981). Ten 12 m × 3 m, 32 mm denier mist nets were deployed in each site 50 m apart in a grid pattern approximately 200 m × 250 m. Each site was sampled for 250 net hours. All birds captured were identified, and then marked to distinguish them from new captures subsequently by cutting the tip of a single rectrix (in the case of residents), or by banding with US Fish and Wildlife numbered bands (in the case of Neotropical migrants). Birds were then released near the point of capture. The identity of all species was established using field-guides, scientific keys and consultation with experts at the University of Nicaragua, Managua and elsewhere.

Mammals were sampled using a 40-m diameter circular trapping array with 15 sample points, 5 equally spaced on an inner 20-m diameter circle, and 10 on the outer 40 m diameter circle. Twenty Sherman folding traps (8 cm × 9 cm × 23 cm) were placed in pairs at each of the inner 5 points and at 5 alternate points of the outer circle, one on the ground and one approximately 1.5 m up in a tree or shrub. Small (15 cm × 15 cm × 48 cm) Tomahawk folding traps were placed at the remaining 5 outer sampling points. Care was used to minimize human scent, and all traps were placed as firmly as possible on the substrate and concealed with brush and leaves. Large (25 cm × 30 cm × 80 cm) Tomahawk folding traps were placed at the center of the 5 sections bounded by the inner and outer sampling rings. Finally, 5 1-m radius circular unbaited tracking stations were established per site by digging up and removing the ground litter and vegetation, and then smoothing the soil so that tracks could be observed and recorded.

Pitfall traps approximately 30 cm deep were placed between each of the inner sampling points. Each pitfall had 31-m long, 15 cm high vertical fences sunk into the ground radiating out from the pitfall traps at equal angles. Pitfall buckets were filled with water 5 cm deep and had 3-mm diameter holes drilled into the sides 5 cm from the bottom to keep them from overflowing in case of rain. In addition, searches for amphibians and reptiles were conducted during the 7 days of each trapping session.

Live traps were baited with bananas and peanut butter and checked every 24 h. Any animals captured were identified, marked by cutting a distinctive pattern in the pelage with scissors, and released, with the exception that voucher specimens were retained as needed in areas outside the biosphere reserve. Track stations were checked daily during the week-long trapping session and the presence and identity of tracks recorded. In cases where tracks could not be identified, sketches, notes and measurements were made at the site, and the soil smoothed out after each check. Pitfall traps were checked daily, and except for species which could be readily identified at the site, their contents was preserved in 10% alcohol and labeled for later identification.

On some of the sites, vegetation-sampling points were established on a random bearing 10 m from each of the 10 nets or the 10 outermost Sherman traps in a random direction. The number of times vegetation contacted a 3-m pole held vertically at this random point was recorded in 3 1-m height classes. In addition, the distance from that random point to the nearest tree that is part of the canopy was measured, as well as the species of tree.

Data analysis

Capture rates for each bird, mammal, amphibian, and reptile species were calculated by habitat as number of birds per 250 net hours, or captures per 7-day trapping period for mammals, amphibians, and reptiles.

Because species richness varies with sampling intensity, we analyzed species richness using rarefaction constructed from the means of 1,000 randomizations of sample order with the program "EcoSim" (Gotelli and Entsminger 2001). Expected values for richness and abundance for each level of sampling intensity were combined to produce a sample-based rarefaction curve in which the x axis was the number of individuals based on the number of samples chosen and the y axis was the corresponding number of species that were sampled. Rarefaction analyses were conducted separately for all bird species combined, and resident and Neotropical migrants separately. Species richness was compared among habitats by assessing the overlap of 95% confidence intervals at an intermediate level of sampling intensity (4 sites for bird data, 5 for mammals, 3 for reptiles and amphibians).

In addition rarefaction analyses we used Jaccard's coefficient to measure the similarity of habitat pairs. Where similarity = $2c/(a+b+c)$, and c is the number of species shared by two habitats, and a and b the total number of species in each habitat. Values of this index vary from 0 to 1; 0 indicates that assemblages differ totally, and 1 that they are identical. In addition to the calculation of similarity indices, we used multidimensional scaling (MDS) to calculate the similarity among samples based on their species composition and abundance. MDS is a procedure for fitting a set of points in a space such that the distance between points on the MDS plot corresponds to the dissimilarity among sampled sites. The fit of configuration distances to the original data was evaluated by calculating stress using Kruskall's stress formula, with values near 0 indicating a better fit. In addition, we examined Shepard diagrams (plots of the distance between points in the final plot with observed dissimilarities in the original data) to verify that they appeared as straight lines or smooth curves, also indicating good fit with the original data.

Vegetation variables were averaged for each plot, and compared among habitats using one-way ANOVA with Tukey's post hoc comparisons.

Results

Birds

We captured 1,433 individuals of 140 bird species during the 2 years of the study (Table 1). Of these, 89% were tropical resident species, and the remainder were

Table 1 Average capture rates (per 250 h) for birds in primary forest (n = 6), secondary forest (n = 5), allspice (n = 6) and pasture (n = 6) in north-central Nicaragua 2003–2004

Species	Status	Primary forest	Secondary forest	Allspice	Pasture	n
Variable Seedeater *Sporophila aurita*	r	1.333	2.600	6.833	8.167	111
White-collared Manakin *Manacus candei*	r	6.167	8.800	0.833	0.000	86
Scarlet-rumped Tanager *Ramphocelus passerinii*	r	1.000	2.800	4.000	6.833	85
Ochre-bellied Flycatcher *Mionectes oleagineus*	r	6.000	5.000	0.833	0.333	68
Long-tailed Hermit *Phaethornis superciliosus*	r	6.167	4.800	0.833	0.167	67
Blue-throated Goldentail *Hylocharis eliciae*	r	1.667	2.200	4.167	2.000	58
Blue-black Grassquit *Volatinia jacarina*	r	0.167	0.400	4.000	4.000	51
Buff-throated Saltator *Saltator maximus*	r	1.167	2.400	3.167	1.667	48
Yellow-faced Grassquit *Tiaris olivacea*	r	0.333	0.200	4.667	2.167	44
Blue-black Grosbeak *Cyanocompsa cyanoides*	r	2.333	3.400	0.333	0.000	33
Thick-billed Seed Finch *Oryzoborus funereus*	r	0.833	0.400	2.000	2.167	32
Red-headed Manakin *Pipra mentalis*	r	3.833	1.400	0.333	0.000	32
Little Hermit *Phaethornis longuemareus*	r	2.333	3.000	0.000	0.167	30
Orange-billed Sparrow *Arremon aurantiirostris*	r	1.500	3.000	0.000	0.000	24
Red-throated Ant-tanager *Habia fuscicauda*	r	2.000	1.600	0.333	0.000	22
Buff-throated Foliage-gleaner *Automolus ochrolaemus*	r	1.000	2.000	0.667	0.000	20
Wedge-billed Woodcreeper *Glyphorhynchus spirurus*	r	2.667	0.600	0.167	0.000	20
Band-tailed Barbthroat *Threnetes ruckeri*	r	0.667	2.400	0.500	0.000	19
Grey-headed Tanager *Eucometis penicillata*	r	0.667	1.400	1.167	0.000	18
House Wren *Troglodytes aedon*	r	0.000	0.000	2.167	0.833	18
Wood Thrush *Hylocichla mustelina*	m	1.000	1.600	0.500	0.000	17
Northern Bentbill *Oncostoma cinereigulare*	r	1.500	1.400	0.000	0.167	17
Stripe-headed Sparrow *Arremonops conirostris*	r	0.333	0.600	0.667	1.167	16
Groove-billed Ani *Crotophaga sulcirostris*	r	0.000	0.000	0.667	2.000	16
Indigo Bunting *Passerina cyanea*	m	0.167	0.200	0.833	1.333	15
White-ruffed Manakin *Corapipo leucorrhoa*	r	1.167	1.000	0.000	0.333	14
Yellow Bellied Elaenia *Elaenia flavogaster*	r	0.667	0.000	1.167	0.500	14
Ovenbird *Seiurus aurocapillus*	m	0.167	1.600	0.667	0.167	14
Clay-colored Robin *Turdus grayi*	r	0.667	0.800	1.000	0.000	14
Kentucky Warbler *Oporornis formosus*	m	0.333	1.400	0.667	0.000	13
Golden-crowned Spadebill *Platyrinchus coronatus*	r	2.000	0.200	0.000	0.000	13
Bronzy Hermit *Glaucis aenea*	r	0.500	0.400	1.000	0.000	11

Table 1 continued

Species	Status	Primary forest	Secondary forest	Allspice	Pasture	*n*
Stub-tailed Spadebill *Platyrinchus cancrominus*	r	1.000	1.000	0.000	0.000	11
Chestnut-sided Warbler *Dendroica pensylvanica*	m	0.333	0.200	0.833	0.333	10
Ruddy-tailed Flycatcher *Terenotriccus erythrurus*	r	1.333	0.400	0.000	0.000	10
Blue Ground-Dove *Claravis pretiosa*	r	0.000	0.000	0.833	0.667	9
Long-billed Gnatwren *Ramphocaenus melanurus*	r	0.333	1.00	0.167	0.167	9
Rufous-bellied Hummingbird *Amazilia tzacatl*	r	0.000	0.000	1.333	0.000	8
Dusky Antbird *Cercomacra tyrannina*	r	0.833	0.600	0.000	0.000	8
Spotted Antbird *Hylophylax naevioides*	r	1.000	0.400	0.000	0.000	8
Scaly-breasted Hummingbird *Phaeochroa cuvierii*	r	0.000	0.600	0.833	0.000	8
Wilson's Warbler *Wilsonia pusilla*	m	0.500	0.800	0.000	0.167	8
Tennessee Warbler *Vermivora peregrina*	m	0.000	0.000	1.00	0.333	8
Steely-vented Hummingbird *Amazilia saucerrottei*	r	0.000	0.000	1.167	0.000	7
Yellow Tyrannulet *Capsiempis flaveola*	r	0.000	0.000	1.000	0.167	7
Bicolored Antbird *Gymnopithys leucaspis*	r	0.500	0.800	0.000	0.000	7
Worm-eating Warbler *Helmitheros vermivorus*	m	0.500	0.800	0.000	0.000	7
White-breasted Woodwren *Henicorhina leucosticta*	r	0.833	0.400	0.000	0.000	7
Royal Flycatcher *Onychorhynchus coronatus*	r	0.333	0.800	0.167	0.000	7
Hooded Warbler *Wilsonia citrina*	m	0.667	0.600	0.000	0.000	7
Green Honeycreeper *Chlorophanes spiza*	r	0.000	0.000	1.000	0.000	6
White-necked Jacobin *Florisuga mellivora*	r	0.333	0.000	0.667	0.000	6
Blue Grosbeak *Guiraca caerulea*	m	0.000	0.200	0.000	0.833	6
Dotted-wing Antwren *Microrhopias quixensis*	r	0.833	0.200	0.000	0.000	6
Slaty Antwren *Myrmotherula schisticolor*	r	0.333	0.800	0.000	0.000	6
Rose-breasted Grosbeak *Pheucticus ludovicianus*	m	0.000	0.000	0.167	0.833	6
Barred Antshrike *Thamnophilus doliatus*	r	0.000	0.400	0.500	0.167	6
Black-faced Grosbeak *Caryothraustes poliogaster*	r	0.500	0.400	0.000	0.000	5
Plain-brested Ground Dove *Columbina minuta*	r	0.000	0.000	0.833	0.000	5
Barred Woodcreeper *Dendrocolaptes certhia*	r	0.833	0.000	0.000	0.000	5
Magnolia Warbler *Dendroica magnolia*	m	0.000	0.000	0.500	0.333	5
Mountain Elaenia *Elaenia frantzii*	r	0.000	0.600	0.000	0.333	5
Black-and-white Warbler *Mniotilta varia*	m	0.500	0.000	0.333	0.000	5
Great-crested Flycatcher *Myiarchus crinitus*	m	0.000	0.000	0.667	0.167	5

Table 1 continued

Species	Status	Primary forest	Secondary forest	Allspice	Pasture	*n*
Social Flycatcher *Myiozetetes similis*	r	0.000	0.000	0.500	0.333	5
Violet-crowned Woodnymph *Thalurania colombica*	r	0.667	0.200	0.000	0.000	5
Plain Xenops *Xenops minutus*	r	0.167	0.200	0.333	0.167	5
White-bellied Emerald *Amazilia candida*	r	0.333	0.200	0.167	0.000	4
Long-billed Starthroat *Heliomaster longirostris*	r	0.000	0.000	0.167	0.500	4
Sulphur-rumped Flycatcher *Myiobius sulphureipygius*	r	0.500	0.200	0.000	0.000	4
Ocellated Antbird *Phaenostictus mcleannani*	r	0.500	0.000	0.167	0.000	4
American Redstart *Setophaga ruticilla*	m	0.167	0.200	0.167	0.167	4
Blue-gray Tanager *Thraupis episcopus*	r	0.000	0.000	0.667	0.000	4
Mistletoe Tyrannulet *Zimmerius vilissimus*	r	0.000	0.000	0.667	0.000	4
Common Tody Flycatcher *Todirostrum cinereum*	r	0.000	0.000	0.667	0.000	4
Ruddy Foliage-gleaner *Automolus rubiginosus*	r	0.000	0.600	0.000	0.000	3
Common Ground Dove *Columbina passerina*	r	0.000	0.000	0.333	0.167	3
Shining Honeycreeper *Cyanerpes lucidus*	r	0.000	0.000	0.333	0.167	3
Blue-crowned Antbird *Gymnocichla nudiceps*	r	0.000	0.600	0.000	0.000	3
Spotted-crowned Woodcreeper *Lepidocolaptes affinis*	r	0.333	0.000	0.167	0.000	3
Buff-rumped Warbler *Phaeothlypis fulvicauda*	r	0.000	0.000	0.500	0.000	3
Eye-ringed Flatbill *Rhynchocyclus brevirostris*	r	0.500	0.000	0.000	0.000	3
Spotted-breasted Wren *Thryothorus maculipectus*	r	0.167	0.400	0.000	0.000	3
Violaceous Trogon *Trogon violaceus*	r	0.500	0.000	0.000	0.000	3
Golden-winged Warbler *Vermivora chrysoptera*	m	0.000	0.200	0.333	0.000	3
Blue-chested Hummingbird *Amazilia amabilis*	r	0.000	0.000	0.333	0.000	2
Yellow-billed Cacique *A mblycercus holosericeus*	r	0.000	0.400	0.000	0.000	2
Ruby-throated Hummingbird *Archilochus culibris*	m	0.333	0.000	0.000	0.000	2
Bright-romped Attila *Attila spadiceus*	r	0.333	0.000	0.000	0.000	2
Red-footed Plumeleteer *Chalybura urochrysia*	r	0.167	0.000	0.167	0.000	2
Short-billed Pigeon *Columba nigrirostris*	r	0.000	0.000	0.000	0.333	2
Tropical Pewee *Contopus cinereus*	r	0.000	0.000	0.333	0.000	2
Keel-billed Motmot *Electron carinatum*	r	0.333	0.000	0.000	0.000	2
Acadian Flycatcher *Empidonax virescens*	m	0.000	0.000	0.333	0.000	2
Gray-crowned Yellowthroat *Geothlypis poliocephala*	r	0.000	0.000	0.333	0.000	2
Lesser Greenlet *Hylophilus decurtatus*	r	0.000	0.000	0.333	0.000	2
White-whiskered Puffbird *Malacoptila panamensis*	r	0.333	0.000	0.000	0.000	2

Table 1 continued

Species	Status	Primary forest	Secondary forest	Allspice	Pasture	*n*
Checker-throated Antwren *Myrmotherula fulviventris*	r	0.333	0.000	0.000	0.000	2
Painting Bunting *Passerina ciris*	m	0.000	0.000	0.000	0.333	2
Crimson-collared Tanager *Phlogothraupis sanguinolenta*	r	0.000	0.000	0.000	0.333	2
White-throated Spadebill *Platyrinchus mystaceus*	r	0.000	0.400	0.000	0.000	2
Russet Antshrike *Thamnistes anabatinus*	r	0.333	0.000	0.000	0.000	2
Tropical Kingbird *Tyrannus melancholicus*	r	0.000	0.000	0.167	0.167	2
Green-breasted Mango *Anthracothorax prevostii*	r	0.000	0.000	0.167	0.000	1
Olive Sparrow *Arremonops rufivirgatus*	r	0.000	0.000	0.167	0.000	1
Rufous Motmot *Baryphthengus martii*	r	0.000	0.200	0.000	0.000	1
Northern Beardless-tyrannulet *Camptostoma imberbe*	r	0.000	0.000	0.000	0.167	1
Swainson's Thursh *Catharus ustulatus*	m	0.000	0.000	0.000	0.167	1
Ruddy Ground Dove *Columbina talpacoti*	r	0.000	0.000	0.000	0.167	1
Olive-sided Flycatcher *Contopus borealis*	m	0.167	0.000	0.000	0.000	1
Fasciated Antshrike *Cymbilaimus lineatus*	r	0.167	0.000	0.000	0.000	1
Blue Dacnis *Dacnis cayana*	r	0.000	0.000	0.167	0.000	1
Black-banded Woodcreeper *Dendrocolaptes picumnus*	r	0.000	0.000	0.167	0.000	1
Yellow Warbler *Dendroica petechia*	m	0.000	0.000	0.167	0.000	1
Yellow-bellied Flycatcher *Empidonax flaviventris*	m	0.000	0.200	0.000	0.000	1
Magnificient Hummingbird *Eugenes fulgens*	r	0.000	0.000	0.167	0.000	1
Gray-breasted Wood-wren *Henicorhina leucophrys*	r	0.000	0.200	0.000	0.000	1
Yellow-breasted Chat *Icteria virens*	m	0.000	0.200	0.000	0.000	1
Black-cowled *OrioleIcterus dominicensis*	r	0.000	0.000	0.167	0.000	1
Streaked-headed Woodcreeper *Lepidocolaptes souleyetti*	r	0.000	0.200	0.000	0.000	1
Gray-chested Dove *Leptotila cassinii*	r	0.167	0.000	0.000	0.000	1
Black-crested Coquette *Lophornis helenae*	r	0.000	0.000	0.167	0.000	1
White-fronted Nunbird *Monasa morphceus*	r	0.000	0.200	0.000	0.000	1
Dusky-capped flycatcher *Myiarchus tuberculifer*	r	0.000	0.000	0.167	0.000	1
Sulphur-bellied Flycatcher *Myiodynastes luteiventris*	r	0.000	0.000	0.000	0.167	1
Golden-olive Woodpecker *Piculus rubiginosus*	*r*	0.000	0.000	0.000	0.167	1
Hepatic Tanager *Piranga flava*	r	0.000	0.000	0.000	0.167	1
Summer Tanager *Piranga rubra*	m	0.000	0.200	0.000	0.000	1
Black-headed Saltator *Saltador atriceps*	r	0.167	0.000	0.000	0.000	1
Grayish Saltator *Saltator caerulescens*	r	0.000	0.200	0.000	0.000	1

Table 1 continued

Species	Status	Primary forest	Secondary forest	Allspice	Pasture	*n*
Thrushlike Manakin *Schiffornis turdinus*	r	0.000	0.200	0.000	0.000	1
Tawny-throated Leaftosser *Sclerurus mexicanus*	r	0.000	0.200	0.000	0.000	1
Lousiana Waterthrush *Seiurus motacilla*	m	0.000	0.200	0.000	0.000	1
Olivaceous Woodcreeper *Sittasomus griseicapillus*	r	0.167	0.000	0.000	0.000	1
Ruddy-breasted Seedeater *Sporophila minuta*	r	0.000	0.000	0.000	0.167	1
White-shouldered Tanager *Tachyphonus luctuosus*	r	0.000	0.200	0.000	0.000	1
Golden-hooded Tanager *Tangara larvata*	r	0.000	0.000	0.000	0.167	1
Great Antshrike *Taraba major*	r	0.000	0.200	0.000	0.000	1
Banded Wren *Thryothorus pleurostictus*	r	0.000	0.000	0.167	0.000	1
Black-throated Trogon *Trogon rufus*	r	0.000	0.200	0.000	0.000	1

Neotropical migrants. Overall bird species richness was highest in primary forest, secondary forest and allspice, and lowest in pasture (Fig. 1a), however, patterns of species richness differed between resident and migrant birds. Resident species exhibited similar richness in primary forest, secondary forest, and allspice, and significantly lower richness in pasture than all other habitats (Fig. 2a). In contrast, species richness of migrants was significantly lower in primary forest than secondary forest, and did not differ among primary, allspice and pasture, or among secondary forest allspice and pasture (Fig. 2b).

Bird species composition also differed among habitats. The similarity in both resident and migrant bird species composition between primary forest and secondary forest was greater than the similarity between allspice and either primary or secondary forest, and allspice and pasture were more or less equally similar to all other habitats (Fig. 3). This pattern was true for both resident and migrant birds, however the similarity values between allspice and primary forest were generally lower for residents than migrants Fig. 3). Similarly, the results of the multidimensional scaling showed considerable overlap of secondary forest, allspice and pasture, but less overlap between primary forest and allspice and pasture (Fig. 4). Finally, 18 bird species were captured only in primary forest, 23 only in secondary forest, 23 only in allspice, and 11 only in pasture (Table 1). There was no difference in the percentage of individual birds that were migrants among primary forest (5.2%), secondary forest (11.4%), allspice (9.7%) and pasture (11.1%; $F_{(3,22)} = 0.97$, $P = 0.43$).

Mammals

We captured 116 individuals of 28 mammal species during the 2 years of the study (Table 2). Overall mammal species richness was significantly higher in primary forest than all other habitats, did not differ between secondary forest and allspice, and was significantly lower in pasture than all other habitats (Fig. 5a). Mammal communities of primary forest had relatively low similarity to secondary forest (40%), allspice (44%) or pasture (20%) in contrast to the relatively high similarity between secondary forest and allspice (55%; Fig. 5b). Pasture had relatively low similarities to other habitats (≤25%). In contrast to the similarity analyses, the multidimensional scaling analyses indicated that there was substantial overlap in mammal species composition between primary and secondary forests and between pasture and allspice (Fig. 6). Nevertheless, the multidimensional scaling analyses indicated that there was less overlap between primary forest and allspice and pasture. Six mammal species were captured only in primary forest, 3 only in secondary forest, and 4 only in allspice (Table 2). No species were encountered only in pasture. Finally, we encountered two species of non-native rodent in allspice plantations (black rat, and house mouse), and one species in secondary forest (Norway rat).

Reptiles and amphibians

We encountered 128 individuals of 32 reptile and amphibian species during the 2 years of the study (Table 3). Rarefaction analysis indicated that species richness was highest in primary forest, secondary forest and allspice, and lower in pasture, however this difference was not significant (Fig. 7a). The similarity in species composition between primary forest and other habitats was low (≤17%), secondary

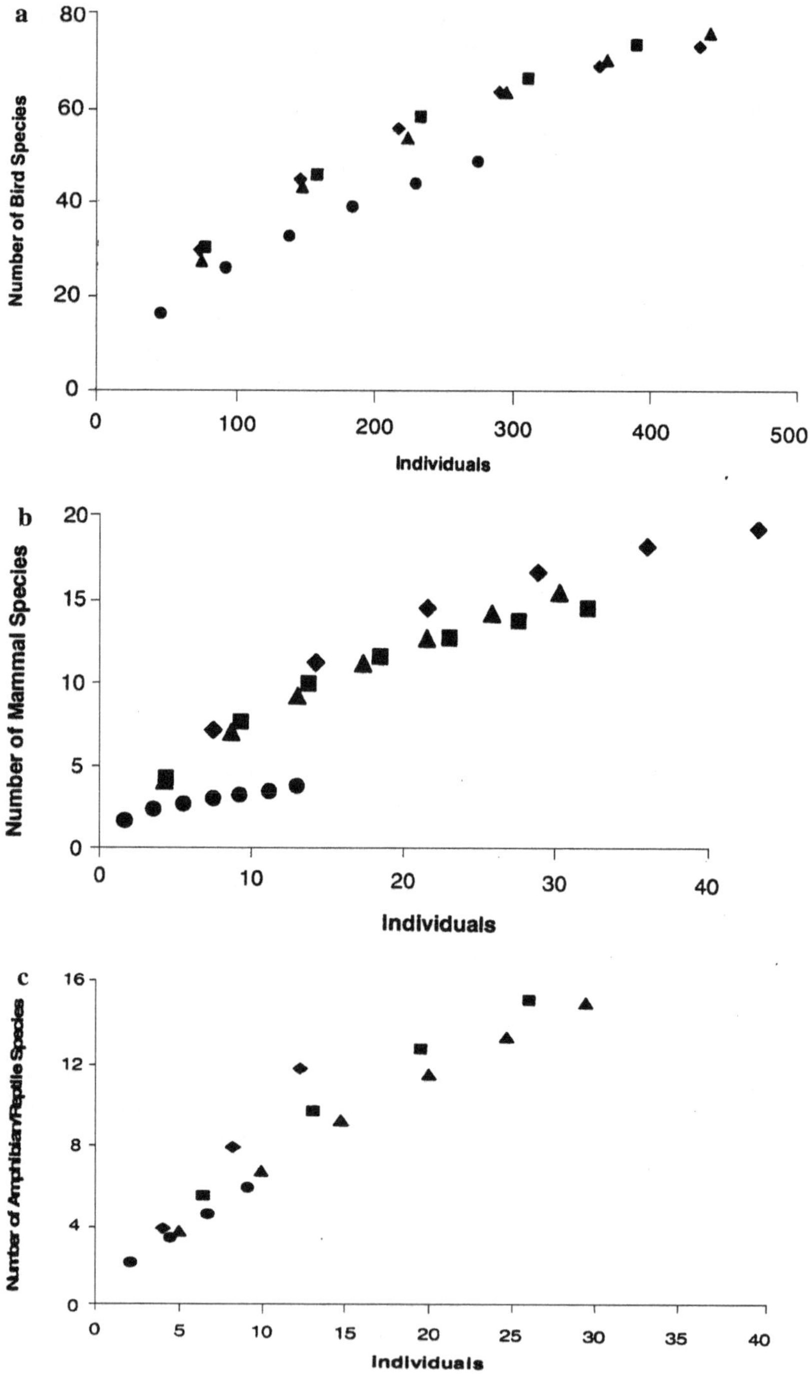

Fig. 1 Rarefaction curves for birds (**a**), mammals (**b**) and herpetofauna (**c**) in primary forest (◆), secondary forest (■), allspice (▲), and pasture (●)calculated from data collected in northcentral Nicaragua, 2002–2004. Primary forest, secondary forest and allspice generally have higher diversity of all taxa relative to pasture

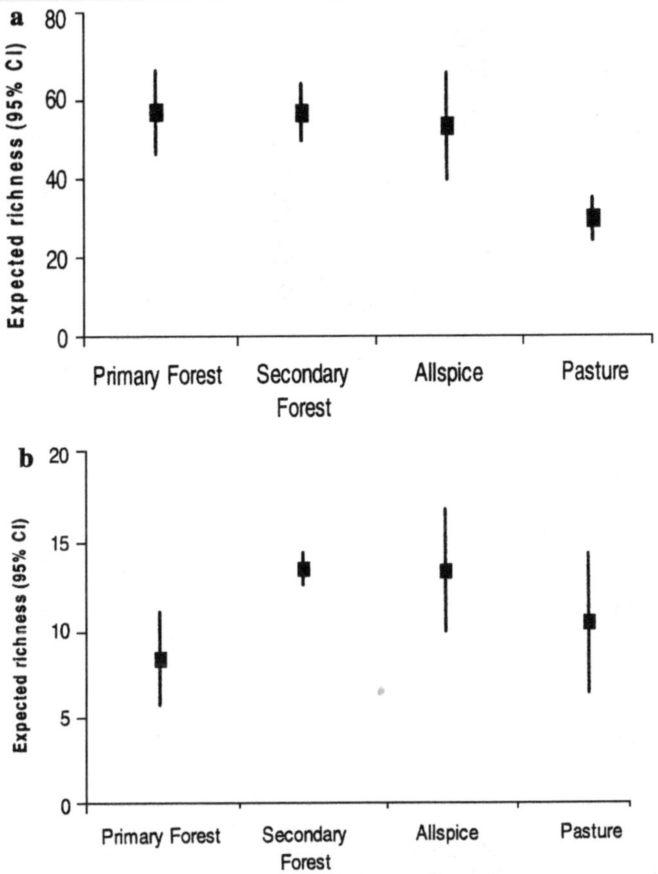

Fig. 2 Mean expected number of resident birds (**a**) and migrant birds (**b**) compared among habitats using rarefaction of data from 23 sites in north-central Nicaragua 2002–2004. Resident and migrant birds exhibit different patterns of species richness among habitats

forest and allspice were more similar (37%), and pasture was most similar to allspice (27%), less similar to secondary forest (20%), and least similar to primary forest (12%; Fig. 7b). Multidimensional scaling analyses indicated that there was substantial overlap between secondary forest, allspice and pasture, (Fig. 8) and less overlap between primary forest and pasture. Six species were captured only in primary forest, 4 only in secondary forest, 6 only in allspice, and 2 only in pasture (Table 3).

Habitat

Tree circumference and tree height were greater in primary forest than other habitats (Table 4). There were no significant differences among habitats in structure in any of the three strata, although structure tended to be higher between 0-1 m and lower 2-3 m above ground in pasture than other habitats. One-hundred-twenty-nine plant species were recorded in allspice plantations.

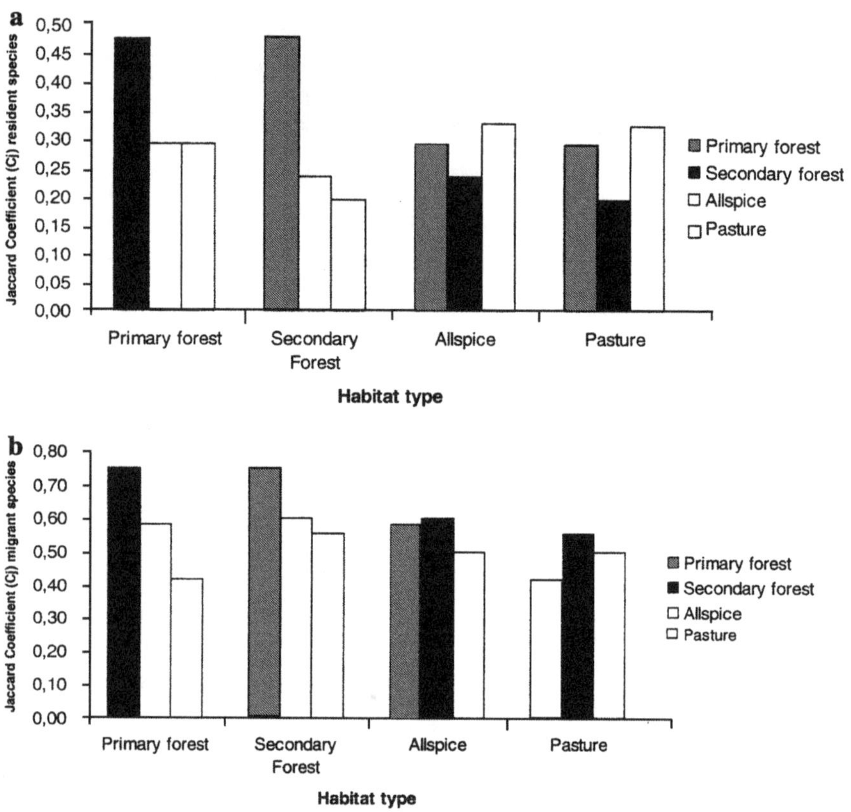

Fig. 3 Jaccard similarity coefficients of resident birds (**a**) and migrant birds (**b**) among habitats using data from 23 sites in north-central Nicaragua 2002–2004. For each habitat on the horizontal axis, the height of the three columns indicates the percentage of species shared with the other three habitats

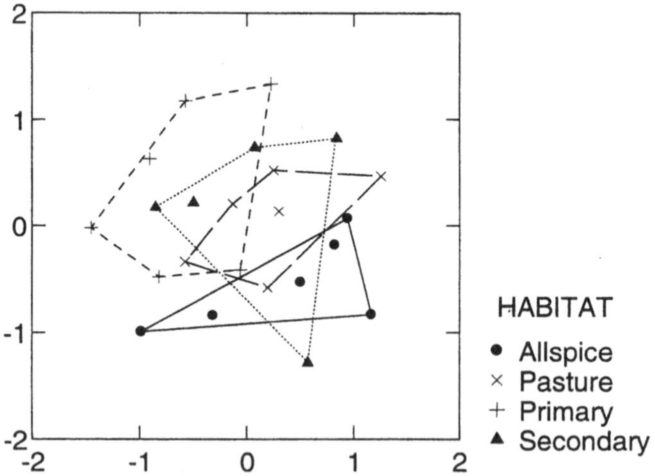

Fig. 4 Multi-dimensional scaling ordination of the abundances of bird species among habitats based on mist net captures of birds at 23 sites in north-central Nicaragua 2003–2004. Stress of final configuration was 0.01, proportion of variance (RSQ) was 0.69. The distance between points in this diagram illustrates the difference between sites in terms of their bird species composition

Table 2 Average capture rates (per 7 days) for mammals in primary forest (n = 7), secondary forest (n = 6), allspice (n = 7) and pasture (n = 7) in north-central Nicaragua 2002–2004

Species	Primary forest	Secondary forest	Allspice	Pasture	*n*
Common gray four-eyed opossum *Philander opossum*	0.5	0.57	0.57	0.86	17
Cotton rat *Sigmodon hispidus*	0.33	0.14	0.86	0.71	14
Common opossum *Didelphis marsupialis*	0.83	0.43	0.29	0	10
Big-eared climbing rat *Ototylomys phyllotis*	0.83	0.43	0.14	0	9
Dusky rice rat *Melanomys caliginosus*	0.5	0.14	0.43	0.14	8
Central American agouti *Dasyprocta punctata*	0.33	0.43	0.29	0	7
Virginia opossum *Didelphis virginiana*	0.33	0.43	0.29	0	7
Mexican mouse opossum *Marmosa mexicana*	0	0.57	0.29	0	6
Paca *Agouti paca*	0.5	0.14	0.14	0	5
Northern tamandua *Tamandua mexicana*	0.5	0.14	0	0	4
White-tailed deer *Odocoileus virginianus*	0.33	0.29	0	0	4
Striped hog-nosed skunk *Conepatus semiestratus*	0.33	0	0.14	0	3
Tayra *Eira barbara*	0.33	0	0.14	0	3
Brown four-eyed opossum *Metachirus nudicaudatus*	0.33	0	0	0	2
Northern raccoon *Procyon lotor*	0	0.29	0	0	2
White-nosed Coati *Nasua narica*	0	0.29	0	0	2
Central American Spiny Rat *Proechimys semispinosus*	0	0.14	0.14	0	2
Black rat *Rattus rattus*	0	0	0.14	0	1
Norway rat *Rattus norvegicus*	0	0.14	0	0	1
Mt. Pirri Isthmus Rat *Isthmomys pirrensis*	0.17	0	0	0	1
Big Pocket Gopher *Orthogeomys matagalpae*	0.17	0	0	0	1
Desmarest's spiny pocket mouse *Heteromys desmarestianus*	0.17	0	0	0	1
House mouse *Mus musculus*	0	0	0.14	0	1
White-throated rice rat *Oryzomys albigularis*	0.17	0	0	0	1
Nine-banded armadillo *Dasypus novemcinctus*	0	0	0.14	0	1
Two-toed sloth *Choloepus hoffmanni*	0	0	0.14	0	1
Jaguar Panthera onca	0.17	0	0	0	1
Unknown wild cat	0.17	0	0	0	1

Discussion

Differences in richness among habitats

We observed that species richness in allspice plantations was comparable to species richness in primary forest for some taxa. These results are similar to patterns reported in studies of vertebrates in other types of mixed agroforestry in the tropics. For example, coffee, cacao and other countryside habitats harbor a large number of native species (Wunderle and Latta 1996; Estrada et al. 1997; Greenberg et al. 1997a, b, 2000, Reitsma et al. 2001; Hughes et al. 2002; Daily et al. 2003; Perfecto et al. 2003; Tejada-Cruz and Sutherland 2004). These studies have been viewed in a positive light by many, and taken at face value, suggest that crops cultivated in a mixed productive system can ameliorate the effects of the destruction of primary forests on native forest biodiversity. The high species richness, as well as the high number of species reported only in allspice, is particularly striking considering that the oldest of these allspice plantations was only 5 years old. As time progresses, the structure of the habitat will become more similar to primary forest, and will likely support even more species.

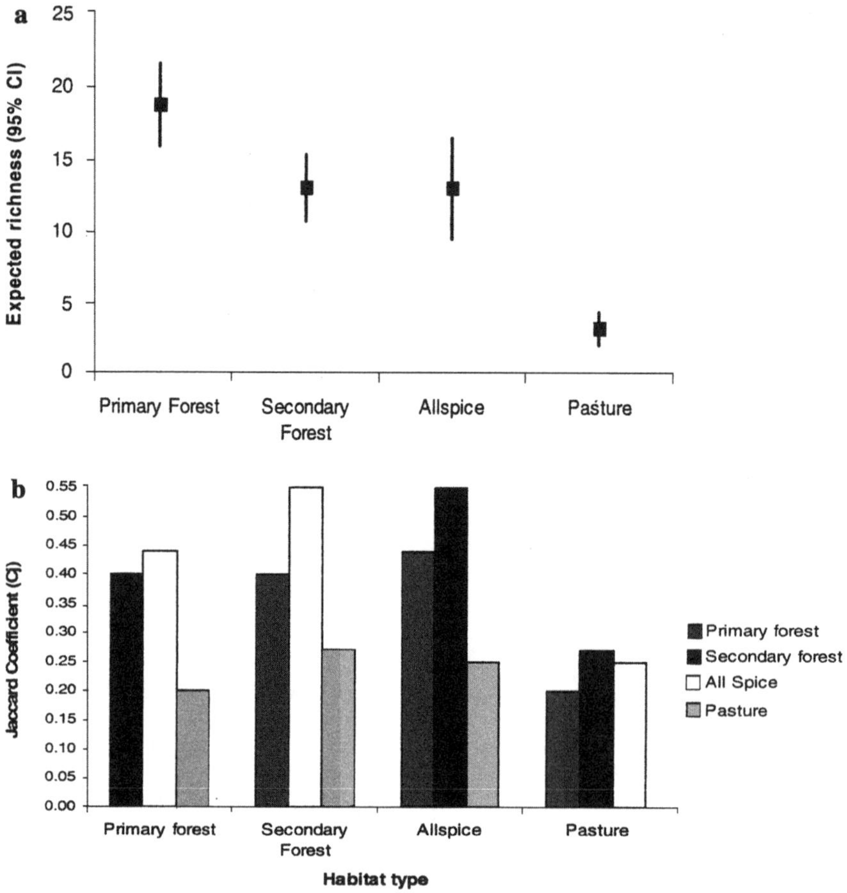

Fig. 5 Mean expected number of mammal species compared among habitats using rarefaction (**a**) and similarity in mammal species composition among habitats (**b**) using data from 27 sites in north-central Nicaragua 2002–2004

The abundance of Neotropical migrants in allspice is particularly notable given the numerous studies documenting declines in populations of these species (Rappole 1995), as well as the high species richness of these species in agroforestry systems. As in other studies, we observed that Neotropical migrant species richness was lowest in primary habitats and higher in habitats such as secondary forests, agroforestry systems (allspice) and pasture (Greenberg et al. 1997a, b, 2000, Reitsma et al. 2001; Hughes et al. 2002; Tejeda-Cruz and Sutherland 2004). Although their presence in human modified habitats may provide a first-order indication of their suitability for these species, survival rates of some migrants can be lower in human modified habitats due to predation (Rappole et al. 1989). Thus, the existence of these species in modified habitats might signify that preferred habitats are saturated, resulting in the exclusion of some individuals from preferred habitats (Winker et al. 1990; Marra et al. 1993). Furthermore, it is likely that we underestimated the differences between primary forest and other habitats for both migrants and residents because mist nets sample decreasing amounts of the avifauna [migrant as well as resident] with increasing canopy height (Rappole et al. 1998). More detailed studies of survival

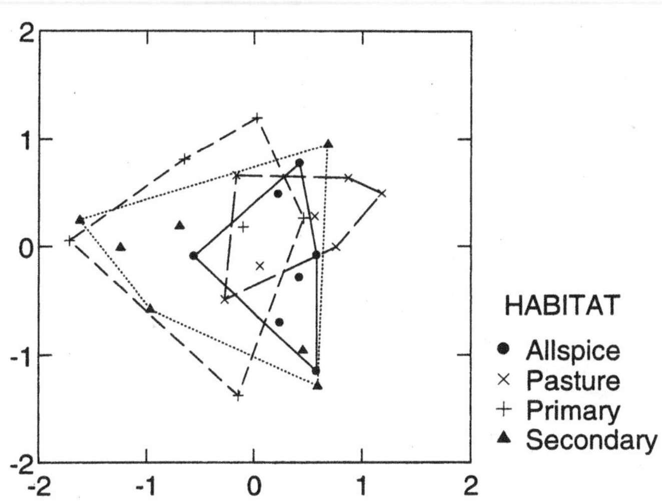

Fig. 6 Multi-dimensional scaling ordination of the abundances of mammal species among habitats based on trapping at 27 sites in north-central Nicaragua 2003–2004. Stress of final configuration was 0.02, proportion of variance (RSQ) was 0.80

rates of migrants in these different habitats are needed before we can conclude that allspice and other agroforestry systems are actually high quality habitat for the migrant species that use them (Greenberg et al. 2000; Reitsma et al. 2001; Tejada-Cruz and Sutherland 2004; Rappole et al. 2003a).

Allspice and vertebrate conservation

Despite the fact that species richness is comparable between allspice and primary forest for some taxa, there were substantial differences in species composition between primary forest and the other habitats. This is consistent with the findings of studies of other types of agroforestry systems (Heinen 1992; Greenberg et al. 2000; Reitsma et al. 2001; Tejada-Cruz and Sutherland 2004), and the limitations of these systems in conserving native biodiversity (Rappole et al. 2003; Naidoo 2004). Nevertheless, there are other ways in which allspice, as it is cultivated at our sites, can contribute to conservation of biodiversity beyond its direct habitat value. First, all of the sites used in this study were sites converted from intensive agriculture to allspice through plantings. Thus, at the most elementary level, the cultivation of allspice provides a more complex habitat that will support more native species than the pasture it is replacing, as well as potentially provide a buffer area for adjacent primary forest areas (Moguel and Toledo 1999; Tejada-Cruz and Sutherland 2004). Other types of agroforestry, such as coffee and cacao are also preferable to intensive agriculture because of relatively high species richness (Greenberg et al. 2000; Reitsma et al. 2001); however in many cases, these plantations have occurred at the expense of native forest (Rice and Greenberg 2000; Donald 2004). In contrast, the program encouraged by MDI involves the reclamation of formerly intensively cultivated agricultural land.

In addition to the value of allspice as habitat for tropical species, the cultivation of allspice will create economic incentives that will further contribute to the conservation of biodiversity. The cultivation of allspice will reduce the economic incentive

Table 3 Average encounter rates (per 7 days) for amphibians and reptiles in primary forest (n = 3), secondary forest (n = 5), allspice (n = 7) and pasture (n = 4) in north-central Nicaragua 2003–2004

Species	Primary forest	Secondary forest	Allspice	Pasture	n
Central American Whiptail *Ameiva festiva*	0.00	1.43	1.33	1.10	55
Speckled Racer *Drymobius margaritiferus*	0.00	0.50	0.29	0.25	7
Mexican Vine Snake *Oxybelis aeneus*	0.00	0.40	0.43	0.25	6
Red Eyed Tree Frog *Agalychnis callidryas*	0.33	0.40	0.29	0.00	5
Green Basilisk *Basiliscus plumifrons*	0.00	0.40	0.29	0.00	4
Smooth Skinned Toad *Bufo haematiticus*	0.33	0.40	0.00	0.00	4
Purple Caecilian *Gymnopis multiplicata*	0.33	0.20	0.14	0.25	4
Cukra Climbing Salamander *Bolitoglossa striatula*	0.33	0.20	0.14	0.00	3
Eyelasah viper *Brothiechis schlegelii*	0.00	0.40	0.00	0.00	3
Cane Toad *Bufo marinus*	0.00	0.00	0.43	0.00	3
Neotropical Green Anole *Norops biporcatus*	0.00	0.20	0.14	0.00	3
Rain Forest Frog *Rana vaillanti*	0.00	0.20	0.29	0.00	3
Tropical Rat Snake *Spilotespullatus*	0.00	0.60	0.00	0.00	3
Common Basilisk *Basiliscus basiliscus*	0.00	0.00	0.29	0.00	2
Helmeted Iguana *Corytophanes cristata*	0.00	0.20	0.14	0.00	2
Neotropical Racer *Drymobius chloroticus*	0.00	0.00	0.14	0.00	2
Tropical Water Snake *Hydromorphus concolor*	0.00	0.40	0.00	0.00	2
Bicolored Coral Snake *Micrurus multifasciatus*	0.33	0.00	0.14	0.00	2
Warschewitsch's Frog *Rana warszewitschii*	0.33	0.20	0.00	0.00	2
Misfit Leaf Frog *Agalychnis saltator*	0.00	0.00	0.14	0.00	1
Jumping Pit Viper *Atropoides nummifer*	0.33	0.00	0.00	0.00	1
Striped Basilisk *Basiliscus vittatus*	0.00	0.00	0.00	0.25	1
Fer de Lance *Bothrops asper*	0.33	0.00	0.00	0.00	1
Mussurana *Clelia clelia*	0.00	0.20	0.00	0.00	1
Mimicking Rain Frog *Eleutherodactylus mimus*	0.00	0.20	0.00	0.00	1
Purple Caecilian *Gymnopis multiplicata*	0.00	0.00	0.14	0.00	1
Common cat-eyed snake *Leptodeira annulata*	0.33	0.00	0.00	0.00	1
Central American Coral Snake *Micrurus nigrocinctus*	0.33	0.00	0.00	0.00	1
Red coffee snake *Ninia sebae*	0.00	0.00	0.14	0.00	1
Anolis sp. *Norops sp.*	0.00	0.00	0.00	0.25	1
Masked tree frog *Smilisca baudinii*	0.33	0.00	0.00	0.00	1
Turnip-tailed Gecko *Thecadactylus rapicauda*	0.33	0.00	0.00	0.00	1

for the felling of mature allspice trees in areas of virgin forest, as was the previous practice. Previously allspice was exported in its whole form, and thus, the fruits had little value added. With the introduction of the technology for the production of essential oils, wild allspice trees will increase in value, which will discourage Nicaraguans from harvesting allspice by cutting down fruiting trees. Furthermore, the semi-permanence of allspice plantations will yield repeated harvests from the same area, which will reduce the necessity of clearing additional virgin forest (Donald 2004), for which the protection of water supplies is an added incentive. Finally, the social and economic stability derived from the production of essential oils will result in the discouragement of indiscriminant deforestation by squatters, similar to that observed in cacao farming regions of eastern Brazil (Donald 2004).

The occurrence of non-native native species can be a problem in some kind of plantations through predation or competition with native species, or the transmission of disease (Daily and Erlich 1996; Laurance and Cochrane 2001). We encountered two species of non-native rodents in allspice plantations (black rat and house mouse). These species occurred in relatively low numbers, accounting for only 7% of

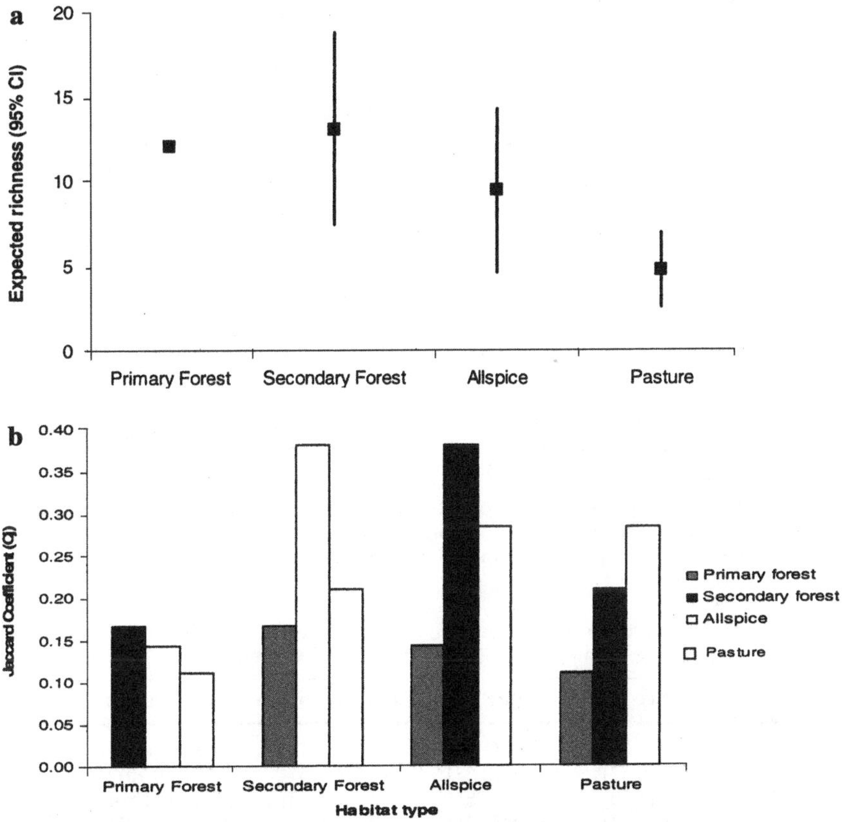

Fig. 7 Mean expected number of amphibian and reptiles compared among habitats using rarefaction (**a**) and similarity in herpetofaunal species composition among habitats (**b**) using data from north-central Nicaragua 2002–2004

species captured in allspice plantations. Thus, it appears unlikely that allspice plantations support numbers of non-native species sufficient to cause significant impacts on the native fauna, although this potential risk merits further investigation.

We conclude that the potential for allspice to contribute to biodiversity conservation by increasing the biodiversity of areas of former pasture and to reducing incentives for the clearing of additional primary forest is substantial. Care should be exercised, however, in indiscriminately promoting its cultivation on the basis of its biodiversity benefits. Shade coffee has been touted as a panacea for the conservation of tropical biodiversity, however it has become clear that, in the absence of rigorous certification programs based on scientifically established criteria combined with a detailed analysis of the effects of economic incentives associated with its promotion, it might in fact increase the loss biodiversity (Rappole et al. 2003a, b). As long as sufficient attention is given to the effect of the economic incentives on the agricultural practices in the region, allspice cultivation can improve the living standards of the human populace, and also improve the prospects for the native fauna of the region.

Although we found evidence of large mammals in the surrounding forest, such as droppings of tapirs (*Tapirus bairdii*), with the exception of white-tailed deer, larger

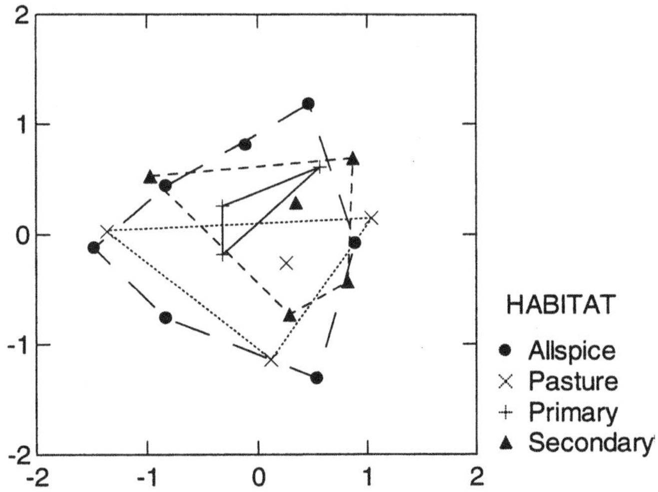

Fig. 8 Multi-dimensional scaling ordination of the abundances of reptile and amphibian species among habitats based on timed searches at 15 sites in north-central Nicaragua 2003–2004. Stress of final configuration was 0.005, proportion of variance (RSQ) was 0.84

Table 4 Average (standard error) habitat variables compared among primary forest, secondary forest, allspice and pasture at sampling sites in north-central Nicaragua 2003–2004

Habitat	Allspice	Pasture	Primary forest	Secondary forest	*P*-value
Circumference	41.8 (11.6)[b]	4.17 (1.23)[b]	141.1 (29.1) [a]	108.9 (20.0)[b]	*P*=0.005
Altura	7.06 (0.57)[b]	6.66 (2.53)[b]	23.7 (2.25) [a]	12.2 (0.70)[b]	*P*<0.002
Contacts 0–1	7.38 (1.70)	11.9 (0.48)	6.07 (1.63)	8.03 (1.56)	*P*=0.06
Contacts 1–2	5.53 (1.34)	4.88 (1.21)	4.10 (0.36)	5.33 (1.72)	*P*=0.88
Contacts 2–3	3.45 (1.31)	1.15 (0.79)	4.90 (0.90)	4.77 (0.68)	*P*=0.08

Common superscripts indicate means that did not differ significantly

[a]number of contacts of vegetation with a 3-m pole held vertically

mammals were notably absent from our study sites. This is probably due to the relative small scale of our sampling effort, which was constrained to relatively few, small trapping arrays. Wide ranging species dependent on forest should have high conservation priority (Daily et al. 2003), and might be particularly sensitive to habitat fragmentation, and we suggest that efforts to survey these species would be a valuable supplement to our investigation.

Conclusions and implications for conservation

Allspice cultivated in a mixed productive system provides habitat for a diverse community of resident and migrant birds, mammals, reptiles and amphibians. The habitat value of the multistoried, structurally complex habitats provided by allspice cultivation is only one way in which this form of agroforestry can contribute to the conservation of biodiversity. The cultivation of allspice will reduce the economic incentive for the felling of mature allspice trees in areas of virgin forest, as was the previous practice. In addition, the semi permanence of allspice plantations will yield repeated harvests from the same area, which will reduce the necessity of clearing additional virgin forest. Finally, it is expected that the social and economic stability derived from the production of essential oils will result in the discouragement of indiscriminant deforestation by squatters.

Acknowledgements Funding for this project was provided through the Global Environmental Facility and the Environmentally & Socially Sustainable Development Sector Management Unit, Latin America and the Caribbean Region of the World Bank. The authors would like to thank the members of CoopeSiuna Cooperative and the Programa Campesino a Campesino of La Unión Nacional de Agricultores y Ganaderos (UNAG) for field help and logistical support. The manuscript benefited from the comments of J. Wunderle and S. Sillett. R. Chaverri assisted with data summary and analysis.

References

Daily GC, Erlich PR (1996) Global change and human susceptibility to disease. Ann Rev Energy Environ 21:125–144
Daily GC, Ceballos G, Pacheco J (2003) Countryside biogeography of neotropical mammals: conservation opportunities in agricultural landscapes of costa rica. Conserv Biol 17:1814–1826
Donald PF (2004) Biodiversity impacts of some agricultural commodity production systems. Conserv Biol 18:17–37
Estrada A, Coates-Estrada R, Meritt DA (1997) Anthropogenic landscape changes and avian diversity at Los Tuxtlas, Mexico. Biod Conserv 6:19–43
Global Environment Facility (1997) Atlantic Biological Corridor Project, Nicaragua. Report No: 16535-NI World Bank, Washington, DC
Gotelli NJ, Entsminger GL (2001) EcoSim: Null models software for ecology. Version 7.0. Acquired Intelligence Inc. & Kesey-Bear. http://www.homepages.together.net/~gentsmin/ecosim.htm. Accessed July 2004.
Greenburg R, Bichier P, Angon AC, Reitsma R (1997a) Bird populations in shade and sun coffee plantations in central Guatemala. Conserv Biol 11:448–459
Greenburg R, Bichier P, Sterling J (1997b) Bird populations in rustic and planted shade coffee plantations in eastern Chiapas, Mexico. Biotropica 29:501–514
Greenberg R, Bichier P, Angon AC (2000) The Conservation value for birds of cacao plantations with diverse planted shade in Tabasco, Mexico. Anim Conserv 3:105–112
Heinen JT (1992) Comparisons of the leaf litter herpetofauna in abandoned cacao plantations and primary rain forest in Costa Rica: some implications for faunal restoration. Biotropica 24:431–439
Hughes JB, Daily GC, Ehrlich PR (2002) Conservation of tropical forest birds in countryside habitats. Ecol Lett 5:121–129
Karr JR (1981) Surveying birds with mistnets. In: Ralph CJ, Scott JM (eds) Estimating numbers of terrestrial birds. Studies in Avian Biology 6, pp 62–67
Kamaljit SB, Seidler R (1998) Natural forest management and conservation of biodiversity in tropical forests. Conserv Biol 12:46–55
Laurance WF, Cochrane MA (2001) Synergistic effects in fragmented landscapes. Conserv Biol 15:1488–1489
Marra PP, Sherry TW, Holmes RT (1993) Territorial exclusion by a long-distance migrant warbler in Jamaica—a removal experiment with American Redstarts (*Setophaga ruticilla*). Auk 110:565–572
Moguel P, Toledo VM (1999) Coffee cultivation and biodiversity conservation. Conserv Biol 13:11–21
Naidoo R (2004) Species richness and community composition of songbirds in a tropical forestagricultural landscape. Anim Conserv 7:93–105
Perfecto I, Mas A, Dietsch T, Vandermeer J (2003) Conservation of biodiversity in coffee agroecosystems: a tri-taxa comparison in southern Mexico. Biod Conserv 12:1239–1252
Rappole JH (1995) The ecology of migrant birds: a Neotropical perspective. Smithsonian Institution Press, Washington, DC
Rappole JH, Ramos MA, Winker K (1989) Winter wood thrush movements and mortality in southern Veracruz. Auk 106:402–410

Rappole JH, Winker K, Powell GVN (1998) Migratory bird habitat use in southern Mexico: mist nets versus point counts. J Field Ornithol 69:635–643
Rappole JH, King DI, Vega Rivera JH (2003a) Coffee and conservation. Conserv Biol 17:334–336
Rappole JH, King DI, Vega Rivera JH (2003b) Coffee and conservation III; a reply to Philpott and Dietsch. Conserv Biol 17:1847–1849
Reitsma R, Parrish JD, Mclarney W (2001) The role of cacao plantations in maintaining forest avian diversity in southeastern Costa Rica. Agroforest Syst 53:185–193
Rice RA, Greenberg R (2000) Cacao cultivation and the conservation of biological diversity. Ambio 29:167–173
Smith JH (2003) Environmental assesment: land-cover assessment of conservation and buffer zones in the BOSAWAS natural reserve of Nicaragua. US EPA Landscape Characterization Branch E243-05; DOI: 10.1007/s00267-002-2774-8
Tejeda-Cruz C, Sutherland WJ (2004) Bird responses to shade coffee production. Anim Conserv 7:169–179
Winker K, Rappole JH, Ramos MA (1990) Population dynamics of the Wood Thrush in southern Veracruz, Mexico. Condor 92:444–460
Wunderle JM, Latta SC (1996) Avian abundance in sun and shade coffee plantations and remnant pine forests in the Cordillera Central, Dominican Republic. Ornitol Neotrop 7:19–34

TOPICS IN BIODIVERSITY AND CONSERVATION

1. David L. Hawksworth and Alan T. Bull (eds.): *Arthropod Diversity and Conservation.* 2006 ISBN 978-1-4020-5203-0
2. David L. Hawksworth and Alan T. Bull (eds.): *Forest Diversity and Management.* 2006 ISBN 978-1-4020-5207-3
3. David L. Hawksworth and Alan T. Bull (eds.): *Human Exploitation and Biodiversity Conservation.* 2006 ISBN 978-1-4020-5282-0
4. David L. Hawksworth and Alan T. Bull (eds.): *Marine, Freshwater, and Wetlands Biodiversity Conservation.* 2006 ISBN 978-1-4020-5733-4
5. David L. Hawksworth and Alan T. Bull (eds.): *Vertebrate Conservation and Biodiversity.* 2007 ISBN 978-1-4020-6319-0

SPRINGER NATURE

GPSR Compliance

The European Union's (EU) General Product Safety Regulation (GPSR) is a set of rules that requires consumer products to be safe and our obligations to ensure this.

If you have any concerns about our products, you can contact us on ProductSafety@springernature.com

In case Publisher is established outside the EU, the EU authorized representative is:

Springer Nature Customer Service Center GmbH
Europaplatz 3
69115 Heidelberg, Germany